73rd Highway Geology Symposium (HGS 2024)

Lawrence, Kansas, USA
9-12 September 2024

ISBN: 979-8-3313-1289-3

Printed from e-media with permission by:

Curran Associates, Inc.
57 Morehouse Lane
Red Hook, NY 12571

CURRAN ASSOCIATES INC. proceedings.com

Some format issues inherent in the e-media version may also appear in this print version.

Printed with permission by Curran Associates, Inc. (2025)

For permission requests, please contact Highway Geology Symposium (HGS)
at the address below.

Highway Geology Symposium (HGS)
C/O John Pilipchuk
1589 Mail Service Center
Raleigh, NC, 27699-1589

Phone: (919) 707-6851
Fax: (919) 250-4237

jpilipchuk@ncdot.gov

Additional copies of this publication are available from:

Curran Associates, Inc.
57 Morehouse Lane
Red Hook, NY 12571 USA
Phone: 845-758-0400
Fax: 845-758-2633
Email: curran@proceedings.com
Web: www.proceedings.com

Tuesday, September 10

Thursday, September 12

No.	Time	Title	Primary Author	
17	8:00 AM - 8:20 AM	West Virginia Highway 340, An Intensive Rockfall Mitigation in Harpers Ferry National Park	David Crotsley	277
18	8:20 AM - 8:40 AM	Case Histories of Roadside Rockfall Barrier Applications	Mike Koutsourais	301
19	8:40 AM - 9:00 AM	Utah HWY 189 Hazard Rock Removal using AutoStem™ - A Case Study in Provo Canyon	Jack Fitzgerald	318
20	9:00 AM - 9:20 AM	Rockfall Hazard Mitigation along I-68 at Sideling Hill, Washington County, Maryland	Brian K. Banks	331
21	9:20 AM - 9:40 AM	Precision Presplitting: Redefining Accuracy in Rock Blasting	Anthony Konya	352
22	9:40 AM - 10:00 AM	Role of Instrumentation and Monitoring in a Geotechnical Asset Management Program	Darren Beckstrand	369
23	10:40 AM - 11:00 AM	Modern Borehole Logging & Database Management - A Geodata Lifecycle	Declan Vanderhor	390
24	11:00 AM - 11:20 AM	SGAM - Smart Geotechnical Asset Management	Niccolò Belcecchi	399
25	11:20 AM - 11:40 AM	Spatial Analytics, Remote Sensing and Advances in Technology as They Apply to Highway Geological Investigations and Critical Infrastructure Asset Management	Robert R. Hendricks	410
26	11:40 AM - 12:00 PM	Advanced Multi-Sensing Techniques for Geotechnical Asset Management	Rakesh Salunke	421

73rd Highway Geology Symposium
BETTER HIGHWAYS
THROUGH APPLIED GEOLOGY
LAWRENCE, KANSAS 2024

73RD HIGHWAY GEOLOGY SYMPOSIUM

2024 HGS Proceedings

Sept. 9-12, 2024 | Double Tree Hotel | Lawrence, KS

Kansas Geology:
More than Meets the Eye

Kansas geology is marked by its diverse sedimentary formations, largely a product of the state's history as a shallow inland sea during the Paleozoic and Mesozoic eras. The state features extensive layers of limestone, shale, and sandstone, rich with marine fossils, particularly from the Carboniferous and Permian periods. These ancient seabeds are best exemplified by the Flint Hills, where resistant chert (flint) layers create rolling hills distinct from the surrounding plains.

The western part of Kansas showcases the Niobrara Chalk, a remnant of the Cretaceous Period, characterized by dramatic formations like Monument Rocks. Central Kansas is notable for its vast salt deposits formed during the Permian Period, which are economically significant today.

Additionally, the Pleistocene glaciations influenced the state's geology, as glaciers extended into Northeast Kansas they left behind erratic's and deposited vast amounts of sediment, shaping the river valleys and influencing soil composition.

Overall, Kansas geology is a testament to its dynamic past, from ancient oceans and rich fossil beds to distinctive sedimentary landscapes and significant mineral resources.

Cover photos and inside cover: Monument Rocks, Gove County

Dedicated to John Duffy

John Duffy at the Igor Paramassi test facility in Meano, Italy in 2007. Photo by Tom Badger.

This year's Highway Geology Symposium proceedings are dedicated to John D Duffy. John was active in HGS and the Transportation Research Board Engineering Geology Committee for decades. He was instrumental in advancing rockfall analysis and design in the US and began doing so 36 years ago.

John's own description of how he became involved in rockfall technology: "My first HGS meeting was in Park City, Utah back in 1988. My supervisor at the time, Marvin McCauley, had been a long time HGS participant and encouraged me to attend. In my relatively new job, I had been assigned to develop an understanding on the subject of rockfall. The theme of that years HGS was "rockfall" and largely centered on the development of the new Colorado Rockfall Simulation Program (CRSP). Bob Barrett (Colorado DOT) was showcasing the program and its attributes."

At that meeting he met Robert Thommen (GeoBrugg) and together they developed field testing for new rockfall barriers being produced. A year later John managed to obtain funding and began field testing fences - proving their efficacy - and, as they say, "the rest is history." For the next 35 years John was involved with development of rockfall barrier technology, pioneered in-house rock slope evaluation and mitigation programs including on-rappel work procedures and helped other DOTs set up their own practices of rockfall mitigation. His work and contributions were recognized nationally and internationally, and known for their practical, common-sense approach.

Throughout his career he advanced rockfall technology, mentored people he worked with as well as consultants, contractors and manufacturers, contributed to the HGS and TRB, and authored/co-authored many papers and several chapters of the rockfall textbooks published by TRB. I remember meeting John in an airport once, with Mike Vierling as the three of us were heading to a Highway Geology Symposium. He was walking with a cane and Mike and I were curious as to why.

John explained that he had an accident on a site months before, when a Caterpillar D9 backed over him. He explained that the track had pushed him down into some mud and that his femur had snapped in the process. He said he was well on the way to being fully mended - but had to share his amusement at the looks he got parking his car in a handicapped space at Pismo Beach, grabbing his surfboard from the roof rack and hobbling down to the water for his PT.

Schedule at a Glance

73rd Highway Geology Symposium

Lawrence, KS | September 9-12, 2024

	Monday, September 9	Location
9:00 AM - 5:00 PM	Registration	Regency Preconvene
9:00 AM - 5:00 PM	Exhibitor Setup	Regency Preconvene & Ballroom
1:00 PM - 5:00 PM	**Field Day:** Drilling & Sampling Techniques	Offsite - KS Geological Survey
5:00 PM - 6:30 PM	National Steering Committee Meeting	Regency DII & DIII
6:00 PM - 8:00 PM	Icebreaker Reception - *Sponsored by GeoStabilization International/Access Limited*	Regency Preconvene & Ballroom
	Tuesday, September 10	**Location**
7:00 AM - 8:00 AM	Breakfast	Regency Ballroom
7:00 AM - 5:00 PM	Registration	Regency Preconvene
7:00 AM - 5:00 PM	Exhibits Open	Regency Preconvene & Ballroom
9:00 AM - 5:00 PM	Companion Activities Departs - *Preregistration required*	Offsite
8:00 AM - 8:10 AM	HGS Welcome & Opening Remarks – Kyle Halverson	
8:10 AM - 8:30 AM	Dedication – Bill Webster	
8:30 AM - 9:00 AM	Kansas Geologic Survey - KGS State Geologist – Jay Kalbas	Brazilian Ballroom
9:00 AM - 9:20 AM	Kansas DOT State Transportation Engineer – Greg Schieber	
9:20 AM - 10:00 AM	**Session 1** - Young Authors	
10:00 AM - 10:30 AM	Mid-Morning Break - *Sponsored by Central Mine Equipment Company*	Regency Preconvene & Ballroom
10:30 AM - 12:00 PM	**Session 2** - Young Authors	Brazilian Ballroom
12:00 PM - 1:00 PM	Lunch	Regency Ballroom
1:00 PM - 1:40 PM	**Session 3** - Young Authors	Brazilian Ballroom
1:40 PM - 3:00 PM	**Session 4** - Foundations/Ground Improvements	
3:00 PM - 3:40 PM	Afternoon Break - *Sponsored by Hager-Richter Geoscience, Inc.*	Regency Preconvene & Ballroom
3:40 PM - 4:20 PM	**Session 5** - Landslide/Research & Technology	
4:20 PM - 5:00 PM	**Session 6** - Case Study/Geophysics	Brazilian Ballroom
5:00 PM - 5:15 PM	Field Trip Preview	
5:30 PM - 9:30 PM	Complimentary Shuttles to Downtown Lawrence	Hotel Lobby (every 30 mins)
6:00 PM - 8:00 PM	**TRB Mid-Year Meeting:** Workflow Choices & Project Outcomes	Brazilian Ballroom
	Wednesday, September 11	**Location**
7:30 AM - 8:00 AM	Grab n' Go Breakfast	Regency Ballroom
8:00 AM	Board Buses for HGS Field Tour - *Transportation Sponsored by Maccaferri*	Hotel Lobby
8:00 AM - 6:00 PM	HGS Field Trip	Offsite
12:00 PM	Lunch: Coronado Heights - *Sponsored by Geobrugg North America*	Coronado Heights
6:00 PM	Field Trip Drop Off *(subject to vary)*	Hotel Lobby
Evening	Dinner on Own	
6:30 PM - 10:30 PM	Complimentary Shuttles to Downtown Lawrence	Hotel Lobby (every 30 mins)
	Thursday, September 12	**Location**
7:00 AM - 8:00 AM	Breakfast	Regency Ballroom
7:00 AM - 5:00 PM	Registration	Regency Preconvene
7:00 AM - 1:00 PM	Exhibits Open	Regency Preconvene & Ballroom
8:00 AM - 9:40 AM	**Session 7** - Rockfall	Brazilian Ballroom
9:40 AM - 10:00 AM	**Session 8** - Asset Management I	
10:00 AM - 10:40 AM	Mid-Morning Break - *Sponsored by Hayes Drilling*	Regency Preconvene & Ballroom
10:40 AM - 12:00 PM	**Session 8** - Asset Management II	Brazilian Ballroom
12:00 PM - 1:00 PM	Lunch	Regency Ballroom
1:00 PM - 5:00 PM	**Field Day:** Geophysical Techniques	Offsite - KS Geological Survey
1:00 PM - 5:00 PM	Exhibitor Move Out	
6:00 PM - 9:00 PM	HGS Banquet & Awards - *Entertainment sponsored by Ameritech Slope Contractors*	Brazilian Ballroom

***Subject to change**

Highway Geology Symposium History, Organization, and Function

Inaugural Meeting

Established to foster a better understanding and closer cooperation between geologists and civil engineers in the highway industry, the Highway Geology Symposium (HGS) was organized and held its first meeting on March 14, 1950, in Richmond Virginia. Attending the inaugural meeting were representatives from state highway departments (as referred to at that time) from Georgia, South Carolina, North Carolina, Virginia, Kentucky, West Virginia, Maryland, and Pennsylvania. In addition, a number of federal agencies and universities were represented. A total of nine technical papers were presented.

W.T. Parrott, an engineering geologist with the Virginia Department of Highways, chaired the first meeting. It was Mr. Parrott who originated the Highway Geology Symposium.

It was at the 1956 meeting that future HGS leader, A.C. Dodson, began his active role in participating in the Symposium. Mr. Dodson was the Chief Geologist for the North Carolina State Highway and Public Works Commission, which sponsored the 7th HGS meeting.

Symposium Locations

Since the initial meeting, 69 consecutive annual meetings have been held in 33 different states. Between 1950 and 1962, the meetings were east of the Mississippi River, with Virginia, West Virginia, Ohio, Maryland, North Carolina, Pennsylvania, Georgia, Florida, and Tennessee serving as host state.

In 1962, the symposium moved west for the first time to Phoenix, Arizona where the 13th annual HGS meeting was held. Since then, it has alternated, for the most part, back and forth from the east to the west. The Annual Symposium has moved to a different location as shown on the next page.

Organization

Unlike most groups and organizations that meet on a regular basis, the Highway Geology Symposium has no central headquarters, no annual dues, and no formal membership requirements. The governing body of the Symposium is a steering committee composed of approximately 20 - 25 engineering geologist and geotechnical engineers from state and federal agencies, colleges and universities, as well as private service companies and consulting firms throughout the country.

Steering committee members are elected for three-year terms, with their elections and re-elections being determined principally by their interests and participation in and contribution to the Symposium. The officers include a chairman, vice chairman, secretary, and treasurer. all of whom are elected for a two-year term. Officers, except for the treasurer, may only succeed themselves for one additional term.

A number of three-member standing committees conduct the affairs of the organization. The lack of rigid requirements, routing and relatively relaxed overall functioning of the organization is what attracts many participants.

Meeting sites are chosen two to four years in advance and are selected by the Steering Committee following presentations made by representatives of potential host states. These presentations are usually made at the steering committee meeting, which is held during the Annual Symposium. Upon selection, the state representative becomes the state chairman and a member of the Steering Committee.

The symposia are generally scheduled for two and one-half days, with a day-and-a-half for technical papers plus a full day for the field trip. The Symposium usually begins with a TRB session and an evening Ice-Breaker the first day, a full day of technical presentations the second day, a field trip on the third day followed by the annual banquet that evening, and a half day of technical presentations on the final day.

The Field Trip

The field trip is the focus of the meeting. In most cases, the trips cover approximately 150 to 200 miles, provide for six to eight scheduled stops, and require about eight hours. Occasionally, cultural stops are scheduled around geological and geotechnical points of interests. To cite a few examples: in Wyoming (1973), the group viewed landslides in the Big Horn Mountains; Florida's trip (1976) included a tour of Cape Canaveral and the NASA space installation; the Idaho and South Dakota trips dealt principally with mining activities; North Carolina provided stops at a quarry site, a dam construction site, and a nuclear generation site; in Maryland, the group visited the Chesapeake Bay hydraulic model and the Goddard Space Center. The Oregon trip included visits to the Columbia River Gorge and Mount Hood; the Central mine region was visited in Texas; and the Tennessee meeting in 1981 provided stops at several repaired landslide in Appalachia regions of East Tennessee.

In Utah (1988) the field trip visited sites in Provo Canyon and stopped at the famous Thistle Landslide, while in New Mexico, in 1990, the emphasis was on rockfall treatments in the Rio Grande River canyon and included a stop at the Brugg Wire Rope headquarters in Santa Fe.

Mount St, Helens was visited by the field trip in 1994 when the meeting was in Portland, Oregon, while in 1995 the West Virginia meeting took us to the New River Gorge Bridge that has a deck elevation of 876 feet above the water. In Cody, Wyoming the 1996 field trip visited the Chief Joseph Scenic Highway and the Beartooth Uplift in northwest Wyoming. In 1997 the meeting in Tennessee visited the newly constructed future I-26 highway in the Blue Ridge of East Tennessee. The Arizona meeting in 1998 visited the Oak Creek Canyon near Sedona and a mining ghost town at Jerome, Arizona. The Virginia meeting in 1999 visited the "Smart Road" Project that was under construction. This was a joint research project of the Virginia Department of Transportation and Virginia Tech University. The Seattle Washington meeting in 2000 visited an ancient lahar in the Mount Rainier area. A stop during the Maryland meeting in 2001 was the Sideling Hill road cut for I-68 which displayed a tightly folded syncline in the Allegheny Mountains.

The California field trip in 2002 provided a field demonstration of the effectiveness of rock netting against rock falls along the Pacific Coast Highway. The Kansas City meeting in 2004 visited the Hunt Subtropolis which is said to be the "world's largest underground business complex". It was created through the mining of limestone by way of the room and pillar method. The Rocky Point Quarry provided an opportunity to search for fossils at the North Carolina meeting in 2005. The group also visited the US-17 Wilmington Bypass Bridge which was under construction. Among the stops at the Pennsylvania meeting were the Hickory Run Boulder Field, the No.9 Mine and Wash Shanty Museum, and the Lehigh Tunnel.

Garden of the Gods, Colorado

The New Mexico field trip in 2008 included stops at a soil nailed wall along US-285/84 north of Santa Fe and a road cut through the Bandelier Tuff on highway 502 near Los Alamos where rockfall mesh was used to protect against rockfalls. The New York field trip in 2009 included the Niagara Falls Gorge and the Devil's Hole Trail. The Oklahoma field trip in 2010 toured the complex geology of the Arbuckle Mountains in the southern part of the state along with stops at Tucker's Tower and Turner Falls.

In the bluegrass state of Kentucky, the 2011 HGS field trip included stops at Camp Nelson which is the site of the oldest exposed rocks in Kentucky near the Lexington and Kentucky River Fault Zones. Additional stops at the Darby Dan Farm and the Woodford Reserve Distillery illustrated how the local geology has played such a large part in the success of breeding prized Thoroughbred horses and made Kentucky the "Birthplace of Bourbon".

Portland, Maine

In Redding, California, the 2012 field trip included stops at the Whiskeytown Lake, which is one in a series of lakes that provide water and power to northern California. Additional stops included Rocky Point, a roadway construction site containing Naturally Occurring Asbestos (NOA), and Oregon Mountain where the geology and high rainfall amounts have caused Hwy 299 to experience local and global instabilities since first constructed in 1920.

The 2013 field trip of New Hampshire highlighted the topography and geologic remnants left by the Pleistocene glaciation that fully retreated approximately 12,000 years ago. The field trip included stops at various overlooks of glacially-carved valleys and ranges; the Old Man of the Mountain Memorial Plaza, which is a tribute to the famous cantilevered rock mass in the Franconia Notch that collapsed on May 3, 2003; the lacustrine deposits and features of the Glacial Lake Ammonoosuc; views of the Presidential Range; bridges damaged during Tropical Storm Irene in August 2011; and the Willey Slide, located in the Crawford Notch where all members of the Willey family were buried by a landslide in 1826.

The 2014 field trip presented a breathtaking tour of the geology and history of southeast Wyoming, ascending from the high plains surrounding Laramie at 7000 feet to the Medicine Bow Mountains along the Snowy Range Scenic Byway. Visible along the way were a Precambrian shear zone, and glacial deposits and features. From the glacially carved Mirror Lake and the Snowy Range Ski Area, the path wound east to the Laramie Mountains and the Vedauwoo Recreational Area, a popular rock climbing and hiking area before returning to Laramie.

In Sturbridge, MA, the 2015 field trip focused on the Connecticut Valley, a Mesozoic rift basin that signaled the breakup of Pangea, and the Berkshires, which represents the collision and amalgamation of an island arc system with the North American Laurentian margin.

The field trip in 2016 was an urban setting along the western edge of Colorado Springs and around Manitou Springs. Stops included the Pikeview Quarry, Garden of the Gods Visitor Center, and several other locations where rockfall and debris flow mitigation, post-flooding highway embankment repair, and a nonconformity in the rock records that spans 1.3 billion years were observed.

The 2017 field trip provided an opportunity to view the geology of northern Georgia. Stops included the Bellwood Quarry, which, at one time was run by the City of Atlanta and also served as a prison labor camp. It will eventually serve as a 2.4 billion-gallon water storage facility for the City of Atlanta upon completion of a tunnel to connect the quarry to two water treatment plans and three pump stations. Additional stops included the Buzzi Unicem Cement Plant to get a close up view of the Clairmont Melange, The Cooper Furnace near the Allatoona Dam, and the New Riverside Ochre-Emerson Barite mine.

The 2018 field trip in Portland Maine provided a good overview of the geology of coastal Maine. Field trip stops included a stop at the Sherman Salt Marsh near Newcastle which was recently restored to its natural state after the dam that carried US Highway 1 washed out during a 2005

Columbia River Gorge, Oregon

Mount Rainier, Washington

storm. Additional stops included the site of the 1996 landslide near Rockland Harbor that consumed several homes and the rock slope remediation project at the Penobscot Narrows Bridge near Prospect Maine. A lobster lunch along the shore of Penobscot Bay was one of several highlights of the field trip

The 2019 field trip in Portland Oregon travelled the Columbia River Gorge west. Starting at the Crown Point Vista House and Portland Women's Forum State Scenic Viewpoint above the gorge to learn about the river highway. Descending into the gorge, we stopped at scenic Multnomah Falls and Benson Bridge, and saw flexible rockfall fence installed to protect the lodge and historic Columbia River Highway. Other stops included lunch at Cascade Locks, Bonneville Landslide and rockfall areas along the highway.

The 2022 field trip in the Ashville area took us through Ordovician (500 my) to Precambrian (1.2 by) migmatized ortho and paragneisses, metamorphosed intrusives, thrust faults and contacts representing three orogenies and complex sequences of basement and terranes. We crossed the Brevard Fault zone several times, which is a structure that has been studied and interpreted for 100 years. Various attempts to define the structure have been made, especially in the pre-plate tectonic era. It has been theorized that these structures were as high, or higher than the Rockies at formation. 200 million years of rifted erosion leave us with an exposed look at deep orogenic roots of multiple thrust events. Precipitation in the area is between 60-100" per year. There are deep ancient colluvial deposits, complex mineralization and weathering profiles, and non-linear/planar discontinuities. These deposits and precipitation make for distinct issues within the state. Deep foundations rarely present problems. We traveled over I-26 and the Blue Ridge Escarpment where they highway is being widened. Stops included the I-26 Old Howard Gap Slide Area, the US 74 Gerton Slide, a shallow landslide barrier on I-40 W, and the Buckner Gap Cut.

The 2023 Field trip in Tacoma, WA traveled to Mt Rainer where HGS goers were able to take various hikes around the park. Additional highlights included driving down into the Ohop Valley outwash channel, a drive by of Alder Lake, a drive along Copper Creek Forest Road 59, a view of Nisqually Glacier from Nisqually Rive and a final stop at Ricksecker Point.

List of Highway Geology Symposium Meetings

No.	Year	HGS Location	No.	Year	HGS Location
1st	1950	Richmond, VA	38th	1987	Pittsburg, PA
2nd	1951	Richmond, VA	39th	1988	Park City, UT
3rd	1952	Lexington, VA	40th	1989	Birmingham, AL
4th	1953	Charleston, WV	41st	1990	Albuquerque, NM
5th	1954	Columbus, OH	41st	1991	Albany, NY
6th	1955	Baltimore, MD	43rd	1992	Fayetteville AR
7th	1956	Raleigh, NC	44rd	1993	Tampa, FL
8th	1957	State College, PA	45th	1994	Portland, OR
9th	1958	Charlottesville, VA	46th	1995	Charleston, WV
10th	1959	Atlanta, GA	47th	1996	Cody, WY
11th	1960	Tallahassee, FL	48th	1997	Knoxville, TN
12th	1961	Knoxville, TN	49th	1998	Prescott, AZ
13th	1962	Phoenix, AZ	50th	1999	Roanoke, VA
14th	1963	College Station, TX	51st	2000	Seattle, WA
15th	1964	Rolla, MO	52nd	2001	Cumberland, MD
16th	1965	Lexington, KY	53rd	2002	San Luis Obispo, CA
17th	1966	Ames, IA	54th	2003	Burlington, VT
18th	1967	Lafayette, IN	55th	2004	Kansas City, MO
19th	1968	Morgantown, WV	56th	2005	Wilmington, NC
20th	1969	Urbana, IL	57th	2006	Breckinridge, CO
21st	1970	Lawrence, KS	58th	2007	Pocono Manor, PA
22nd	1971	Norman, OK	59th	2008	Santa Fe, NM
23rd	1972	Old Point Comfort, VA	60th	2009	Buffalo, NY
24th	1973	Sheridan, WY	61st	2010	Oklahoma City, OK
25th	1974	Raleigh, NC	62nd	2011	Lexington, KY
26th	1975	Coeur d'Alene, ID	63rd	2012	Redding, CA
27th	1976	Orlando, FL	64th	2013	North Conway, NH
28th	1977	Rapid City, SD	65th	2014	Laramie, WY
29th	1978	Annapolis, MD	66th	2015	Sturbridge, MA
30th	1979	Portland, OR	67th	2016	Colorado Springs
31st	1980	Austin, TX	68th	2017	Marietta, GA
32nd	1981	Gatlinburg, TN	69th	2018	Portland, ME
33rd	1982	Vail, CO	70th	2019	Portland OR
34th	1983	Stone Mountain, GA	71st	2022	Asheville, NC
35th	1984	San Jose, CA	72nd	2023	Tacoma, WA
36th	1985	Clarksville, TN	73rd	2024	Lawrence, KS
37th	1986	Helena, MT	74th	2025	Morgantown, WV

73RD HIGHWAY GEOLOGY SYMPOSIUM

IN THIS SECTION:

Field Day Information
Mid-Year Meeting
Field Trip Schedule
Medallian Awards
Emeritus Members
Committee Officers & Members
Contact List
Young Author Awards

Coronado Heights,
McPherson County

HGS Field Day Events

The 73rd Annual HGS Conference is proud to present two separate field day events, organized to demonstrate common geotechnical investigation practices. These events offer participants a "boots on the ground" experience with geotechnical drilling, sampling, and geophysics techniques, providing valuable insights into subsurface investigations.

Field Day Event #1
Monday, September 9

1:00 PM – 5:00 PM

Drilling and Sampling:

The Drilling and Sampling Field Day will offer participants an opportunity to see various drilling and sampling methods that are commonly used or emerging technologies on transportation projects. Drilling and Sampling are the backbone of geotechnical site investigations, however for new geo-professionals there is gap in the application and practice. Below is a summary of each method that will be demonstrated.

- **CPT**- A Cone Penetration Test (CPT) is commonly used to determine the subsurface stratigraphy in situ (in place) and to estimate geotechnical parameters of the materials present. Geotechnical engineers typically use a CPT test to determine the necessary construction requirements for infrastructure – roadbeds, bridges, buildings. A demonstration will be done to show you the setup and the practice to pushing a CPT behind a CPT/Drilling platform. Elements involved are an instrumented probe, push rods, electronics, earth anchors, logging computer and a hydraulic ram-set (drill rig). The CPT "push" will be done using a CPT Probe that meets an ASTM standard for construction and will be advanced at a controlled push rate following the ASTM practice.
- **Drive Point-** The Drive Point is a simple geotechnical testing method developed by the Wyoming Department of Transportation. Using a hydraulic 140 pound automatic hammer, an AW rod with a two inch conical tip is driven into the ground, with the operator recording blows per foot. The test helps WYDOT Geologists determine relative density within the subsurface, and aids in determining compressible layers, potential scour depths, and H-Pile refusal among others.
- **Rock Coring-** The Kansas Department of Transportation Geology Section will be performing a demonstration in rock core sampling. The demonstration will explain the procedures for drilling, retrieving, and analyzing rock core samples. For this presentation, a truck mounted CME 55 rotary drill rig will be used to retrieve NQ sized core samples. This process and information obtained is important to properly design road and bridge foundations.
- **SPT/Shelby Tube**- The Shelby tube/SPT demo, will drill down and push a Shelby tube. KDOT Geotech Engineers will demonstrate measuring recovery, pocket penetrometer readings, and capping the tube. Further discussion of what the Shelby tube sample can be used for in the laboratory and how it is considered an undisturbed sample. Additional drilling will advance down where an SPT test will be performed. Discussion of how SPT blow counts are measured and where the N-value comes from, how SPT data can be used and what testing can be done on the sample that is recovered.
- **MWD (Measurement While Drilling)**-The MWD demonstration will be performed to show participants real time data acquisition, along with associated samples. A continuous sample will be done to show subsurface stratigraphy and how MWD data parameters can be associated. Drilling will continue down where rock coring will be performed. This portion of the demonstration will illustrate MWD parameters that are useful during rock coring as well as visualization of the underlying bedrock. Discussion opportunities during this demonstration will allow for participants to hear from MWD experts.

Field Day Event #2
Thursday, September 12
1:00 PM – 5:00 PM

Geophysics:

The Geophysics Field Day will offer participants an opportunity to see various geophysical methods used on transportation projects. The application of Geophysics on transportation projects has historically been an underutilized method to determine subsurface conditions, however with the Federal Highway Administrations (FHWA) development of the "A-Game" these methods have been proven to be very beneficial. Below is a summary of each method that will be demonstrated.

- **Electrical Resistivities-** Electrical resistivity is a common method for helping determine the shallow stratigraphy of highway projects. It creates a profile of the subsurface showing the difference in resistivity of earth materials. Current is injected through two electrodes into the ground, and the resulting voltage is measured through other electrodes. Modern systems can measure the voltage at several electrodes at once and automatically switch the current and voltage locations through numerous configurations. The resistivity method can help determine top of bedrock and fractures within bedrock, changes in soil and rock type, location of groundwater, and can detect voids.
- **Downhole SPT-Seismic Refraction-** The demonstration will showcase a new SPT-seismic testing method for 3D characterization of a large volume of soil/rock properties with a single standard penetration test (SPT). Specifically, the seismic wavefields generated during a conventional SPT are recorded by a 2D grid of geophones on the ground surface without interfering with SPT crew. The seismic data are then analyzed by a 3D full-waveform inversion (3D FWI) to extract material properties. Leveraging SPT-induced wavefields predominantly composed of body wave components within soil/rock masses, this method provides new imaging capabilities for subsurface soil/rock with enhanced accuracy and resolution at depths. It enables characterization at foot-pixels across a large 3D volume (up to 60 ft around SPT boring). Requiring only a single boring for 3D imaging, the method is cost-effective and efficient for site characterization, especially for imaging of deep voids within weathered and karst rock.
- **Downhole Televiewer-** Hager-Richter Geoscience (HRGS) will provide a presentation and field demonstration on the use of borehole geophysical logging methods to characterize subsurface conditions. The methods presented are non-invasive borehole geophysical logging methods that provide in-situ subsurface conditions as part of geotechnical and environmental investigations by characterizing subsurface geology (overburden, bedrock, groundwater flow) as well as in-situ conditions of man-made structures such as drilled shafts and other foundation elements. Borehole geophysical logging methods can be used as part of a subsurface investigation program to supplement other methods (drilling, sampling/lab testing, surface geophysics) and provide high resolution in-situ results to characterize subsurface conditions encountered in boreholes throughout many stages of a project lifecycle. The presentation will include information on how borehole geophysical logging results can be used as part of geologic/hydrogeologic and geo-engineering investigations, specific logging methods will be discussed along with limitations of the methods, logging deliverables will be presented, and a demonstration will be provided giving those attending the opportunity to see borehole geophysical logging equipment including the logging winch, control unit, and a multitude of individual logging probes.
- **GPR-** Schnabel Engineering will be providing both conventional pulse-based and step-frequency ground penetrating radar field demonstrations. This is the opportunity for hands on experience with two GPR systems. The conventional GPR system is typically used to image geologic, karst, or void features, map utilities, or for structural evaluations such as locating reinforcement steel concrete. The step-frequency (also referred to as 3D-GPR) system will be attached to a vehicle and is typically used to scans roads, runways, or large relatively flat areas. The step-frequency system has multiple sensors, allowing for near continuous coverage with each pass of the 5-ft wide antenna. We will showcase the

mechanisms of the equipment and how they are able to collect geophysical data that is then, in turn, interpreted and displayed. We will also show multiple data sets and visualizations for several markets including transportation, geotechnical, dams, and environmental.

- **Seismic Refraction and MSAW-** Jackson State University will demonstrate the Multichannel Analysis of Surface Waves (MASW) method, which is an advanced geophysical method utilized to assess subsurface conditions by analyzing the propagation of surface waves. This technique involves generating seismic waves using a source such as a sledgehammer or specialized seismic equipment, which are then recorded by an array of geophones placed along the ground surface. The setup of the equipment will include a seismograph, 24 geophones (for Vibration collection), a 12-volt battery, a spread cable (to connect the geophones), a strike plate, a triggering cable, and a striking hammer. The geophones will be placed in the soil at specific intervals (based on the investigation depth) and connected with the spread cable. Seismic waves will be generated by striking the plate with a sledgehammer while the seismograph records the data from the geophones.

  The recorded data captures the dispersion characteristics of surface waves, where different frequencies travel at varying velocities depending on the subsurface material properties. By analyzing these dispersion curves, MASW provides a detailed shear wave velocity profile, revealing the stiffness and layering of the subsurface materials. In the field demonstration, data processing using software to create a profile showing the subsurface layers and their properties will be presented. This step-by-step demonstration will help to understand how MASW works and its practical applications in assessing subsurface conditions for highway, levees and other infrastructure projects. Besides, a comparison of the MASW results with other ERI profiles on the highway embankment will be presented.

- **Magnetometer-** JCollier Geophysics will provide a demonstration of UAV-Magnetometry survey. This geophysical method will demonstrate the capabilities of acquiring magnetic data suspended from a drone. A processed section of the area of interest will be presented which will showcase the result of this area. The capabilities of using a MAG survey are to detect buried magnetic objects and/or to map certain geologic features. Geologic applications of the MAG method include finding fault structures and lava tubes. Near surface applications include searching for ferrous metal drums, ferrous metal waste deposits, ferrous metal pipelines and utilities, Unexploded Ordnance (UXO), as well as old foundations or ancient site characterization if present.

  The magnetometry (MAG) method is a non-invasive geophysical method used to measure the strength and/or direction of magnetic fields produced by magnetized objects. Many rocks and minerals are magnetized by induction of the Earth's magnetic field and cause spatial perturbations or "anomalies" in the Earth's main field.

  The polarization and strength of an object's magnetic field depends on its magnetic susceptibility, which is the ratio between the object's magnetization and the strength of the inducing field. Man-made objects composed of magnetic minerals containing iron are highly susceptible and can cause large anomalies thousands of times greater than geologic anomalies.

  With the use of drones and flight planning software, we are able to make precise and altitude conscious paths for land and water operations.

Prairie Trail Scenic Byway, McPherson County

Tuesday, Sept. 10, 6 p.m.

TRB Midyear Meeting

2024 TRB Standing Committee on Geotechnical Instrumentation and Modeling (AKG60) Midyear Meeting

Workflow Choices and Project Outcomes

At the 73rd Highway Geology Symposium, Lawrence

The committee meeting is open to all HGS attendees.

The midyear meeting will be a hybrid session (in-person and online) and feature:

Updates on AKG60 workshops at the 2025 Annual Meeting in Washington DC

Brief discussion on Research Needs Statements (RNS) and Webinar ideas

Presentations and Discussion related to "Geotechnical Field Tools and Techniques for Phased Site Investigations."

Presentation and Discussion related to "Geotechnical Risk and Reliability: How Investigation, and Analysis Choices Impact Outcomes."

Presentation and Discussion related to "Geotechnical Modeling, Geotechnical Baseline Reports, and Digital Project Delivery."

This event will complement the HGS field demonstrations and explore how geophysical and in-situ direct-push, boring, and sampling tools and techniques can be used in a complimentary fashion to perform efficient phased site investigations for assessment of geotechnical parameters, locating areas of additional investigation interest, assessment of variability, and creation of geotechnical design models. Phased investigations, with exploration geophysics in the lead, are becoming increasingly common to make better use of drilling and sampling resources, increase productivity, decrease site disruption, save time, and reduce cost.

We then begin to explore the value of different site characterization practices and how project risk is influenced by choices throughout the geotechnical workflow- desk review, site investigation, lab testing, instrumentation/monitoring, design analysis, reporting, visualization, and deliverables.

HGS Field Trip Schedule

Time	Activity Information
7:30 AM	Grab n' Go Breakfast-Hotel Lobby
8:00 AM	Depart from Hotel
9:30 AM	Stop #1 Tallgrass Prairie National Preserve
10:30 AM	Depart from Tallgrass Prairie National Preserve
12:00 PM	Stop #2 Coronado Heights and Lunch
1:30 PM	Depart from Coronado Heights
2:15 PM	Stop #3 Mushroom Rock State Park
3:15 PM	Depart from Mushroom Rock State Park
6:00 PM	Arrive back at Hotel

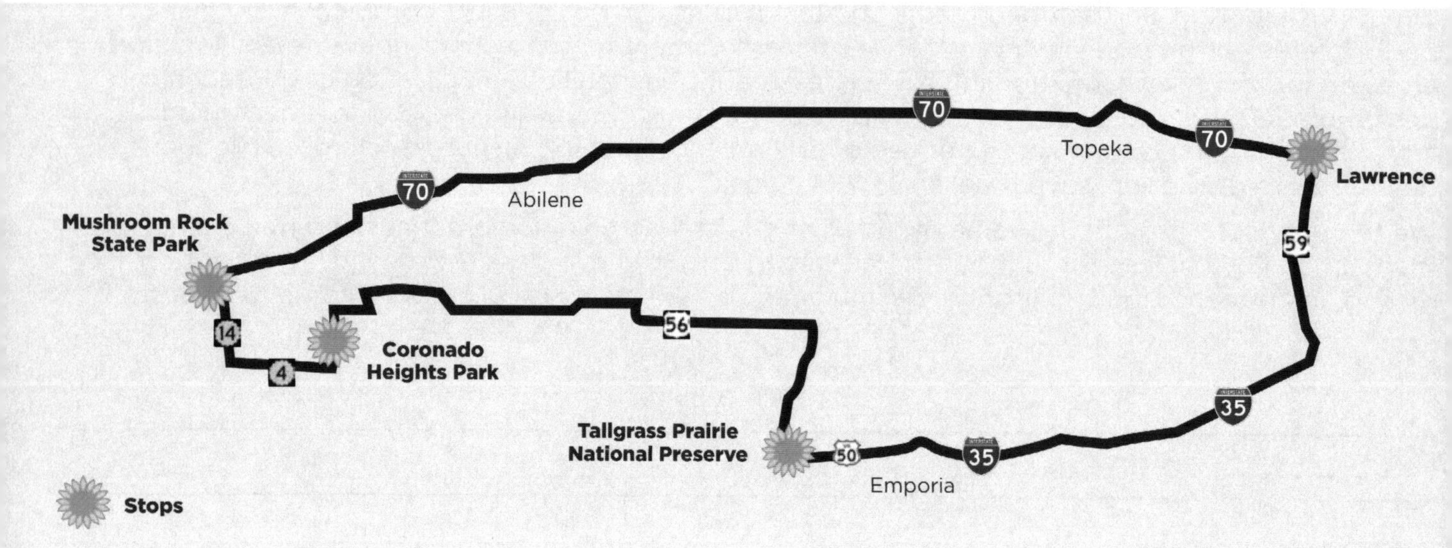

Features Along the way

- US-59 Rock Cut
- Transition from Osage Cuestas to Flint Hills Physiographic Region
- Transition from Flint Hills to Smoky Hills Physiographic Region
- Kanopolis State Park
- I-70: First Segment of Interstate Open

Scan Here for Info

Medallion Award Recipients

Technical Sessions

At the technical sessions, case histories and applied state-of-the-art papers are most common; with highly theoretical papers the exception. The papers presented at the technical sessions are published in the annual proceedings. All proceedings are available to download from www.HighwayGeologySymposium.org

Member Recognition

A Medallion Award was initiated in 1970 to honor those persons who have made significant contributions to the Highway Geology Symposium. The selection was and is currently made from the members of the national steering committee of the HGS.

Name	Year	Name	Year
Hugh Chase	1970	Earl Wright	1997
Tom Parrott	1970	Russell Glass	1998
Paul Price	1970	Harry Ludowise	2000
K.B. Woods	1971	Sam Thornton	2000
R.J. Edmondson	1972	Bob Henthorne	2004
C.S. Mullin	1974	Mike Hager	2005
A.C. Dodson	1975	Joseph A. Fischer	2007
Burrell Whitlow	1978	Ken Ashton	2008
Bill Sherman	1980	A. David Martin	2008
Virgil Burgat	1981	Michael Vierling	2009
Henry Mathis	1982	Dick Cross	2009
David Royster	1982	John F. Szturo	2010
Terry West	1983	Christopher Ruppen	2012
Dave Bingham	1984	Jeff Dean	2012
Vernon Bump	1986	John Pilipchuk	2015
C.W. "Bill" Lovell	1989	Peter Ingraham	2016
Joseph A. Gutierrez	1990	Richard Lane	2017
Willard McCasland	1990	Steve Sweeny	2018
W.A. "Bill" Wisner	1991	John Duffy	2018
David Mitchell	1993	Krystle Pelham	2018
Harry Moore	1996	Marc Fish	2023

Emeritus Members of the National Steering Committee

Emeritus Members

A number of past members of the national steering committee have been granted Emeritus status. These individuals, usually retired, resigned from the HGS Steering Committee, or are deceased, have made significant contributions to the Highway Geology Symposium. A total of 43 persons have been granted Emeritus status.

R. F. Baker	George S. Meadors, Jr.
John Baldwin	Willard MaCasland
David Bingham	Henry Mathis
Vernon Bump	David Mitchell
Virgil E. Burgat	Harry Moore
Robert G. Charboneau	W. T. Parrot
Hugh Chase	Nicholas Priznar
Jim Coffin	Paul H. Price
Dick Cross	David L. Royster
A. C. Dodson	Bill Sherman
Tom Eliassen	Willard L Sitz
Walter Fredericksen	Mitchell Smith
John "Brandy" Gilmore	Jim Stroud
Russell Glass	Steve Sweeney
Robert Goddard	Sam Thornton
Joseph Gutierrez	Berke Thompson
Mike Hager	Mike Vierling
Rich Humphries	Burrell Whitlow
Charles T. Janik	W. A. "Bill" Wisner
Richard Lane	Earl Wright
John Lemish	Ed J. Zeigler
Bill Lovell	

Dedications

Several proceedings volumes have been dedicated to past HGS Steering Committee members who have passed away. The 36th HGS Proceedings were dedicated to David L. Royster (1931 - 1985, Tennessee) at the Clarksville, Indiana Meeting in 1985. In 1991 the Proceedings of the 42nd HGS held in Albany, New York were dedicated to Burrell S. Whitlow (1929 - 1990, Virginia). The 64th HGS Proceedings were dedicated to Earl Wright (1931 – 2012) at the North Conway, New Hampshire meeting. The 65th proceedings were dedicated to Nicholas Priznar (1952 – 2014) at the Laramie, Wyoming meeting. The 76th HGS held at Colorado Springs, Colorado dedicated the proceedings to Vern McGuffy (1934 – 2016). The proceedings for the 68th HGS held in Marietta, Georgia were dedicated to Richard (Dick) Cross (1944 - 2016). The proceedings for the 69th HGS are dedicated to Dave Bingham (1932-2018) and Joe Gutierrez (1926-2018). The Proceedings of the 71st HGS are dedicated to Vernon (Vern) Bump. The Proceedings of the 73rd HGS are dedicated to John Duffy.

HGS National Steering Committee Officers and Members

HGS National Steering Committee Officers

Bill Webster - Chairman

CalTrans
5900 Folsom Blvd.,
Sacramento, CA 95819
916-662-1183
bill_webster@dot.ca.gov

Kyle Halverson - Vice-Chairman

Kansas Department of Transportation Bureau of Structures and Geotechnical Services
700 SW Harrison St.. Topeka, KS 66603
785-600-8165
kyle.halverson@ks.gov

Marc Fish - Secretary

WSDOT Geotechnical Office
1655 S. 2nd Ave.,
Tumwater, WA 98512
360-709-5498
fishM@wsdot.wa.gov

John Pilipchuk - Treasurer

NCDOT Geotechnical Engineering Unit (Retired)
1512 US Highway 70 East
Hillsborough, NC 27278
919-622-3381
johnpilipchuk@yahoo.com

HGS National Steering Committee Members

Ken Ashton

West Virginia Geological Survey
1 Mont Chateau Rd.,
Morgantown, WV 26508
304-594-2331
ashton@wvgs.wvnet.edu

Vanessa Bateman

Engineering & Construction Headquarters U.S. Army Corps of Engineers
441 G St.,
NW Washington DC, 20314-1000
202-761-7423
vanessa.c.bateman@usace.army.mil

Simon Boone

Access Limited Construction
1102 Pike Ln.,
Oceano, CA 94335
540-420-2678
simon@alcinc.com

James Dahill

Wyoming DOT
5300 Bishop Blvd.,
Cheyenne, WY 82009
307-777-4205
james.dahill@wyo.gov

Jeff Dean

ODOT & Terracon (Retired)
2412 Cedar Oak Drive
Edmond, OK 73013
405-503-1463
jeff4txaggs@aol.com

Bob Henthorne

PEC
400 S. Kansas Ave.,
Topeka, KS 66603-3754
785-233-8300
bob.henthorne@pec1.com

Peter Ingraham

Scarptec Inc.
19 Lord Jeffrey Dr.,
Amherst, NH 03031
603-785-0262
peter@scarptec.com

Jody Kuhne

North Carolina DOT (Retired)
78 Flint St.
Asheville, NC 28801
828-779-9482
jkuhne@ncdot.gov

HGS National Steering Committee Members (continued)

Sarah McInnes
PA DOT
District 6-0
7000 Geerdes Blvd.,
King of Prussia, PA 19406
610-205-6544
smcinnes@pa.gov

Krystle Pelham
New Hampshire Dept. of Transportation Bureau of Materials and Research
PO Box 483, 5 Hazen Dr.,
Concord, NH 03302-0483
603-271-1657
krystle.pelham@dot.nh.gov

Victoria Porto
PA DOT (Retired)
10 Pine Lake Dr.,
Carlisle, PA 17015
717-805-5941
vamporto@aol.com

Erik Rorem
Geobrugg North America, LLC (Retired)
20483 Whistle Punk Rd.,
Bend, OR 97702
505-690-7144
erik.rorem6522@gmail.com

Christopher A. Ruppen - Young Author Committee
GeoStabilization International
3808 Sunflower RD.,
New Brighton, PA 15066
724- 272-7532
chris.ruppen@gsi.us

Tim Shevlin, R.G.
Geobrugg North America LLC
Salem, OR 97302
503-423-7258
tim.shevlin@geobrugg.com

Deana Sneyd
Petrologic Solutions, Inc.
3997 Oak Hill Rd.,
Douglasville, GA 30135
678-313-4147
dsneyd@gmail.com

John F. Szturo
HNTB Corporation
715 Kirk Dr.,
Kansas City, MO 64105
816-527-2275
jszturo@hntb.com

Terry West - Medallion, Emeritus
Earth and Atmospheric Science Dept. Purdue University
West Lafayette, IN 47907-1297
765-494-3296
trwest@purdue.edu

Past, Present, and Future Symposium Contact List

Year	State	Member	Contact Information	
2013	New Hampshire	Krystle Pelham	603-271-1657	Krystle.Pelham@dot.state.nh.us
2014	Wyoming	Jim Coffin	307-777-4205	Jim.coffin@wyo.gov
2015	Massachusetts	Peter Ingraham	603-688-0880	peter_ingraham@golder.com
2016	Colorado	Ty Ortiz	303-921- 2634	Ty.ortiz@state.co.us
2017	Georgia	Deana Sneyd	678-313-4147	Dsneyd61@gmail.com
2018	Maine	Krystle Pelham	603-271-1657	Krystle.pelham@dot.state.nh.us
2019	Oregon	Scott Burns	503-725-3389	BurnsS@pdx.edu
2022	North Carolina	John Pilipchuk	919-707-6851	jpilipchuk@ncdot.gov
		Jody Kuhne	828-250-3285	jkuhne@ncdot.gov
2023	Washington	Marc Fish	360-485-5825	fishm@wsdot.wa.gov
2024	Kansas	Kyle Halverson	785-600-8165	Kyle.halverson@ks.gov
2025	West Virgina	Ken Ashton	304-594-2331	ashton@wvgs.wvnet.edu

Young Author Award

Young Author Award

The Highway Geology Symposium has always encouraged participation of Young Professionals, realizing that Young Professionals are the future of the Organization. This participation was taken formal in 2014, with the formation of an annual National Young Author Competition, where Young Authors have the opportunity to prepare papers and present their work. To participate, Young Authors must be up to 35 years old or younger, the principal author of the paper and the sole presenter of the paper at the Symposium. Papers are reviewed and judged based on Technical Presentation of the Paper (including Geology), Originality of the Work, Applicability of the Work to Others and Paper Layout. One Young Author is selected each year to receive the coveted Young Author Award, with presentation of the award conducted at the annual Symposium banquet.

Year	Award Recipient and Paper Titles
2014	Simon Boone, "Performance of Flexible Debris Flow Barriers in a Narrow Canyon"
2015	Cory Rinehart, "High Quality H20: Utilizing Horizontal Drains for Landslide Stabilization"
2016	Todd Hansen, "Geologic Exploration for Ground Classification: Widening of the I-70 Veterans Memorial Tunnels"
2017	James Arthurs, "Construction of Transportation Infrastructure in Weathered Volcanic Ash Soils"
2018	Brian Felber, "Geotechnical Challenges for Bridge Foundations & Roadway Embankment Design in Peats and Deep Glacial Lake Deposits"
2019	Anya Brose, "The Assessment and Remediation of Wabasha St. Rock Fall"
2022	Christopher Mayer, "Using Geophysics to Evaluate the Results of a Grouting Program in Karstic Geology"
2023	Cody Chaussee, "Bolt Creek Fire: Post-Wildfire Debris Flow Risk Assessment and Barrier Design on US 2, Near Grotto, WA

KANSAS GEOLOGY FACT

Kansas was part of the Tri-State Mining district in the extreme southeast part of the state. It was active for over one-hundred years (1850-1950). This area produced 50% of the zinc and 10% of the lead in the United States at that time.

Mushroom Rock, Ellsworth County

73rd Highway Geology Symposium
BETTER HIGHWAYS THROUGH APPLIED GEOLOGY
LAWRENCE, KANSAS 2024

THANK YOU TO OUR SPONSORS!

The following companies have graciously contributed to the success of the Symposium.
We gratefully appreciate the support of our sponsors.

Platinum Level Sponsors

BRUGG
Geobrugg
Safety is our nature

Field Trip Lunch Wednesday

Geobrugg North America, LLC

Tim Shevlin
tim.shevlin@geobrugg.com
503-423-7258
www.geobrugg.com

Geobrugg is the global leader in the supply of high-tensile steel wire, safety nets and meshes. 65 years of experience have made Geobrugg the answer to reliable solutions against natural hazards.

MACCAFERRI

Field Trip Transportation Wednesday

Maccaferri, Inc.

Michael Koutsourais
m.koutsourais@maccaferri.com
301-641-8072
www.maccaferri.com

Pioneers in the industry in rockfall protection systems and natural hazard mitigation, Maccaferri offers a wide range of systems to stabilize rock faces, soil slopes and snow masses, reducing risks to people, buildings, and critical infrastructure. Maccaferri's wide variety of systems offer the highest level of performance in the industry.Maccaferri's product offerings include the SteelGrid and RockMesh simple and pinned drapery systems, High Energy Absorption (HEA) cable nets, Dynamic Rockfall Barriers, Hybrid Barriers and Attenuators, Debris Flow Barriers, Rockfall Embankments, and Snow Avalanche Protection systems. Maccaferri's design software suite includes the MacRO1 and MacRO2 software for simple and pinned draperies, and the Mac S-Design for surface strengthening in soil nailing applications.Feel free to visit our website – www.maccaferri.com

Gold Level Sponsors

Ameritech Slope Constructors INC

Geohazard Mitigation

Thursday Closing Banquet Entertainment

Ameritech Slope Constructors, Inc.

Roger Moore
rmoore@ameritech.pro
828-633-6352
www.ameritech.pro

Ameritech Slope Constructors, Inc. is a geohazard mitigation contractor who provides rockfall mitigation, slope stabilization, shotcrete installation and related rope access services for stabilizing soil and rock slopes.

GSI GeoStabilization International®

Monday Ice Breaker

GeoStabilization International / Access Limited

Paige Barnett
paige.barnett@gsi.us
425-758-4757
www.geostabilization.com/accesslimited.com

Our passion is to develop and install innovative solutions that protect people and infrastructure from the dangers of geohazards. GeoStabilization International® has combined the most experienced and responsive geohazard mitigation professionals in North America. GeoStabilization specializes in design/build/warranty landslide repair, rockfall mitigation, excavation shoring, grouting, and GRS-IBS abutment construction. With over 4800 completed geohazard repairs in the last 19 years, our combined experience and nationwide emergency response capability are unmatched.

Silver Level Sponsors

TabLogs

TabLogs

David Adcock
david@tablogs.com.au
0-402-333-783
www.tablogs.com

TabLogs empowers your geotechs to log consistently and to nationally recognized US standards by featuring standardized workflows. But TabLogs isn't just for logging boreholes! Using our intuitive desktop application, you can scope and design investigations, create customisable reports and import/export your existing database of borehole logs!

Bronze Level Sponsors

Apex Rockfall Mitigation

Sarah Walton
swalton@apexrfm.com
925-503-7078
www.swalton@apexrfm.com

Apex Rockfall Mitigation is a leader in the rockfall and geohazard stabilization industry. Our experience in restricted and limited access locations is unsurpassed. Apex continues to create and deploy some of the most hi-tech equipment, materials, techniques and tools available to the industry. Apex Rockfall Mitigation has wide ranging ability allowing us to provide rock solid services throughout the country.

BOREDM

Louis Aaron
louis@boredmlogs.com
602-492-3076
www.boredmlogs.com

BoreDM is the modern standard for geotechnical and environmental subsurface data management. Created by a team of software developers with input from hundreds of engineers, BoreDM has rapidly become the favorite at 12 state departments of transportation and leading firms across the United States, Australia, and Canada. BoreDM's technology stack enables an intuitive user experience, unmatched support, rock solid data security protocols, and the ongoing integration of cutting-edge capabilities.

Hager-Richter Geoscience, Inc.

Jeffrey Reid
jeff.reid@hager-richter.com
603-370-7518
www.hager-richter.com

HAGER-RICHTER GEOSCIENCE, INC. is a well-established small business that specializes in Surface and Borehole Geophysics for Engineering applications. The firm has been in business since 1984, has earned a national reputation, and has a nationwide practice. Hager-Richter is headquartered in New Hampshire and has had a fully staffed and equipped New York/New Jersey Regional Office in New Jersey since 2001. Hager-Richter has extensive experience in providing high resolution geophysical services to support transportation infrastructure projects throughout the United States.

Tuesday AM Break

Central Mine Equipment Company

Brad Jouppi
info@cmeco.com
800-325-8827
www.cmeco.com

CME manufactures auger, core and rotary drilling rigs mounted on trucks, skids, tracks, and rubber-tire all-terrain carriers. We also manufacture hollow stem augers, conventional flight augers, sampling and drilling tools to support our machines.

HAYES DRILLING, INC.

Thursday AM Break

Hayes Drilling, Inc.

Luke Schuler
mluke.schuler@gmail.com
816-918-7553
www.hayesdrillinginc.com

Hayes Drilling, Inc. is recognized as the leading drilled shaft and excavation support contractor in MId-America. Since 1958, we have been providing high quality, cost effective and innovative foundation solutions to our customers. Our list of core services include drilled shafts, design-build earth retention, micropiles, soldier beams and lagging, soil/rock nailing, permanent rock/soil anchors and secant piles. We are also experienced in performing drilled shaft load testing and drilled shaft post-grouting. We welcome the opportunity to work with you in developing the right solution to your next foundation project.

HNTB 100+ YEARS OF INFRASTRUCTURE SOLUTIONS

HNTB

John Szturo
jszturo@hntb.com
816 527-2275
https://www.hntb.com/

HNTB is celebrating its 110th anniversary as a civil engineering firm. HNTB consists of over 7000 employee owners in over 80 offices across the United States. HNTB has over 70 geologists, geotechnical engineers, and construction inspectors, providing support for simple to complex subsurface investigations, roadway design, pavement, bridge foundations, retaining walls, rock slope and landslide work for multiple DOT's and railroads. Please visit HNTB.com for career opportunities.

FOUNDATION TESTING & CONSULTING, LLC
Knowledge To Build On®

FTC

William Jones
cj@ftandc.com
913-626-8499
www.FTandC.com

FTC provides a variety of construction phase testing for bridge and other infrastructure foundations. Testing services include cross-hole sonic logging (CSL), thermal integrity profiling (TIP), dynamic pile testing (PDA), sonic echo/impluse response testing (SE/IR), and surface geophysical surveys.We also provide a suite of data analytic services through our PileTrac service that captures history pile installation data and test results to improve design and construction of future pile supported bridge projects.

Keller NA

Kert Thielen
kert.thielen@keller-na.com
469-634-8415
www.keller-na.com

Keller is a Design-Build geotechnical construction company offering a wide and market-leading portfolio of geotechnical construction solutions including: Ground Improvement, Grouting, Deep Foundations, Earth Retention, and Groundwater control. Solutions incorporate one or more of the following geotechnical techniques:Vibro Stone Columns, Rigid Inclusions, Soil Mixing, Vibro Piers, Jet Grouting, Permeation Grouting, Injection Systems, Auger Cast Piles, Drilled Shafts, Driven Piles, Helical Piles, Micropiles, Underpinning Assemblies, Diaphragm Walls, Tieback Anchors, Soil Nailing, Soldier Piles and Lagging, Secant/Tangent Piles, as well as many other techniques.

NHAZCA S.r.l

Miriana Bottigoni
miriana.bottigoni@nhazca.com
+39 069-506-5820
www.nhazca.it/en/

NHAZCA (Natural HAZards Control and Assessment), Startup of "Sapienza" University of Rome, is a services and consultancy company, leader in the analysis and monitoring for the management and mitigation of risks of natural hazards and large infrastructures. As a result of a constant effort in the innovation and optimization of the newest remote sensing technologies, NHAZCA provides to its customers cutting-edge solutions for the management, control and monitoring of natural and man-induced hazards, supporting the construction and administration of large infrastructures and natural resources exploitation projects. NHAZCA is an international leader in monitoring services through innovative technologies such as the Satellite (A-DInSAR) and Terrestrial (TInSAR) SAR Interferometry. NHAZCA develops also the PhotoMonitoring technology to monitor with extreme precision changes and surface movements of structures and terrain using images acquired from different platforms (terrestrial, aerial, satellite) and sensors (optical, multispectral, infrared, radar).

Voss Signs

Tom Tenerovicz
tom@vosssigns.com
315-682-6418
www.vosssigns.com

Since 1965, Voss Signs, LLC has produced custom and stock signs for various customers that include: Forestry Professionals, Land Owners, State and Federal Government Agencies.

JEAN LUTZ North America

Michel Lariau
contact@jeanlutzna.com
330-702-1476
www.jeanlutzna.com

Manufacturer of MWD datalogger, sensors and software for mwd data processing.

KESCO, Inc. and Fragmentation Efficiency Services, Inc.

dconn@kesco-fes.com
724-548-2811
www.kescoblasting.com
fesconsult.com

KESCO, Inc., and Fragmentation Efficiency Services, Inc. (FES) have been engaged in explosive industry for over 60 years. We are an engineering and blasting firm that applies data to solve unique challenges using custom-controlled blasting designs.We also partner with Geotechnical Engineering Firms and State Transportation Agencies to provide economical and safe solutions for their challenging projects.

Professional Engineering Consultants

Ashley Smith
ashley.smith@pec1.com
316-641-2708
www.pec1.com/

Established as a professional association in 1965 after the merger of three engineering design firms, Professional Engineering Consultants (PEC) has been delivering comprehensive engineering solutions for more than 58 years.Headquartered in Wichita, Kansas, with additional locations in Kansas, Missouri, Oklahoma and Colorado, our staff of more than 400 hold professional engineering licenses in all 50 states. Our work takes us across the nation and around the world to design an increasing variety of projects from highways to hospitals — water towers to wind turbines — stadiums to sidewalks — and everything in between.

Contributing Sponsor

ROCK ENGINEERING SOLUTIONS

Field Trip Beverages

Scarptec, Inc.

Peter Ingraham
peter@scarptec.com
www.scarptec.com

KANSAS GEOLOGY FACT

According to the Kansas Geological Survey, of the three main rock types (igneous, sedimentary and metamorphic) sedimentary is the most common found in Kansas.

Little Jerusalem Badlands State Park

Photo Credit: Nick Abt

Exhibitors

Acker Drill Co., Inc.

John Lang
jlang@Ackerdrill.com
800-752-2537
www.ackerdrill.com

Acker Drill Company was founded in 1916 and has been exporting its products worldwide since 1960. Today, we are recognized as a world leader in the manufacture of drill rigs and tooling for the geotechnical, environmental, mineral exploration, mining, and civil engineering industries.

Collier Geophysics LLC.

Rhett Gore
rgore@colliergeophysics.com
254-968-8741
www.colliergeophysics.com

Collier Geophysics is a service-disabled, veteran-owned, small business founded in 2018 with offices in Texas, California, Colorado, Georgia, Massachusetts, North Carolina, Oklahoma, Tennessee, and Wisconsin. We offer a full spectrum of expert surface, borehole, airborne, and marine geophysical solutions for today's needs throughout the U.S. and around the globe. If you have problems or challenges in any of our service areas, let us bring you solutions.

Delve Underground

Allison Halvorson
halvorson@delveunderground.com
925-300-5774
www.delveunderground.com

Delve Underground is a leader in heavy civil engineering, serving the transportation, water, wastewater, and energy industries. Specializing in tunnel design, we provide innovative solutions to the most challenging underground problems. We offer comprehensive design, construction management, and dispute resolution capabilities. Our expertise includes civil, structural, and geotechnical engineering. Founded in 1954, as Jacobs Associates, Delve Underground is an employee-owned firm with 25 offices and 350 team members throughout the United States, Canada, Australia, and New Zealand.

GEOPROBE SYSTEMS®

Lori Christensen
chril@geoprobe.com
785-404-1160
www.geoprobe.com

Launched in 1987, Geoprobe Systems® manufactures innovative drilling rigs and tooling - engineered to simplify fieldwork - empowering drilling companies around the globe to succeed as productive and profitable industry leaders. We incorporate customer feedback to continuously develop solutions to make drillers' jobs faster, easier, and safer, across water well, geothermal, cathodic protection, geotechnical, environmental, exploration and construction drilling industries. Recognized for providing superior service support, Geoprobe® keeps you in the field via live phone support by our expert service technicians based at several locations across the United States.

Gilson Company, Inc.

Abbi Smith
asmith@gilsonco.com
740-548-7298
www.globalgilson.com

Gilson Company, Inc. manufactures material testing equipment.

Harrison Western Construction Corporation

Joey Vaupel
jvaupel@harwest.com
720-552-0490
www.harrisonwestern.com

Harrison Western Construction Corporation was founded in Denver, Colorado and has provided superior geotechnical, mining, tunneling, design and engineering services for over 55 years. We have completed over $2.5 billion of geotechnical, underground, and heavy civil projects. Our extensive and diverse experience is complemented by strong leadership and hands-on management, ensuring projects are delivered on budget and on time. Our clients include Local, State and Federal entities, as well as corporate and private owners. These clients range from some of the largest private operators in the world to public entities like cities and counties, departments of transportation, water and sewer authorities, state departments of natural resources and the Federal Government.

KANE GeoTech, Inc.

Joseph McElhany
joey.mcelhany@kanegeotech.com
209-472-1822
www.kanegeotech.com

KANE GeoTech, Inc specializes in difficult civil and geotechnical engineering problems. Our services include rockfall, debris flow and shallow landslide mitigation, as well as geotechnical instrumentation, and geotechnical analysis and design. We build partnerships with our clients and affiliates, taking time to learn project needs and how we can deliver the best solutions, most effective products and fastest service.

Kansas Geological Survey

Blair Schneider
blair.schneider@ku.edu
785-864-0663
www.kgs.ku.edu

The Kansas Geological Survey (KGS) is a research and service division of the University of Kansas that investigates and provides information about the state's geologic and groundwater resources.

SIMCO Drilling Equipment

Ryan Gross
rgross@simcodrill.com
515-490-3868
www.simcodrill.com

Drill Rig Manufacturer.

Solmax

Maggie Kooi
mkooi@solmax.com
706-693-1803
www.solmax.com/en/ams

Solmax is a world leader in sustainable construction solutions for civil and environmental infrastructure. Its pioneering products separate, contain, filter, drain and reinforce essential applications in a more sustainable way - making the world a better place. The company was founded in 1981 and has grown through the acquisition of GSE, TenCate Geosynthetics and Propex. It is now the largest geosynthetics company in the world, empowered by more than 2,000 talented people. Solmax is headquartered in Quebec, Canada, with subsidiaries and operations across the globe.

Trumer Schutzbauten

Ahren Bichler
a.bichler@trumer.cc
604-732-0325
www.trumer.cc

TRUMER SCHUTZBAUTEN is a world leader in the design and manufacturing of geohazard mitigation structures that protect against rockfall, debris flow, avalanche and unstable slopes. From slope mesh to massive barriers, we protect life, buildings and infrastructure where failure is not an option.

Williams Form Engineering Corp.

Dan Jeffers
djeffers@williamsform.com
616-866-0815
www.williamsform.com

Williams Form Engineering Corporation has been providing threaded steel bars and accessories for rock, soil and concrete anchors, post-tensioning systems, and concrete forming hardware systems in the construction industry for over 100 years.Our rock and soil anchor product line includes our Spin-Lock mechanical rock anchors, polyester resin anchors, multiple corrosion protection anchors, soil nails, strand anchors, mechanical soil anchors, Geo-Drill Hollow-Bar anchors, and micropiles. For concrete anchoring we offer Spin-Lock anchors, undercut anchors, reusable anchors and cast-in-place anchors. We also have a full line of All-Thread Rebar for tiebacks, micropiles and post-tensioning.

73RD HIGHWAY GEOLOGY SYMPOSIUM

Thank You!

With grateful acknowledgment in making this event possible:

Ken Ashton

Bob Henthorne

Luke Metheny

Blair Schneider

John Szturo

John Barker

John Pilipchuk

HGS Steering Committee

Kansas Geological Survey

Delaney Event Management

Kansas Department of Transportation

See you in **West Virginia!**

Notes

Crawling through Steep Slopes:

Highlighting the Use of Walking Excavators through the Highway 14 Slope Repair Project in Santa Clarita, California

Nolan T. Lescalleet, G.I.T.
Access Limited
1102 Pike Lane, Oceano, CA 93445
(805) 592-2230
Nolan.Lescalleet@alccinc.com

Shane Nia
Access Limited
1102 Pike Lane, Oceano, CA 93445
(805) 592-2230
Shane@alccinc.com

Prepared for the 73rd Highway Geology Symposium, September, 2024

Acknowledgements

For example: The author(s) would like to thank the individuals/entities for their contributions in the work described:

Jon Simpson – Access Limited
Kevin Wiesman – Access Limited
Simon Boone, P.G. – Geostabilization International
Sukut Construction
California Department of Transportation, Design Branches A and B

Disclaimer

Statements and views presented in this paper are strictly those of the author(s), and do not necessarily reflect positions held by their affiliations, the Highway Geology Symposium (HGS), or others acknowledged above. The mention of trade names for commercial products does not imply the approval or endorsement by HGS.

ABSTRACT

Walking excavators, also commonly referred to as spider excavators, are redefining construction methods and increasing jobsite and public safety. These machines present efficient solutions for accessing and operating in challenging, steep, and unstable terrains where traversing with traditional machinery is deemed unsafe. The defining characteristic of these machines, when compared to traditional tracked or wheeled excavator equipment, is their four independently operating telescopic legs, which enable the machines to perform a crawling or spider-like movement. Walking excavators perform a range of construction services, such as excavating and grading, difficult drilling programs, maintenance of landscapes, and emergency response construction projects, making them indispensable in the geohazard mitigation industry.

A slope stabilization project is nearing completion along a portion of Highway 14 (SH-14) in Santa Clarita, California. The California Department of Transportation designed an extensive solution to improve the soil stability along an approximately 2,275-ft long and nearly 200-ft tall slope. This solution includes excavating and regrading over 200,000 cubic yards of material, installing up to four tiers of erosion control benches, drilling 5,607 soil nails in both soil and rock, and installing approximately 554,000 square feet of pinned, or anchored, double twisted wire mesh and turf reinforcement mat.

This paper explores the history of the walking excavator before demonstrating key features and advantages these machines can bring to construction projects through Access Limited's scope of work performed at the SH-14 project.

INTRODUCTION

The British Geological Survey (BGS) defines geological hazards, or geohazards, as "the natural geological processes that present a direct risk to people or an indirect risk by impacting development." The BGS further defines geohazards into two categories: earth hazards, such as earthquakes, volcanoes, and tsunamis; and shallow geohazards including landslides, sinkholes, rockfall-related events, karst weathering, and Quaternary-attributed processes (BGS Research, 2024). Earth hazards are generally inevitable as "Mother Nature" has had millions to billions of years to develop these processes; however, shallow geohazards are the result of both natural processes and human development. Whether man has placed an asset or roadway at the bottom of a precariously to perpetually unstable slope or in the line of a potential rockfall trajectory, the advancement of geohazard mitigation engineering and complex designs is a function of the need for new technologies, and vice versa.

Mobile walking excavators, commonly referred to as Spyder or spider excavators, are invaluable machines in the geohazard mitigation industry. These machines fall into the construction family of excavators and include some similar components to traditional excavators like a chassis and frame, boom and arm, cabin equipped with controls for the operator, an engine, and various attachments. However, dissimilar from a traditional excavator, a walking excavator is equipped with four telescopic legs that operate independently of one another, similar to the legs of a four-legged animal, as opposed to being simply equipped with four mounted rubber tires or two all-terrain vehicle (ATV) tracks. Images 1 and 2 illustrate two projects where Access Limited spider excavators were utilized for visual reference. These defining legs allow the operator of the machine to navigate efficiently and safely through and over variable terrains, including

Image 1 – Image depicts a walking excavator crawling down an approximately 45-degree slope at a ski resort project in Idaho.

Image 2 – Image depicts two walking excavators drilling soil nails to facilitate installation of an avalanche mitigation system at a ski resort in Utah.

moderate-to-high angle grades (up to and sometimes exceeding 50 degrees), rocky and talus-covered debris fields, and loose or unconsolidated piles of fine-grained sediments, such as coal tailing piles and beach sands.

Between 2022 and 2024, a slope repair and stabilization project along Highway 14 (SH-14) in Santa Clarita, California required using one of these machines to aid in repairing the slope as traditional heavy machinery proved to be insufficient and unsafe in navigating the slope grades. As such, this paper expands on the history of walking excavators, and presents a general background of geohazard mitigation construction designs and technologies before analyzing the use of these machines at the SH-14 slope stabilization project. This case study aims to highlight the significance these machines can bring to the geohazard mitigation industry.

HISTORY & DEVELOPMENT OF THE WALKING EXCAVATOR

Walking excavators have been utilized in the construction industry dating back to the mid-1960s. Ernst Menzi and Joseph Kaiser developed the first walking excavator in 1966 when they identified a continuous need for a piece of machinery that was able to safely traverse and perform in the steeper slopes of their mountainous homeland in Switzerland. Together, Menzi and Kaiser manufactured the MUK 2000, a single-axle machine that moved through use of the boom and excavator bucket rather than drivetrain-powered tracks or wheels. The MUK 2000 performed its first job in 1967, which included a river channel cleanout project in Schaanwald, Liechtenstein. Various debris had reportedly dammed the river causing flooding upstream of the blockage and throughout the river channel. Accordingly, the MUK 2000 was mobilized and was able to traverse through the channel and shallow waters to clean out the blockage. Implementing the MUK 2000 proved to be a resounding success with project representatives stating that the one "machine [could] achieve the work of at least 12 to 15 men" (Kaiser AG, 2015).

Menzi and Kaiser ultimately parted ways later in 1967 with Menzi controlling the existing company and Kaiser moving on to partner with an Italian-based company called Moro. Kaiser and Moro shortly thereafter developed the KAMO 3X machine (Richardson, 2024). Other walking excavator companies began to emerge in the late 1970s, including the Italian company Euromach and the German company Karl Schaeff KG coming on the scene in 1977 and 1978, respectively, although Kaiser and Menzi continued to and have remained the household names for the walking excavator industry.

Walking excavators became common equipment for most excavation contractors in Europe. By the end of 1979, Kaiser's sales alone included selling and distributing approximately 1,800 walking excavators to contractors in mountainous European countries like Switzerland and Austria. By the end of 1980, walking excavators were present on four continents.

The machines proved to be versatile and efficient machines with the ability to access steeper terrains than was accessible with traditional machinery, and climb over short elements including boulders and retaining or knee walls. In 1981, one project reported that the use of these machines reduced the overall construction cost by 2/3s. An article in the Liechtensteiner Volksblatt (a former Liechtenstein newspaper) further stated that these machines "can operate on grades as steep as 70 degrees and in water up to 8 ft deep" (Kaiser AG, 2015).

Unfortunately for the walking excavator, the 1990s saw a large boom in the manufacturing and versatility of the tracked mini excavators, which were far less expensive than walking excavators. The adaptation of the mini excavator gave these machines the ability to traverse low angles while having a significant advantage in torque and lifting capacity when compared to the walking excavator. As a result, walking excavator sales diminished over the next decade. This decline in sales fueled the need for innovations to the walking excavator design, as well as attachments to further showcase the use of these machines in construction.

Innovations to these machines between the 1990s and today have included the introduction of all-wheel drive and four-wheel steering, articulating booms similar to a traditional excavator, increased down pressure and lifting capacity, and continued modifications to the telescoping legs to lower ground pressure and ground disturbance (Kaiser AG, 2015). Walking excavators today can be equipped with a wide array of attachments, including various widths and types of digging and grading buckets, component drills, mulchers and vegetation masticating heads, clam shell buckets, winches, tree cutters, and more. These machines come in a variety of weight classes generally ranging from 7 to 16-ton machines, each performing efficiently for specific tasks, yet all continuing to showcase their abilities in steep slope construction projects.

These machines continue to gain popularity in the United States in regions with high frequencies of geohazards like California, the Pacific Northwest, and Pennsylvania. Although expensive machines, the introduction of these machines into construction methodologies has proven to have positive project implications, such as improved safety, enhanced designs, refined construction feasibility, and overall reductions in project cost.

Image 3 – Historical image of a KAMO 3X walking excavator being piloted by Joseph Kaiser in 1970 (Kaiser AG, 2015).

Image 4 – Historical image of MUK 3000 crawling onto truck by itself in 1966 (Kaiser AG, 2015)

GEOHAZARD CONSTRUCTION METHODOLOGIES

Traditional machinery may be effective in most scenarios; however, when access to construction areas is limited and involves unique topographic terrains, steep slope construction methods are generally necessary. One example of steep slope construction is using certified ropes access technicians, which involves laborers tying themselves and equipment off at the top of the slope and rappelling down to the construction targets. Tie off points can include natural features like trees and boulders or manually installed permanent elements, such as metal posts and wire rope anchors. It is important to note that tying off to natural features requires experienced ropes access technicians to determine if a natural feature is safe for construction, whereas manually installed tie-off points can be and should be tested to ensure the system should not fail when fully loaded with personnel and equipment.

Where steep slope construction projects call for larger pieces of machinery than just manpower, construction durations and costs can become long and expensive if adequate access is not easily attainable with traditional methods. For example, envision a 50-ft tall soil or rock slope adjacent to a highway. The preliminary construction of said rock slope and highway likely included purchasing the land necessary to develop the roadway and adjacent slopes, delineating temporary construction easements to facilitate mobilizing machinery necessary to construct the roadway asset, traversing heavy machinery to the top of a hillslope, excavating overburden, and, where rock is present, performing drill-and-blast processes to fragment and remove the bedrock. Many of the highways that line the country today were built in the better part of the mid- to late-20th century. Accordingly, many of these rock slopes are now in deteriorating states, having been exposed to years of weathering and natural processes. State Department of Transportation agencies are tasked today with creating asset management programs for these rock slopes in an effort to prioritize construction projects, form maintenance programs to capital projects, and allocate budgets appropriately.

The following case study looks at a project that Access Limited performed steep slope construction activities on that required the use of both specialized steep slope machinery and certified ropes access technicians to install a slope mitigation system.

HIGHWAY 14 (SH-14) SLOPE EROSION REPAIR PROJECT

Project Background

SH-14 includes a multi-lane highway beginning in Los Angeles, California and extends to the northeast through Los Angeles and San Bernadino Counties. Based on conversations with the California Department of Transportation (Caltrans), the highway corridor construction started in 1963 and was completed in 1975. Caltrans, Branch A, designed a slope repair project along the southbound (SB) side of SH-14 between approximate postmiles 29.0 and 29.5, specifically between Stations (Sta.) 1532+20 and 1554+94. Refer to Figure 1 for the project location. Existing cut slopes from the original construction of SH-14 in the late 1960s showed significant fatigue including surficial failures and rills throughout the slope face. The slope repair includes an approximately 2,274-ft long portion of slope that crests at over 100 ft tall, as measured from roadway grades, and slopes down to the roadway at an approximate 2H:1V slope

Figure 1 – The figure locates the project site on the project locus in the upper right hand corner, and further depicts the project location and slope limits along Highway 14. The site is located immediately northwest of the Highway 14 southbound travel lanes, as shown.

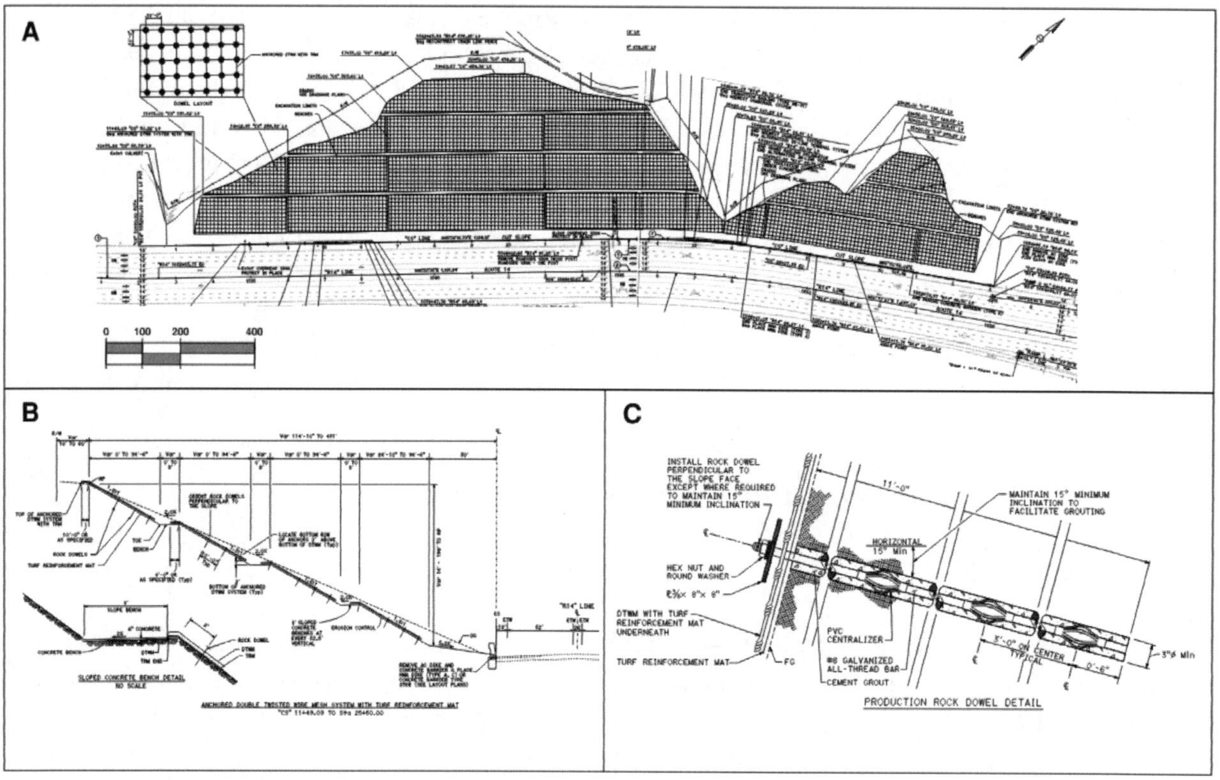

Figure 2 – Snapshots from the SH-14 project construction plans dated for approval on December 3, 2021. (A) Depicts the proposed limits of the pinned mesh solution including dowel layout and proposed grading. (C) Provides a snapshot of a typical section of the slope, which depicts the proposed 1.8H:1V slope grades and three erosional control benches between slope cuts. (D) A typical construction detail for a production soil nail.

(approximately 27-degrees). Based on historical images, the roadway alignment appears to have been constructed prior to 1985.

Based on a 2020 Geotechnical Design Report (GDR) for the project drafted by Caltrans, slope stability analyses suggest that the slope should be regraded from a 2H:1V to a 1.8H:1V, approximately 29 degrees (note: slope stability analyses were redacted from the provided GDR and are therefore not included herein), and that erosional control measures should be installed. These erosion control measures consist of creating 8 to 10-ft wide benches every 50 ft up the slope, as measured from above the roadway, and installing a pinned mesh, or anchored mesh, solution to the slope face to increase stability and promote revegetation of the slope. The pinned mesh includes applying a double-twisted wire mesh (DTWM) and turf reinforcing mat (TRM) across the newly graded slope surface and anchoring the DTWM and TRM to the slope with 10 to 15-ft long soil nails, or rock dowels (these are used interchangeably in the contract plans), on a 10x10-ft wide grid pattern (Branch A, 2020). Total quantities for this project include excavating and regrading over 200,000 cubic yards of material, installing three tiers of erosion control benches, drilling 5,607 soil nails in both soil and rock, and installing approximately 554,000 square feet of DTWM and TRM (District 07, 2021). Proposed construction plans and typical sections are provided in Figures 2A through 2C above.

Geologic Background

The project site is located in the Transverse Ranges Province, an area characterized by east-west trending folds and faults that define the alignments of the mountain ranges. Placerita Canyon is located about one mile south of the project area.

Bedrock in the vicinity of the site is mapped as the Saugus Formation, which consists of light gray to light reddish-brown pebble-cobble conglomerate, sandstone, and siltstones. Figure 3 provides a geologic map of the project area. Orientations of bedding planes appear to be heavily influenced by the San Gabriel Fault, an east-southeast to west-northwest trending right-lateral strike slip fault, which strikes to the northwest through the project area. Bedding planes to the northeast of the SGF generally dip down at low to moderate angles (7 to 63 degrees) towards the fault, and bedrock strata to the southwest of the SGF dip down either to the northwest and west at sub-horizontal to low angles (5 to 22 degrees) or to the southwest at low to moderate angles (33 to 35 degrees). Within the Mint Canyon Quadrangle, several antiforms and synforms are mapped approximately one mile east of the site and trending roughly parallel to the SGF (Dibblee, Jr., 1996). However, the SGF and these antiforms and synforms trend to the southeast into the San Fernando Quadrangle, where the SGF is shown characterizing the structure and alignment of the San Gabriel Fault Zone, which is mapped approximately 2 to 3 miles southeast of the project area (Dibblee, Jr., 1991).

Surficial soils in the vicinity of the project site generally consist of alluvium and high terrace deposits characterized by sands and gravels (Dibblee, Jr., 1991). According to the Caltrans 2021 GDR, two test borings, designated RC-19-001 and RC-19-002, were advanced at the top of the slope to termination depths of 165 and 90 feet below ground surface (ft bgs), respectively. Materials encountered and logged are briefly described in the bullets below (Branch A, 2020). Refer to Table 1 for depths and thicknesses of the units at each boring location.

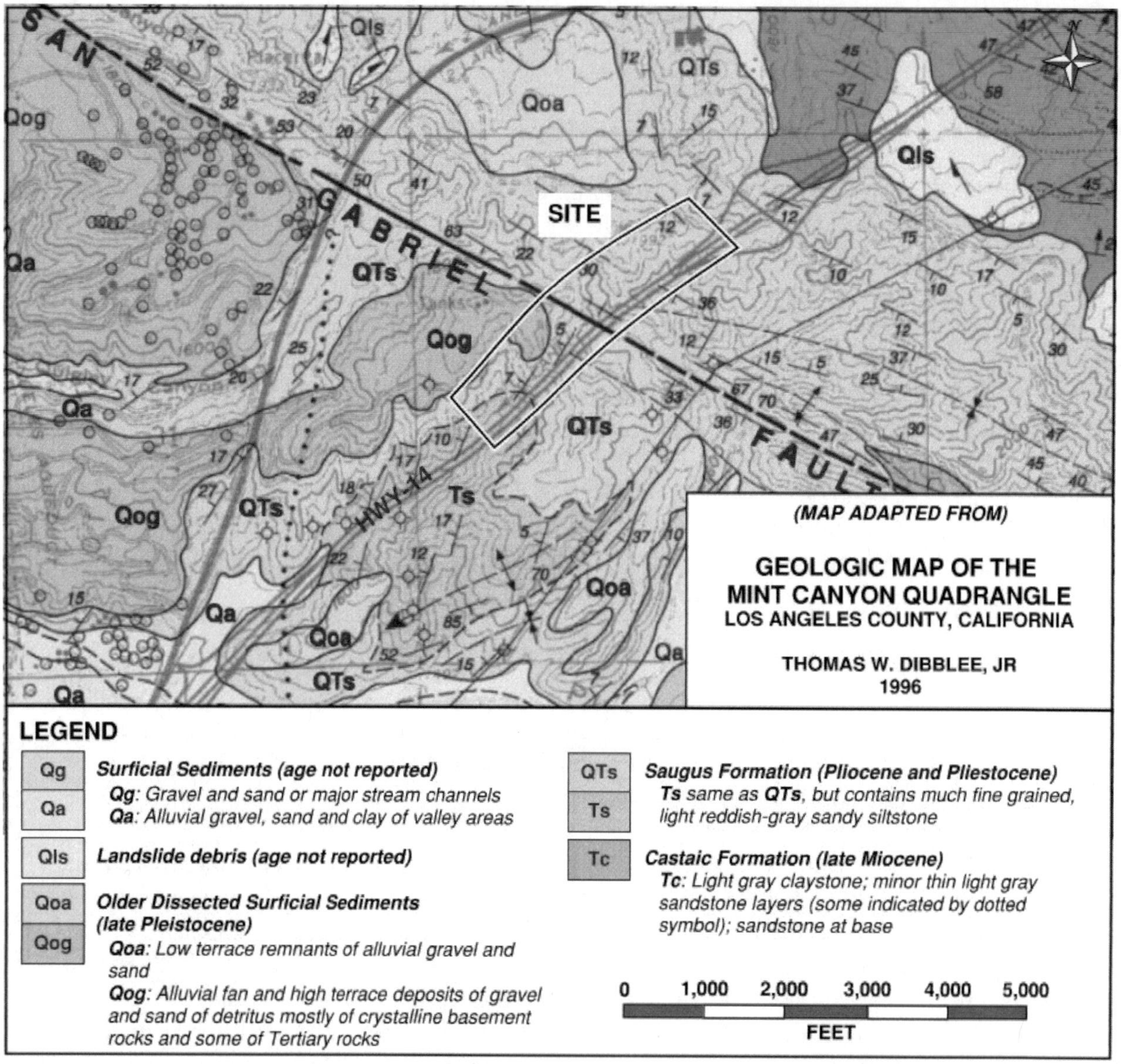

Figure 3 – An adapted geologic map of the project area taken from the 1996 Geologic Map of the Mint Canyon Quadrangle, drafted by Thomas W. Dibblee, Jr. The project area is delineated along Highway 14 on the map for clarity.

Table 1 – Summary of Test Boring Exploration Program

Boring ID	Ground Surface		Overburden			Bedrock		
			Bottom of Unit / Top of Rock		Thickness	Bottom of Exploration		Rock Cored
	(ft bgs)	(El., NAVD)	(ft bgs)	(El., NAVD)	(ft)	(ft bgs)	(El., NAVD)	(ft)
RC-19-001	0	1938.2	99.5	1838.7	99.5	165.5	1772.7	66
RC-19-002	0	1906.9	NE	NA	0	90	1816.9	90

Notes:

ft / ft bgs: feet below ground surface

El., NAVD: Elevations reference the North American Vertical Datum

- Well-graded to poorly-graded SANDS with varying amounts of silt and gravel above approximate. Gravel and boulder concentrations appear to increase with depth until bedrock is encountered.

- When encountered, bedrock at depth was generally described as a light brown to reddish-brown to gray, poorly indurated Sandstone with minor units including thin beds of conglomerate, siltstone, and claystone.

Groundwater was not encountered in the test borings (Branch A, 2020).

According to the California Geological Survey's Landslide Inventory online mapper, no landslides have been reported on the hillslope. However, numerous landslides, including shallow surficial failures to deep rockslides have been reported and mapped within the project region. As such, the presence of these landslides combined with the active and complex tectonic structure of this region further supports the need for further slope stabilization of this Highway 14 rock slope (CA.gov, 2024).

Construction

In 2022, Sukut Construction (Sukut) won the bidding process for the project and mobilized to the site in September 2022. Between September of 2022 and today (this project is currently under and nearing the end of construction), Sukut has been using traditional heavy machinery (i.e., excavators, bulldozers, skid steers, etc.) and construction methods to excavate and regrade the slope to the recommended 1.8H:1V slope grades and create the erosional control mid-slope benches. Sukut also installed an 8-ft tall temporary construction barrier along the north side of the southbound travel lane, which consisted of a fence mounted atop a K-rail, to prevent loose materials from rolling down the slope and hitting the SH-14 travel lanes.

In August 2023, Sukut contracted Access Limited as a subcontractor to install the pinned mesh soil stabilization solution. Accordingly, Access Limited mobilized their crews and equipment to the site in May of 2023. A breakdown of Access Limited's crew and Machinery necessary to complete the scope of work is summarized in Table 2 below.

Table 2 – Crew & Equipment List

Scope of Work	Crew Size		Equipment List	
	Title	No.	Equipment	No.
Install Double-Twisted Wire Mesh & Erosion Control Mat	Superintendent	1	Work Trucks	5
	Ropes access technician	6	LF of Air Hose	100
	Walking excavator operator	1	Forklifts	2
			Komatsu PC228 Excavators	2
			Equipment Trailer including Soil Nail Testing Equipment	1
			Grout Mixer	1
			Kaiser S2 Walking Excavators	1 to 2

Methods for Pinned Mesh Install

Access Limited is performing this scope of work by piggy-backing their construction processes and crews, working their way from the southern treatment area to the north, and top of the slope to bottom. Specifically, the following steps have been and continue to be carried out (refer to Figure 4):

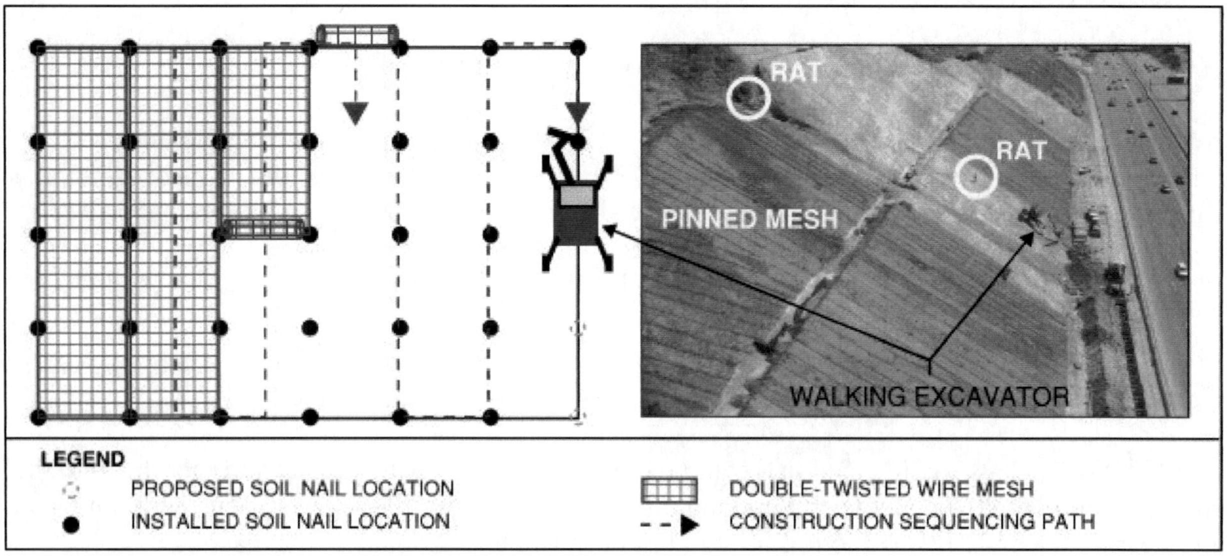

Figure 4 – The figure above attempts to present a visual representation of Access Limited's construction sequencing. The left figure provides a conceptual schematic of the construction sequencing path, and the photo on the right is a snapshot of the project area taken in July 2024 via drone.

1. The spider excavator, operator, and one RAT traverse to the top of a soil nail column line. The spider crawls down to the desired soil nail elevation and drills a borehole to design embedment (between 10 and 15 ft) and generally perpendicular to the slope face, yet must maintain at least 15-degrees of inclination to facilitate grouting. Once the borehole is drilled, the spider moves down the slope in the same column to drill the next soil nail. Once a column line is completed, the machine crabwalks sideways to the next column line and begins the process again.

2. After a borehole is complete, RATs rappel down to the borehole location and install the soil nail, equipped with centralizers, and tremie grout the dowel into place. Additional RATs are responsible for mixing grout at the top of the slope and pumping it down to the dowel locations using the air hoses. Dowels are installed ensuring that a 6 to 12-in portion of the dowel is sticking out of the ground to later be locked off onto the pinned mesh solution. Dowels are installed in a manner similar to the spider excavator's movement mentioned above, traversing down column lines to maximize efficiency.

3. Once the soil nails in a column line are installed, RATs will roll out the TRM over the treatment area from the top of the slope, adjusting accordingly to avoid the soil nail stickups. Once the TRM is installed, the DTWM can be rolled out over the TRM and soil nails. RATs

then lock off the soil nails to the mesh and TRM using 8 x 8 x 3/8-in spike plates, washers, and hex nuts.

4. Load tests are additionally being performed on soil nails to ensure the design load is achieved. Specifically, 5% of the total production soil nails are being tested to the minimum allowable design load of 5 tons (approximately 11 kips), and a total of two verification tests on soil nails are being completed along the total slope area (District 07, 2022).

Image 5 – Site photo depicts an Access Limited Kaiser S2 walking excavator and operator drilling a soil nail along the SH-14 project slope. Two ropes access technicians are shown providing ground support to the operator, as needed.

Results & Performance Takeaways

As construction activities for the slope repair project are still progressing, the results and discussion provided herein draw on the construction progress and methods to date.

Access Limited's construction process is proving to be quite beneficial. Utilizing the walking excavator and RATs has allowed for several hundred linear feet of dowel to be installed per day. Access Limited initially mobilized some Komatsu excavators to install the lower elements, however the slope grade proved to be too steep for traditional, tracked machines. The RATs at this project also set a new internal company production rate record for the amount of mesh installed in one day using ropes access methods (please note that actual production rates were omitted from this paper for protection of Access Limited's bids against their competitors).

Assessment of Additional Construction Alternatives

The walking excavator has proven to be a pivotal piece of equipment in the construction process. Without the use of this machine, Sukut and/or Access Limited would have to find other methods of installing the pinned mesh elements. Some other alternatives to installing the pinned mesh solution are discussed below.

1. Sukut and/or Access Limited could have mobilized long-reach heavy machinery to install the soil nails. Although often an effective solution, in this scenario, the constructed bench widths would not be wide enough to mobilize and safely mount a long-reach piece of machinery to install the dowels necessary. To reiterate, the slope crest in some areas is approximately 200 ft above the roadway elevation and 400 ft away from the paved CA-14 shoulder, when you include the three to four tiers of 8-10-ft wide benches and the newly regraded 1.8H:1V slopes.

2. Access roads could have been installed for traditional heavy machinery to reach the soil nail locations. However, this would only increase the amount of excavation and regrading needed on the project and may further require temporary design-build solutions to be developed to stabilize the access roads (i.e., temporary shoring, retaining walls, etc.).

Figure 5 – Figure provides an aerial imagery comparison of pre-construction conditions and conditions nearing completion of construction. Imagery is provided by Google Maps. (A) Image depicts pre-construction slope conditions from June 2023. (B) Image depicts slope conditions during construction from March 2024.

3. The slope grades and soil nails could have been constructed and installed in lifts. This would have allowed for at least the drilling portion of the pinned mesh solution to be completed during the excavation and regrading phase. The problem with this alternative is that the slope surfaces and rock dowel locations would be left open to the weathering. There were instances during the construction process where severe rains caused surficial slides on the slope before the full pinned mesh solution could be installed. Accordingly, equipment had to be remobilized to these locations to regrade the recent slide areas. If the soil nails had also been installed and grouted in place, the nails may have been impacted and would also have had to be removed and reinstalled. Furthermore, the mesh would have to be installed after all of the lifts were completed, so it is ultimately more pragmatic for the soil nails and DTWM/TRM to be installed under the same mobilization.

4. RATs could have manually drilled each soil nail using wagon drills and/or pluggers. Although possible, manually drilling all of the proposed soil nails would likely result in one of two scenarios: (a) either the same two-person crew size would not be able to achieve the same amount of production per day as the walking excavator; or, (b) a larger crew size could be mobilized to the site with the correct amount of equipment necessary to achieve the same production, although the labor cost of the project would drastically increase. In Access Limited's experience and depending on the cost, this inflated labor cost can double to triple the cost of the walking excavator installation process.

   Furthermore, ropes access construction methods are inherently taxing on the bodies of laborers and often puts persons in non-ergonomic positions on slope. Ropes access equipment can be extremely heavy while performing construction activities, with laborers generally carrying thirty to fifty pounds of gear and ropes on their person. It can be a very powerful construction method in steep slope terrains, but machinery should be relied upon where feasible to keep laborers safe and healthy.

Image 6 – Image depicts conditions of the project slope taken from a drone as of July 2024, looking west.

CONCLUSIONS

Walking excavators have been in use since the 1960s and are growing in popularity for the geohazard mitigation industry. They are efficient tools that can access a variety of precarious slope conditions, enhance jobsite and laborer safety, improve production rates, and decrease unnecessary construction costs and methods, such as the need for access roads and working platforms for other pieces of traditional machinery. For project design teams, it is important to engage with the general and specialty contractors early in the design phase to further understand construction feasibility of engineered designs.

The walking excavator has proven to be a key piece of machinery needed to efficiently complete the pinned mesh installation scope at the SH-14 project. Access Limited's project schedule is currently right on track and scheduled to be completed in mid-August 2024.

REFERENCES

1. BGS Research. *Shallow geohazards.* British Geological Survey. https://www.bgs.ac.uk/geology-projects/shallow-geohazards/. Accessed July 16, 2024.

2. Branch A. *Geotechnical Design Report (GDR) for Slope Erosion Repair on Southbound Route 14 Postmiles 29.0 to 29.5.* State of California Department of Transportation, March 11, 2020.

3. CA.gov. *Landslide Inventory (Beta).* California Geological Survey of the California Department of Conservation. https://maps.conservation.ca.gov/cgs/lsi/. Access July 18, 2024.

4. Dibblee, Jr., Thomas W. *Geologic Map of the Mint Canyon Quadrangle, Los Angeles County, California.* Santa Barbara Museum of Natural History. 1996.

5. Dibblee, Jr., Thomas W. *Geologic Map of the San Fernando and North 1/2 Van Nuys Quadrangles, Los Angeles County, California.* Santa Barbara Museum of Natural History. 1991.

6. District 07. *Notice to Bidders and Special Provisions for Construction on State Highway in Los Angeles County, in Santa Clarita from 0.9 mile north of Placerita Canyon Road undercrossing to 0.2 mile south of Golden Valley Road overcrossing in District 07 on Route 14.* State of California Department of Transportation, May 2, 2022.

7. District 07. *Project Plans for Construction on State Highway Construction on State Highway in Los Angeles County, in Santa Clarita from 0.9 mile north of Placerita Canyon Road undercrossing to 0.2 mile south of Golden Valley Road overcrossing.* State of California Department of Transportation, December 3, 2021, pp. 1-61.

8. Kaiser AG. An excavator for the mountains – development of the Kaiser mobile walking excavator. *50 Years (1965 – 2015) of KAISER Mobile Walking Excavators,* 2015, pp. 126- 169.

9. Richardson, K. *Everything You Need to Know About Walking "Spider" Excavators.* Machines4U. Spider Excavators. machines4u.com.au/mag/spider-excavators/. Accessed July 16, 2024.

Post-Wildfire Effects on Geotechnical Assets

Katelyn Card
Washington State Department of Transportation
310 Maple Park, Transportation Building
Olympia, Washington 98504
(360) 789-8334
katelyncard@wsdot.wa.gov

Prepared for the 73rd Highway Geology Symposium, September 2024

Acknowledgements

The author would like to thank the individuals/entities for their contributions in the work described:

Samuel Johnston – Washington State Department of Transportation
Marc Fish – Washington State Department of Transportation
Cody Chaussee – Washington State Department of Transportation
Nishanthi Perera – Washington State Department of Transportation
Sebastian Dirringer – Landslide Technology

Disclaimer

Statements and views presented in this paper are strictly those of the author(s), and do not necessarily reflect positions held by their affiliations, the Highway Geology Symposium (HGS), or others acknowledged above. The mention of trade names for commercial products does not imply the approval or endorsement by HGS.

Copyright Notice

ABSTRACT

The Pacific Northwest has experienced a significant increase in wildfires over recent years, driven by factors like prolonged periods of hot and dry weather, combined with an accumulation of fuel due to past forest management practices. Wildfires have not only become more frequent but also more intense, posing threats to communities, wildlife habitats, and air quality across the region. The Washington State Department of Transportation has also observed damage to constructed geotechnical assets across the state following these wildfires. This paper examines two case studies: Bolt Creek Wildfire of 2022 and the Sourdough Wildfire of 2023 and the impacts of those wildfires to constructed geotechnical assets and provides a foundation for developing guidelines to protect and rehabilitate these assets following wildfires.

The first case study focuses on the Bolt Creek Wildfire in 2022 near Skykomish, WA, where an approximate 14,766 acres set fire and damaged a cable net slope protection system along US 2. The second case study investigates impacts to a wire mesh slope protection system on SR 20 following the Sourdough Wildfire in 2023, which burnt about 6,360 acres within the North Cascades National Park. Future implementation for post-wildfire assessments of geotechnical assets should be considered following these wildfires and establishing guidelines for this within regional asset management plans. These guidelines will enhance the reliability of infrastructure to wildfire-induced hazards, ultimately improving the safety and resilience of transportation systems in wildfire-prone regions.

INTRODUCTION

The magnitude and frequency of wildfires in the Pacific Northwest has surged in recent years. Driven by factors such as prolonged droughts and the accumulation of fuel due to poor forest management practices, the amplification in the intensity of these fires poses substantial threats to communities, wildlife habitats, and air quality across the region. Not only do wildfires have immediate impacts but they can cause long term impacts to critical infrastructure. Geotechnical assets that play a vital role in the stability and safety of transportation systems can also be affected by wildfires.

The Washington State Department of Transportation (WSDOT) observed several wildfires over the past 5 years that have impacted known unstable slopes along the highway. Some of these slopes have been previously mitigated as part of the state's Geotechnical Asset Management Plan. This paper will examine post-wildfire impacts of geotechnical assets and WSDOT's response using two case studies (Figure 1), outline guidelines for post-wildfire assessments, and provide recommendations for enhancing infrastructure resilience.

The first case study focuses on the Bolt Creek Wildfire near Skykomish, WA, which scorched approximately 14,766 acres and damaged a cable net slope protection system along US 2. The second case study investigates the Sourdough Wildfire within the North Cascades National Park, which burned about 6,360 acres and impacted a wire mesh slope protection system on SR 20. These case studies will highlight the specific damages incurred and the subsequent rehabilitation efforts, offering valuable insights into the challenges and best practices for managing geotechnical assets post-wildfire.

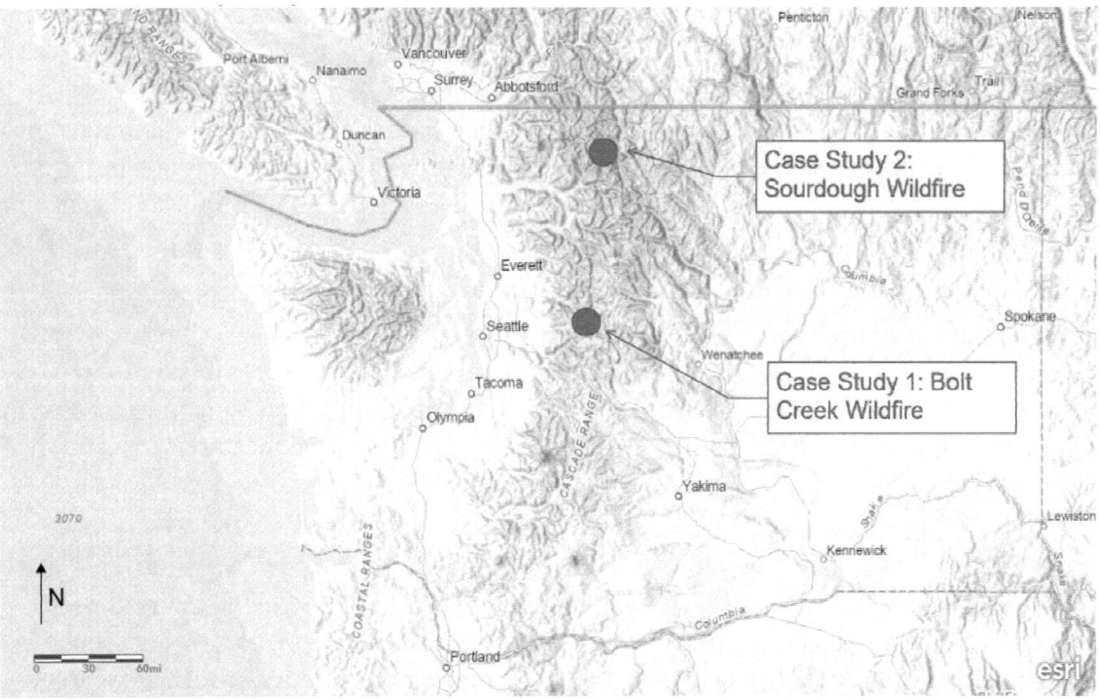

Map for WSDOT Geotechnical field input

Esri, USGS | King County, WA State Parks GIS, Esri, TomTom, Garmin, FAO, NOAA, USGS, Bureau of Land Management, EPA, NPS, USFWS | VaLoy Briscoe, 10/15/2015 | Washington State Department of Transportation | King County, WA State Parks GIS, Esri, TomTom, Garmin, FAO, NOAA, USGS, Bureau of Land Management, EPA, NPS, USFWS

Figure 1: Site map showing the location of the case studies

BACKGROUND

The population growth across the state of Washington has caused greater consequential risk of severe wildfires. Residential areas are now encroaching on forested lands that may benefit from periodic wildfires or controlled burns. This has resulted in the accumulation of fuel which causes fires to burn at higher intensity and cover larger areas. These wildfires have had negative impacts on communities, wildlife habitats, and air quality. For example, wildfire smoke can cause long term heath impacts to people in surrounding areas. While wildlife can typically outrun wildfires, their habitats are being destroyed by the expansion of residential areas. The massive fires throughout the state then destroy their remaining habitats. Wildfire smoke can also cause long term health impacts to wildlife. The ignition source of wildfires varies from anthropogenic causes (out of control campfires or fireworks) to natural causes (lightning strikes).

The Washington State Department of Transportation (WSDOT) has made a commitment to maintaining, preserving, and improving transportation assets for current and future generations. As part of this goal, WSDOT's Geotechnical Asset Management Plan works to reduce risk and build climate preparedness. The program mitigates risks of known unstable slopes and constructed geotechnical assets before they reach an elevated level of risk that would require intervention by outside stakeholders. These objectives are obtained by performing routine inspections on the known unstable slopes and for constructed geotechnical assets. This allows the Unstable Slope Management System (USMS) to keep track of the assets through routine data collection and ensure known unstable slopes are being mitigated before the involvement of WSDOT executive management or the Governor's Office due to an emergency. The unstable slopes are rated based on their level of risk, which is evaluated by the likelihood of failure and the consequence if failure occurs. Unstable slopes that have a high or very high-risk level are considered to be in poor condition and are programmed to be mitigated first. The consistent inspections of geotechnical assets also ensure that the assets are remediated as close to their annualized lowest life cycle cost as possible, and/or when they are no longer functioning as designed. Geotechnical assets are given a rating between A through C, where: an A rating means the asset is in poor condition and no longer functioning as designed, B is in fair condition, and C is in good condition and functioning as designed (*7*).

In recent years as wildfires have become more common and are impacting unstable slopes and geotechnical assets along WSDOT right-of-way, there has been a significant push towards implementing a post-wildfire inspection protocol to ensure the resilience of our transportation infrastructure. The Bolt Creek Wildfire and the Sourdough Wildfires are two major catalysts for this work and examples of how the Agency has dealt with these scenarios in the past couple of years.

CASE STUDY 1: BOLT CREEK WILDFIRE (2022)

The first case study presents the Bolt Creek Wildfire of 2022, which occurred near Skykomish, Washington. This case study will discuss the initial site reconnaissance, specific recommendations made for post-fire damage, and the key lessons learned from the incident. Through examining the impacts on the cable net slope protection system along US 2 and subsequent response to the fire, we aim to highlight critical insights that can inform better preparation and resilience strategies for geotechnical assets in fire-prone areas.

Description of the Bolt Creek Wildfire

At this location, the South Fork Skykomish River meanders down the valley floor and flows westward to the Puget Lowland. US 2 runs parallel with the South Fork Skykomish River. The South Fork Skykomish River Valley has steep slopes on the valley walls that range from 45 degrees to vertical. The primary vegetation consists of Douglas-fir, western hemlock, Pacific silver fir, and mountain hemlock (*3, 10*).

The Bolt Creek Wildfire occurred in September 2022 burning the slopes on the northeast side of US 2 in Washington State from the shoulder of the highway to the crest of Grotto Mountain (Figure 2). The Bolt Creek Fire burned over 12,000 acres of local watersheds, including tributaries to the South Fork Skykomish River (*3, 6, 10*). It was reported as of October 31, 2022, that a total of 14,766 acres had been affected by the wildfire (*4*).

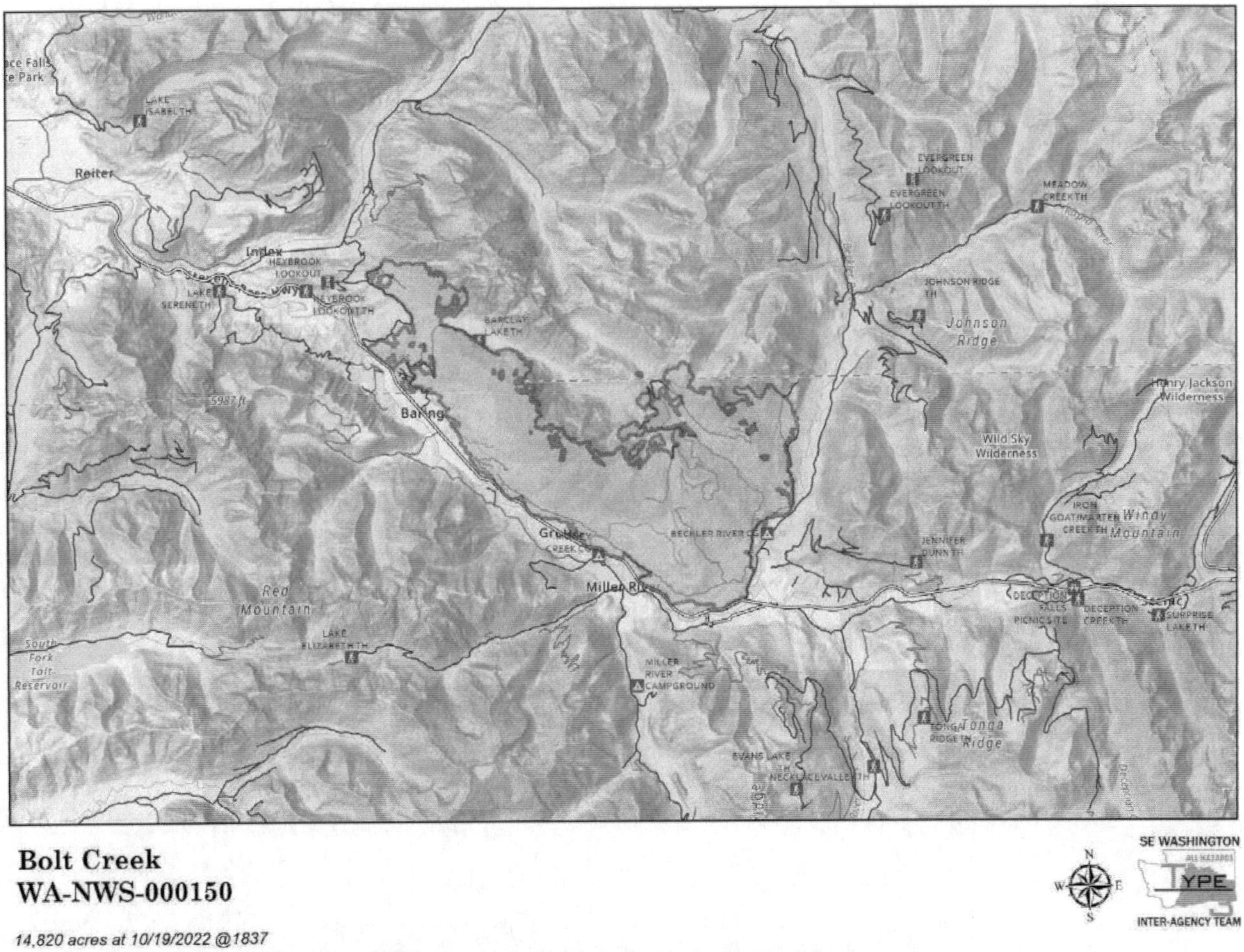

Figure 2: Bolt Creek Wildfire Map (*4*)

Leading up to the wildfire, weather conditions were hot and dry. Burn bans were in effect for the area at the time of the wildfire. Due to strong winds across the region, the fire was able to rapidly spread to the limits of the containment areas (*3*).

The wildfire was reported to have been caused by humans and continued to spread through the naturally accumulated fuel in the surrounding forest. The primary fuel sources consisted of timber litter and understory, closed timber litter, and hardwood litter. The heavy amounts of fuels continued to burn hot until a sudden change in weather pattern brought heavy precipitation to the area (*4*). The fire was reported to burn down to US 2, where firefighters worked to isolate it from crossing over the highway.

Impacts on Geotechnical Assets

Within the burned area along US 2 at the approximate milepost 46.05 is a rock cut that has been previously identified in WSDOT's USMS as Slope #571. The slope varies in height from approximately 50 to 75 feet and is composed of small to large blocky, fractured granodiorite. This slope has a high-risk rating and was programmed to be mitigated. Mitigation of this slope took place in 2005 which included the installation of a cable net slope protection system. The cable net slope protection system was inspected in September 2021 as part of the Geotechnical Asset Management Plan and given a "C" rating, which means the asset was still functioning as designed with little to no repairs needed (*11*).

After the Bolt Creek fire, the Regional WSDOT Office reported fire related damage to the asset. Engineering geologists from the WSDOT State Geotechnical Office responded by conducting a site reconnaissance of the reported damage. The damaged section of cable net was at approximately 20 feet in height. WSDOT Maintenance reported a large tree (approximately 4 feet in diameter) fell upslope of this location, slid down the slope and landed on the highway (*11*) (Figure 3).

Figure 3: Image showing damaged section of cable net slope protection system

During the site visit, the cable net system appeared to be intact, but as the tree slid over the asset it damaged an upper 10-foot section along the crest of the slope. WSDOT engineering geologists identified a failed anchor and associated grout block hanging below the damaged section of the cable net. Despite the failed anchor, the top rope was still above the crest of the slope as designed (*11*).

Although vegetation grew through many places in the slope protection system, but the fire did not make it down to the rock cut. Vegetation above the crest of the slope was scorched resulting in additional trees being dislodged down towards the highway and possibly causing additional damage to the slope protection system. At the time of the inspection, US 2 remained closed because the fire was still active.

Response and Rehabilitation Efforts

Following the initial site visit, the State Geotechnical Office submitted recommendations for repairs to mitigate the damage caused the wildfire. The recommendations included cutting the wire rope anchor that was attached to the suspended grout block and to remove the grout block from the slope. The State Geotechnical Office recommended removing hazardous scorched trees from the crest of the slope and remove additional debris caught in the slope protection system (*11*).

Typically, a failed anchor would need to be replaced to keep the asset in a good state of repair. Upon reviewing the design plans and as-builts of the cable net slope protection system the spacing between anchors was 25 feet. With the failed anchor, the spacing between anchors was now 50 feet. According to WSDOT general special provisions, 50 feet between anchors is within the specifications for the height of this slope so the failed anchor did not need to be replaced, as the horizontal top rope was still above the slope crest and the system overall was still functioning as designed. (*11*).

To remediate this asset, WSDOT maintenance removed all identified hazard trees above the crest of the slope surrounding the cable net slope protection system. Loose debris was removed, and the wire rope anchor was cut so it did not add additional strain to the system. This work was completed during the closure of the highway. It took approximately 5 days to remove the debris from behind the cable net and hazard trees. The highway re-opened on September 24, 2022.

Lessons Learned and Key Insights

It was not necessary to replace the failed anchor because the anchor spacing between the remaining functional anchors was 50 feet, which still complied with WSDOT's general special provisions for cable net slope protection systems of that height, and the system was still functioning as designed.

A key lesson from this experience is the potential benefit of reducing anchor spacing in fire-prone areas. By decreasing the distance between anchors, the system may tolerate individual anchor failures while remaining functional and minimizing the need for extensive repairs. Implementing such adjustments in anchor spacing could enhance the resiliency of slope protection systems in regions susceptible to wildfires, ensuring continued effectiveness and reducing maintenance costs following fire-related damage.

Another impactful mitigation is to proactively remove hazardous trees and understory vegetation along WSDOT right-of-way in fire-prone areas. Regularly clearing potentially dangerous debris and vegetation can reduce the risk of these material affecting assets during a wildfire. This is a preventative measure that can help maintain the integrity and functionality of these systems, further protecting WSDOT's critical infrastructure from wildfire damage.

CASE STUDY 2: SOURDOUGH WILDFIRE (2023)

The second case study presents the Sourdough Wildfire of 2023. This case study will discuss the emergency rockfall response following the fire, the State Geotechnical Office's reconnaissance of the reported damage of geotechnical assets on SR 20, specific recommendations for repairs and enhancements to the damaged geotechnical assets, and the key lessons learned from the incident.

Description of the Sourdough Wildfire

At this location, SR 20 runs parallel with Gorge Lake and Diablo Lake. These lakes were formed by the Gorge Dam and Diablo Dam as part of the Skagit River Hydroelectric Project (*1*) Steep terrain and valley walls are located on either side of the highway ranging from 45 degrees to vertical at some rock cuts. The vegetation consists of mixed-age mixed conifer and some shrub species including vine maple, ocean spray, and birch (*5*).

The Sourdough Wildfire occurred near Newhalem, Washington on July 29, 2023. The fire was located within North Cascades National Park Service Complex in Ross Lake National Recreation Area (Figure 4). The fire burned at very steep inaccessible terrain on the northwest side of SR 20. The fire was contained by October 1, 2023, after burning approximately 6,369 acres (*5*).

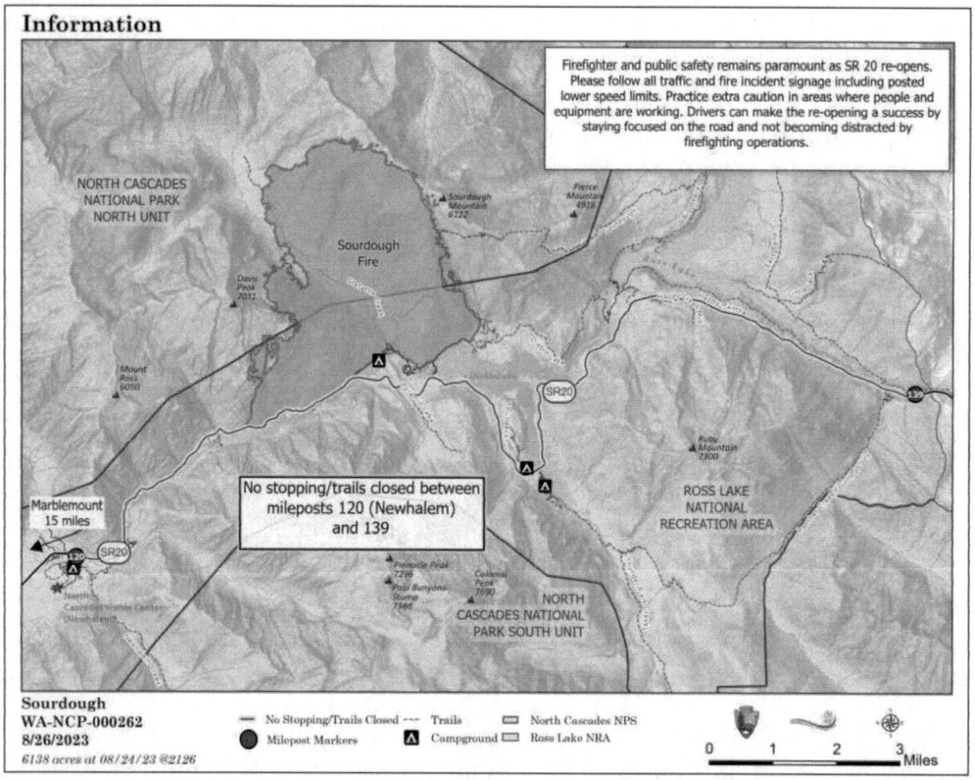

Figure 4: Map of Sourdough Wildfire (*5*)

Weeks leading up to the wildfire consisted of warm temperatures and low humidity. The fire was reportedly caused by a lightning strike. Reports suggest that the terrain and deep timber fuel and understory caused spreading up and down the terrain towards the drainage bottom and then moving back up hill (*5*). The weather continued to remain warm with low humidity as fire response crews were working to contain the wildfire. At the time the fire began, SR 20 was closed due to the hazards presented by the wildfire.

Impact on Geotechnical Assets

WSDOT previously identified the rock cut on SR 20 as an unstable slope in the USMS. This slope is known as Slope #504 and the deficiency description is "rockfall". This slope was previously mitigated with a wire mesh slope protection system in 1981. There were two subsequent contracts for repair work on the wire mesh, the first was in 1987 and the second was in 1992. The wire mesh slope protection system is approximately 400 feet in length along the highway, and is approximately 40 to 100 feet in height, extending from approximately 45 feet above the ditch to approximately 40 feet above the crest of the slope. This wire mesh was installed to reduce the risk of rockfall from the slope. The slope is comprised of moderately to highly fractured banded gneiss and the overburden above the rock cut consists of forest duff. The asset was last inspected in 2019 with a "C" rating, meaning it was functioning as designed with little to no repairs needed (*9*).

WSDOT Maintenance notified the WSDOT State Geotechnical Office of an increase in rockfall along this section of SR 20 following the wildfire. WSDOT contracted Landslide Technology to assess rock slope conditions and evaluate rockfall potential along SR 20 between milepost 124 and 126 where the wildfire intersected the highway. Landslide Technology informed the WSDOT State Geotechnical Office of potential damage to the western portion of the wire mesh slope protection system and unknown conditions of ground anchors. Landslide Technology also reported woody debris in the ditch and trees leaning at the crest of the slope. The report states that "large volume events have impacted this area in the past two years and there is an elevated concern for the risk of additional rockfall following the wildfire" (*8*) (Figure 5).

Figure 5: Photo of reported damage from Landslide Technology Report (*8*)

The WSDOT State Geotechnical Office conducted a detailed site reconnaissance of the damage associated with the Sourdough Wildfire and the wire mesh slope protection system. From the highway, some deformation of the wire mesh was visible, but a more thorough review of the condition of the anchors and wire mesh required rope access techniques. The top rope and lateral anchors had loose nuts on the anchor plates. The intermediate rope termination anchor appeared deformed and impacted by burnt debris (Figure 6). There was a significant accumulation of burnt debris on the mid-slope bench. Three of the wire mesh panels on the western end of the system had significant zinc oxide corrosion from the heat of the fire. An anchor was also observed to have failed due to a large volume rockfall event following the wildfire. This event severed the wire rope tagline that extended from the anchor to the wire mesh. This caused the top rope to sag 15 to 20 feet below the freshly exposed brow (Figure 7) (*9*).

Figure 6: Image showing intermediate rope damage

Figure 7: Image showing top rope sag

On the east side of the system, where it transitions into a taller section of the wire mesh slope protection system, tag lines had burnt debris lying on top and were applying additional stress to the system. This section had been repaired in the 1991 contract and replaced with cable net. Another ground anchor had been pulled out in this area due to the fire, though surrounding anchors appeared to be intact (Figure 8). The anchor spacing at this location is 25 feet between anchors and with the failed anchor the spacing is 50 feet between intact anchors (*9*).

Figure 8: Image showing the compromised anchor on the east side of the system

SR 20 remained closed following the Sourdough Wildfire through the winter. The highway is typically closed to the public during the winter; however, the highway is still maintained by WSDOT Maintenance for Seattle City Light to access the Gorge Dam. The WSDOT State Geotechnical Office determined that there was still a large rockfall risk in the area due the fire damage. The Regional Office requested that the geotechnical asset be repaired as soon as possible to minimize the risk of rockfall to Seattle City Light workers during the winter and to the traveling public once the highway opened in the following Spring.

Response and Rehabilitation Efforts

Following the site visit, the State Geotechnical Office submitted recommendations for repairs to the wire mesh slope protection system. The State Geotechnical Office recommended clearing loose burnt woody debris from the system and clearing and grubbing along the crest of the slope to remove the brush that had grown through the wire mesh. It was also recommended that rock slope safety scaling be performed within the limits of the wire mesh slope protection system to remove the loose rocks trapped within the wire mesh.

Typically, the existing system would be repaired which would include replacing the damaged panels and failed anchors. However, due to the increase in rockfall from upslope associated with burnt vegetation and forest duff, it was determined that it would be appropriate to replace the western end of the system with an attenuated slope protection system (post-supported system) (Figure 9). This would ensure that the additional debris coming from upslope would not roll over the wire mesh and would provide additional containment for the rockfall impacting the highway.

Figure 9: Showing the proposed replacement with the attenuator system

The eastern portion of the slope protection system had far less zinc oxide corrosion than the western portion. It was determined that this section's failed anchor could be replaced and then wire mesh slope protection be connected with the attenuator slope protection installed to the west. Due to the height of the slope protection system on the east side (100 feet), 25-foot anchor spacing is required for WSDOT general special provisions (*9*).

The Region moved quickly to execute an emergency contract to conduct the repairs as recommended by the WSDOT State Geotechnical Office. Repair and replacement work began February 27, 2024. The State Geotechnical Office provided geotechnical construction support during the project. An engineering geologist was onsite and advised the proposed layout of the attenuator posts and proposed extending the system to ensure that the maximum amount of rockfall would be contained by the system. The work was completed in the last week of March 2024.

Lessons Learned and Key Insights

During the rehabilitation process following the Sourdough Wildfire, the detailed inspection revealed significant damage to the wire mesh slope protection system. The fire burned away vegetation, leading to increased rock exposure and subsequent rockfall. To mitigate this elevated risk, the western portion of the system was replaced with an attenuator that can catch debris from further upslope. This strategic decision was crucial as the increased rockfall volume heightened the risk to the highway. The work was completed quickly, which allowed WSDOT to reopen the highway in the Spring as planned to ensure minimal disruption to the transportation network.

The WSDOT representative from the State Geotechnical Office provided geotechnical construction support throughout the project and made critical decisions to extend the boundary of the attenuator to capture additional rockfall from upslope. This proactive measure highlighted the importance of adaptive and responsive decision-making during rehabilitation efforts.

Key insights gained from this project include recognizing how wildfires can alter failure dynamics of already unstable slopes which can increase rockfall frequencies, requiring updated designs for slope protection systems. The successful implementation of the attenuator demonstrates the effectiveness of such adaptive measures in mitigating post-wildfire geotechnical risks and ensuring the resilience of transportation infrastructure.

FUTURE IMPLEMENTATION FOR POST-WILDFIRE ASSESSMENTS

This section discusses the importance of implementing post-wildfire assessments for impacted geotechnical assets. Conducting these inspections directly following a wildfire, when it is safe to do so, is essential to ensure the integrity and safety of WSDOT infrastructure. Typical protocols for these assessments and the new guidelines developed may be used for future asset inspections, especially post-wildfire. Additionally, future advanced technologies to determine structural damage to the steel (the main component in geotechnical assets) will be proposed as future work. As wildfires become increasingly frequent due to climate change, characterized by warming temperatures and prolonged droughts in the Pacific Northwest, adopting these practices is becoming critical and necessary.

Importance of Post-Wildfire Assessments

As part of our Geotechnical Asset Management Plan, the WSDOT State Geotechnical Office routinely inspects our geotechnical assets to verify that they remain well-maintained and that they operate as designed. Doing these inspections ensures that the traveling public remains safe by reducing the risk of slope failures that could impede the highway systems throughout the state. Due to the increased risk of wildfire related damage, we may have to consider changes to our designs to remain resilient to wildfires.

These case studies highlight the potential for a wildfire to damage ground anchors, to oxidize steel, and to increase rockfall frequency and dynamics of failures. In the Bolt Creek Wildfire case study, a tree slid down from up slope and damaged a ground anchor. Had the spacing not already been adequate without that anchor, this tree hazard would have resulted in a costly repair and emergency contract. Future implementation of shorter anchor spacing in wildfire-prone areas may assist in preserving the system's intended functionality of these types of slope protection systems if anchors fail due to wildfires.

In the Sourdough Wildfire case study, we determined that the wire mesh slope protection system had significant zinc oxide damage to the steel. This caused a significant reduction in strength of the steel. While not quantified, we inferred that this damage led to holes in the panels and an overall reduction in strength by visual observation.

During a thorough site investigation, it was determined that rockfall originating from upslope was significantly increased due to fire destroying the vegetation. This led to our decision to install the attenuator system as a replacement for the wire mesh slope protection system on the west end. Without the inspections, the damaged geotechnical asset and unstable slope posed a significant risk to the traveling public along SR 20.

Proposed Guidelines

While the State Geotechnical Office is already conducting regular inspections of our known unstable slopes and constructed geotechnical assets, it is imperative that we also conduct additional inspections following a wildfire. The following steps include a protocol and guidelines for conducting inspections of geotechnical assets following the wildfire and how to proceed with the gathered information.

1. Safety: Ensure that the wildfire-affected area is safe for inspection. Coordinate with local relevant authorities to obtain clearance and necessary safety protocols.

2. Initial Visual Inspection: Conduct a visual inspection to determine any potential signs of damage to the asset. This may include but are not limited to damage to netting, dislodged components, deformation of the system, exposed areas that may have previously been protected, and material degradation. Identify if zinc oxide corrosion is present. This may impact the integrity of the strength of the steel and cause damage if additional forces are applied to the system.

3. Structural Integrity Evaluation: Assess the structural integrity by examining the anchoring systems, support structures, and material condition. It is important to verify the current state of the system against the as-built records and count the number of components to check that they are all still intact on the slope.

4. Stability Assessment: Evaluate the stability of the slopes and embankments protected by the geotechnical assets. Look for signs of erosion, slope instability, or potential rockfall risk as the fire may have exacerbated the risk of failure. It is important to note changes in vegetation, visible tension cracks, debris flow transportation paths, and increases rockfall accumulation in the ditch. It is important to monitor water flow in the area and note any changes due to the increased risk of post-wildfire debris flow hazards (*2*).

5. Instrumentation Inspection: If instrumentation is installed within the vicinity of a geotechnical asset, check the functionality of the instruments, and collect a reading to ensure no significant changes have occurred. Ensure that monitoring equipment such as

inclinometers, piezometers, and ground movement sensors are operational and accurately recording data as needed.

6. Documentation and Reporting: Document all findings from the assessments including photographs of reported damage with measurements and any observations of slope instability. This report may include geotechnical recommendations for the observed damage if applicable. If necessary, highlight the need for enhanced mitigation due to the increase in failure frequency due to the wildfire. Any structural components should be repaired to keep the asset functioning as designed and minimize the risk of slope failures inundating the highway.
7. Implementation of Mitigation Measures: Based on the assessment and recommendations, implement appropriate mitigation measures to repair, reinforce, or replace damaged geotechnical assets. Prioritize actions that enhance the resilience to future wildfire events, such as adding anchors to ensure anchor spacing is appropriate in the case of a failure. These actions should promote the long-term stability of the system and ensure the resiliency of transportation infrastructure.
8. Monitoring and Maintenance: Continue regular monitoring of the geotechnical assets to ensure that repairs continue to function as designed. WSDOT generally inspects A-rated assets annually, B-rated assets every 3 years, and C-rated assets every 5 years. Following a wildfire, the frequency of the asset should increase for the following 3 years. The asset and slope should be inspected after 1 year and again 2 years later to determine that the mitigation is continuing to function as designed and that no additional repairs or rehabilitation of the slope are required.

Future Work

Future work to understand how geotechnical assets are affected by wildfires will involve conducting experiments to comprehensively understand how the strength of steel is impacted by high temperatures. This will include performing in-house testing where steel samples are subjected to temperatures equivalent to those experienced during a wildfire, followed by tension and compressive strength tests to determine the viability of the steel post-heat treatment. It will be important to make sure the testing regime will include temperatures reported during specific wildfires. This data will serve as a baseline for understanding the effects that wildfire can have on geotechnical assets. The results of the test can be compared to the reported temperatures during specific wildfires to assess if the steel behaved similarly in the lab as it did in the field. This type of comprehensive testing will give us a complete understanding of the reduction of the strength of geotechnical assets due to wildfire hazards.

Continuing this research is crucial, as the Pacific Northwest faces the growing threat of increased wildfires. Insights gained from these tests will be instrumental in the planning and rehabilitation of geotechnical assets following wildfires. Understanding how wildfires affect the structural integrity of steel components can help us better plan for repairs of our geotechnical assets.

CONCLUSION

Increasing frequency and intensity of wildfires in the Pacific Northwest present significant challenges to the resilience and safety of highway infrastructure and constructed geotechnical assets. Through the examination of the two case studies, Bolt Creek Wildfire and Sourdough Wildfire, this paper highlights the serious impacts of wildfires on slope protection systems and provides valuable insights for future planning of emergency response and rehabilitation efforts of these systems.

A key lesson learned from these case studies is the necessity of incorporating design changes to geotechnical asset plans in wildfire-prone areas. This will ensure that if the components fail, the asset will continue to function as designed. Additionally, it is important to be flexible in the implementation of these mitigation measures during construction.

It is imperative to adopt comprehensive post-wildfire assessment protocols to accurately assess structural damage to geotechnical assets. The guidelines provide a steppingstone for a resilient transportation system. Future research will provide key information on the structural integrity of steel components following wildfires. The proposed research will provide essential baseline data that can be used to understand the exact effects of wildfires on geotechnical assets. This understanding will help in future planning and repairing of geotechnical assets in wildfire-prone regions.

As climate change continues to exacerbate wildfire conditions, these measures will be vital in ensuring the ongoing safety and functionality of transportation infrastructure. By integrating these practices into regional asset management plans, we can enhance the resilience of our geotechnical systems and better protect communities and critical infrastructure from wildfires.

REFERENCES:

1. City of Seattle. Skagit River Hydroelectric Project. Skagit River Hydroelectric Project - City Light. https://www.seattle.gov/city-light/energy/power-supply-and-delivery/hydroelectric-resources/skagit-river-hydroelectric-project. Accessed July 20, 2024.
2. De Graff, J. V. Improvement in quantifying debris flow risk for post-wildfire emergency response. Geoenvironmental Disasters, vol. 1. 2014. pp. 1-10.
3. Forest Service, Burned-Area Report for the Bolt Creek Fire, Report #FS-2500-8 (2/20). United States Department of Agriculture, 2022.
4. InciWeb. Bolt Creek Fire Incident Information. InciWeb Wildfire. https://inciweb.wildfire.gov/incident-information/wanws-bolt-creek-fire. Accessed July 11, 2024.
5. InciWeb. Sourdough Fire Incident Information. InciWeb Wildfire. https://inciweb.wildfire.gov/incident-information/warlp-sourdough-fire. Accessed July 11, 2024.
6. Mickelson, K., Allen, M. Wildfire-Associated Landslide Emergency Response Team Report – Bolt Creek, Suiattle River, Boulder Lake, and Lake Toketie Fires, King and Snohomish Counties, Washington State Department of Natural Resources, Washington Geological Survey. 2022.
7. Washington State Department of Transportation, Geotechnical Asset Management Plan, 2024.
8. Washington State Department of Transportation, Landslide Technology, SR 20 MP 124.5 Sourdough Fire Complex, Rock Slope Review – Emergency Response, Geotechnical Site Visit Report, 2023.
9. Washington State Department of Transportation, SR 20 MP 125.25 Wildfire Damage to Wire Mesh Slope Protection System, Geotechnical Site Visit Report, 2023.
10. Washington State Department of Transportation, US 2 MP 40 to 50 Bolt Creek Fire Debris Flow Assessment & Recommendations, Geotechnical Memorandum, 2022.
11. Washington State Department of Transportation, US 2 MP 46 Damaged Cable Net, email communication, 2022.

Intermediate Geomaterials In Kansas:

Challenges And An Approach

Jason Kolb
Kansas Department of Transportation
District Five/Area Two- Geology Office
733 SW Purity Springs Drive
El Dorado, KS 67042
(316) 251-6596
Jason.Kolb@ks.gov

Prepared for the 73rd Highway Geology Symposium, September, 2024

Disclaimer

Statements and views presented in this paper are strictly those of the author(s), and do not necessarily reflect positions held by their affiliations, the Highway Geology Symposium (HGS), or others acknowledged above. The mention of trade names for commercial products does not imply the approval or endorsement by HGS.

Copyright Notice

ABSTRACT

Intermediate geomaterials (IGMs) can present some unique challenges to the geologist seeking to provide engineering recommendations for highway improvements. In southcentral and southwestern Kansas, areas blessed (or cursed?) with an abundance of IGM, the Kansas Department of Transportation (KDOT) Geology Section has wrestled with some of these difficulties firsthand. In this paper, the challenges presented to the geologist by these IGMs will be examined. These include the heterogeneous nature of the IGM present in these areas and the lack of an ideal sampling, testing, or interpretive method for understanding and characterizing these IGMs. Also explored will be an approach utilized by KDOT Geology when working with these IGMs. The approach is flexible in that it varies depending on the IGM being dealt with and is integrative in that it combines many sampling, testing, and interpretive methods together and views their results in the light of past geologic experience. While the one 'secret weapon' or 'trick' for dealing with IGMs in southcentral and southwestern Kansas has yet to be discovered, experience has shown that working with them can be more bearable when utilizing something like the flexible, integrative approach outlined in this paper.

INTRODUCTION

IGM Definition

Intermediate geomaterials (IGMs) are geomaterials that exist at the transition between soil and rock. IGMs tend to possess some of the characteristics of both soils and rocks simultaneously, and their behavior can be soil-like, rock-like, or somewhere in between. Mokwa and Brooks noted in 2008 (*1*) that material strength is sometimes used to distinguish IGMs and unconfined compression values in the range of 12.5 to 260 ksf or SPT N-values greater than 50 blows per 1 foot have been identified for this purpose. IGM is a more recent term, and historically other terms were more common such as soft rock or weak rock.

General Challenges

Investigating project sites that contain IGMs can be a real challenge for field personnel. Many commonly used sample collecting methods are geared either toward sampling soil, such as sampling via thin walled tubes (Shelby tubes) or split barrel samplers while performing Standard Penetration Tests (SPTs); or geared toward sampling rock, such as sampling via rock coring. Sampling IGMs by these common methods can be difficult. For example, an IGM may be too dense and hard for a thin-walled tube (Shelby tube) to be advanced into it. Simultaneously, the same IGM may not core well due to a low recovery and a poor Rock Quality Designation (RQD) during the rock coring process. Even when samples can be collected it has been observed by Mokwa and Brooks in 2008 (*1*) that "currently available sampling methods are not adequate for obtaining reasonably undisturbed samples of IGMs, and laboratory testing methods cannot fully correct for disturbances that inevitably occur when an IGM deposit is sampled using either soil or rock sampling techniques.".

The geologist or geotechnical engineer working with IGMs in the design phase of the project can experience difficulties as well. As previously mentioned, the data collected during the investigation of the project site may not be great in terms of how much there is and how reliable it is. Additionally, there is not always a consensus on design methods to use for foundation elements in IGMs. Most familiar design methods are for foundation elements in either a type of soil or rock. Sometimes it is recommended to pigeonhole IGMs into a soil or rock classification during analysis. These factors can lead to poor design. Even something as simple as identifying rock excavation versus common excavation on cross sections for the project plans can also be made painful when the geomaterials at the project site are IGMs. IGMs do not always organize easily and neatly into the binary common excavation versus rock excavation.

Objective

The above list of general challenges posed to field personnel, geologists, and geotechnical engineers by IGMs is not exhaustive. The Kansas Department of Transportation (KDOT) Geology Section has wrestled with some of the above challenges as well as some unique challenges brought forth by some of the IGMs present in central and western Kansas. In what follows, some of the unique challenges posed to KDOT Geology by some Kansas IGMs will be examined. An approach utilized by KDOT Geology for dealing with these IGMs will then be presented. The approach is flexible in that it varies depending on the IGM being dealt with and is integrative in that it combines many sampling, testing, and interpretive methods together and views their results in the light of past geologic experience. It is hoped that other agencies may be able to learn something from this approach that can help them in their own difficulties with IGMs.

IGMS OF KANSAS

Classification

According to Mokwa and Brooks in 2008 (*1*) a common way to classify IGMs is by differentiating them based on properties such as cohesiveness. In Kansas, IGMs are generally classified and differentiated by KDOT Geology based upon the geologic formation in which they occur. For example, an IGM encountered within the Wellington Formation while working on a project in the city of Wichita would immediately classify it as something like a 'Wellington Formation IGM'. This classification would then allow KDOT Geology to keep in mind specific challenges that have come up in the past while working with Wellington Formation IGMs. Methods that have worked well while dealing with Wellington Formation IGMs could also be referenced. Two commonly encountered Kansas IGMs that represent well the types of IGMs present in the state are presented below along with some of the unique characteristics and challenges they pose.

Wellington Formation IGMs

In Kansas the Permian aged Wellington Formation is present near the ground surface in a band running from north to south through the central part of the state. The city of Wichita is founded upon alluvial materials overlying the Wellington Formation and because of this the Wellington is frequently dealt with by KDOT Geology. Zeller in 1968 (*2*) referencing Swineford in 1955 (*3*) provides a general description of the Wellington Formation in the following quote:

> *"The Wellington is predominantly shale with minor amounts of limestone and dolomite, siltstone, and gypsum and anhydrite (Swineford, 1955). The shales are chiefly gray and greenish-gray, with some red, maroon, and purple shale. The limestones and dolomites are generally light colored and argillaceous. Thick beds of salt are present in the subsurface. The Wellington includes marine and brackish- and fresh-water deposits."*

When it is encountered by KDOT Geology most of the geomaterial within the Wellington Formation can be classified as an IGM. More specifically, as clayey and silty shale makes up a large part of the Wellington Formation, Wellington Formation IGMs can safely be thought of as cohesive IGMs. Outcrops of the Wellington Formation are extremely rare. Sampling Wellington Formation IGMs via rock coring can be difficult due to the fragile nature of the material. Rock coring these IGMs for core samples (see Figure 1) is still a preferred sampling method of KDOT Geology and sample sized pieces can usually be collected.

Figure 1- Wellington Formation IGM Core Samples

Specific Challenges

Wellington Formation IGMs are commonly interbedded with gypsum and anhydrite. This gypsum and anhydrite occur within the Wellington Formation in various thicknesses, from very thin veins all the way up to layers multiple feet thick. When it occurs in veins the gypsum and anhydrite seem to be more irregular in how they are oriented. When it is present in thicker layers the gypsum and anhydrite are bedded in more regular, consistent layers. As the gypsum and anhydrite generally possess a higher shear strength than the Wellington Formation IGMs it lends a heterogeneous character to the Wellington Formation as a whole that can complicate the project. For example, irregular veins of gypsum were noted in core samples at a bridge site. Will these irregular veins of gypsum affect pile penetration during driving? Or maybe a thick layer of gypsum is present under a bridge site. Will this thick layer of gypsum be encountered by proposed drilled shafts and is that a problem due to potential solubility of this gypsum?

Drilled shaft design recommendations have been particularly difficult in Wellington Formation IGMs. The main problem KDOT Geology has encountered during design is the low axial resistances and high settlements calculated using drilled shaft design methods. This is primarily due to the very low q_u numbers from sampled rock cores of Wellington Formation IGM. Shafts must be designed with a very large diameter and very deep to build up any kind of axial resistance and prevent excessive settlement.

The construction of the drilled shaft foundations within Wellington Formation IGMs has been problematic as well. As previously mentioned, shafts must be designed deeper and larger diameter than usual to work within Wellington Formation IGMs. Drilling shafts larger diameter and deeper can be more of a challenge for drilling contractors. Further complicating matters is the fact that the geomaterial itself does not lend itself to a clean shaft once drilled. Even with concrete placement immediately after the shaft has been drilled and cleaned out, anomalies are often picked up during crosshole sonic logging (CSL) testing which can be costly and take time to fix.

Ogallala Formation IGMs

The Miocene to Pliocene aged Ogallala Formation exists near the ground surface in large swathes of western Kansas. The Ogallala Formation derives from streams and rivers carrying eroded material from the Rocky Mountain uplift in the west. Zeller from 1968 (*2*) again with the description:

> *"The Ogallala is massive to cross-bedded, generally arkosic gravel, sand, and silt, locally cemented with calcium carbonate. It is greenish-gray, pink, red, tan, and ash-gray in color. It contains limestone, volcanic ash, diatomaceous marl, opaline, sandstone, and bentonitic clay. It contains diagnostic vertebrate and plant fossils."*

During our investigations KDOT Geology commonly finds the Ogallala Formation made up of a mixture of cohesionless sand and gravel and weakly to moderately cemented sand and gravel (see Figure 2). These weakly to moderately cemented sand and gravel layers and zones are generally referred to as 'mortar beds' and can be classified as IGMs. The orientation of these Ogallala Formation IGM 'mortar beds' usually varies both vertically and horizontally. Rarely well cemented, dense sandstone 'mortar beds' are encountered within the Ogallala Formation, and these can be classified as rock.

Figure 2- Ogallala Formation IGM 'mortar bed' outcrop

Specific Challenges

As previously mentioned, the orientation of these Ogallala Formation IGM 'mortar beds' usually varies both vertically and horizontally. This makes determining the exact extent of these Ogallala Formation IGM 'mortar beds' below ground difficult. Degrees of cementation within these Ogallala Formation IGM 'mortar beds' can also vary significantly so that infrequent or non-continuous sampling often doesn't capture the true character of these deposits. These qualities make Ogallala Formation IGMs challenging to deal with in a couple ways.

Firstly, it makes identifying rock excavation versus common excavation on project plans tricky. Buried nodules or pockets of cemented Ogallala Formation IGM 'mortar beds' can remain undiscovered during investigations due to their being present in between borings. This can lead to these buried nodules or pockets of Ogallala Formation IGM 'mortar beds' being misclassified as common excavation when in fact there should be shown as rock excavation on the project plans.

It also makes predicting how driven pile will behave when driven into Ogallala Formation very difficult. Of the traditional sampling methods utilized by KDOT Geology; rock coring, Shelby Tubes, and SPTs, only SPTs can reliably sample the Ogallala Formation. As these SPTs are not continuous but only occur every 5 feet, the variable nature of the Ogallala Formation is not captured. This means a dense Ogallala Formation IGM 'mortar bed' could be missed by this type of sampling. Problems can also occur when drilling occurs adjacent to proposed pile driving locations instead of right on them. Even if the material is well understood and captured where drilling occurs the variability within the Ogallala Formation means the material could be quite different at the adjacent location where pile is to be driven. This is important because if pile refuse shallower than anticipated during construction due to an unforeseen denser 'mortar bed' it may mean that pile has not advanced deep enough to support the lateral loads of the bridge. Or, on the other hand, if pile must be advanced deeper than anticipated during construction it may mean costly change orders.

APPROACH

Basic Structure

The basic structure of the general approach utilized by KDOT Geology for dealing with project sites containing IGMs of the state is shown below-

- Investigate the project site thoroughly by performing a high volume of various kinds of sampling and testing techniques at pertinent locations.
- Perform multiple analytical methods pre-construction for foundation design.
- Identify 'sister projects'/projects in the same general area with similar geology and see what sampling, testing, and analytical methods worked best at characterizing IGMs and determining how they would behave during construction. Put a greater emphasis on methods that have worked well and emphasize less methods that have not worked well.
- During construction, visit the site to monitor work and perform testing such as dynamic analyses with the Pile Driving Analyzer (PDA) and the Case Pile Wave Analysis Program (CAPWAP) analyses for driven pile and CSL for drilled shafts.

Each step of this approach will be expanded on in what follows.

Site Investigation

The investigation at project sites containing IGMs should include performing a higher frequency of borings and tests to characterize the geology. This is due to the commonly variable, heterogeneous nature of the IGMs themselves. A poor understanding of the IGM will be had with infrequent borings and testing because all the characteristics of the IGM will not be captured. Performing borings and tests at pertinent locations as close as possible to all proposed foundation locations and cut sections for excavation is also important. This is due again to the variable, heterogeneous character of the IGM. KDOT Geology has found sampling often and in important locations is especially important while working in the Ogallala Formation IGMs due to their extremely heterogeneous character.

Sampling and Testing methods

A list of sampling and testing methods with a brief word about each is provided (see Table 1). As general rule of thumb, if the sampling or testing method is possible in the given IGM and provides even a little useful data then it is performed. Relying on only one sampling or testing method to characterize an IGM doesn't work well in our experience because no one sampling or testing method alone provides one with an adequate understanding of the IGM.

Table 1 – Sampling and Testing methods for IGMs	
Method	Remarks
Coring (see Figure 3)	Coring IGMs is always conducted where possible. It provides the geologist/geotech with a continuous undisturbed sample for inspection. Samples can be tested via unconfined compression tests for q_u quantities which are valuable for many design procedures.
SPTs	A versatile/easily performed test in IGMs. Can be utilized in many design procedures though KDOT Geology has had less success with these SPT based design procedures.
Shelby Tubes	IGMs are typically too hard and dense to be successfully sampled via Shelby Tube. It could be an option in very soft IGMs though.
Drive Points	Drive Point tests are performed by driving small diameter steel rods continuously into the ground via the autohammer of the drill rig while recording the blows/foot. Though a new testing method for KDOT Geology it has shown that it could be useful in anticipating how driven pile will behave in Ogallala Formation IGMs.
CPTs	IGMs are typically too hard and dense to be successfully sampled via KDOT Geology CPT testing equipment. It could be possible in very soft IGMs or with a more heavy-duty CPT set-up.
Seismic Surveys	Seismic Surveys performed by KDOT Geology determine the shear wave velocity along profiles of project sites. It is hoped that correlations can be identified with this testing method and things such as the locations of IGMs and how driven pile will behave in them.
Power Augers	Helpful for determining the extent of IGMs. Quick and easy.

Figure 3- Coring

Some sampling methods are better adapted to sampling certain IGMs than others. For example, coring works well for sampling more cohesive material like Wellington Formation IGMs but is usually unsuitable for sampling more cohesionless or weakly cemented material like Ogallala Formation IGMs. SPTs and Drive Points perform better in Ogallala Formation IGMs.

Analytical Methods

Analytical methods are commonly utilized by KDOT Geology to provide specific foundation recommendations pre-construction. Various static methods are used pre-construction to predict how pile will behave when driven into IGMs. These pre-construction static methods for driven pile in IGMs can be quite inaccurate and are used more to get a rough idea how pile will behave for the project plans. The SHAFT 2023 program is the primary tool utilized to determine drilled shaft recommendations. Space is limited here but shown below is a general idea of how analytical methods are used by KDOT Geology to provide recommendations for driven pile and drilled shafts in IGMs.

Driven Pile

Various static methods are used by KDOT Geology to try and predict pile penetrations and resistances pre-construction. The traditional way to deal with pile driven into IGMs is to input SPT, Shelby Tube, or even rock core sample data in the computer program DrivenPiles (*4*) and analyze. The DrivenPiles program employs the computational methods of the older FHWA DRIVEN (*5*) program with some minor changes. Utilizing DrivenPiles for pile driven to IGMs has been met with limited success. Often during pile driving, pile penetrations and resistances measured while performing Pile Driving Analyzer (PDA) tests and the Case Pile Wave Analysis Program (CAPWAP) (*6*) analyses are much different than those predicted pre-construction by DrivenPiles analysis. Mokwa and Brooks in 2008 (*1*) have shown that CAPWAP capacities of pile driven into IGMs are highly accurate when compared to static load tests. This means that the fault lies with DrivenPiles when DrivenPiles and CAPWAP capacities do not match. Despite its weaknesses DrivenPiles is still commonly utilized to get a rough idea how pile might behave during design.

Other static methods are also utilized by KDOT Geology for predicting pile behavior in IGMs but these static methods don't have as much of a history of use with KDOT Geology and it hasn't really been determined yet if these other methods are more accurate or not for the Section. These include the IDOT static method as described by Long, Hendrix, and Baratta in 2009 (*7*) and the method developed specifically for pile driven into IGMs by Ng, Massud, Oluwatuyi, and Wulff in 2022 (*8*). We anticipate that using these other static methods will supplement DrivenPiles and make predictions more accurate.

Drilled Shafts

KDOT Geology's process for determining drilled shaft recommendations during the pre-construction design phase of the project is a little more straightforward. KDOT Geology primarily utilizes Ensoft Inc's Shaft 2023 program (*9*) for drilled shaft recommendations. In the technical manual of the Shaft 2023 program, users are advised to analyze cohesive IGMs under the program's 'weak rock' designation. This 'weak rock' designation utilizes recommendations from O'Neill et al from 1996 (*10*) for estimating axial resistance and settlement of drilled shafts under axial loading. The only difficulty seen here is the tendency for IGM samples to give low unconfined compression q_u values which can lead to very large shaft diameters and rock socket lengths in some cases. Various shaft diameters and lengths are examined with the Shaft 2023 program, and the most economical and easy to construct configurations are recommended in the report.

If gypsum or anhydrite are present within an IGM, as is commonly the case in Wellington Formation IGMs, KDOT Geology generally ensures rock sockets avoid large pockets or layers of this gypsum or anhydrite. This protects the rock socket of the shaft from the effects of the potential dissolution of gypsum or anhydrite in the future. It should also be noted that a common practice among KDOT Geology is checking the final drilled shaft recommendations by performing hand calculations. For a drilled shaft socketed into an IGM that would mean performing the method outlined by O'Neill et al in 1995 (*10*) by hand, the old-fashioned way. This will confirm that the drilled shaft recommendations were reasonable, and no errors were made while running the Shaft 2023 program.

'Sister Projects'

'Sister projects' are projects that share a similar geographic region and possess similar geology. KDOT Geology identifies 'sister projects' to consider how experiences working in similar areas with similar geology can help inform work on current projects. This can be especially useful when working with certain IGMs. Say a geologist must recommend how steep to set a backslope in an IGM. According to their sampling and testing data the IGM appears resistant enough to set the backslope at a 2:1 slope. They could then look at 'sister projects' containing the IGM in question to see if said projects had backslopes within the IGM that were set at a 2:1 and assess how well those slopes have performed. If the slopes had performed well on the 'sister projects' the geologist could then confidently proceed with the 2:1 recommendation. If the slopes had failures, then reconsidering the recommendation would likely be warranted.

'Sister Projects' for Driven Pile Recommendations

KDOT Geology has found identifying 'sister projects' to be exceptionally useful when providing driven pile recommendations in IGMs specifically. For example, a geologist is working on pile recommendations for pile driven to bear within Wellington Formation IGM on a project in north Wichita. The geologist could recall a large project recently completed in north Wichita that contains PDA testing with CAPWAP analyses for pile driven to bear into a very similar Wellington Formation IGM. The geologist could take this data into account when making pile recommendations on the current project they are working on. This could be as simple as looking at a historic 'sister project' to see what type of pile was recommended and how far pile advanced into the IGM before reaching bearing or as complex as figuring out a correction factor from a group of 'sister projects' and applying it to a static analysis to better predict how pile will perform. KDOT Geology has had such success utilizing this PDA data from driven pile 'sister projects' that they have started working with Foundation Testing and Consulting, LLC (FTC) in building and utilizing a driven pile PDA database called Piletrac (*11*).

Piletrac

Piletrac was developed by Foundation Testing and Consulting, LLC (FTC). Piletrac is a database that allows one to visually see various PDA datasets on a series of online dashboards. The PDA data is currently from roughly 120 projects and includes data from roughly 600 PDA test pile drives. These PDA test piles were performed by FTC in Kansas and Missouri and KDOT in Kansas. The dashboards show data such as the type of pile and hammers used, average pile penetration into bedrock, average pile length, average pile resistance, etc. These can be sorted by things such as the project county or type of pile used. The idea behind the database is having a place where all of this information is able to be quickly accessed and recalled to see what works and what does not work in regards to driven pile in specific regions and specific geologic circumstances.

Monitoring and Testing During Construction

Because IGMs in Kansas are difficult to work with KDOT Geology has always made a point to monitor project sites during construction and require PDA testing for driven pile and CSL testing for drilled shafts.

Site Monitoring

On more difficult projects such as those projects containing IGMs, visiting the project site to monitor how the construction process is going is worthwhile. Relationships can often be developed with field personnel so that when problems arise, they know who they can contact for help. KDOT Geology has put a particular emphasis on being on site when investigative core hole data is collected and when drilled shaft construction is taking place.

Testing

KDOT Geologists perform PDA testing and CAPWAP analyses at project sites where pile is driven into IGMs. The frequency of this testing is commonly a PDA test at every other foundation element of the bridge. The reason for these PDA tests is to confirm that adequate resistance is obtained by the driven pile and to avoid overdriving and pile damage. KDOT Geology has also found that performing these PDA tests in person allows the geologist to gain a better understanding of how specific pile behaves in certain circumstances within certain IGMs and thus improves pile recommendations on future projects.

CSL testing and sometimes Thermal Integrity Profiling (TIP) is required on drilled shafts by KDOT Geology. This testing is done to confirm that no anomalies are present within the drilled shaft. KDOT Geology has found that anomalies can commonly occur in drilled shafts in certain Kansas IGMs, so this testing is very important on projects that have drilled shafts socketed into IGM. The results of CSL testing and TIP are reviewed by the Chief Geologist of KDOT and remediation of the shaft is usually required if significant anomalies are detected.

CONCLUSION

To sum things up, IGMs in Kansas are complex and mysterious. A bare bones investigation will not fully capture the character of the IGM. Taking this minimal data and rushing through the design phase utilizing a cookie cutter method compounds this problem. Not taking the time to reflect on past experiences with projects located in a similar area and possessing similar geology is a missed opportunity to refine one's understanding of the IGM and the design. Basically, to understand the IGM and be able to provide quality engineering recommendations for highway improvements what is needed is to cast a wide net, uncovering and collecting every piece of pertinent information possible and integrating it into what is ideally a deep understanding of the IGM and how it will behave. Something like this thorough detailing and integrating in order to understand is what the approach outlined in this paper is after. It is hoped that by presenting this approach the reader of this paper will be able to come away with something that will aid them in their own struggles with IGMs.

REFERENCES:

1. Mokwa, R. L., and Brooks, H. *Axial Capacity of Piles Supported on Intermediate Geomaterials.* Montana Department of Transportation, 2008.

2. Zeller, D. E., ed. *The Stratigraphic Succession in Kansas.* Kansas Geological Survey Bulletin 189, 1968.

3. Swineford, Ada. *Petrography of upper Permian rocks in south-central Kansas.* Kansas Geological Survey Bulletin 111, 1955.

4. DrivenPiles. Version 1.4.0. Computer software by Multidimensional Software Creations, LLC (MDSC).

5. DRIVEN. Version 1.2. Computer software by the Federal Highway Administration.

6. CAPWAP. Version 2014 14.3.6. Computer software by Pile Dynamics, Inc.

7. Long, J. H., J. Hendrix, and A. Baratta. *Evaluation/Modification of IDOT Foundation Piling Design and Construction Policy*. Univ. of Illinois, Illinois Center for Transportation Research Report No. FHWA-ICT-09-037, 2009.

8. Ng, K., Massud, N., Oluwatuyi, O., and Wulff, S.S. *Development of LRFD Recommendations of Driven Piles on Intermediate Geomaterials.* Mountain-Plains Consortium, 2022.

9. SHAFT. Version 2023. Computer software by Ensoft, Inc.

10. O'Neill, M. W., Townsend, F.C., Hassan, K.M., Buller, A. and Chen, P.S. *Load Transfer for Drilled Shafts in Intermediate Geomaterials.* U.S. Department of Transportation Publication No. FHWA-RD-95-172, 1996.

11. Piletrac. Version 2023. Online datasets and dashboards by Foundation Testing and Consulting, LLC.

Commit to Clean:

Reducing the Risk of Invasive Species Spread and Introduction via Field Equipment and Gear

Sara A. Ricklefs
Invasive Species Action Network
215 East Lewis Street, Suite 202
Livingston, MT 59047
(785) 207-7624
sara@stopais.org

Jennifer Riddle
Invasive Species Action Network
215 East Lewis Street, Suite 202
Livingston, MT 59047
(406) 222-7270
jennifer@stopais.org

Erin Bjorklund
Invasive Species Action Network
215 East Lewis Street, Suite 202
Livingston, MT 59047
(406) 222-7270
erin@stopais.org

Prepared for the 73rd Highway Geology Symposium, September, 2024

Acknowledgements

The authors would like to thank the following individuals for their contributions in reviewing earlier drafts of this paper:

Bob Henthorne – Geotechnical Engineering, PEC
Jason Allen – Maintenance Noxious Weed Coordination, Montana Department of Transportation

Disclaimer

Statements and views presented in this paper are strictly those of the author(s), and do not necessarily reflect positions held by their affiliations, the Highway Geology Symposium (HGS), or others acknowledged above. The mention of trade names for commercial products does not imply the approval or endorsement by HGS.

ABSTRACT

Terrestrial and aquatic invasive species have detrimental impacts on ecological and economic systems. These include, but are not limited to the reduction of biodiversity, diminished water quality and/or soil health, fragmented habitat, and infrastructure damage. Prevention is the best tool available for managing new invasions since containment and eradication efforts generally come at great monetary cost and time investment. Often, prevention techniques center around simple measures that individuals can easily implement in their daily routines or procedures. Invasive species prevention campaigns are often directed at recreational communities. However, numerous industries (including, most notably, military branches and wildfire crews) have diligently adopted and reported upon invasive species prevention measures for gear and highly mobile equipment. The risk of invasive species transport on gear and equipment is heightened in the geotechnical industry as equipment is commonly utilized across multiple job sites and in varying areas and systems (including upland, riparian, urban, and rural). This is especially critical in the spring, summer, and early fall when the height of seed release/spawning seasons coincide with an increase in field investigations and construction efforts. Due to soil disturbance at job sites, the risk of one or more invasive species establishing a population is high. Adopting preventative techniques within departmental or agency procedures will not only reduce the risk of invasive species introduction but will also help maintain equipment and gear. Additionally, preventative actions will reduce liability and limit post-project investments from final users, partner firms, neighboring landowners, and governmental agencies. Herein, we discuss preventative measures that can be implemented, the legality of project site requirements, and the benefits of such preventative efforts.

INTRODUCTION

Invasive species are one of the leading causes of biodiversity loss across the globe. Invasive species can contribute to habitat degradation, change natural regimes (e.g., fire, flood), alter nutrient cycling, and degrade or remove ecosystem functions and services. Furthermore, they threaten the public by negatively impacting natural resources, trade, commerce, property values, agriculture, infrastructure, and health (*1*).

Immense research focus has been placed on identifying pathways of introduction to better establish preventative actions and policies (*2*, *3*, *4*, *5*, *6*, *7*, *8*). Unfortunately, introductions are often exacerbated or even solely facilitated by human activity (i.e., humans acting as vectors). One high profile example of accelerated introduction due to human interference would be zebra mussels (*Dreissena polymorpha*) via recreational (*9*, *10*) and commercial boat (*11*, *12*) movement. While dispersal via natural pathways (e.g., birds) is possible, it remains low risk and at a smaller scale than human-caused pathways (*13*). Comparatively, the introduction of the Burmese python (*Python molurus bivittatus*) to southern Florida can be solely attributed to the release of individuals (either accidentally or intentionally) and is likely tied to the pet trade (*14*). The introduction and eventual establishment of these species has had tremendous economic and ecological impact, and they are just two species among many whose invasions (as a result of human interference) are responsible for detrimental impacts. Notable human-caused pathways include recreational (e.g., bait release, movement of seeds on clothing and gear, movement of insects within wood products) and industrial (e.g., domestic and international trade, movement of organisms on equipment, movement of organisms within food and agricultural systems) based activities. Furthermore, species invasions are expected to rise at exponential rates with increased global transport and human movement (*1*, *15*) as well as climatic changes (*16*). This threat crosses all taxa and ecosystems, including species that have not historically been privy to introduction (*17*). These changes require new management methods and outreach for invasive species detection.

The United Nations Intergovernmental Platform on Biodiversity and Ecosystem Services (*1*) estimates that more than 37,000 invasive species are contributing to costs of more than $400 billion annually for monitoring, management, and eradication efforts. Since 1960, researchers have concluded that the United States has spent more than $1.22 trillion on management, damages, or loss caused by invasive species (*18*). Each of these studies (*1, 18*) claim that estimated costs related to invasive species are likely to be severely underestimated as they are reliant upon reporting. While the issue is clearly global in nature, invasive species have monumental impacts on local systems and communities, up to and including native species extinctions and reduced climate resilience. Thus, individual actions worldwide are essential to combat impacts at both a global and regional level.

Actions associated with species invasions (Figure 1) fall into four stages: 1) prevention; 2) eradication; 3) containment; and 4) long-term management or asset-based protection. Prevention of introduction and establishment remains the most cost-effective measure for species invasions. Within this stage, managers largely focus on public outreach, monitoring, and establishing preventative techniques and reporting avenues for voluntary or mandatory employment. Once a species has been introduced, there is a short period of time where eradication may be possible. This length of time is dependent upon early detection and may be impacted by a species' natural history (e.g., fecundity). If a small population is able to establish itself and avoid eradication from management efforts, or is detected too late, containment efforts may be deployed for that species. The goal of which is to reduce the likelihood of an infestation spreading. If containment is not achieved, managers will then work toward long-term management efforts. Each of these stages is less cost-effective than the previous stage.

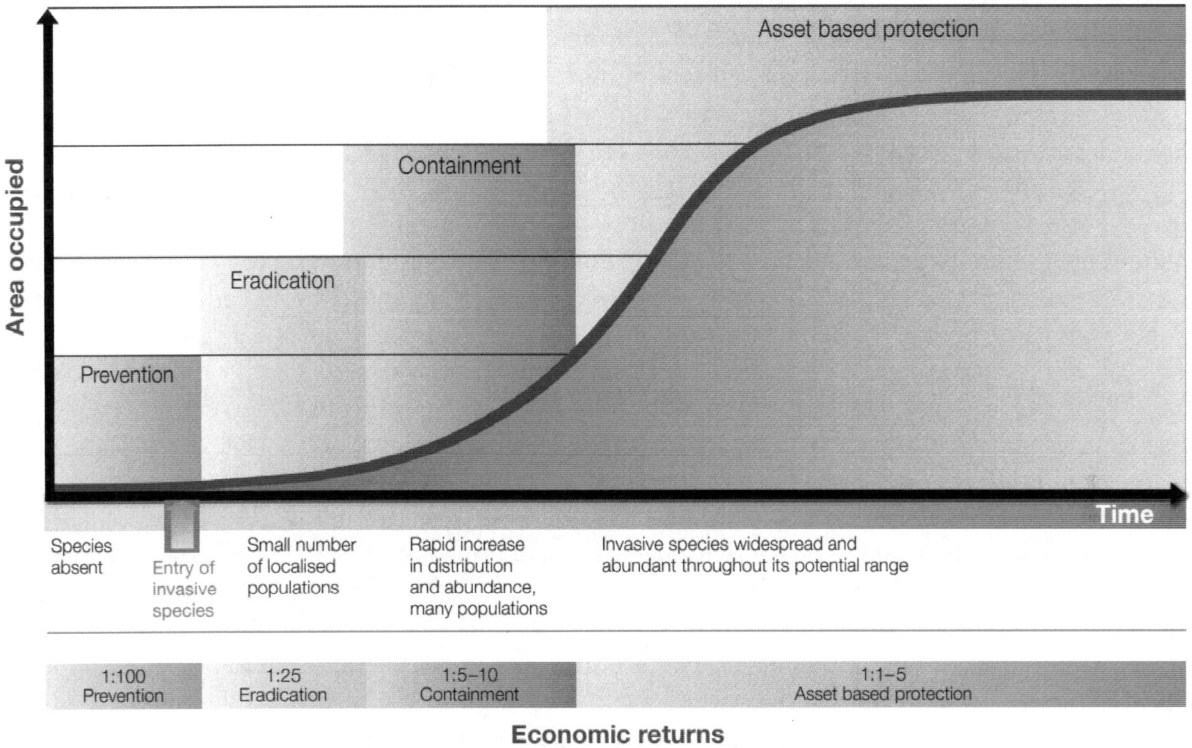

Figure 1 – The invasion curve (directly from *19*) demonstrates cost over time based on infestation size. Preventing invasive species introductions and eventual establishment is key to reducing not just long-term investment in management or control, but also costs associated with damages and loss.

Industry Implications

Many industries, such as wildland firefighting, military service, and transportation, have great potential to propel the spread of invasive species. This risk is in part due to the use of highly mobile crews (many of which are working in both interstate and/or international capacities), large numbers of staff on the ground, and diverse equipment (*20*, *21*, *22*). While wildland fire crews (*23*, *24*, *25*, *26*) and military forces (*27*, *28*) have taken great strides to incorporate invasive species prevention into their operations, the highway transportation industry faces unique challenges and barriers to prevention. One such challenge for construction efforts is the seasonality of work. Major highway construction projects often coincide with spawn and seed release and cause high rates of soil disturbance within project sites and rights-of-way (ROWs). Areas like these construction sites, with widespread disturbance and high rates of development, provide premier habitat for invasive species establishment. Additionally, the high rates of soil disturbance within project sites and ROWs present prime opportunities for invasive species to spread quickly upon introduction. Roadways are intended to connect areas; in that, their purpose also provides opportunities for invasive species to quickly spread along corridors (*29*, *30*). When paired with greater population densities and greater rates of introduction, urban and suburban areas prove to be hot spots for invasive species (*15*). Developing industries, such as the highway construction industry, therefore, can contribute greater influence as invasive species vectors (i.e., individuals who facilitate the spread of invasive species) simply due to the nature and location of their work.

Invasive species' impacts are pervasive, and transportation infrastructure (e.g., highways, railroads, ports, bridges) is not immune. Often, impacts to infrastructure are intertwined with those affecting natural systems and local communities (*31, 32*). For instance, Japanese knotweed (*Reynoutria japonica*) is known to grow through pavement and asphalt (*32*), rapidly weakening structures and subsurfaces. It can also increase erosion, particularly in areas prone to flooding and where knotweed is found in great densities (*33*). Outside of steep investments to repair infrastructure affected and safety hazards to those traveling on damaged infrastructure, rapid monocultures of knotweed can also cause cascading impacts to communities (*34*). Research has found that invertebrate and plant diversity (*35*) decrease with Japanese knotweed encroachment, homogenizing and degrading habitats for other taxa. Some invasive species, such as kudzu (*Pueraria montana*), Japanese knotweed, and giant reed (*Arundo donax*), can be a safety hazard once established in dense stands along roadways by blocking sightlines at intersections and along shoulders (*31, 32*). Consequently, if treated without careful planning, previously dense stands may result in large areas of bare soil and increased erosion (*32*). Lastly, annual invasive grasses [e.g., cheatgrass (*Bromus tectorum*), wild oats (*Avena fatua*)] and tumbleweeds [e.g., kochia (*Kochia scoparia*), Russian thistle (*Salsola tragus*)] may increase fire hazard and fuel sources along roadways (*32, 36, 37*), posing a risk to travelers, maintenance crews, maintenance equipment, and roadside infrastructure.

Additionally, invasive species prevention may be regulated and enforced by public policy. Entitled *Safeguarding the Nation from the Impacts of Invasive Species*, Executive Order 13112 (*38*) provides that it is "the policy of the United States to prevent the introduction, establishment, and spread of invasive species, as well as to eradicate and control populations of invasive species that are established." The Order was written and amended in accordance with and for consistent implementation with the following laws:

- National Environmental Policy Act (*39*)
- Nonindigenous Aquatic Nuisance Prevention and Control Act (*40*)
- Plant Protection Act (*41*)
- Lacey Act (*42*)
- Endangered Species Act (*43*)
- Noxious Weed Control and Eradication Act (*44*)

Each federal agency is required to adhere to this Order, including the prevention, early detection, rapid response, and accurate and reliable monitoring of invasive species within their actions. Additionally, they must work to restore native ecosystems, support research, and promote education and outreach related to invasive species. The Order calls for each agency to collaborate with partners (e.g., states, territories, tribal entities) and ensure that strong frameworks are in place for management of invasive species. The framework set forth in E.O. 13112 (*38*) is both supported and required by the Federal Highway Administration. To gain a comprehensive understanding of regulation in accordance with the transportation industry will also require independent research at the local and state levels. It is the responsibility of practitioners to set forth policies and plans that meet regulatory requirements at all levels. Doing so not only ensures compliance, but also reduces organizational liability and impacts to public health and safety.

Lastly, the transportation industry utilizes diverse equipment and fixed assets. Since invasive species prevention activities often involve regular cleaning (*21*), adoption of these techniques is likely to result in lower maintenance and replacement costs. The upfront time investment for adopting invasive species prevention best practices is minimal in comparison to the potential cost of species invasions' impacts to public health and safety, native ecosystems, eradication or management efforts, transportation infrastructure, organizational liability, and equipment maintenance and replacement. Herein, we provide recommendations specifically for the highway transportation and construction industry to be considered for incorporation into future project planning.

PREVENTION TECHNIQUES

Comprehensive invasive species plans should address prevention, monitoring, and management. Best practices range from individual actions within project-based procedures to department- or agency-wide policies. Project managers and designers should focus on two objectives: 1) prevent invasive species introduction or spread; and 2) reduce disturbance as much as possible. Taking preventative measures at every stage from field investigation through construction will help reduce management actions and costs for end users, neighboring landowners, and maintenance crews. Reducing soil disturbance as much as possible will help prevent post-project spread, as presence of disturbed or bare soil can increase the facilitation of species already present in the seed bank and heightens the establishment risk of those that may be introduced. We've provided recommendations for general consideration, as well as those specific to the planning/design, construction, and post-project phases. These were developed through consultation of federal policy and written state and federal agency protocols, combined with our experiences in invasive species prevention within the outdoor recreational community. Our recommendations do not ensure compliance with or satisfaction of applicable local, state, or federal regulations. The recommendations provided are intended to be a tool for organizations that may be helpful in the development of best management practices for their departments, agencies, or contracting partners.

General Recommendations

1. Collaborate with relevant local, state, and federal agencies to remain informed on current and potential infestations with emphasis on state listed, high priority noxious weeds.

2. Train staff on noxious weed and invasive species identification and reporting mechanisms. Include information on known impacts and associated costs (e.g., cost effectiveness in relation to adoption of preventative measures in comparison to long-term management, damage, and maintenance).

3. Utilize applications or other online tools for reporting (e.g., iNaturalist). Ideally, choose an application that includes geospatial recognition for mapping purposes, photo upload for record confirmation, and is shared with and/or utilized by partnering agencies and/or organizations.

4. Inspect and clean clothing and equipment of all mud, soil, seeds, plant debris, and invertebrates. If water is used to clean, dry items or equipment thoroughly. Perform these

activities at the site to reduce the movement of any existing invasive species off-site. If additional cleaning is performed off-site, be mindful of runoff and drainage to reduce introductions to new environments. Helpful tools include portable wash stations, hand-pumped pressurized water sprayers, pressurized air, handheld brushes, and vacuums.

5. If a watercraft is used, clean the equipment of all plant debris, mud, organisms, and standing water; drain the craft of all water; and dry it completely before launching in a new waterbody. For thorough cleaning with pressurized water, consider taking the watercraft to a car wash after completing the previous steps on-site. If applicable, have it inspected and/or decontaminated at a watercraft inspection and decontamination station.

6. Train staff and implement effective control methods in rights-of-way and project sites. Methods and timing of management efforts are likely to be species-specific.

7. Develop species-specific plans for regionally common invasive species and noxious weeds. Collaborate with relevant local, state, and federal agencies to identify management best practices, including timing of management actions, targeted herbicides or pesticides, short-term and long-term monitoring plans, and utilization of appropriate cultural management practices (e.g., mowing, grazing, mulching, biocontrol). Include relevant decontamination, disinfectant, and disposal processes for excavated materials or debris (e.g., drying, burying, burning) where applicable.

8. Participate in and/or support public outreach for invasive species awareness and prevention.

Planning/Design Recommendations

1. Address and quantify invasive species introduction risk in project planning and design processes. Considerations may include proximity to fragile natural sites (e.g., ecological preserves, critical habitats) or high-risk systems (e.g., marshes, rivers), gaining an understanding of equipment's recent exposure to invasive species or noxious weeds, and unavoidable soil disturbance.

2. Identify and map existing invasive species populations within the project site. Report as necessary, avoid disturbance in these areas, and consider management action prior to construction when necessary. Utilize flagging where appropriate to denote populations for construction crews with emphasis on high priority invasive species and state listed noxious weeds.

3. Minimize soil disturbance.

4. Use potable water or water from the same drainage for geotechnical investigations. Clean all drilling equipment and tanks of all plant debris, mud, organisms, and standing water; drain equipment of all water; and dry it completely prior to moving to the next project. If drainage is infested with Dreissenid mussels, use potable water or take steps to decontaminate all equipment (*21*).

5. Incorporate use of Certified Weed-Free products (e.g., forages for mulching, gravels for shoulders, seeds for revegetation), if available. If Certified Weed-Free products are unavailable, monitor the site for invasive species emergence to enable early detection and rapid response protocols.

Construction Recommendations

1. Refer to project plans to avoid any flagged or reported invasive species within the project site. If new populations are located, report to agency or departmental authority for proper management action or avoidance.

2. Reduce soil disturbance as much as possible.

3. Water used during construction should be potable water or sourced from the same drainage. Clean all water tanks and pumps of all plant debris, mud, organisms, and standing water; drain equipment of all water; and dry it completely prior to moving to the next project. If drainage is infested with Dreissenid mussels, use potable water or take steps to decontaminate all equipment (*21*).

4. Plant native seed mixes (utilize Certified Weed-Free, if available) as soon as possible to reduce occurrences of bare soil. If utilizing straw, hay, gravel, etc. for mulching purposes, incorporate Certified Weed-Free products, if available. Be mindful of local, state, and federal regulations when developing seed mixes and always seek native species. If Certified Weed-Free products are unavailable, monitor the site for invasive species emergence to enable early detection and rapid response protocols.

5. Excavated materials removed from project sites containing invasive species should not be utilized at new project sites without decontamination. It is best to utilize excavated materials within the same project site or extent of infestation whenever possible. If certain high-risk invasive species (dependent upon region) are located within the project site, excavated materials including plant debris may need to be buried [e.g., Japanese knotweed (*Reynoutria japonica*) and giant knotweed (*R. sachalinensis*) at least five feet below grade (*45*)]. During transport, cover any invasive species-infested materials.

Post-Project Recommendations

1. Develop plans to monitor project site in accordance with authoritative agency's regulations. The plan should include monitoring known invasive species populations (and their potential spread) as well as potential invasive species emergence. Prioritize management actions based on aggressiveness of species invasion and according to relevant regulations (e.g., state noxious weed policies and enforcement).

2. Implement management actions based on species-specific management plans with careful consideration of seed release timing and spawning seasons to not further influence the spread of the species.

3. Create a long-term invasive species management plan for project sites. Consider density, aggression, and early detections when prioritizing management actions.

CONCLUSION

Invasive species present one of the greatest threats to biodiversity today. Their capacity for dispersal, adaptation, reproduction, and growth can impair ecosystem function. Invasive species degrade habitat quality by disrupting soil health, water quality, and biodiversity. Some invasions have resulted in the alteration of natural fire (*36*, *46*) or flood regimes (*47*, *48*, *49*) and the extirpation or even extinction of native species (*50*, *51*, *52*, *53*).

Invasive species also pose threats to public health and safety. For example, the emerald ash borer (*Agrilus planipennis*) and other forest pests have reduced climate resiliency in communities by diminishing shade and tree cover (*54*). Other pests have significant negative impacts on the world's food supply (*55*). Research has also shown that numerous invasive species contribute considerably to the spread of pathogens and novel diseases (*56*). Lastly, species invasions cause great damage to municipal infrastructure, including power, water, housing, and transportation systems (*31*).

Often, introduction and subsequent establishment occurs rapidly causing detection and management to be difficult. Invasive species spread is largely facilitated through human activities, such as global trade and development. Prevention of the spread and introduction of invasive species remains the most cost-effective strategy available to managers to combat their impacts. Researchers estimate an annual global investment of over $400 billion in invasive species-related costs (*1*) with the United States alone having spent an estimated $1.22 trillion since 1960 (*18*). Climate change and increased global trade are expected to further aid the introduction and spread of invasive species in coming years, resulting in even greater investments for management and response efforts. Therefore, taking systematic steps to inhibit their spread is imperative. Focusing on preventative measures also reduces the need for herbicide and pesticide applications, which come at great cost both economically and ecologically.

Adoption of preventative measures across many user groups, including developing industries, outdoor recreators, trade, and agricultural and food production, helps to ensure that communities and ecosystems are protected from greater harm. Outreach programs, such as Work Clean Go administered by the North American Invasive Species Management Association (NAISMA, *57*), have proven to be useful resources for industries interested in incorporating invasive species prevention and messaging into their practices. Industries that have adopted invasive species prevention programs, such as Work Clean Go, not only protect the environment but also gain economic, safety, and municipal advantages. It is imperative that industry professionals employ invasive species prevention efforts to reduce impacts to ecosystems, ensure public health and safety, adhere to public policy, protect infrastructure, and reduce organizational liabilities and investments.

REFERENCES

1. Intergovernmental Platform on Biodiversity and Ecosystem Services (IPBES). *Thematic Assessment Report on Invasive Alien Species and their Control of the Intergovernmental Science-Policy Platform on Biodiversity and Ecosystem Services*. Roy, H.E., Pauchard, A., Stoett, P., and Renard Truong, T. (eds). IPBES Secretariat, Bonn, Germany, 2023. DOI: 10.5281/zenodo.7430682.

2. Molnar, J.L., Gamboa, R.L., Revenga, C., and Spalding, M.D. Assessing the Global Threat of Invasive Species to Marine Biodiversity. *Frontiers in Ecology and the Environment*, 6 (9), 2008, pp. 485-492. DOI: 10.1890/070064.

3. Hulme, P.E. Trade, Transport, and Trouble: Managing Invasive Species Pathways in an Era of Globalization. *Journal of Applied Ecology*, 46, 2009, pp. 10-18. DOI: 10.1111/j.1365-2664.2008.01600.x.

4. Sanchirico, J.N., Albers, H.J., Fischer, C., and Coleman, C. Spatial Management of Invasive Species: Pathways and Policy Options. *Environmental and Resource Economics*, 45, 2010, pp. 517-535. DOI: 10.1007/s10640-009-9326-0.

5. Padayachee, A.L., Irlich, U.M., Faulkner, K.T., Gaertner, M., Procheş, Ş., Wilson, J.R.U., and Rouget, M. How Do Invasive Species Travel To and Through Urban Environments? *Biological Invasions*, 19 (12), 2017, pp. 1-14. DOI: 10.1007/s10530-017-1596-9.

6. Saul, W., Roy, H.E., Booy, O., Carnevali, L., Chen, H., Genovesi, P., Harrower, C.A., Hulme, P.E., Pagad, S., Pergl, J., and Jeschke, J.M. Assessing Patterns in Introduction Pathways of Alien Species by Linking Major Invasion Databases. *Journal of Applied Ecology*, 54, 2017, pp. 657-669. DOI: 10.1111/1365-2664.12819.

7. Turbelin, A.J., Diagne, C., Hudgins, E.J., Moodley, D., Kourantidou, M., Novoa, A., Haubrock, P.J., Bernery, C., Gozlan, R.E., Francis, R.A., and Courchamp, F. Introduction Pathways of Economically Costly Invasive Alien Species. *Biological Invasions*, 24, 2022, pp. 2061-2079. DOI: 10.1007/s10530-022-02796-5.

8. Lieurance, D., Canavan, S., Behringer, D.C., Kendig, A.E., Minteer, C.R., Reisinger, L.S., Romagosa, C.M., Flory, S.L., Lockwood, J.L., Anderson, P.J., Baker, S.M., Bojko, J., Bowers, K.E., Canavan, K., Carruthers, K., Daniel, W.M., Gordon, D.R., Hill, J.E., Howeth, J.G., Iannone III, B.V., Jennings, L., Gettys, L.A., Kariuki, E.M., Kunzer, J.M., Laughinghouse IV, H.D., Mandrak, N.E., McCann, S., Morawo, T., Morningstar, C.R., Neilson, M., Petri, T., Pfingsten, I.A., Reed, R.N., Walters, L.J., and Wanamaker, C. Identifying Invasive Species Threats, Pathways, and Impacts to Improve Biosecurity. *Ecosphere,* 14, 2023, e4711. DOI: 10.1002/ecs2.4711.

9. Johnson, L.E. and Padilla, D.K. Geographic Spread of Exotic Species: Ecological Lessons and Opportunities from the Invasion of the Zebra Mussel *Dreissena polymorpha*. *Biological Conservation,* 78, 1996, pp. 23-33.

10. De Ventura, L., Weissert, N., Tobias, R., Kopp, K., and Jokela, J. Overland Transport of Recreational Boats as a Spreading Vector of Zebra Mussel *Dreissena polymorpha. Biological Invasions*, 18, 2016, pp. 1451-1466. DOI: 10.1007/s10530-016-1094-5.

11. O'Neill, C.R., and Dextrase, A. The Introduction and Spread of the Zebra Mussel in North America. *In* Proceedings of the 4th International Zebra Mussel Conference, Madison, Wisconsin. March 1994, pp. 433-446.

12. Oliver, M.E. Linking Zebra Mussel Invasion and Waterborne Commerce in the USA. *Water Policy*, 16 (3), 2014, pp. 536-556. DOI: 10.2166/wp.2013.067.

13. Johnson, L.E. and Carlton, J.T. Post-Establishment Spread in Large-Scale Invasions: Dispersal Mechanisms of the Zebra Mussel. *Ecology*, 77 (6), 1996, pp. 1686-1690.

14. Wilson, J.D., Dorcas, M.E., Snow, R.W. Identifying Plausible Scenarios for the Establishment of Invasive Burmese Pythons (*Python molurus*) in Southern Florida.

15. Pyšek, P., Hulme, P.E., Simberloff, D., Bacher, S., Blackburn, T.M., Carlton, J.T., Dawson, W., Essl, F., Foxcroft, L.C., Genovesi, P., Jeschke, J.M., Kühn, I., Liebhold, A.M., Mandrak, N.E., Meyerson, L.A., Pauchard, A., Pergl, J., Roy, H.E., Wingfield, M.J., and Richardson, D.M. Scientists' warning on invasive alien species. *Biological Reviews* 95, 2020, pp. 1511-1534.

16. Walther, G., Roques, A., Hulme, P.E., Sykes, M.T., Pyšek, P., Kühn, I., and Zobel, M. Alien Species in a Warmer World: Risks and Opportunities. *Trends in Ecology and Evolution* 24 (12), 2009, pp. 686-693. DOI: 10.1016/j.tree.2009.06.008.

17. Seebens, H., Blackburn, T.M., Dyer, E.E., Genovesi, P., Hulme, P.E., Jeschke, J.M., Pagad, S., Pyšek, P., van Kleunen, M., Winter, M., Ansong, M., Arianoutsou, M., Bacher, S., Blasius, B., Brockerhoff, E.G., Brundu, G., Capinha, C., Causton, C.E., Celesti-Grapow, L., Dawson, W., Dullinger, S., Economo, E.P., Fuentes, N., Guénard, B., Jäger, H., Kartesz, J., Kenis, M., Kühn, I., Lenzner, B., Leibhold, A.M., Mosena, A., Moser, D., Nentwig, W., Nishino, M., Pearman, D., Pergl, J., Rabitsch, W., Rojas-Sandoval, J., Roques, A., Rorke, S., Rossinelli, S., Roy, H.E., Scalera, R., Schindler, S., Štajerová, K., Tokarska-Guzik, B., Walker, K., Ward, D.F., Yamanaka, T., and Essi, F. Global Rise in Emerging Alien Species Results from Increased Accessibility of New Source Pools. *PNAS* 115 (10), 2018, pp. E2264-E2273. DOI: 10.1073/pnas.1719429115.

18. Fantle-Lepczyk, J.E., Haubrock, P.J., Kramer, A.M., Cuthbert, R.N., Turbelin, A.J., Crystal-Ornelas, R., Diagne, C., and Courchamp, F. Economic Costs of Biological Invasions in the United States. *Science of the Total Environment* 806 (3), 2022, 151318. DOI: 10.1016/j.scitotenv.2021.151318.

19. Victorian Government. *Invasive Plants and Animal Policy Framework*. DPI Victoria, Melbourne, 2010.

20. Marsh, A.S., Hayes, D.C., Klein, P.N., Zimmerman, N., Dalsimer, A., Burkett, D.A., Huebner, C.D., Rabaglia, R., Meyerson, L.A., Harper-Lore, B.L., Davidson, J.L., Emery, M.R., Warziniack, T., Flitcroft, R., Kerns, B.K., and Lopez, V.M. Sectoral Impacts of Invasive Species in the United States and Approaches to Management. *In* Invasive Species in Forests and Rangelands of the United States, Chapter 9, 2021, pp. 203-229. DOI: 10.1007/978-3-030-45367-1_9.

21. Bureau of Reclamation. *Inspection and Cleaning Manual for Equipment and Vehicles to Prevent the Spread of Invasive Species*. U.S. Department of the Interior, Bureau of Reclamation. 2021.

22. Santini, A., Maresi, G., Richardson, D.M., and Liebhold, A.M. Collateral Damage: Military Invasions Beget Biological Invasions. *Frontiers in Ecology and the Environment*, 21 (10), 2023, pp. 469-478. DOI: 10.1002/fee.2640.

23. National Wildfire Coordinating Group (NWCG). *Beware Hitchhikers in Fire Pants*. https://fs-prod-nwcg.s3.us-gov-west-1.amazonaws.com/s3fs-public/2023-06/iss-beware-hitchhikers-in-fire-pants-highlights.pdf. Accessed 26 June 2024.

24. Watters, A. Decontamination Protocols for Watercraft and Wildland Firefighting Equipment in Preventing the Spread of Invasive Quagga (*Dreissena rostriformis bugensis*) and Zebra (*Dreissena polymorpha*) mussels. University of Nevada, Las Vegas Theses, Dissertations, Professional Papers, and Capstones, 2311, 2014, pp. 1-143. DOI: 10.34917/7048630.

25. Shaefer, C., Wu, S., Sanchez, A., and Laufman, J. *Fire Equipment Foot Valve Study Investigating Leakage Potential for Aquatic Invasive Species (AIS) Transfer*. United States Department of Agriculture, Forest Service, Washington, DC, 2016.

26. National Wildfire Coordinating Group (NWCG). *Guide to Preventing Aquatic Invasive Species Transport by Wildland Fire Operations*. PMS 444, 2017.

27. Confrancesco, Jr., A.F., Reaves, D.R., and Averett, D.E. *Transfer of Invasive Species Associated with the Movement of Military Equipment and Personnel*. ERDC/EL TR-07-8. U.S. Department of Defense, Legacy Resource Management Program, Washington, D.C., 2007.

28. Weldy, T. *Conserving Biodiversity on Military Lands: A Guide for Natural Resource Managers 3rd Edition*. U.S. Department of Defense, Environment, Safety, and Occupational Health Network and Information Exchange (DENIX), 2022.

29. Joly, M., Bertrand, P., Gbangou, R.Y., White, M-C, Dubé, J., and Lavioe, C. Paving the Way for Invasive Species: Road Type and the Spread of Common Ragweed (*Ambrosia artemisiifolia*). *Environmental Management*, 48, 2011, pp. 514-522. DOI: 10.1007/s00267-011-9711-7.

30. Benedetti, Y. and Morelli, F. Spatial Mismatch Analysis Among Hotspots of Alien Plant Species, Road, and Railway Networks in Germany and Austria. *PLOS One*, 12 (8), 2017, e0183691. DOI: 10.1371/journal.pone.0183691.

31. Invasive Species Advisory Committee (ISAC). *Invasive Species Impacts on Infrastructure.* U.S. Department of the Interior, Office of the Secretary, Washington, D.C., 2016.

32. Moore, V. *Invasive Species Impacts on Transportation Infrastructure*. FHWA-HEP-23-021. Federal Highway Administration, Washington, DC, 2021.

33. Matte, R., Boivin, M., and Lavoie, C. Japanese Knotweed Increases Soil Erosion on Riverbanks. *River Research and Applications*, 38 (3), 2022, pp. 561-572. DOI: 10.1002/rra.3918.

34. Lavoie, C. The Impact of Invasive Knotweed Species (*Reynoutria* spp.) on the Environment: Review and Research Perspectives. *Biological Invasions*, 19, 2017, pp. 2319-2337. DOI: 10.1007/s10530-017-1444-y.

35. Gerber, E., Krebs, C., Murrell, C., Moretti, M., Rocklin, R., and Schaffner, U. Exotic Invasive Knotweeds (*Fallopia* spp.) Negatively Affect Native Plant and Invertebrate Assemblages in European Riparian Habitats. *Biological Conservation,* 141 (3), 2008, pp. 646-654. DOI: 10.1016/j.biocon.2007.12.009.

36. Lambert, A.M., D'Antonio, C.M., and Dudley, T.L. Invasive Species and Fire in California Ecosystems. *Fremontia*, 38 (2), 2010, pp. 29-36.

37. Gray, M.E., and Dickson, B.G. Applying Fire Connectivity and Centrality Measures to Mitigate the Cheatgrass-Fire Cycle in the Arid West, USA. *Landscape Ecology*, 31, 2016, pp. 1681-1696. DOI: 10.1007/s10980-016-0353-2.

38. Safeguarding the Nation from the Impacts of Invasive Species. E.O. 13112, 1999, as amended.

39. National Environmental Policy Act. 42 U.S.C. 4321 *et seq.*, 1969, as amended.

40. Nonindigenous Aquatic Nuisance Prevention and Control Act. 16 U.S.C. 3371-3378 *et seq.*, 1990.

41. Plant Protection Act. 7 U.S.C. 7701 *et seq.*, 2000.

42. Lacey Act. 18 U.S.C. 42, 16 U.S.C. 3371-3378 *et seq.*, 1990, as amended.

43. Endangered Species Act. 16 U.S.C. 1531 *et seq.*, 1973, as amended.

44. Noxious Weed Control and Eradication Act. 7 U.S.C. 7781 *et seq.*, 2004.

45. Pennsylvania Department of Transportation. *Invasive Species Best Management Practices.* Publication 756. Pennsylvania Department of Transportation, 2014.

46. Brooks, M.L., D'Antonio, C.M., Richardson, D.M., Grace, J.B., Keeley, J.E., DiTomaso, J.M., Hobbs, R.J., Pellant, M., and Pyke, D. Effects of Invasive Alien Plants on Fire Regimes. *BioScience*, 54 (7), 2004, pp. 677-688. DOI: 10.1641/0006-3568(2004)054[0677:EOIAPO]2.0.CO;2.

47. Stromberg, J.C., Lite, S.J., Marler, R., Paradzick, C., Shafroth, P.B., Shorrock, D., White, J.M., and White. M.S. Altered Stream-Flow Regimes and Invasive Plant Species: The *Tamarix* Case. *Global Ecology and Biogeography*, 16, 2007, pp. 381-393. DOI: 10.1111/j.1466-8238.2007.00297.x.

48. Havel, J.E., Kovalenko, K.E., Thomaz, S.M., Amalfitano, S., and Kats, L.B. Aquatic Invasive Species: Challenges for the Future. *Hydrobiologia*, 750, 2015, pp. 147-170. DOI: 10.1007/s10750-014-2166-0.

49. Thomaz, S.M. Propagule Pressure and Environmental Filters Related to Non-Native Species Success in River-Floodplain Ecosystems. *Hydrobiologia*, 849, 2022, pp. 3679-3704. DOI: 10.1007/s10750-021-04624-8.

50. Mooney, H.A. and Cleland, E.E. The Evolutionary Impact of Invasive Species. *PNAS*, 98 (10), 2001, pp. 5446-5451. DOI: 10.1073/pnas.091093398.

51. Donlan, C.J. and Wilcox, C. Diversity, Invasive Species and Extinctions in Insular Ecosystems. *Journal of Applied Ecology,* 45, 2008, pp. 1114-1123. DOI: 10.1111/j.1365-2664.2008.01482.x.

52. Sax, D.F. and Gaines, S.D. Species Invasions and Extinction: The Future of Native Biodiversity on Islands. *PNAS*, 105 (1), 2008, pp.11490-11497. DOI: 10.1073/pnas.0802290105.

53. McCreless, E.E., Huff, D.D., Croll, D.A., Tershy, B.R., Spatz, D.R., Holmes, N.D., Butchart, S.H.M., and Wilcox, C. Past and Estimated Future Impact of Invasive Alien Mammals on Insular Threatened Vertebrate Populations. *Nature Communications*, 7, 2015, 12488. DOI: 10.1038/ncomms12488.

54. Marini, L., Ayres, M.P., and Jactel, H. Impacts of Stand and Landscape Management on Forest Pest Damage. *Annual Review of Entomology*, 67, 2022, pp. 181-199. DOI: 10.1146/annurev-ento-062321-065511.

55. Heeb, L., Jenner, E., and Cock, M.J.W. Climate-smart Pest Management: Building Resilience of Farms and Landscapes to Changing Pest Threats. *Journal of Pest Science*, 92, 2019, pp. 951-969. DOI: 10.1007/s10340-019-01083-y.

56. Chinchio, E., Crotta, M., Romeo, C., Drewe, J.A., Guitian, J., and Ferrari, N. Invasive Alien Species and Disease Risk: An Open Challenge in Public and Animal Health. *PLOS Pathogens*, 16 (10), 2020, e1008922. DOI: 10.1371/journal.ppat.1008922.

57. North American Invasive Species Management Association. *WorkCleanGo*. https://naisma.org/naisma-resources/workcleango/. Accessed 26 June 2024.

Diagnosing Mechanically Stabilized Earth (MSE) Wall Using Spectral Analysis of Surface Waves (SASW) Method

Md Fahimuzzaman Khan
Jackson State University
1400 J.R. Lynch Street, JSU Box 17068, Jackson, MS 39217-0168
j01012191@students.jsums.edu

Sadik Khan, PhD, P.E.
Jackson State University
1400 J.R. Lynch Street, JSU Box 17068, Jackson, MS 39217-0168
j01014295@students.jsums.edu

Ian LaCour P.E.
Mississippi Department of Transportation
412 E. Woodrow Wilson Ave. Jackson, MS 39216
ilacour@mdot.ms.gov

Avipriyo Chakraborty
Jackson State University
1400 J.R. Lynch Street, JSU Box 17068, Jackson, MS 39217-0168
sadik.khan@jsums.edu

Sayedul Kibria
Jackson State University
1400 J.R. Lynch Street, JSU Box 17068, Jackson, MS 39217-0168
j00986721@students.jsums.edu

Prepared for the 73rd Highway Geology Symposium, September 2024

Acknowledgements

The authors sincerely acknowledge the Mississippi Department of Transportation (MDOT) for supporting the study of Diagnosing Mechanically Stabilized Earth (MSE) Wall Using Spectral Analysis of Surface Waves (SASW) Method.

Disclaimer

Statements and views presented in this paper are strictly those of the authors, and do not necessarily reflect positions held by their affiliations, the Highway Geology Symposium (HGS), or others acknowledged above. The mention of trade names for commercial products does not imply the approval or endorsement by HGS.

ABSTRACT

Mechanically Stabilized Earth (MSE) wall is a common earth-retaining structure in the transportation infrastructure network that is used to construct retaining walls, bridge abutments, and other structures. The significance of the MSE wall is high due to its affordability, adaptability, and strength, making it a vital element. In this study, a failure in the MSE wall at exit 99 on the I-55 North highway is analyzed, and the condition was assessed using the Spectral Analysis of Surface Waves (SASW) method. SASW is a non-destructive geophysical method that analyzes the dispersion of surface waves to evaluate subsurface conditions, which is particularly useful in identifying inconsistencies like voids or weaknesses without disturbing the structure. By applying the SASW method at different points along the MSE wall, a dispersion curve was generated for each location, providing crucial data on the subsurface properties. Subsequently, an average surface velocity contour graph was plotted, integrating the data comprehensively. This evaluation identified the potential void location within the wall, demonstrating how SASW can serve as an effective tool in diagnosing and mitigating issues in MSE structures contributing to safer and more reliable infrastructure.

INTRODUCTION

Mechanically Stabilized Earth (MSE) walls have emerged as a cornerstone in modern civil engineering, particularly within the transportation infrastructure sector (1). These structures have revolutionized the construction of retaining walls, bridge abutments, embankments, and other essential components of highways and roadways (2). The widespread adoption of MSE walls is primarily driven by their numerous advantages over traditional retaining wall systems, including cost-effectiveness, ease of construction, superior performance, and adaptability to various site conditions.

MSE walls are composite structures consisting of three main components: a foundation, a facing element, and soil reinforcement. The foundation provides the base support, which is crucial for the overall stability of the structure (3). The facing element, typically made of precast concrete panels or modular blocks, serves the dual purpose of retaining the backfill material and providing an aesthetically pleasing exterior. The soil reinforcement, usually composed of metallic strips, geosynthetic materials, or a combination of both, enhances the overall stability and load-bearing capacity of the wall. The principle behind MSE walls lies in the interaction between the reinforcement and the soil, creating a cohesive mass that can withstand significant lateral earth pressures (4). This interaction allows for the construction of steeper slopes and higher walls compared to unreinforced soil structures, making MSE walls particularly suitable for areas with limited space or challenging topography.

Compared to traditional reinforced concrete walls, MSE walls offer several advantages. They require less material and labor, resulting in significant cost savings depending on the project specifics. The modular nature of MSE wall components allows for quicker and more efficient construction, which is particularly beneficial in large-scale infrastructure projects with tight deadlines (5). Furthermore, the flexibility of MSE walls in terms of design and aesthetics enables them to blend seamlessly with the surrounding environment, enhancing the visual appeal of transportation corridors.

Despite their numerous advantages and widespread use, MSE walls are not immune to failures. Several factors can contribute to the instability and eventual failure of these structures, necessitating a thorough understanding of potential failure modes and their causes (6). Poor backfill material quality is one of the primary factors that can compromise the structural integrity of an MSE wall. The use of backfill with high fine content or inadequate drainage properties can lead to increased pore water pressure, reduced shear strength, and, ultimately, wall instability. Inadequate reinforcement, either in terms of length, strength, or spacing, is another critical factor that can lead to MSE wall failures. Insufficient reinforcement can result in excessive deformation of the wall face, internal instability, or even global failure of the structure.

Foundation stability is another crucial factor in the performance of MSE walls. Unstable foundation soil, due to factors such as poor compaction, high moisture content, or underlying voids, can lead to differential settlement and wall instability. Thorough geotechnical investigations and appropriate foundation preparation are essential to mitigate these risks. These factors underscore the importance of proper material selection and protective measures in the design and construction of MSE walls, especially in harsh environmental conditions.

Given the potential for failure and the critical role MSE walls play in transportation infrastructure, it is imperative to employ reliable methods for assessing their condition throughout their service life. Non-destructive evaluation (NDE) techniques have gained significant attention in recent years due to their ability to detect internal anomalies without causing damage to the structure (7). Among these techniques, Spectral Analysis of Surface Waves (SASW) has emerged as a particularly effective tool for evaluating the integrity of MSE walls and their foundation soils.

SASW is a geophysical method that analyzes the propagation of surface waves to evaluate subsurface conditions. The technique involves generating surface waves using an impact source and recording their propagation using receivers placed on the surface (8). By analyzing the dispersion characteristics of these waves, it is possible to infer the properties of the subsurface materials, including their stiffness and layering. This non-invasive approach makes SASW particularly suitable for assessing existing structures where minimal disturbance is desired (9).

The effectiveness of SASW in detecting subsurface anomalies, such as voids or zones of weakness, has been demonstrated in various geotechnical applications(10). In the context of MSE walls, SASW can provide valuable insights into the condition of the backfill material, the integrity of the reinforcement zone, and the properties of the foundation soil. This information is crucial for identifying potential issues before they manifest as visible signs of distress or failure.

This study focuses on the analysis of a failed MSE wall located at exit 99 on the I-55 North highway in Jackson, Mississippi. The wall exhibited signs of instability, prompting a detailed investigation using the SASW method. The primary objective was to identify potential voids or weaknesses within the wall that could have contributed to its failure. By generating and analyzing dispersion curves at various points along the MSE wall, it was possible to create a comprehensive contour map of surface wave velocities, highlighting areas with potential anomalies.

The application of SASW in this case study demonstrates its potential as a diagnostic tool for MSE wall assessment. By providing detailed information about the subsurface conditions, SASW can complement visual inspections and other traditional assessment methods, leading to more comprehensive and accurate evaluations of MSE wall integrity (11). The findings from this investigation not only provide valuable insights into the specific case study but also contribute to the broader understanding of MSE wall behavior and failure mechanisms. Furthermore, this research highlights the importance of integrating advanced non-destructive testing techniques, such as SASW, into routine inspection and maintenance programs for transportation infrastructure (12).

As the use of MSE walls continues to grow in transportation and other civil engineering applications, the need for effective assessment and monitoring techniques becomes increasingly critical. This study aims to demonstrate the potential of SASW as a valuable tool in this context, paving the way for more widespread adoption of advanced non-destructive testing methods in infrastructure management and maintenance strategies.

METHODOLOGY

This study focused on the analysis of MSE walls exhibiting signs of instability at exit 99 on the I-55 North highway in North Jackson, Mississippi. The investigation site comprised a rectangular pavement section measuring 35 meters in length and 9.6 meters in width, consisting of both concrete (6 meters) and asphalt (29 meters) surfaces. Visual inspection revealed significant structural issues, including bulging in the western MSE wall panels and spalling of fill material in the eastern MSE wall panels, as shown in Figure 1.

Figure 1 (a) Site location and study area (b) Instability in the MSE wall.

To conduct a comprehensive, non-destructive evaluation of the MSE walls and their underlying conditions, we employed the Spectral Analysis of Surface Waves (SASW) method. This technique was chosen for its ability to provide detailed information about subsurface conditions without causing damage to the existing structure, aligning with the objectives outlined in the introduction. A battery-powered Seismic Geophysical Testing Platform was utilized for the SASW testing. The system consisted of a hammer for generating acoustic energy (surface waves), two accelerometers for recording surface wave propagation, a rugged, field-ready portable computer for data acquisition, and BNC cables for connecting the accelerometers to the computer (13). The accelerometers were capable of functioning effectively with up to 3.6 meters of spacing, allowing for flexibility in the testing setup. Surface wave velocities were recorded using WinSW software, which displayed the velocities in time domain intervals. A schematic of the SASW setup is presented in Figure 2.

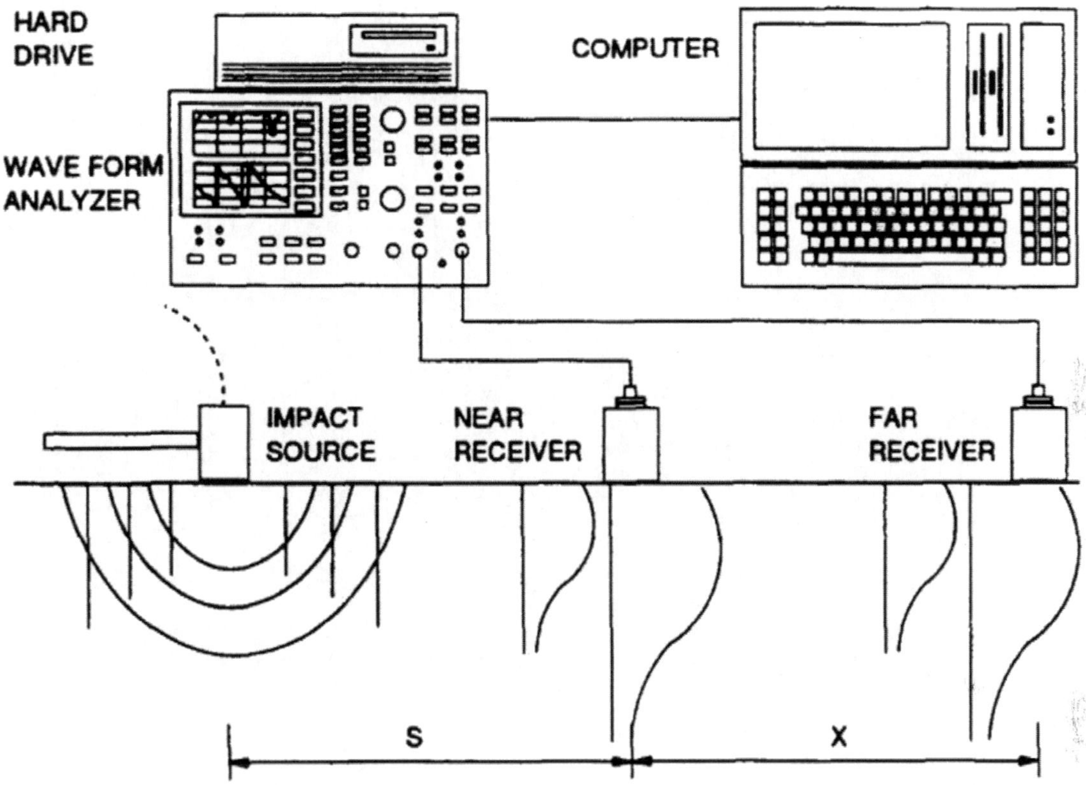

Figure 2 Schematic Diagram of SASW Setup (14)

To ensure systematic and comprehensive data collection, the pavement section was divided into a grid system. The site was divided into 24 grids along its 35-meter length with a spacing of 1.5 meters and 9 grids along its 9.6-meter width with a spacing of 1.2 meters. SASW data was collected at each grid intersection, resulting in 216 data collection points across the site, as illustrated in Figure 3. This detailed grid system allowed for high-resolution analysis of the subsurface conditions, which is crucial for identifying potential anomalies or weaknesses within the MSE wall structure.

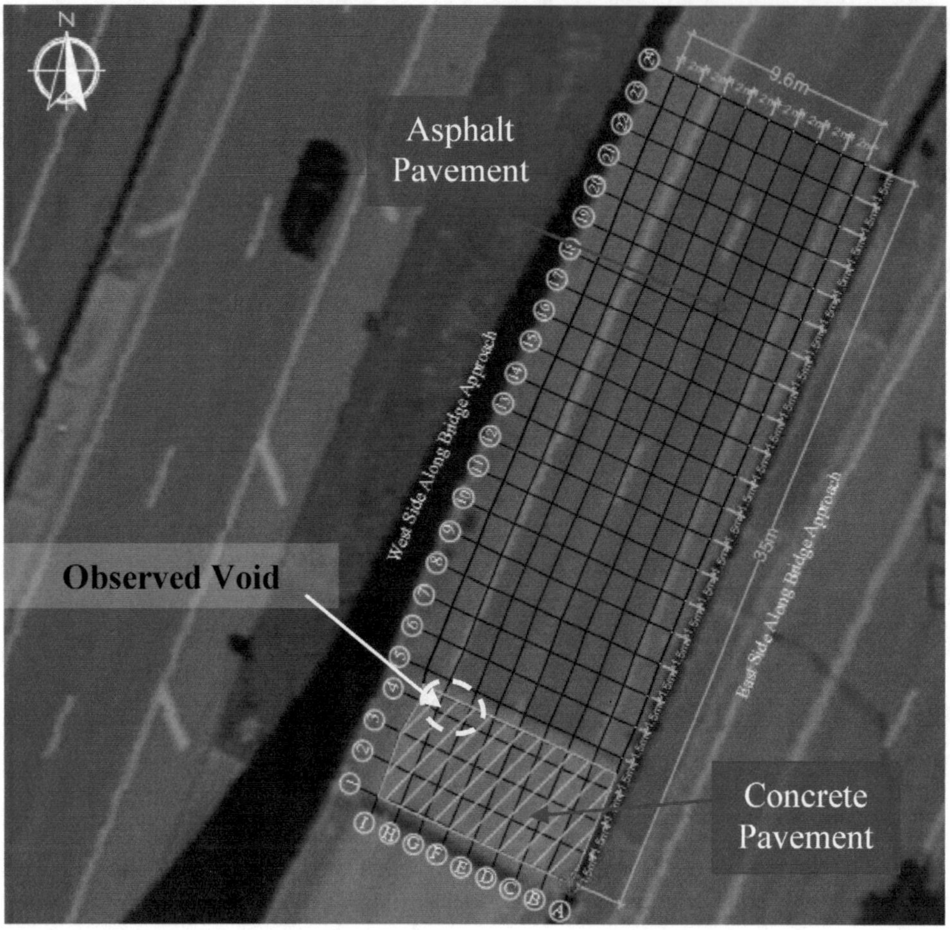

Figure 3 Grid System for Data Collection

To complement the SASW data and provide additional context for the site conditions, drone imagery was captured. This imagery was used to create a Digital Elevation Model (DEM) and an orthomosaic model of the site. These high-resolution images from an aerial perspective aided in accurate mapping and analysis of the site, allowing for a comprehensive understanding of the surface conditions in relation to the subsurface data collected through SASW.

Figure 4 Data Collection for SASW Method

At each of the 216 grid points, two accelerometers were installed on the pavement surface. Surface waves were generated using the hammer source, and wave propagation data was recorded using the WinSW software (Figure 4). This systematic approach ensured consistency in data collection across the entire site, allowing for reliable comparison and analysis of subsurface conditions at different locations.

The collected SASW data was processed to generate individual dispersion curves for each of the 216 grid points and a contour map of the average surface wave velocity across the entire site. The dispersion curves provided information about the variation of surface wave velocity with frequency at each point, allowing for the inference of subsurface material properties and layering. The contour map offered a comprehensive visualization of the subsurface conditions across the entire site, highlighting areas of potential concern or anomalies.

By integrating the SASW results with visual observations and drone imagery, a thorough assessment of the MSE wall condition was done, identifying potential voids, weaknesses, or other factors contributing to the observed instability. This multi-faceted approach aligns with the growing need for advanced, non-destructive evaluation techniques in infrastructure assessment and maintenance, as highlighted in the introduction. Through this methodology, we aimed to demonstrate the effectiveness of SASW in evaluating MSE wall integrity and to contribute to the broader understanding of MSE wall behavior and failure mechanisms, as outlined in our study objectives.

RESULTS

The data analysis process focused on identifying potential voids or weaknesses within the MSE walls, which could explain the observed structural issues. To support SASW, collected drone images were processed in software to generate DEM and orthomosaic map, which are shown in Figure 5. This picture gives an overall outlook of the study area.

A key finding from the SASW analysis was the correlation between surface wave velocities and subsurface conditions. Higher surface wave velocity values were observed in areas suspected of having voids or inconsistencies. This phenomenon can be attributed to the confinement of surface waves to the pavement surface in regions with voids, resulting in higher measured velocities. Conversely, areas where surface waves could disperse through underlying materials exhibited lower velocities, suggesting more stable subsurface conditions.

The data processing involved generating coherence, phase difference, and velocity dispersion curves for each grid point using WinSW software (Figure 6). This analysis facilitated the identification of velocity dispersion curves, from which average velocities were calculated for each grid point. MATLAB was then employed to generate a contour map, visualizing the variation in velocity across the site and highlighting potential anomalies.

(a) (b)

Figure 5 Drone image of site area (a) Orthomosaic Model (b) DEM

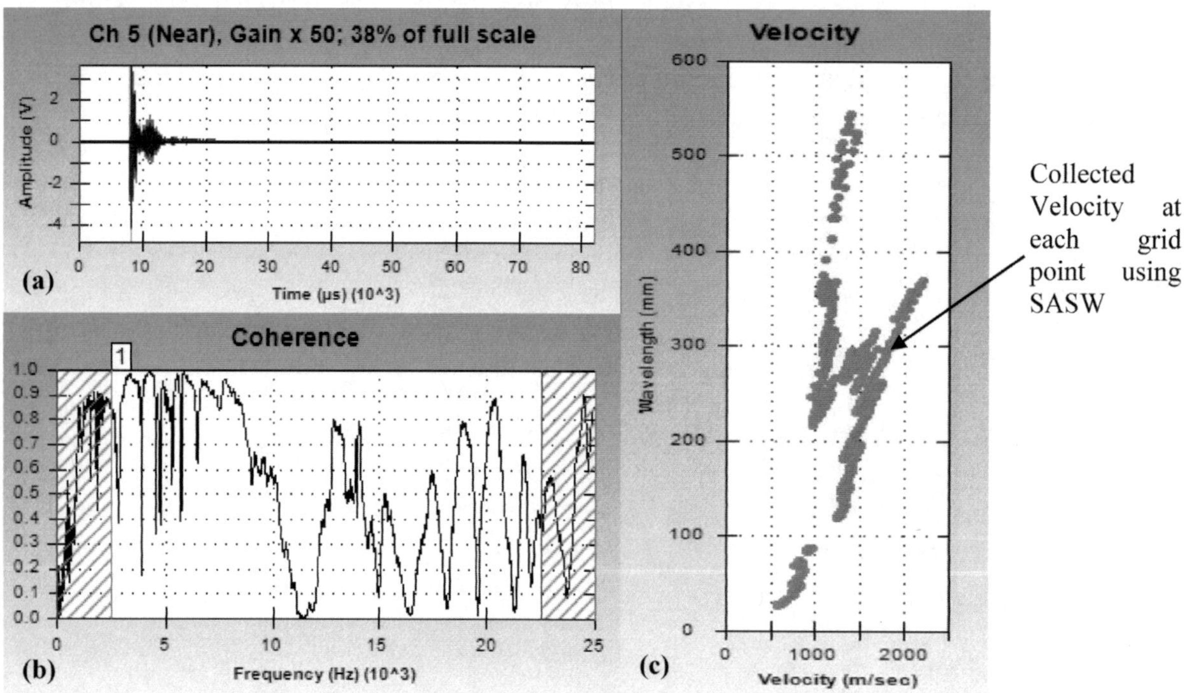

Figure 6 Typical Data Analysis Procedure of SASW points

For each of the 216 grid points, dispersion curves were generated and analyzed to determine surface wave velocities. To understand better, dispersions curved were combined based on every grid line (A-I). In Figure 7 and Figure 8, two combined dispersion curves are shown for Grid C and Grid G, respectively. From the grid results, the difference in the velocities can be observed in grid C and grid G. Grid G mostly sits in the failure zone, whereas grid C was not in the failure zone. These individual velocities were then averaged to create a comprehensive velocity contour map of the entire site. This map served as a crucial tool in highlighting areas of potential structural concern within the MSE walls.

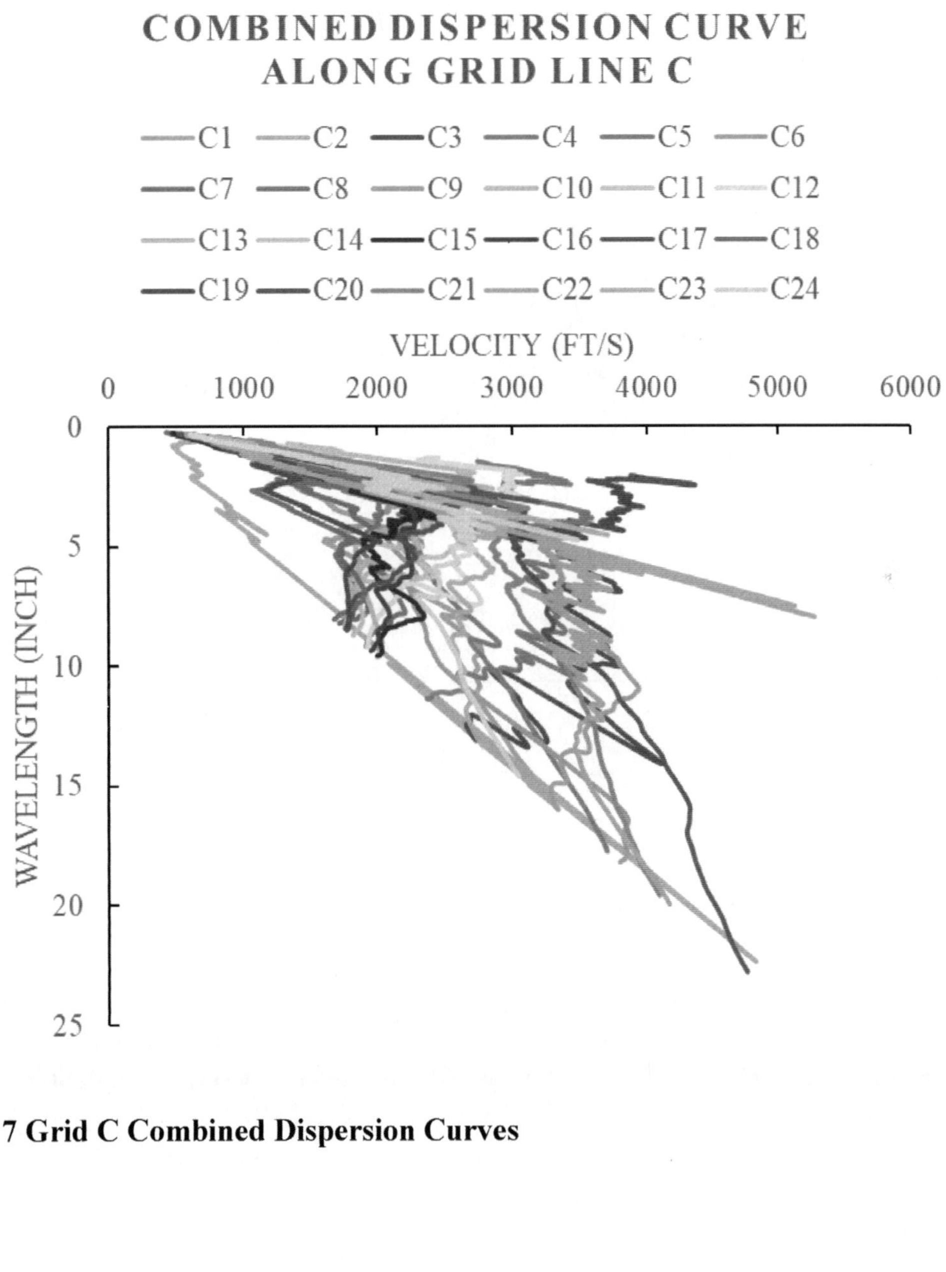

Figure 7 Grid C Combined Dispersion Curves

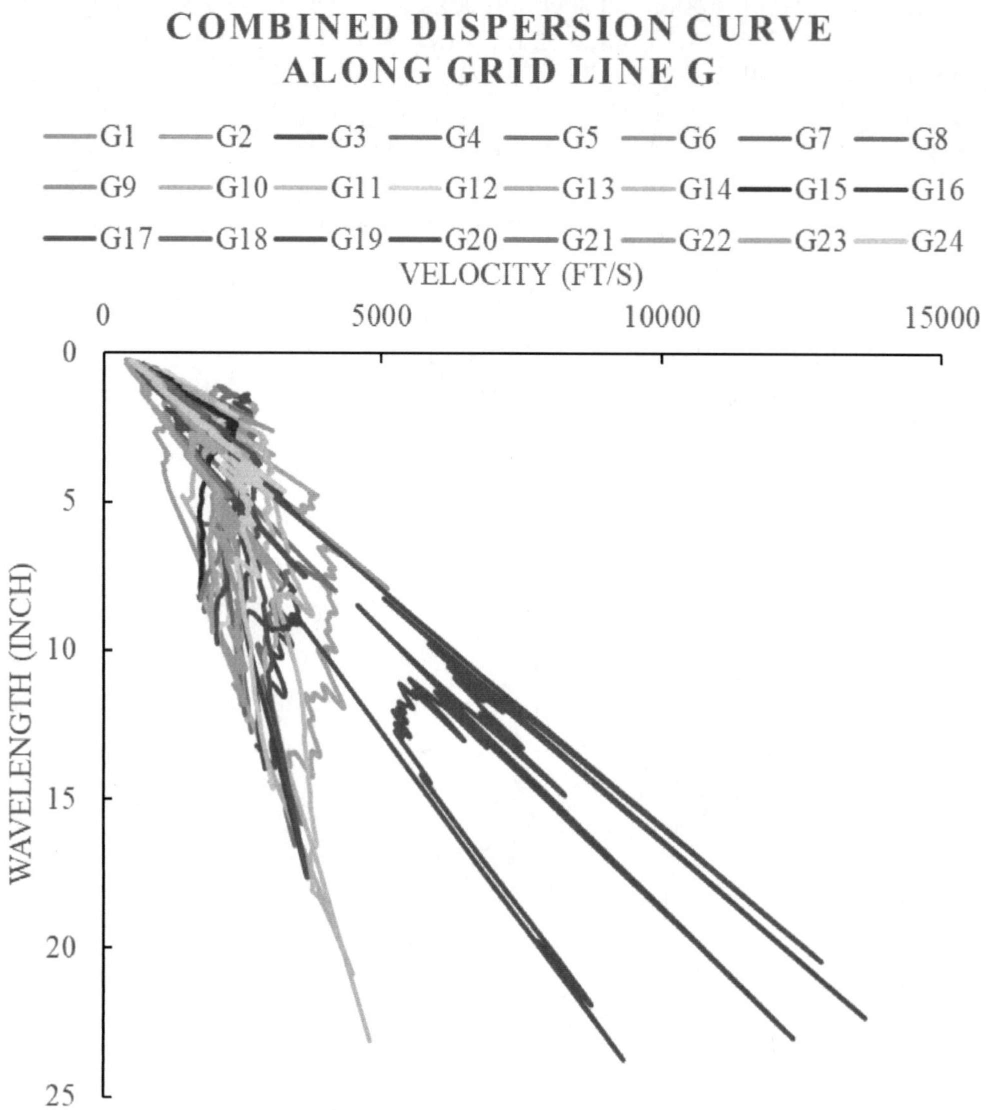

Figure 8 Grid G Combined Dispersion Curves

Significant variations in velocity were observed across different locations of the pavement, corresponding to the underlying material properties and potential void locations (Figure 9). Consistently higher velocity values were recorded in areas with suspected voids compared to locations without anomalies. In these void locations, surface wave velocities were confined to the pavement surface, resulting in higher measurements. In contrast, regions devoid of anomalies exhibited a decrease in surface wave velocity as the waves dispersed across the underlying materials.

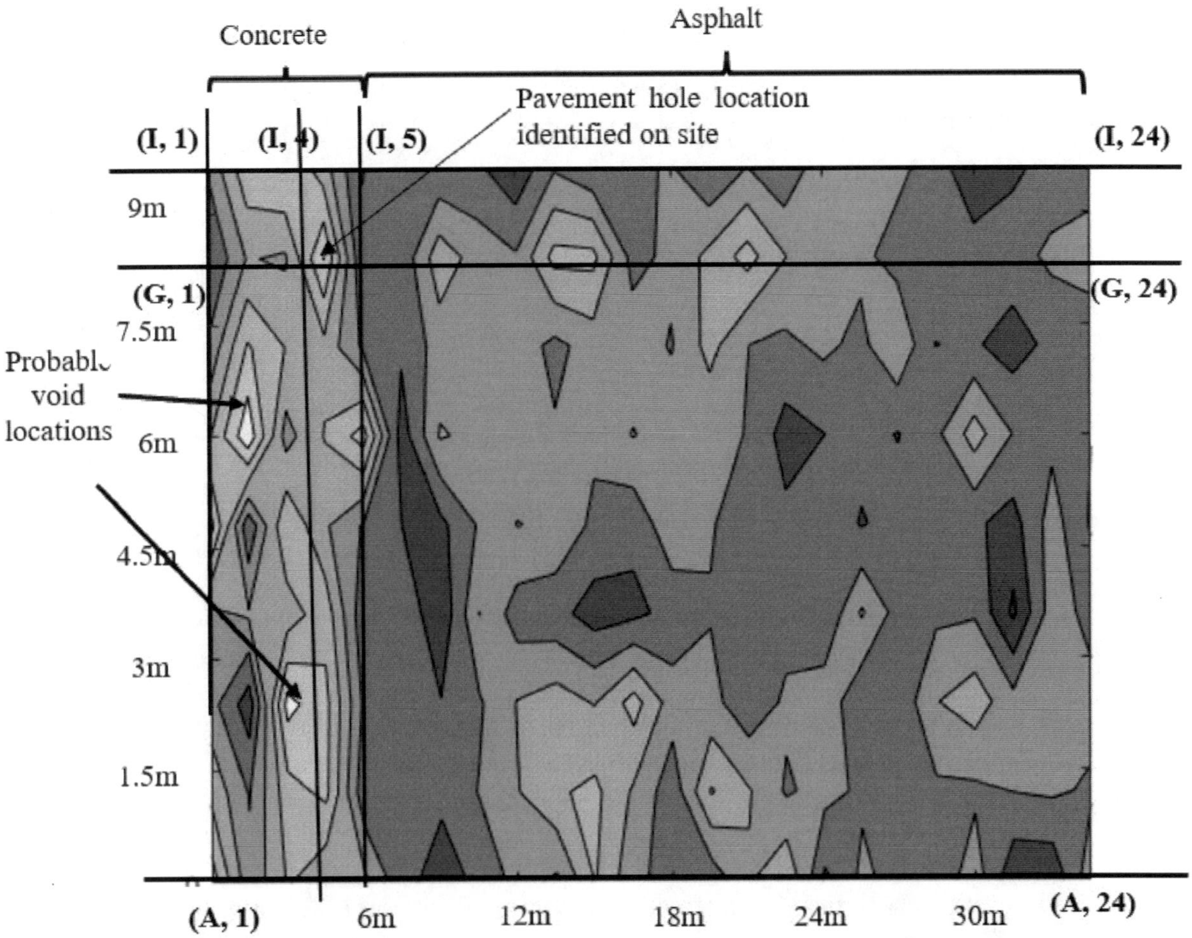

Bend Color	Range of Velocity	Bend Color	Range of Velocity
	<152m/s		610m/s-<762 m/s
	152m/s-<304 m/s		762m/s-<914 m/s
	304m/s-<457 m/s		914m/s-<1067 m/s
	457m/s-<610 m/s		1067m/s- <1219m/s

Figure 9 Contour Map from SASW Results

The SASW results provided critical insights into the subsurface conditions of the MSE walls, which were essential for diagnosing the causes of the observed bulging and spalling. The areas of higher velocity in the contour map closely corresponded with the visually observed structural distress in the MSE walls, particularly the bulging in the western panels and spalling in the eastern panels.

These findings demonstrate the capability of the SASW method to detect and map subsurface anomalies that may not be apparent from visual inspection alone. By identifying potential voids and other structural issues, the SASW analysis has provided valuable data for recommending appropriate mitigation measures to ensure the stability and safety of the MSE walls.

In summary, the SASW method proved to be a reliable and non-destructive approach for evaluating the integrity of MSE walls. The results from this analysis contribute to a safer and more reliable infrastructure by providing detailed insights into subsurface conditions and identifying potential structural issues before they manifest into more significant problems.

DISCUSSION

The SASW method provided insights into the subsurface conditions of the MSE wall at exit 99 on the I-55 North highway. The results from this non-destructive evaluation technique have significant implications for understanding the condition and potential failure mechanisms.

The dispersion curves generated for each of the 216 grid points revealed variations in subsurface material properties across the site area. These variations correlated strongly with the visible signs of distress observed in the MSE wall, particularly the bulging in the western panels and spalling in the eastern panels. This correlation suggests that the surface deformations are indeed indicative of underlying structural issues rather than mere superficial damage.

The contour map of average surface wave velocities provided a comprehensive view of the internal condition. Areas of significantly lower velocities, particularly those behind the wall face, likely indicate zones of inadequate compaction or material degradation. These zones may be contributing to the observed wall instability by reducing the overall strength and cohesion of the reinforced soil mass. The presence of such weakened zones aligns with common failure modes in MSE walls, where inadequate compaction or material degradation can lead to excessive deformation and potential collapse. Abrupt changes in surface wave velocities over short distances, as observed in several areas of the wall, which should be the area of concern. These rapid transitions could indicate the presence of voids, areas of differential settlement, or zones where the reinforcement has lost contact with the surrounding soil.

The correlation between areas of low surface wave velocities and visible wall distress provides strong evidence for the effectiveness of SASW in detecting and mapping subsurface anomalies in MSE walls. This non-destructive method has successfully identified potential problem areas that may not have been apparent from visual inspection alone, demonstrating its value as a diagnostic tool for infrastructure assessment. However, it is important to note that while SASW provides valuable insights into the condition of the wall, it should be used in conjunction with other assessment methods for a comprehensive evaluation.

The findings also highlight the importance of regular monitoring and maintenance of MSE walls, particularly in critical infrastructure applications. The ability to detect subsurface anomalies before they manifest as visible wall deformations could allow for earlier intervention, potentially preventing more severe failures and reducing long-term maintenance costs.

CONCLUSION

This study effectively demonstrated the utility of the SASW method in evaluating the structural integrity of MSE walls. By generating dispersion curves and creating a velocity contour map, the SASW method identified potential voids and inconsistencies within the MSE walls. These findings were validated through field observations, which show the accuracy and reliability of the method.

Mississippi Department of Transportation (MDOT) repaired that section, addressing structural issues that were present earlier. This result could be considered effective and applied as an effective tool for diagnosing MSE walls. The non-destructive nature of SASW helped to identify the anomalies in a short period of time and without creating massive destructive tests.

In conclusion, the SASW method proved to be a strong tool for diagnosing subsurface irregularities in MSE walls. Its application in this study highlights the importance of advanced non-destructive testing techniques in maintaining infrastructure health and safety. Future infrastructure maintenance strategies should incorporate such methodologies to ensure the resilience and reliability of critical transportation systems.

REFERENCES

1. Hulagabali AM, Solanki CH, Thrupthi C, Sushma N, Suokhrie R, Sudarshan S. Pull-Out and Rupture Behavior of Geogrid Reinforcements in MSE Wall Subjected to Seismic Conditions. In: Patel D, Kim B, Han D, editors. Innovation in Smart and Sustainable Infrastructure. Singapore: Springer Nature; 2024. p. 231–49.

2. Chau TL, Nguyen TH, Pham VN. Physical and Numerical Models of Mechanically Stabilized Earth Walls Using Self-Fabricated Steel Reinforcement Grids Applied to Cohesive Soil in Vietnam. Applied Sciences. 2024 Jan;14(3):1283.

3. Yildirim IZ, Turkel B, Guler E. Seismic Response of MSE Walls with Various Reinforcement Configurations: Effect of Input Ground Motion Frequency. Buildings. 2024 Feb 13;14(2):508.

4. Imran H, Al-Jeznawi D, Al-Janabi MAQ, Bernardo LFA. Assessment of Soil–Structure Interaction Approaches in Mechanically Stabilized Earth Retaining Walls: A Review. CivilEng. 2023 Sep 1;4(3):982–99.

5. Jayakrishnan PV, Thiyyakkandi S, Mottadelli L, Halasa González M. Experience from design & construction of a stable feature MSE wall in Dubai. Amer OM, El-Sherbiny R, editors. E3S Web Conf. 2023;368:02020.

6. Alimohammadi H, Memon AA. A Case Study of Mechanically Stabilized Earth (MSE) Retaining Wall Failure in the State of Tennessee; Recommendations for Future Design and Constructions [Internet]. OSF; 2023

7. Benedetto A, Al-Qadi IL, Alani AM, Ciampoli LB, Loizos A, Tosti F. Systematic use of transport infrastructure non-destructive assessment and remote sensing. In: Eleventh International Conference on the Bearing Capacity of Roads, Railways and Airfields. CRC Press; 2022.

8. Schevenels M, Lombaert G, Degrande G, François S. A probabilistic assessment of resolution in the SASW test and its impact on the prediction of ground vibrations. Geophysical Journal International. 2008;

9. Kim G, Hwang S. Shear Wave Velocity Determination of a Complex Field Site Using Improved Non-destructive SASW Testing. Sensors. 2024 Jan;24(10):3231.

10. Gucunski N, Woods RD. Numerical simulation of the SASW test. Soil Dynamics and Earthquake Engineering. 1992;

11. Lew M, Stokoe II KH, Davis CA, Ponnaboyina H, Hudson MB, Perry DL. A Comparison of SASW Survey Results with In Situ Field Investigation Methods. 2012 Jun 20;2726–35.

12. Alonso-Díaz A, Solla M, Rodríguez JL, Elseicy A. Integrated approach to improve Road Infrastructure Resilience through complementary Non-Destructive Techniques. ISPRS Annals of the Photogrammetry, Remote Sensing and Spatial Information Sciences. 2024 Jun 27.

13. Roesset J, Chang D, Stokoe K, Aouad M. Modulus And Thickness Of The Pavement Surface Layer From Sasw Tests. Transportation Research Record [Internet]. 1990

14. Balasubramani DP, Gopalakrishnan V. Subsurface object detection and characterization using Ground Penetrating Radar. Innovative Infrastructure Solutions. 2020;

Investigation of Perched Water Dynamics in Highway Slopes: Using Electrical Resistivity Imaging (ERI) and Reversed Induced Polarization (IP) Methods.

Rahul Biswas

Graduate Research assistant

Jackson State University

1400 J.R. Lynch Street, JSU Box 17068, Jackson, MS 39217-0168
(601)-572-1678
j01017311@students.jsums.edu

Avipriyo Chakrabarty

Graduate Research assistant

Jackson State University

1400 J.R. Lynch Street, JSU Box 17068, Jackson, MS 39217-0168
(601)-767-7935

avipriyo.chakraborty@students.jsums.edu

Sadik Khan, PhD, P.E.
Jackson State University
1400 J.R. Lynch Street, JSU Box 17068, Jackson, MS 39217-0168
sadik.khan@jsums.edu

Prepared for the 73rd Highway Geology Symposium, September, 2024

Acknowledgements

The authors want to acknowledge Jackson State University and Mississippi Department of Transportation (MDOT) for providing necessary site access and investigation.

Disclaimer

Statements and views presented in this paper are strictly those of the authors, and do not necessarily reflect positions held by their affiliations, the Highway Geology Symposium (HGS), or others acknowledged above. The mention of trade names for commercial products does not imply the approval or endorsement by HGS.

ABSTRACT

Perched water, a common occurrence in sloping terrains, poses significant risks to infrastructure stability and transportation safety. The development of perched water (PW) has had a significant impact on the service life of roadway slopes built on highly plastic and expansive clay due to shrink swell potential and formed desiccation crack which can eventually reduce the soil shear strength and cause slope failure. Electric Resistivity method can identify perched water condition on subsoil. However, only relying on electric resistivity may sometimes misinterpret the saturated soil with perched water as in both conditions the value of resistivity can be very low. Induced polarization is another type of geophysical investigation method which focuses on investigating the perched water condition based on the ability of subsurface materials to store and release electrical charge. Current study will investigate the perched water condition based electric resistivity and induced polarization (IP) testing. As a part of the study, electric resistivity testing will be done two sites in Mississippi with high plasticity expansive clay. Initially, electric resistivity will be done on the soil slope for subsurface investigation. Later, The Reversed IP approach was then used to validate the ERI data by determining the polarization properties of the subsurface materials. This study findings identify the presence and accumulation of perched water at varying depths and extents. The integration of ERI and reversed IP data made it possible to precisely map perched water zones within roadway slopes, allowing for more informed judgments about slope stability assessments and mitigating actions.

INTRODUCTION

Perched water zones are areas in the subsurface of slopes under heavy rainfall or above low-permeability soil layers, causing decreased shear strength and increased loading on the slope. These zones are located above the regional water table and result in saturation and positive pore water pressure building up at shallow depths. In expansive clay soil due to high shrink swell behavior desiccation cracks form at the surface which pave the way for water infiltration. These infiltrated water create perched water zone unidentified from the top (Khan et al., 2017 ; Khan et al., 2019; Nobahar et al., 2021; Khan et al., 2023; Robinson et al., 2005; Wu et al., 1999).

Electrical resistivity imaging (ERI) has been widely used in various geophysical studies to assess subsurface conditions and geological features. Yamakawa et al., 2010 discussed the use of a combined penetrometer-moisture probe with geophysical methods to survey hydrological properties of natural slopes which indicates the potential for ERI to provide valuable insights into slope conditions. Similarly, (Williams et al., 2017) utilized time-lapse ERI to image hydrological processes showing the effectiveness of ERI in monitoring water movement in different areas. (Crawford et al., 2018; Ismail et al., 2019) both studied the application of 2-D ERI in landslide investigations, highlighting its ability to characterize landslide types, locate failure zones and identify the area perched water. Additionally, (Awang et al., 2021) utilized ERI to investigate the groundwater table under a rock slope surface, emphasizing the importance of ERI in mapping subsurface conditions and determining the weathering in rock slopes. Overall, ERI is a valuable tool to investigate the perched water conditions on slopes providing insights into geological structures, hydrological processes, and landslide investigations.

Induced Polarization (IP) is a geophysical imaging technique used to identify the chargeability of subsurface materials. IP refers to the ability of soil to store and release electrical charges which can vary based on different ground conditions similar to electrical resistivity. The collaboration of Electrical resistivity imaging (ERI) and Induced polarization (IP) has been widely used for geophysical investigation to provide valuable insight into subsurface soil conditions. (Riddell et al., 2010) used ERI and IP to delineate hydro-geomorphic controls in wetland which studies the importance of understanding the controls to maintain equilibrium of wetlands. (A. Aziz et al., 2013) demonstrates the efficiency of 2D ERI and IP in differentiating the clayey soil layers and emphasized the importance of combining these methods for accurate interpretation of field data. (Attwa et al., 2011) evaluated DC resistivity, FDEM and SIP methods for Imaging perched saltwater and a shallow channel within coastal flat sediments. (Kumar et al., 2021) used ERI and IP data to delineate aquifers contaminated with saltwater. (Mekkawi et al., 2021) integrated ERI and IP in the exploration of iron ore deposits, showing the effectiveness of these method in identifying the lateral and vertical distribution of mineral deposits. However, the previous research studies are not adequate in terms of collaboration of ERI and IP methods for subsurface soil investigation for perched water condition.

This study focuses on identifying the perched water zone with the collaboration of ERI and IP methods in expansive soil like Yazoo clay in Mississippi. With climate variations, particularly rainfall, affect slope stability of highways in Mississippi, where the rainfall intensity is higher than the U.S. average. Identifying perched water zone and the depth of slope failure can enhance repair design, safety and performance of highway slopes. Early detection of failure prone highway slopes is very crucial. The primary aim of this study is to locate the perched water zone using field geophysical surveys that combine Electrical Resistivity Imaging (ERI) and reverse Induced Polarization (IP) methods.

SITE DESCRIPTION

One of the study areas selected for the investigation was Jackson State University premises in Jackson, Mississippi and another investigation area is situated near Terry Road in southwest Jackson, Mississippi (Figure 1). The slope is composed of Yazoo clay, a high plasticity expansive soil, with a weathered upper zone that overlays unweathered clays. The liquid limit of weathered Yazoo clay typically ranges from 70%, resulting in a plasticity index of over 50%. This expansive type of soil can shrink and swell due to moisture variation which can eventually form desiccation cracks. The cracks pave the way for water during rainfall which can create perched water condition inside the soil (Nobahar et al., 2024).

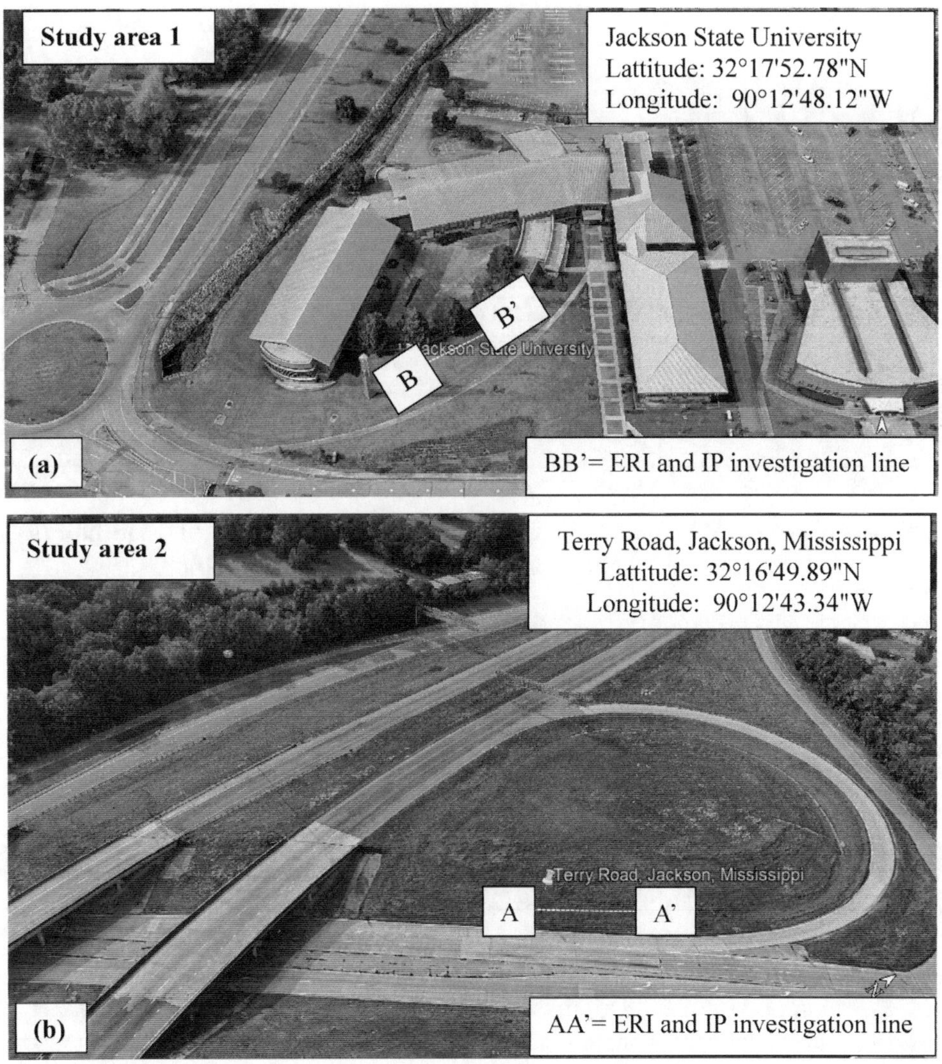

Figure 1. Site Location for Perched Water Condition Detection Using ERI and IP Method (a) Site 1 in Jackson, Mississippi (b) site 2 in Terry Road, Mississippi

METHODOLOGY

Electrical Resistivity Imaging (ERI)

Electrical Resistivity Imaging (ERI), a nondestructive approach, can be used to gather information quickly and easily about the horizontal and vertical characteristics of soil structures and properties. Electrical Resistivity Imaging is a non-invasive and quick method for subsurface investigation which provides a high-resolution image of subsurface features. ERI involves injecting current into the ground through electrodes and measuring the resulting potential differences of electrodes which allows the identification of different materials of subsurface. Power supply/transmitter, current electrodes, multichannel resistivity meter (superstring), switch box, cables and connectors are the key features of electrical resistivity measuring instruments. ERI test was conducted along a 16.5m length of the mid highway slope of Terry Road Jackson Mississippi site for July 2024. 28 number of electrodes were used at a 0.6m spacing for collecting the ERI data.

Figure 2. Investigation of Perched Water Condition Using ERI and IP Method (a) Data Collection b) ERI and IP Investigation Setup Line

Induced Polarization

Polarization (IP) is a geophysical method to identify the subsurface materials by measuring their ability to temporarily hold an induced electric charge. The highway slope of the study area at Terry Road, Jackson consists of Yazoo clay, a highly plastic clay soil. Induced polarization (IP) properties of clayey soil are distinct due to the unique properties of clay materials. Clayey soils typically show high chargeability due to the presence of clay minerals. Clay minerals can hold and release electrical charges efficiently. The fine-grained nature of clay soil provides a large surface area for ion absorption. Also, clays soil may have low resistivity because clay particles can retain water and dissolved ions which conduct electricity. Power supply/transmitter, current electrodes, multichannel resistivity meter (superstring), switch box, cables and connectors are the key features of Induced Polarization (IP) measuring instruments. Induced polarization (IP) test was conducted along a 16.5m length of the mid highway slope of Terry Road Jackson Mississippi site. 28 number of electrodes were used at a 0.6 m spacing for collecting the IP data. 2D software is used in this study as a processing and visualization software of resistivity and induced polarization (IP) data. Earthimager 2D supports various data formats for resistivity and IP surveys. The quality control tool of Earthimager 2D can check and remove any erroneous data points. Earthimager 2D helps to contour colors, overlay information, and display the best output of resistivity/IP surveys of ground soil for further interpretation. Earthimager 2D is used primarily to create 2D surface images based on resistivity and IP surveys.

RESULTS

Figure 4 shows a 2D electric resistance plot that was generated from the collected resistivity data analysis with Earth Imager 2D. This data provides the resistivity values measured in ohmmeters. As resistivity is the quantitative measure of a material's ability to impede the movement of electric current, high moisture content or perched water content increases the electricity reducing resistivity values. In the contour generated from the values from resistivity analysis, regions that are colored blue represent a lower level of resistivity whereas regions that are colored red show a higher level of resistivity indicating the material is less capable of conducting electricity. Similar change has been observed in the IP contours where the value increases from low to high with chromatic change of colors from blue to red. The analysis data is represented with the values of chargeability in milliseconds (ms). IP values refer to the ability of subsurface materials to efficiently store and discharge electrical charge. Regions colored in blue represent lower levels of chargeability whereas regions that are colored red suggest a greater level of chargeability. In the joint effort with resistivity and IP for subsurface investigation, IP method complements resistivity imaging by providing information on the ability of materials to store and release charge. The IP approach enhances resistivity imaging in subsurface exploration by providing insights into the capacity of materials to store and discharge electric charge. Red coloration in resistivity measurements may suggest the presence of dry sands, rocks, or non-conductive materials with high resistance. In IP, the color red may represent materials with a high mineral content or subsurface material that has the ability to polarize the charge including clay, rock or materials with high mineral content. On the other hand, the color blue suggests materials with a lower ability to polarize the charge including clean sands, gravels, non-clayey soils, solid rock formations without metallic content. In perched water condition or in saturated clay, reduced value of IP is observed in presence of water as the IP response is weakened and as the ions become mobile. The ions redistribute themselves more readily, thereby reducing the polarization effect. In Figure 4. blue spots were detected on the uppermost layer of the surface, indicating decreased resistivity values. Contrary to the low IP value shown by the resistivity imaging, the presence of a reduced blue patch in the resistivity zone may indicate the existence of saturated clay rather than a perched water situation in the soil. In areas where perched water is present, both the IP value and resistivity are significantly reduced. Some blue spots detected in the resistivity has high polarization values which may indicate the presence of minerals in the soil resulting in high IP values.

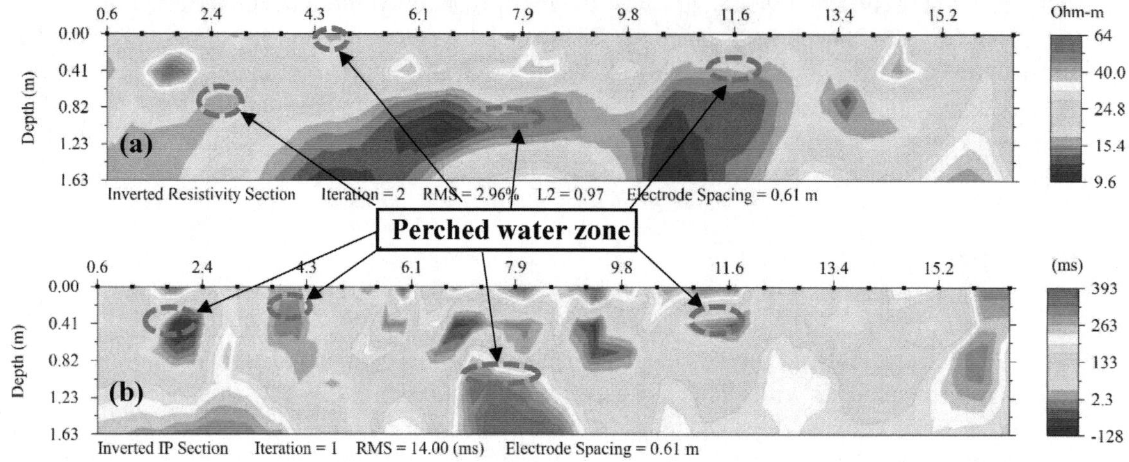

Figure 3. Subsurface Investigation for Perched Water Condition Detection Using 2D Electric Resistivity and Induced Polarization in Site 1 in Jackson, Mississippi (a) Electric Resistivity Imaging (b) Induced Polarization

At Site 2 on Terry Road, a comparable examination of resistivity and IP was conducted. The collected data, obtained from 28 electrodes, was used to construct a contour plot, as shown in Figure 5. A decline in values has been noted in the uppermost layer of soil at a depth of less than 1 meter in resistivity analysis, and a similar decline has also been detected in the IP values. However, the measured IP value does not align with the low resistivity indicated by the contour imaging. This discrepancy suggests that the blue patch observed in the resistivity zone may be caused by the presence of saturated clay rather than perched water. Certain places exhibit elevated resistivity values, which also include an enhanced IP value. This can be suggestive of the presence of dry clay materials with reduced moisture content with mineral content in the soil that results high polarization values.

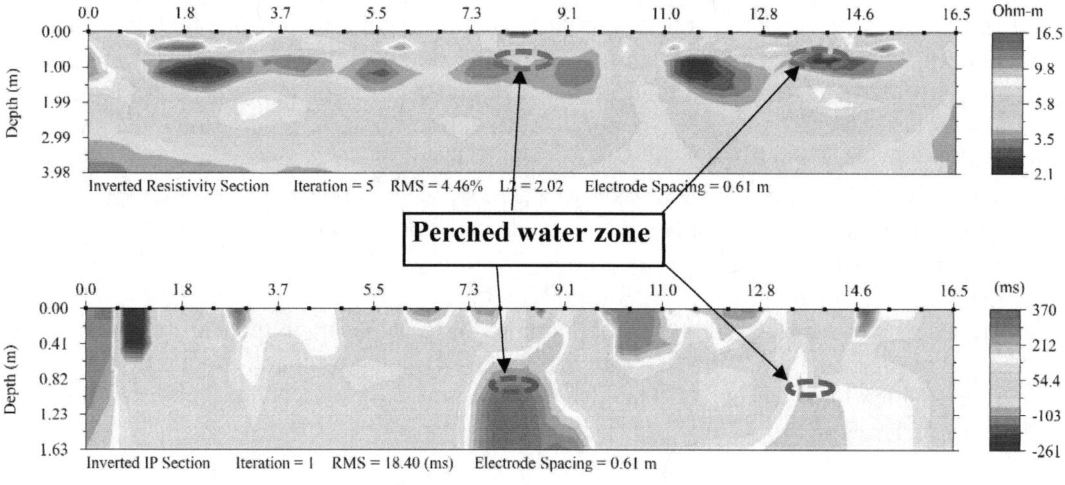

Figure 4. Subsurface Investigation for Perched Water Condition Detection Using 2D Electric Resistivity and Induced Polarization in Site 2 in Terry Road, Mississippi (a) electric resistivity imaging (b) Induced Polarization

DISCUSSION

The current study will examine the electric resistivity and induced polarization (IP) testing based on the perched water condition. Electric resistivity testing will be conducted at two locations in Mississippi that contain high plasticity expansive clay as part of the investigation. Initially, the soil slope will be subjected to electric resistivity testing and later, the Induced Polarization (IP) method was conducted on the slope to investigate the perched water zone inside the soil. It has been seen that both site conditions exhibited a low resistivity value in multiple locations indicating a high saturation level of soil or the presence of a perched water zone. However, the data obtained during IP-based investigation in some of those locations did not indicate a similar decrease in ERI indicating the presence of saturated soil rather than perched water zones. Some locations were detected having low resistivity with high Ip value which may result from the presence of minerals in the soil. Some locations were identified as having a high IP value and low resistivity, which may be attributed to the presence of minerals in the soil.

Electrical resistivity imaging (ERI) is a valuable tool in geophysical studies, providing insights into subsurface conditions and geological features. ERI has already proven handy to detect the soil and moisture condition underneath. Added with ERI, IP inversion is a powerful tool that complements resistivity imaging by helping to detect changes in moisture levels and the presence of perched water beneath the surface. Resistivity values increase in soil conditions with low moisture content or with materials that have low conductivity. In the Induced Polarization method, the values of electric charge increase when there are no materials present that can store and release electric charge. However, these values are highly influenced by the minerals in the soil and their ability to hold electric charge.

Clay minerals possess a substantial surface area and exhibit the ability to attract ions which in general results potentially larger levels of ionic polarization compared to clean sand or gravel. In conditions of saturated soil, both resistivity and IP value can decrease. Nevertheless, the existence of minerals in the moisture and soil might also impact the IP values, leading to deviations from the anticipated outcomes. Like saturated soil, within perched water zones, resistivity values exhibit a substantial drop as a result of the presence of water. Regarding the IP values, the existence of a perched water zone increases the mobility of ions, which reduces the polarization effect and leads to a decrease in the IP value, as observed in the ERI results. However, the value may not decrease as significantly as it would in saturated clay, as perched water contains fewer clay minerals and higher water molecules and ionic concentrations, which results in decreased polarization effects.

Significant risks to transportation safety and infrastructure stability are posed by perched water, a frequent occurrence in sloping terrains. The service life of roadway slopes constructed on highly plastic and expansive clay material has been significantly impacted by the development of perched water inside the soil. This is due to the shrink swell potential and the formation of desiccation cracks, which can ultimately reduce the soil shear strength and result in slope failure. The Electric Resistivity method can detect the presence of suspended water in the subsoil. However, the saturated soil with perched water may occasionally be misinterpreted if electric resistivity is the sole metric used, as the resistivity value in both cases can be extremely low. Induced polarization is an additional geophysical investigation method that concentrates on the perched water condition by examining the capacity of subsurface materials to store and release electrical charge. When the soil is saturated, its resistivity value falls, which can lead to a chance of misinterpretation between saturated soil conditions and perched water zones. Combination of the IP approach with resistivity provide enhances the precision of predicting the perched water zone in subsurface investigations.

SUMMARY AND CONCLUSION

Perched water conditions in soil can be identified using Electrical Resistivity Imaging (ERI) and Reverse Induced Polarization (IP) method. The study examines the presence of perched water in high plasticity expansive clay soil at two location using Electrical Resistivity Imaging (ERI) and Induced Polarization (IP) techniques. Both sites showed low resistivity values, indicating high soil saturation. This information is crucial for understanding the subsurface characteristics of the soil and identifying the potential risks associated with landslides. In perched water zones, resistivity drops due to water presence and IP values decrease due to reduced polarization effects. The study emphasized that combining IP with resistivity improves the precision in detecting perched water zone, although mineral content in soil and moisture can affect the IP values.

DATA AVAILABILITY STATEMENT

All data will be shared from the authors based on reasonable request.

REFERENCES

- Journal

1. A. Aziz, N., R. Al- Qaissy, M., & H. Karim, H. (2013). Differentiating Clayey Soil Layers from Electrical Resistivity Imaging (ERI) and Induced Polarization (IP). *Engineering and Technology Journal*, *31*(17), 3216–3234. https://doi.org/10.30684/etj.31.17A.1
2. Attwa, M., Günther, T., Grinat, M., & Binot, F. (2011). Evaluation of DC, FDEM and IP resistivity methods for imaging perched saltwater and a shallow channel within coastal tidal flat sediments. *Journal of Applied Geophysics*, *75*(4), 656–670. https://doi.org/10.1016/j.jappgeo.2011.09.002
3. Awang, H., Salmanfarsi, A. F., Zaini, M. S. I., Mohamad Yazid, M. A. F., & Ali, M. I. (2021). Investigation of groundwater table under rock slope by using electrical resistivity imaging at Sri Jaya, Pahang, Malaysia. *IOP Conference Series: Earth and Environmental Science*, *682*(1), 012017. https://doi.org/10.1088/1755-1315/682/1/012017
4. Crawford, M. M., Bryson, L. S., Woolery, E. W., & Wang, Z. (2018). Using 2-D electrical resistivity imaging for joint geophysical and geotechnical characterization

of shallow landslides. *Journal of Applied Geophysics*, *157*, 37–46. https://doi.org/10.1016/j.jappgeo.2018.06.009

5. Dey, N., & Sengupta, A. (2018). Effect of rainfall on the triggering of the devastating slope failure at Malin, India. *Natural Hazards*, *94*(3), 1391–1413. https://doi.org/10.1007/s11069-018-3483-9
6. Han, J., Chen, J., Hong, Z., & Shen, S. (2010). Mitigation of levee failures using deep mixed columns and geosynthetics. *Geomechanics and Geoengineering*, *5*(1), 49–55. https://doi.org/10.1080/17486020903452733
7. Ismail, N. E. H., Taib, S. H., & Abas, F. A. M. (2019). Slope monitoring: An application of time-lapse electrical resistivity imaging method in Bukit Antarabangsa, Kuala Lumpur. *Environmental Earth Sciences*, *78*(1), 14. https://doi.org/10.1007/s12665-018-8019-9
8. John A. Cherry, A. R. F. (1979). *Groundwater. Prentice-Hall, Englewood Cliffs, NJ.: Vol. ISBN 0-13-365312-9*. Prentice-Hall, Inc. Englewood Cliffs, New Jersey 07632.
9. Khan, M. S., Hossain, S., Ahmed, A., & Faysal, M. (2017). Investigation of a shallow slope failure on expansive clay in Texas. *Engineering Geology*, *219*, 118–129. https://doi.org/10.1016/j.enggeo.2016.10.004
10. Khan, S., Ivoke, J., & Nobahar, M. (2019). Coupled Effect of Wet-Dry Cycles and Rainfall on Highway Slope Made of Yazoo Clay. *Geosciences*, *9*(8), 341. https://doi.org/10.3390/geosciences9080341

11. Khan, S., Whalin, R. W., Spears, A., & Chakraborty, A. (2023). Bio-Inspired Stabilization of

Levee Slope on Expansive Yazoo Clay at the Maritime and Multimodal Transportation Infrastructure in Mississippi.

12. Kumar, P., Tiwari, P., Singh, A., Biswas, A., & Acharya, T. (2021). Electrical Resistivity and Induced Polarization signatures to delineate the near-surface aquifers contaminated with seawater invasion in Digha, West-Bengal, India. *CATENA*, *207*, 105596. https://doi.org/10.1016/j.catena.2021.105596
13. Mekkawi, M. Mohamed., ElEmam, A. ElKotb., Taha, A. Ismail., Al Deep, M. AbdelSabour., Araffa, S. A. Sultan., Massoud, U. Saad., & Abbas, A. Mohamed. (2021). Integrated geophysical approach in exploration of iron ore deposits in the

North-eastern Aswan-Egypt: A case study. *Arabian Journal of Geosciences*, *14*(8), 721. https://doi.org/10.1007/s12517-021-06964-0

14. Nobahar, M., Khan, M. S., Stroud, M., Amini, F., & Ivoke, J. (2021). Progressive Development of the Perched Water Zone in Highway Slopes Made of Highly Plastic Clay. *Transportation Research Record: Journal of the Transportation Research Board*, *2675*(9), 715–728. https://doi.org/10.1177/03611981211004178

15. Nobahar, M., Worsley, G., Spears, A., Khan, S., & Chakraborty, A. (2024). Climate Adaptive Predictive Approaches for Geotechnical Infrastructure Components in Mississippi. In Geo-Congress 2024 (pp. 482-492).

16. Rahardjo, H., Lee, T. T., Leong, E. C., & Rezaur, R. B. (2005). Response of a residual soil slope to rainfall. *Canadian Geotechnical Journal*, *42*(2), 340–351. https://doi.org/10.1139/t04-101
17. Riddell, E. S., Lorentz, S. A., & Kotze, D. C. (2010). A geophysical analysis of hydro-geomorphic controls within a headwater wetland in a granitic landscape, through ERI and IP. *Hydrology and Earth System Sciences*, *14*(8), 1697–1713. https://doi.org/10.5194/hess-14-1697-2010
18. Robinson, B. A., Broxton, D. E., & Vaniman, D. T. (2005). Observations and Modeling of Deep Perched Water Beneath the Pajarito Plateau. *Vadose Zone Journal*, *4*(3), 637–652. https://doi.org/10.2136/vzj2004.0168
19. Williams, M. R., Buda, A. R., Singha, K., Folmar, G. J., Elliott, H. A., & Schmidt, J. P. (2017). Imaging Hydrological Processes in Headwater Riparian Seeps with Time-Lapse Electrical Resistivity. *Groundwater*, *55*(1), 136–148. https://doi.org/10.1111/gwat.12461
20. Wu, Y. S., Ritcey, A. C., & Bodvarsson, G. S. (1999). A modeling study of perched water phenomena in the unsaturated zone at Yucca Mountain. *Journal of Contaminant Hydrology*, *38*(1–3), 157–184. https://doi.org/10.1016/S0169-7722(99)00015-7
21. Yamakawa, Y., Kosugi, K., Masaoka, N., Tada, Y., & Mizuyama, T. (2010). Use of a Combined Penetrometer–Moisture Probe Together with Geophysical Methods to Survey Hydrological Properties of a Natural Slope. *Vadose Zone Journal*, *9*(3), 768–779. https://doi.org/10.2136/vzj2010.0012

Evaluation of the performance of Vetiver plant using resistivity imaging to repair highway slope failure

Fariha Rahman

Jackson State University

1400 J.R. Lynch Street, JSU Box 17068, Jackson, MS 39217-0168

(601)-291-5578

J00986301@students.jsums.edu

Audrika Nahian

Jackson State University

1400 J.R. Lynch Street, JSU Box 17068, Jackson, MS 39217-0168

(601)-212-2035

j00986152@students.jsums.edu

Sadik Khan, PhD, P.E.

Jackson State University

1400 J.R. Lynch Street, JSU Box 17068, Jackson, MS 39217-0168

sadik.khan@jsums.edu

Ian La Cour, P.E.

Mississippi Department of Transportation

412 E. Woodrow Wilson Ave. Jackson, MS 39216

ilacour@mdot.ms.gov

Shams E Shifat

Jackson State University

1400 J.R. Lynch Street, JSU Box 17068, Jackson, MS 39217-0168

j01033311@students.jsums.edu

Prepared for the 73rd Highway Geology Symposium, September, 2024

Acknowledgements

The authors sincerely acknowledge the National Science Foundation (NSF) CMMI (Award No 2046054) for supporting the study of Vetiver in landslide repair. The authors also appreciate the Mississippi Department of Transportation (MDOT) for supporting the project MDOT State Study 332: Green Landslide Repair Using Deep Rooted Vetiver Grass for MDOT.

Disclaimer

Statements and views presented in this paper are strictly those of the authors, and do not necessarily reflect positions held by their affiliations, the Highway Geology Symposium (HGS), or others acknowledged above. The mention of trade names for commercial products does not imply the approval or endorsement by HGS.

Copyright Notice

ABSTRACT

Vetiver grass is a perennial grass with a dense and long root system. It has applications in slope stability, erosion control, environmental remediation, and many others. The characteristics of Vetiver in surviving very harsh and contaminated environments make it a suitable technology for landslide repair under changed climatic conditions. The objective of the current study is to evaluate the performance of Vetiver grass in reducing landslides in highway slopes. The highway slope along MS 145 near Shubuta, Mississippi, is subject to a creeping failure. Vetiver grass was planted along the slope in October 2023, and the site has been monitored regularly using Electrical Resistivity Imaging (ERI). ERI is a non-destructive subsurface investigation technique that helps to generate the subsurface resistivity profile in a greater depth. The resistivity profile can then be translated to detect any perched conditions or slope movement along the slope. Initial ERI assessments prior to plantation revealed the presence of perched conditions within the slope soil. Subsequent assessments of post-plantation demonstrated a noticeable improvement in the perched zone over time. The findings underscore Vetiver grass's efficacy in alleviating perched conditions on highway slopes constructed on highly expansive clay, a primary trigger for landslides in Mississippi. This nature-based solution offers a cost-effective approach to landslide repair, showcasing the potential of the Vetiver system in climate-adaptive slope repair.

INTRODUCTION

Vetiver grass (Chrysopogon zizanioides) is a perennial grass known for its extensive root system. The long and dense root makes it valuable for soil and water conservation, as well as for its applications in environmental engineering, bioenergy and even medicine. The roots can grow to a depth of up to 3-4 meters in a single year, anchoring the plant firmly into the ground (1). At full maturation, the root diameter of each tiller can reach a width of up to 18 inches. Vetiver grass exhibits resilience to extreme climate variations, adapts to various soil types, tolerates a wide pH range (from 3.3 to 12.5), extreme temperatures from -15°C to +55°C and can thrive in the presence of high levels of micropollutants. Notably, the grass does not require replanting and can be harvested annually (2). The 'Sunshine' vetiver genotype has been recognized by the United States Department of Agriculture (USDA) Natural Resources Conservation Service (NRCS) as a non-invasive and sterile variety suitable for cultivation in the United States. In the United States, vetiver grass has been successfully adapted to USDA Plant Hardiness Zones 9 to 11, encompassing regions like the Pacific Islands, Caribbean Areas, and southern states such as Texas, Louisiana, Mississippi, and Alabama. The adaptability of Vetiver grass to various climatic conditions is another remarkable characteristic. The 'Sunshine' vetiver genotype exhibits the ability to thrive in areas with annual rainfall ranging from 20 to 200 inches and can adapt to a diverse range of soil types, including sand, silt, and clay (3). Earlier study has shown that the Vetiver grass was able to reduce the moisture content of the slope soil even after heavy rainfall events (4). When the Vetiver was in growing, there was a stable moisture trend within the soil system (Figure 1a). However, with the growth of Vetiver, the moisture content started to reduce, with substantial precipitation (Figure 1b).

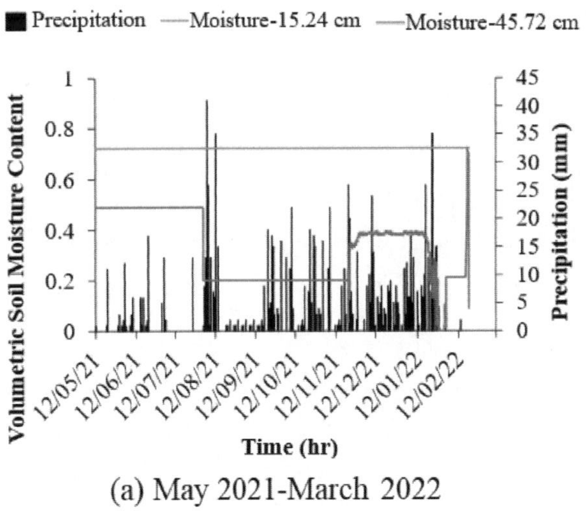

(a) May 2021-March 2022

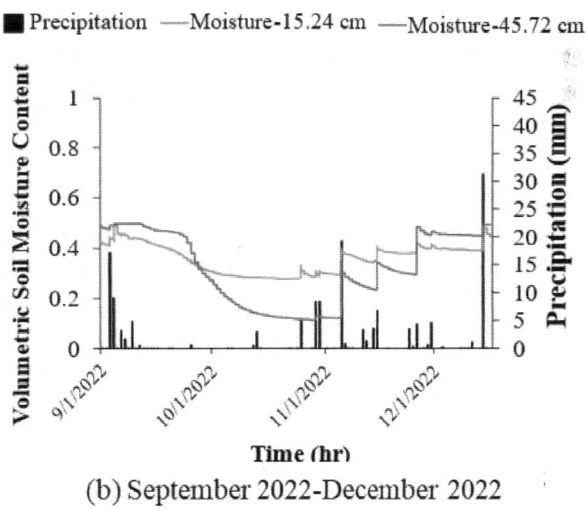

(b) September 2022-December 2022

Figure 1- Changes in Soil Moisture Content (a) During the Growth of Vetiver Root, (b) After the Full Maturation of the Vetiver Root

Climate change significantly impacts geo-infrastructure, which includes essential structures like roads, bridges, railways, airports, dams, and levees. These impacts are largely due to the increased frequency and severity of extreme weather events, temperature variability, sea-level rise, and changes in soil stability and groundwater levels. When heavy rainfall events occur, the rainwater infiltrates into the slope soil and increases the moisture content. Increased soil moisture leads to reduced shear strength (5). Extreme weather events such as heavy rainfall, hurricanes, and storms can cause severe flooding and erosion, overwhelming drainage systems and increasing soil moisture content. The effects lead to costly repairs and increased maintenance needs (6). High winds and storm surges from hurricanes can damage coastal infrastructure, including ports and bridges, necessitating resilient design and construction standards to mitigate these effects (7). Temperature variability, including more frequent extreme heat and cold events, adversely affects geo-infrastructure (8).

Increased precipitation can lead to soil erosion or increase soil moisture content. On the other hand, droughts can cause soil desiccation and shrinkage, affecting structural integrity (9). Rainfall-induced slope failure has become frequent in Mississippi due to the precipitation increase in the region (10). Several studies have been conducted to investigate the factors behind slope failure and its connection to highly expansive clay soil and rainfall. Xu and Zhang, 2010 studied a railway landslide induced by rainfall, aiming to prevent the recurrence of such failures (11). Earlier study shows the effect of rainfall on the stability of unsaturated earth slopes constructed on expansive clay, showing a decrease in the factor of safety after a seven-day rainfall period (12). Ng et al., 2003 highlighted the importance of understanding the performance of unsaturated expansive soil slopes, particularly in the context of major infrastructure projects like the south-to-north water transfer project in China (13). A study conducted by Qi and Vanapalli, 2015 on the hydro-mechanical coupling effect on the stability of unsaturated expansive soil slopes demonstrated that coupled analysis considering swelling leads to different profiles within the surficial layer compared to uncoupled analysis (14). Investigation of a shallow slope failure on expansive clay in Texas found that both a fully softened condition and rainfall played a role in the slope failure (15). Ahmed et al. 2018 monitored moisture variation in the expansive subgrade through field instrumentation and geophysical testing, highlighting the impact of moisture suction on soil properties and pavement performance (16). The impact of rainfall variation on slopes made on expansive Yazoo clay soil in Mississippi was also observed in a study conducted by Nobahar et al., 2020 (17). The study used finite element analysis to investigate the unsaturated hydromechanical behavior of the slopes.

The current study focuses on assessing the efficiency of Vetiver grass in preventing landslides on highway slopes. One highway slope near Shubuta, Mississippi, along MS 145, was selected for the study. Vetiver grass was planted on the slope in October 2023. Since then, regular field investigations have been conducted to monitor the subsurface conditions of the site. Electrical Resistivity Imaging (ERI) profiles were able to capture the changes in soil moisture profiles with time. The ERI profiles showed that the perched condition in the shallower depth of the soil was reduced after the plantation of Vetiver grass. Vetiver grass can improve the subsurface moisture condition of the soil and thus provide additional shear strength to the soil. Therefore, it can be a cost-effective and climate-adaptive solution for highway slope stability.

METHODOLOGY

Site Selection

The slope along the MS 145 highway near Shubuta, Mississippi, was selected for the current study. The coordinates of the site are 31°50'53.8"N 88°41'21.2"W. The Chickasawhay River flows near the slope. It is a 165 ft ×20 ft highway slope. The 3H:1V slope was having creeping slides near the bridge (Figure 2).

Figure 2- Location of the Study Slope

Vetiver Plantation

Around 4500 Vetiver grass was planted on the slope in October 2023. The whole section of the slope was planted with grass. The site was divided into two sections based on the primary field investigation results. Movement was identified near the bridge, and grass spacing of this section (section 1) Vetiver was lower than the other section (section 2) (Figure 3a). The plantation was done in a staggered way (Figure 3b). The growth of the Vetiver was monitored regularly.

(a) (b)

Figure 3- Two Sections of the Slope Repaired with Vetiver Grass

Weather Data

Weather data of the location was obtained from the NASA POWER Data (18). This source provides comprehensive weather data derived from satellite observations and meteorological models. The purpose of obtaining weather data is to incorporate climatic conditions into the analysis of the ERI results. Weather data, like precipitation and temperature, impact soil moisture content, which significantly influences soil resistivity and, consequently, the interpretation of subsurface conditions.

Electrical Resistivity Imaging (ERI)

The Electrical Resistivity Imaging (ERI) technique has proven highly effective in revealing subsurface structures and providing detailed information about the physical properties of rocks for economic, environmental, and engineering applications (19). ERI test aims to determine the subsurface resistivity distribution by taking surface measurements using four electrodes: two current electrodes (A and B) and two potential electrodes (M and N), which can be positioned arbitrarily on the ground surface. Electric currents are injected into the ground through the current electrodes, and the resulting potential differences are measured at the surface using the potential electrodes. Deviations from the expected potential differences in the homogeneous ground reveal information about the subsurface inhomogeneities and their electrical properties. These measurements allow for the estimation of the true resistivity of the subsurface (20). The use of ERI to monitor temporal variations in soil moisture is based on the theory that changes in soil resistivity are due to changes in soil moisture. Specifically, as the soil becomes wetter, its resistivity decreases, and as it dries, its resistivity increases (21).

ERI tests were conducted in the field along the center of the slope in a single line with electrodes spaced 3 feet apart (Figure 4a, Figure 4b, Figure 4c and Figure 4d). During the ERI field investigation, apparent resistivity (r) data were collected using a SuperSting R8/IP device equipped with 56 electrodes. The collected data were subsequently downloaded to a computer and analyzed with EarthImager 2D software.

(a) (b)

(c) (d)

Figure 4- Electrical Resistivity Imaging Tests at the Site

RESULTS

Vetiver Growth Monitoring

The growth of Vetiver is important as this dense root network binds soil particles, reducing the likelihood of slope failures and landslides. The grass's ability to control soil erosion further stabilizes slopes by acting as a physical barrier to surface runoff, slowing water flow and allowing more water to infiltrate the ground. The root system also enhances water infiltration and retention, preventing surface runoff during heavy rainfall and maintaining soil moisture levels. As Vetiver grows, its adaptability to extreme temperatures, high salinity, and varying pH levels ensures it can thrive in diverse climatic conditions and provide long-term stability (22). The growth of Vetiver plants, planted in October 2023, was systematically monitored to ensure their health and suitability for slope stabilization. It was observed that there was a significant improvement in the growth of the grass till June 2024 (Figure 5b).

Figure 5- Site Condition (a) Before Plantation (September 2023) and (b) After Plantation (June 2024)

Weather Data

The weather data included the precipitation and temperature in the site location. These parameters were plotted over time to analyze their seasonal variations (Figure 6). The plot revealed that the highest rainfall occurred between February 2024 and June 2024, with notable spikes on specific days.

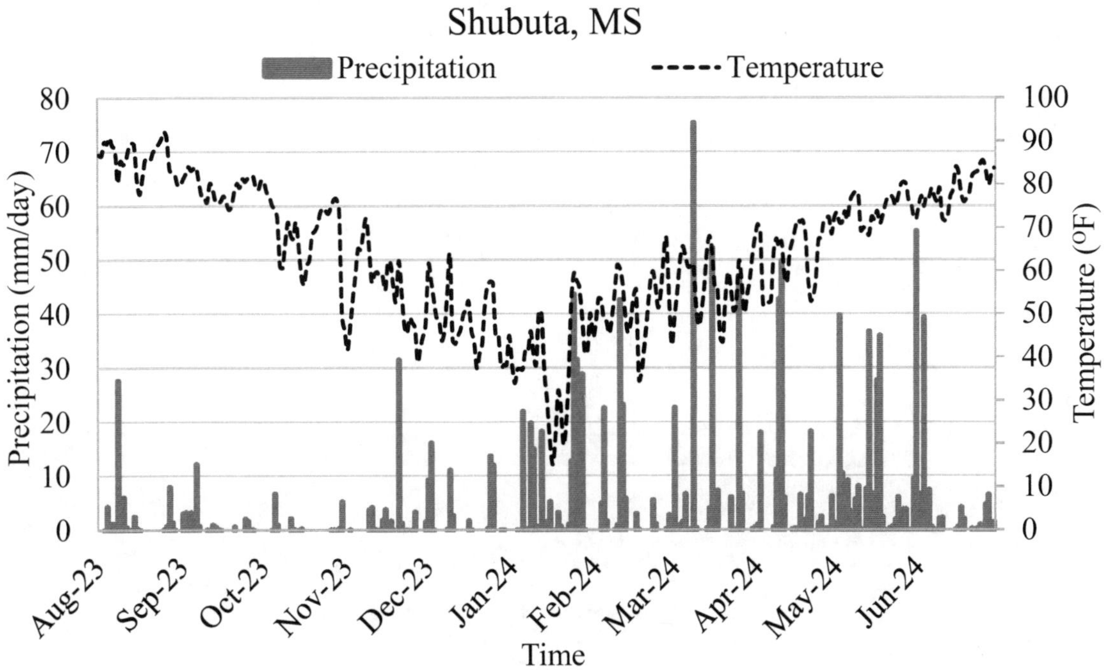

Figure 6- Weather Condition in Shubuta, MS

ERI Profiles

The ERI profiles were obtained from the EarthImager 2D software. In September 2023, the ERI profile identified perched water zones and slope movement. Based on the investigation, a plantation layout was designed. ERI tests were performed after the Vetiver showed significant growth. As soil moisture content increases, the electrical resistivity of the soil decreases. The color gradient from blue to red represents increasing resistivity.

The ERI profiles were obtained from the EarthImager 2D software. In September 2023, the ERI profile identified perched water zones and slope movement. Based on the investigation, a plantation layout was designed. ERI tests were performed after the Vetiver showed significant growth. As soil moisture content increases, the electrical resistivity of the soil decreases (23). The dissolved ions in the soil's pore water facilitate electrical conductivity. When an electric field is applied, the presence of free electrical charges reduces electrical resistivity. Consequently, an increase in soil moisture leads to a decrease in electrical resistivity (24). The color gradient adapted in this study is from blue to red, representing increasing resistivity. Therefore, the blue range color represents a higher moisture content. In Figure 7a, lower resistivity (around 9.1 Ohm-m) represents the presence of a higher moisture zone in September 2023. There was a significant increase in the resistivity value (around 54.8 Ohm-m) in the same zone from March 2024 (Figure 7c). On the other hand, higher resistivity from 90 ft to 144 ft along the ERI line denotes a drier zone or void, as air has higher resistance. High resistivity zones typically correlate with low moisture content, which can indicate dry and brittle material. It was found in an earlier study that high resistivity zones (>1000 Ohm-m) were associated with alluvium or highly weathered rock, which are prone to instability when saturated (25). This zone was, therefore, considered critical during the initial investigation before the Vetiver plantation layout design. Closer Vetiver spacing was used for this section.

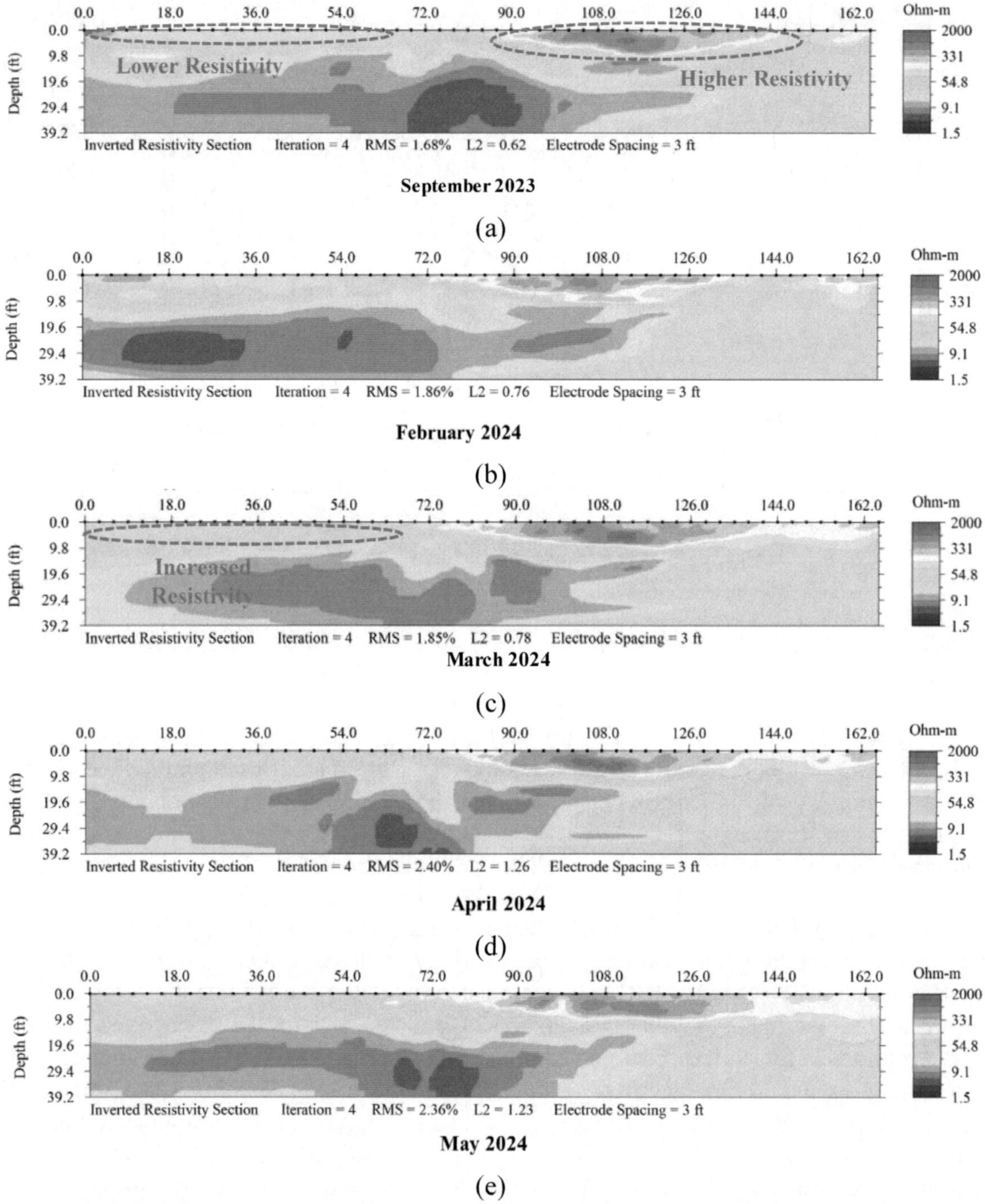

Figure 7- ERI Profiles obtained in (a) September 2023, (b) February 2024, (c) March 2024, (d) April 2024, (e) May 2024

DISCUSSION

After planting Vetiver on the site in October 2023, there was a noticeable increase in resistivity values over time, which can be attributed to the root development and maturation process. Initially, resistivity values were relatively low, indicating higher moisture content. As Vetiver roots began to grow and absorb moisture from the soil from November 2023 to February 2024, resistivity gradually increased, reflecting a reduction in soil moisture. (Figure 8).

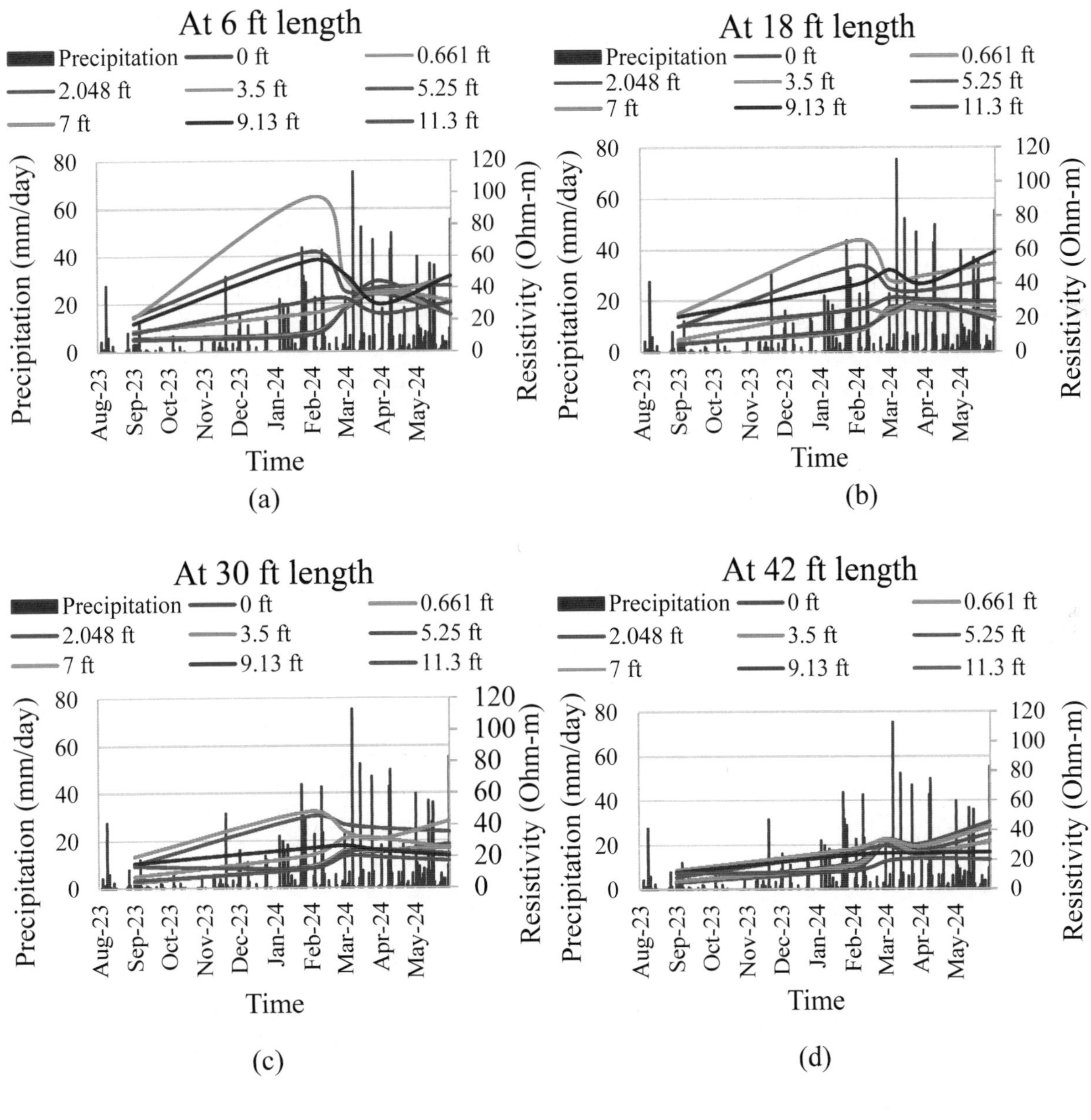

Figure 8- Resistivity Upto 11 ft Depth Along (a) 6 ft Length, (b) 18 ft Length, (c) 30 ft Length and (d) 42 ft Length

By the maturation phase from March 2024 to May 2024, the Vetiver roots had become well-established, significantly drawing moisture from the soil and leading to a notable increase in resistivity values, reaching up to 45 Ohm-m. This increase in resistivity is a clear indication of the soil drying out due to the mature Vetiver roots' enhanced moisture absorption capabilities, demonstrating the relationship between vegetation growth and changes in soil properties.

DISCUSSION AND CONCLUSION

The current study evaluates the performance of Vetiver grass in stabilizing highway slopes built on expansive clay soil using a non-destructive testing method, ERI. The selected slope in Mississippi showed higher moisture content during the early investigation before the plantation. The high moisture content led to lower shear strength. The moisture content of the zone increased over time as the Vetiver roots matured. This indicates that the roots were able to absorb the moisture from the soil. The subsurface investigation provided substantial proof that the Vetiver system was able to reduce the soil moisture content and thus provided additional shear strength to the soil. As Vetiver is a nature-based system that can thrive in harsh environments like heavy precipitation or hot temperatures, it can be a climate-adaptive and sustainable solution for slope stabilization in regions like Mississippi.

REFERENCE

1. Nirola R., M. Megharaj, R. Aryal, and R. Naidu. Screening of Metal Uptake by Plant Colonizers Growing on abandoned Copper Mine in Kapunda, South Australia. *International Journal of Phytoremediation*. vol. 18(4), 2016, pp. 399–405.

2. Meki M. N., J. R. Kiniry, A. W. Worqlul, S. M. Kim, A. S. Williams, J. M. Osorio, J. Reilley. Field and simulation-based assessment of Vetiver grass bioenergy feedstock production potential in Texas. *Agronomy Journal*. vol. 112(4), 2020, pp. 2692–707.

3. Samir S., and S. Khan. Water Balance Final Cover Using Vetiver Grass in Texas. *Proc. Geo-Congress,* Los Angeles, CA, 2023. pp. 182–92. doi:10.1061/9780784484661.019

4. Rahman, F., A. Chakraborty, and S. Khan. A Transformative Approach to Stabilize Highway Slope Using Vetiver Grass. *Proc. ASCE INSPIRE,* Alexandria, VA, 2023. pp. 537-545.

5. Kang Q., Y. Xia, X. Li, W. Zhang, and C. Feng. Study on the Effect of Moisture Content and Dry Density on Shear Strength of Silty Clay Based on Direct Shear Test. *Advances in Civil Engineering*. 2022, pp. 1–9. doi: 10.1155/2022/2213363

6. Transportation Research Board. *Potential Impacts of Climate Change on US Transportation: Special Report* 290. National Research Council, Washington, D.C. 2008.

7. US Army Corps of Engineers. Coastal Risk Reduction Resilience. 2013.

8. Ebi K. L., J. Vanos, J. W. Baldwin, J. E. Bell, D. M. Hondula, N. A. Errett, K Hayes, C. E. Reild, S. Saha, J. Spector and R. Berry. Extreme Weather and Climate Change: Population Health and Health System Implications. *Annu Rev Public Health*. vol. 42, 2021, pp. 293–315.

9. Zeng H., C. S. Tang, Q. Cheng, C. Zhu, L. Y. Yin, and B. Shi. Drought-Induced Soil Desiccation Cracking Behavior with Consideration of Basal Friction and Layer Thickness. *Water Resource Research*. vol. 56(7), 2020. doi: 10.1029/2019WR026948

10. Khan S., F. Amini, M. Nobahar. *Performance Evaluation of Highway Slopes on Yazoo Clay.* FHWA/MDOT-RD-20-286. Mississippi Department of Transportation; 2020.

11. Xu Q., and L. Zhang. The mechanism of a railway landslide caused by rainfall. *Landslides* vol. 7, 2021, pp. 149-156. doi:10.1007/s10346-010-0195-y

12. Hossain J., M. S. Hossain, L. R. Hoyos. Effect of Rainfall on Stability of Unsaturated Earth Slopes Constructed on Expansive Clay. *Proc. Geo-Congress,* San Diego, CA, 2013. pp. 417-425.

13. Ng C. W. W., L. T. Zhan, C. G. Bao, D. G. Fredlund, and B. W. Gong. Performance of an unsaturated expansive soil slope subjected to artificial rainfall infiltration. *Géotechnique* vol. 53(2), 2003, pp. 143–57.

14. Qi S., and S. K. Vanapalli. Hydro-mechanical coupling effect on surficial layer stability of unsaturated expansive soil slopes. *Computers and Geotechnics* vol. 70, 2015, pp. 68–82.

15. Khan M. S., S. Hossain, A. Ahmed, and M. Faysal. Investigation of a shallow slope failure on expansive clay in Texas. *Engineering Geology* vol. 219, 2017, pp. 118–29.

16. Ahmed A., M. S. Hossain, S. Khan, K. Greenwood, and A. Shishani. Moisture Variation in Expansive Subgrade Through Field Instrumentation and Geophysical Testing. In *Conference Proceedings: Soil Testing, Soil Stability and Ground Improvement.* Springer International Publishing, 2018. pp. 45–58.

17. Nobahar M., S. Khan, and J. Ivoke. Combined Effect of Rainfall and Shear Strength on the Stability of Highway Embankments Made of Yazoo Clay in Mississippi. *Geotech Geol Eng* vol. 38(3), 2020, pp. 787–802.

18. National Aeronautics and Space Administration (NASA). *POWER : Data Access Viewer.* NASA, 2024. Accessed Jun. 3, 2024.

19. Keller G. V., and F. C. Frischknecht. Electrical Methods in Geophysical Prospecting. 1996.

20. Karim H. H. Implementation of electrical resistivity imaging (ERI) technique for near surface investigation. *Porc. 2nd Int. Conf. on Buildings, Construction and Environmental Engineering*, Beirut, 2015. pp. 101–108.

21. Schwartz B. F., M. E. Schreiber, and T. Yan. Quantifying field-scale soil moisture using electrical resistivity imaging. *Journal of Hydrology*. vol. 362(3), 2008, pp. 234–46.

22. Grimshaw R. G. Vetiver grass—its use for slope and structure stabilisation under tropical and semi tropical conditions. In *Vegetation and Slopes: Stabilisation, Protection and Ecology.* London: Institute of Civil Engineers; 1994. pp. 26–35.

23. Holmes J., J. Chambers, P. Wilkinson, B. Dashwood, D. Gunn, M. Cimpoiaşu M, M. Kirkham, S. Uhlemann, P. Meldrum, O. Kuras, D. Huntley, S. Abbott, V. Sivakumar and S. Donohue. 4D electrical resistivity tomography for assessing the influence of vegetation and subsurface moisture on railway cutting condition. *Engineering Geology* vol. 307, 2022

24. Hossain S., G. Kibria, and S. Khan. Site Investigation using Resistivity Imaging. Taylor & Francis Group; 2019.

25. Maslinda U., M. M. Nordiana, and A. A. Bery. Porosity determination from 2-D resistivity method in studying the slope failures. *Proc. 1st International Geo-Electromagnetic Workshop*, West Java, Indonesia, 2017.

Developing An Efficient Subsurface Monitoring System with Resistivity Imaging for Railroad Infrastructure

A Q M Zohuruzzaman
Jackson State University
1400 J.R. Lynch Street, JSU Box 17068, Jackson, MS 39217-0168
j01012191@students.jsums.edu

Saimum Hossain
Jackson State University
1400 J.R. Lynch Street, JSU Box 17068, Jackson, MS 39217-0168
j01014295@students.jsums.edu

Mahdi Zulfikar
Jackson State University
1400 J.R. Lynch Street, JSU Box 17068, Jackson, MS 39217-0168
j01014658@gmail.com

Md Fahimuzzaman Khan
Jackson State University
1400 J.R. Lynch Street, JSU Box 17068, Jackson, MS 39217-0168
j00986721@students.jsums.edu

Sadik Khan, PhD, P.E.
Jackson State University
1400 J.R. Lynch Street, JSU Box 17068, Jackson, MS 39217-0168
sadik.khan@jsums.edu

Abby Cisko, Ph.D
Airfields and Pavements Branch,
Geotechnical and Structures Laboratory,
US Army Engineer Research and Development Center (ERDC),
Abby.R.Cisko@erdc.dren.mil

Thomas J (Jeremy) Beasley
Railroad Track Evaluation Program Team Leader,
Geotechnical and Structures Laboratory,
US Army Engineer Research and Development Center (ERDC),
Thomas.J.Beasley@erdc.dren.mil

Prepared for the 73rd Highway Geology Symposium, September 2024

Acknowledgments

The authors sincerely acknowledge the Federal Railwar Authority (FRA) (Contract No: 693JJ623C000011 - P00001) for supporting the study of Detection of Large -Scale Soil Moisture Content, Pore Water Pressure and Matric Suction Using the Electrical Resistivity Imaging Technique.

Disclaimer

Statements and views presented in this paper are strictly those of the authors and do not necessarily reflect positions held by their affiliations, the Highway Geology Symposium (HGS), or others acknowledged above. The mention of trade names for commercial products does not imply the approval or endorsement by HGS.

Copyright Notice

ABSTRACT

Ensuring the stability of extensive railroad tracks is a considerable challenge, largely because ballast and subgrade maintenance are labor-intensive tasks. Impermeable soils, like clay, combined with inadequate drainage, cause water to accumulate beneath the tracks, leading to the migration of ballast into the subgrade. This migration results in the formation of ballast pockets, weakening the subgrade soil's strength and heightening the risk of track failures and derailments. The situation is further complicated by the dynamic nature of soil moisture content, which varies with location and climate, requiring ongoing monitoring to maintain track integrity. This study utilizes an effective subsurface monitoring approach using Electrical Resistivity Imaging (ERI), which measures the apparent resistivity across different soil subgrade layers that can be correlated with the soil moisture levels in the subsurface section. Utilizing IoT-based sensors, the study continuously monitored changes in resistivity attributed to variations in soil moisture. Experiments were conducted on simulated railroad subgrade conditions to observe the changes in resistivity in response to moisture change. The findings aim to pinpoint areas prone to failure, enabling targeted maintenance efforts by utilizing ERI with IoT integration. By doing so, it facilitates the optimized allocation of limited resources, ensuring the durability and safety of railroad infrastructure.

INTRODUCTION:

Railway infrastructure in the 21st century faces multifaceted challenges, ranging from aging assets to increasing operational demands and environmental concerns. Many railway networks, particularly in developed countries, are operating on infrastructure that has exceeded its design life, leading to increased maintenance costs and potential safety risks *(1)*. The American Society of Civil Engineers (2021) reports that 23% of American railway assets are in "poor" condition, underscoring the urgent need for modernization. Climate change presents another significant challenge, with extreme weather events threatening the stability of tracks and embankments *(2)*. Additionally, the increasing demand for higher speed and heavier freight operations puts further stress on existing infrastructure *(3)*.

Central to these challenges is the critical role of ballast and subgrade maintenance in ensuring track stability, safety, and operational efficiency. The ballast layer plays a crucial role in distributing loads (Figure 1), providing drainage, and maintaining track geometry, while the subgrade forms the foundation of the entire track structure *(4)*. Over time, these components degrade due to factors such as traffic loading, environmental conditions, and contamination, leading to track settlement and geometry irregularities *(5)*. Inadequate maintenance of ballast and subgrade can result in increased dynamic loads, accelerated track deterioration, and potential derailments *(6)*.

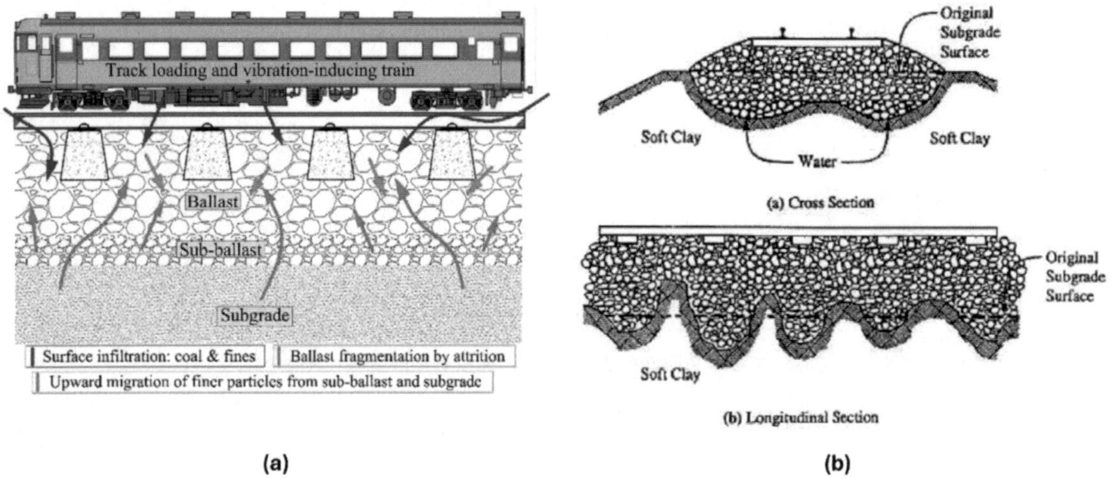

Figure 1 (a) Ballast fouling Mechanisms *(7)*, (b) Typical Formation of Ballast pockets beneath track *(8)*

Water accumulation in railway substructures is a persistent and critical issue that significantly impacts track performance and longevity. When water infiltrates the ballast and subgrade layers, it reduces the shear strength of the soil, leading to decreased bearing capacity and increased susceptibility to deformation under cyclic loading *(6)*. This problem is particularly acute in areas with poor drainage or where the subgrade consists of fine-grained, impermeable soils such as clay *(4)*. Excess water in the track structure can lead to pumping action under dynamic loads, where fine particles are transported upwards, contaminating the ballast layer and further impeding drainage *(9)*.

To address these challenges, this study proposes a continuous Electrical Resistivity Imaging (ERI) system for monitoring railway subgrade conditions. The research aims to develop and evaluate the effectiveness of this system in providing real-time data on subgrade moisture and instabilities. By assessing correlations between ERI measurements and critical subgrade parameters, the study seeks to demonstrate how continuous monitoring can improve maintenance planning and resource allocation.

BACKGROUND

The formation of ballast pockets is a direct consequence of prolonged water accumulation and repeated loading cycles in railway substructures. As the saturated subgrade softens and deforms, ballast particles migrate downward, creating localized depressions or "pockets" filled with a mixture of ballast, subgrade soil, and water *(5)*. These ballast pockets act as water reservoirs, exacerbating drainage issues and accelerating track degradation.

The cyclic nature of train loading further exacerbates the problems associated with water accumulation and ballast pocket formation. Each passing train induces a pumping effect, where water and fine particles are forced upward through the ballast layer, leading to progressive fouling and reduced drainage capacity *(10)*. This process creates a self-reinforcing cycle of degradation, where increased water retention leads to further ballast deterioration and subgrade weakening. Researchers found that the hydraulic conductivity of ballast can decrease by up to two orders of magnitude due to fouling, severely compromising its drainage function *(11)*.

The economic implications of water-induced track degradation are substantial. estimated that water-related issues account for approximately 25% of all track maintenance costs in some European networks *(12)*. Moreover, the presence of ballast pockets and associated track irregularities can lead to increased fuel consumption and wear on rolling stock, adding to the overall operational costs for railway operators *(13)*. From a safety perspective, the Federal Railroad Administration (2018) reported that track geometry defects, many of which can be attributed to substructure issues like water accumulation and ballast pockets, were a contributing factor in 35% of train derailments in the United States between 2014 and 2017.

To mitigate these issues, innovative solutions such as the use of geosynthetics for subgrade separation and reinforcement *(4)*, improved drainage systems *(14)*, and the application of advanced monitoring technologies *(15)* have been proposed. However, the complex and site-specific nature of water-related problems in railway substructures necessitates ongoing research and development of tailored solutions.

Non-destructive testing methods have become increasingly important for monitoring infrastructure stability, including highway slopes and railway subgrades. Among these techniques, Electrical Resistivity Imaging (ERI) has emerged as a particularly powerful and versatile tool. ERI, along with other methods such as LiDAR scanning and Multichannel Analysis of Surface Waves (MASW), offers comprehensive subsurface characterization without disturbing the ground structure *(16)*. The ERI technique involves injecting electrical current into the ground through electrodes and measuring the resulting potential differences at various locations. By systematically altering electrode configurations and spacings, researchers can construct detailed two-dimensional or three-dimensional images of subsurface resistivity distribution *(17)*. ERI's particular strength lies in its sensitivity to variations in moisture content, clay content, and porosity, making it invaluable for assessing groundwater conditions and soil saturation levels in diverse geotechnical applications, including railway subgrade monitoring *(18)*. This non-destructive approach allows for continuous assessment of subsurface conditions, providing crucial data for maintaining the stability and safety of critical infrastructure.

The integration of ERI with Internet of Things (IoT) technology presents a promising approach for continuous monitoring of railway subgrade conditions. This has been found in the case of geotechnical asset management *(19)*. This combination enables real-time data collection on subsurface conditions, particularly soil moisture levels, which are crucial for track stability *(20)*. Recent studies have demonstrated the feasibility of continuous data collection in railway environments using wireless sensor networks *(21)* and integrated multiple sensor types to create comprehensive track health monitoring systems *(22)*.

Advancements in machine learning algorithms applied to monitoring data have opened new avenues for predictive maintenance in railway infrastructure. It was found that the potential of big data analytics in railway maintenance, showing how large-scale data collection could improve the accuracy of degradation predictions (*23*). Building on this work, *(24)* developed a deep learning-based approach for predicting track geometry degradation, utilizing data from various track monitoring systems.

Despite these advancements, challenges remain in the widespread implementation of ERI and IoT-based monitoring systems in railway environments. Issues such as power management for remote sensors, data transmission in areas with poor connectivity, and the integration of multiple data sources need to be addressed *(21)*. Additionally, the interpretation of ERI data in complex geological settings and the development of robust algorithms for automated anomaly detection remain active areas of research *(25)*.

By analyzing the potential impact of continuous ERI monitoring on railway infrastructure's durability, safety, and economic viability, while exploring its integration with existing monitoring techniques, this research aims to contribute to the development of comprehensive subgrade health assessment strategies and improve overall railway infrastructure management. The study seeks to demonstrate how data-driven approaches can enhance maintenance planning, optimize resource allocation, and ultimately lead to more resilient and efficient railway systems.

METHODOLOGY

In the context of railway track monitoring, a comprehensive laboratory study was conducted to simulate and analyze subgrade conditions using advanced sensing technologies and data processing techniques. The experimental setup was designed to replicate a simplified railway track structure, consisting of a 2-inch sandy clay layer compacted at optimum moisture content, overlaid by a 2-inch layer of stone chips, all contained within a 6"x6"x6" box. This controlled environment facilitated precise moisture content manipulation and allowed for experimental repetition when necessary.

The study focused on investigating the properties of sandy and clayey soils typical of railway subgrades. Soil samples collected from various locations were characterized through particle size distribution and plasticity analyses to establish their geotechnical properties. The experimental apparatus incorporated a range of sensors and microcontrollers deployed within the uniformly compacted, moisture-controlled soil boxes.

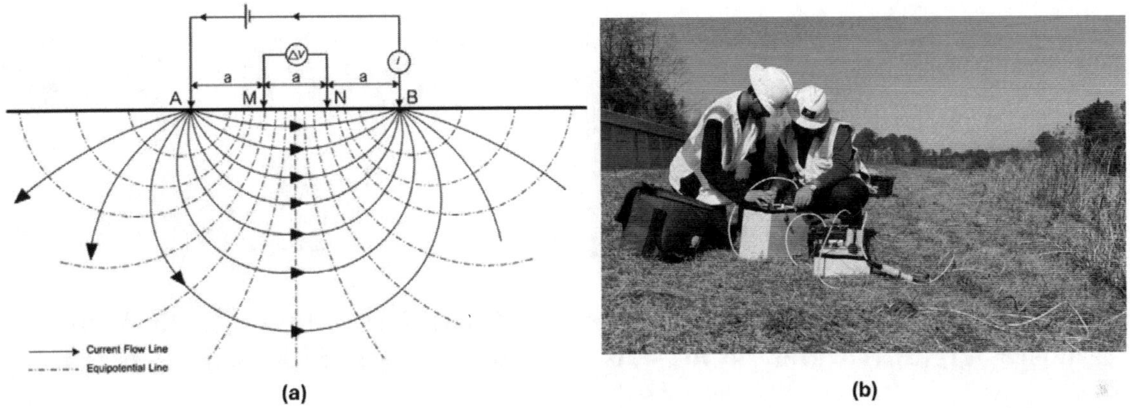

Figure 2 (a) Wenner Array and distribution of electric field underneath (redrawn) (26) **(b) ERI testing in field**

Central to the data acquisition system was the microcontroller, programmed to collect measurements at 15-minute intervals. The parameters monitored included soil resistivity, moisture content, and temperature. Soil resistivity was determined using the Wenner Array method (Figure 2), a widely accepted technique in geophysical surveys. This method involved measuring the voltage between two electrodes (A and B) while injecting current through two other electrodes (M and N), as described by *(26,27)*. Using Teros 12 sensors, Concurrent measurements of moisture content and temperature were obtained, providing a comprehensive dataset for analysis.

The experimental setup leveraged Internet of Things (IoT) technology to facilitate remote monitoring and data collection of railway subgrade conditions. This system architecture integrated various components including microcontrollers, sensors, wireless communication protocols, and cloud computing services, prioritizing open-source solutions where possible to ensure flexibility and cost-effectiveness.

A microcontroller served as the central data acquisition and transmission unit, interfacing with an array of sensors designed to measure soil resistivity, moisture content, and temperature. The sensors were selected based on their ability to provide high-precision measurements under the challenging conditions typical of railway environments.

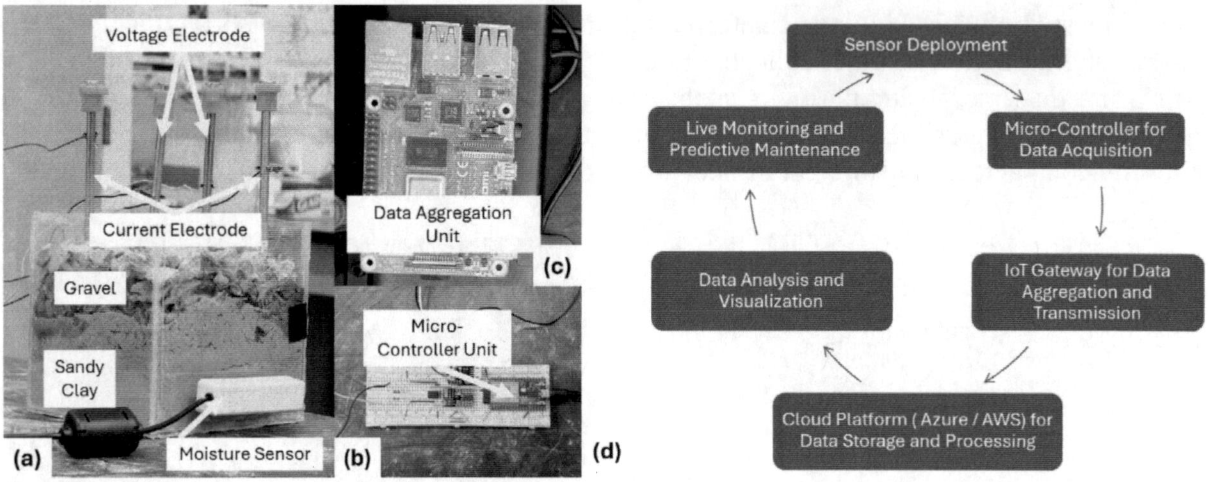

Figure 3 (a) Resistivity Box (b) Resistivity Controller Unit (c) Data Collection Unit (Raspberry Pi) (d) Workflow of IoT-Based Railway Subgrade Monitoring System

Data transmission utilized a lightweight, power-efficient protocol suitable for IoT applications, ensuring reliable communication even in areas with limited connectivity. An IoT gateway was implemented to act as both a local storage hub and a bridge to cloud-based services, providing data redundancy and facilitating remote access.

The data processing flowchart incorporated Python scripts for data cleaning, outlier removal, and statistical analysis. This approach allowed for the identification of trends between soil properties and simulated subgrade conditions. Visualization of the processed data utilized standard Python libraries to generate time-series plots and other relevant graphical representations. A web-based dashboard was developed to present processed data and visualizations, enabling real-time monitoring and analysis. This interface significantly enhanced the ability of researchers and engineers to assess subgrade conditions remotely, reducing the need for frequent site visits.

The integration of cloud computing services into the system architecture ensured secure data storage and remote accessibility. This design choice also positions the setup for future expansions, potentially including integration with other remote sensing technologies to further enhance understanding of geotechnical processes in railway applications. This IoT-enabled monitoring system represents a significant step forward in railway subgrade assessment technology. By providing continuous, high-resolution data collection and real-time analysis capabilities, it offers a powerful tool for proactive maintenance strategies and data-driven decision-making in railway infrastructure management. The system's design addresses key challenges in long-term subgrade monitoring, including power efficiency, data reliability, and ease of integration with existing railway maintenance practices.

RESULTS AND DISCUSSION

The experimental results demonstrate a strong correlation between measured and actual resistance values in the subgrade monitoring system measured and compared with the market standard AGI Supersting results. As illustrated in Figure 4, the linear regression analysis yields an R-squared value of 0.9669, indicating that 96.69% of the variance in measured resistance is explained by the actual resistance. This high coefficient of determination shows the system's reliability and accuracy in assessing subgrade conditions. The fitted line exhibits a positive slope, confirming the expected direct relationship between measured and actual resistance across the range of approximately 15Ω to 45Ω. While minor deviations from the fitted line are observed, particularly for one data point around 25Ω actual resistance, the overall trend suggests robust performance of the measurement system.

The statistical analysis of apparent resistivity and moisture content data from railway subgrade measurements reveals a strong inverse correlation between these parameters. The Pearson correlation coefficient of -0.9146 (p-value = 4.47e-24) indicates a statistically significant relationship, confirming that increases in subgrade moisture content correspond to decreases in apparent resistivity (Figure 4). This strong correlation aligns with established geophysical principles, where increased soil water content typically leads to lower electrical resistivity due to water's conductive properties.

Apparent resistivity measurements exhibited a mean of 0.029978 Ωm-m (SD = 0.000522), ranging from 0.028551 to 0.030984 Ωm-m. The histogram (Figure 5, top left) displays an unimodal distribution, while the box plot (Figure 5, bottom left) shows a narrow interquartile range with few outliers. This consistency suggests a homogeneous subgrade composition in terms of resistivity response. Moisture content data presented a mean of 0.098720 m^3/m^3 (SD = 0.000934), ranging from 0.098 to 0.101 m^3/m^3. The moisture content histogram (Figure 5, top right) reveals a slight positive skew, corroborated by the box plot (Figure 5, bottom right) showing several high-value outliers.

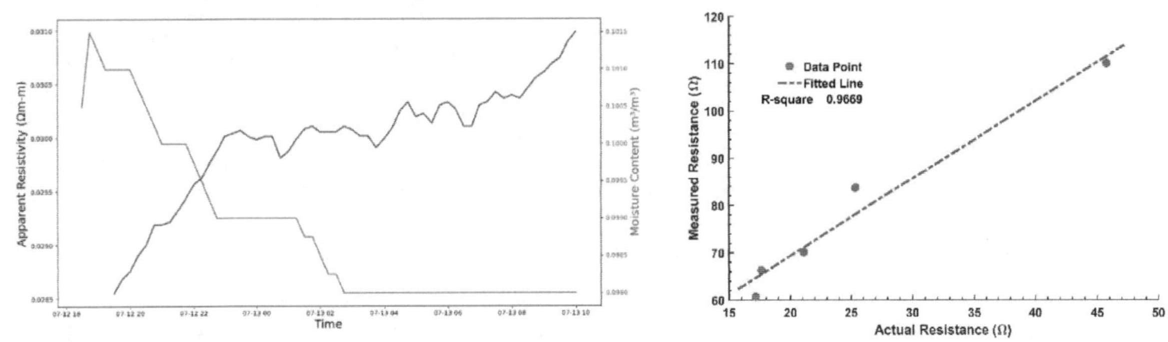

Figure 4 (a) Apparent Resistivity vs Moisture Content (b) Correlation Analysis of Measured and Actual Subgrade Resistance Values

The narrow distribution of both parameters suggests overall consistent subgrade conditions. However, the moisture content outliers require further investigation, as they could signify areas prone to instability or ballast pocket formation. The ability of the Electrical Resistivity Imaging (ERI) method to detect these localized variations demonstrates its value as a non-destructive monitoring tool for subgrade assessment.

The IoT-enabled monitoring system developed in this study offers several advantages, including continuous, high-resolution data collection, remote access, and real-time data analysis. This system can be configured to provide trigger-driven alerts when resistivity values reach specific thresholds, indicating changes in moisture content. Such alerts can help assess the stability of geotechnical infrastructure and facilitate proactive maintenance strategies.

The consistency of resistivity measurements across the study area suggests that the ERI method, when integrated with IoT technology, can provide reliable and reproducible results for subgrade assessment. This reliability is crucial for long-term monitoring applications, where detecting subtle changes over time is essential for proactive maintenance strategies. The ability to collect and analyze real-time data on soil conditions facilitates data-driven decision-making, proactive maintenance, and enhanced risk assessment.

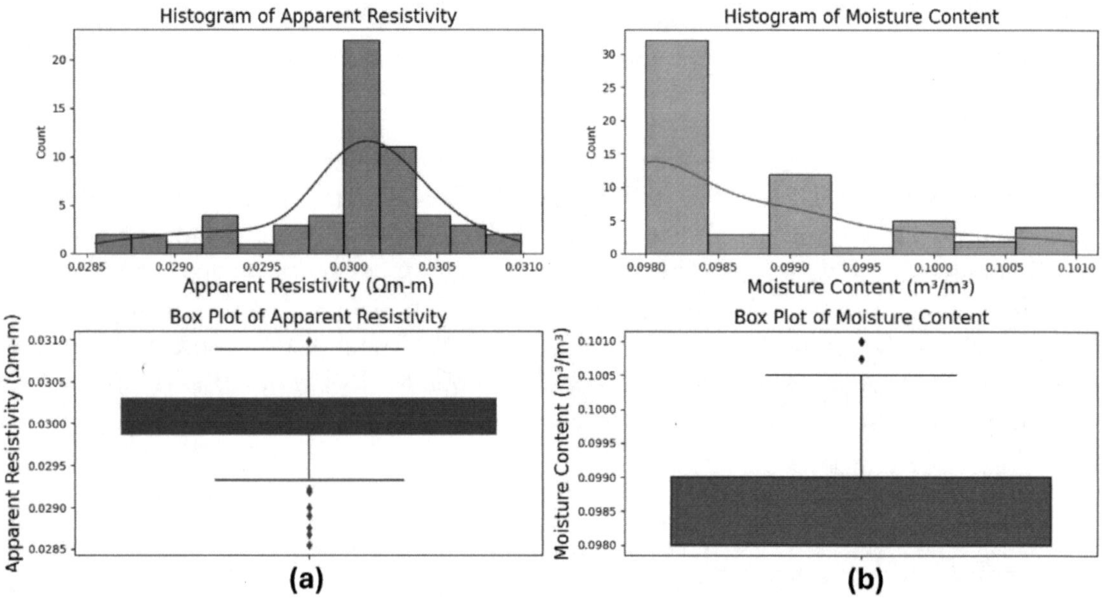

Figure 5 Histogram and Boxplot of (a) Apparent Resistivity, (b) Moisture Content

Despite these promising results, the study acknowledges certain limitations, such as the need for reliable wireless connectivity and the potential for sensor failures in field conditions. These challenges highlight the importance of robust system design and redundancy measures in practical implementations. Nevertheless, the overall findings indicate the significant potential of IoT-enabled ERI systems in advancing geotechnical engineering practices, particularly in the context of railway subgrade monitoring.

CONCLUSION

This study demonstrates the efficacy of Electrical Resistivity Imaging (ERI) integrated with Internet of Things (IoT) technology for non-destructive, continuous railway subgrade monitoring. The strong inverse correlation between apparent resistivity and moisture content provides a reliable means of assessing subgrade conditions remotely. The IoT-based ERI system, utilizing open-source microcontrollers, wireless communication, and cloud computing, could offer a cost-effective solution for real-time subgrade monitoring. This approach significantly reduces manual measurement efforts while providing continuous data streams, enabling early identification of potential issues and facilitating proactive maintenance strategies. For real-world implementation, longer electrodes are necessary to penetrate the ballast layer and make direct contact with the subgrade soil, ensuring accurate measurements unaffected by air voids in the ballast. The consistency of resistivity measurements across the study area indicates that properly implemented ERI can establish a stable baseline for long-term monitoring programs.

In conclusion, the IoT-enabled ERI system for railway subgrade monitoring has the potential to enhance predictive maintenance strategies significantly. By providing continuous, real-time data on subgrade conditions cost-effectively, this approach can improve safety, reduce maintenance costs, and increase the operational efficiency of railway infrastructure. As the railway industry faces challenges of aging infrastructure and climate change, such innovative monitoring techniques will play a crucial role in ensuring the long-term sustainability and reliability of rail networks, representing a significant advancement in railway maintenance practices.

REFERENCES

1. Tzanakakis K. The railway track and its long term behaviour: a handbook for a railway track of high quality. Vol. 2. Springer Science & Business Media; 2013.
2. Dijkstra T, Dixon N. Climate change and slope stability in the UK: challenges and approaches. Quarterly Journal of Engineering Geology and Hydrogeology. 2010;43(4):371–85.
3. Indraratna B, Rujikiatkamjorn C, Salim W. Advanced rail geotechnology–ballasted track. CRC press; 2011.
4. Indraratna B, Rujikiatkamjorn C, Nimbalkar S. Use of geosynthetics in railways including geocomposites and vertical drains. In: Geo-Frontiers 2011: Advances in Geotechnical Engineering. 2011. p. 4733–42.
5. Selig ET, Waters JM. Track geotechnology and substructure management. Thomas Telford; 1994.
6. Spiryagin M, Wolfs P, Cole C, Spiryagin V, Sun YQ, McSweeney T. Design and simulation of heavy haul locomotives and trains. CRC Press; 2016.
7. Bassey D, Ngene B, Akinwumi I, Akpan V, Bamigboye G. Ballast Contamination Mechanisms: A Criterial Review of Characterisation and Performance Indicators. Infrastructures. 2020;5(11).
8. Li D, Selig E. Evaluation of railway subgrade problems. Transportation research record. 1995;1489:17.
9. Ghataora GS, Rushton K. Movement of water through ballast and subballast for dual-line railway track. Transportation Research Record. 2012;2289(1):78–86.
10. Mayuranga HGS, Navaratnarajah SK. Effect of clay fouling on track drainage capacity of railway Ballast: Experimental and numerical study. Construction and Building Materials. 2023 Jan 18;364:129987.
11. Tennakoon N, Indraratna B, Rujikiatkamjorn C, Nimbalkar S, Neville T. The role of ballast-fouling characteristics on the drainage capacity of rail substructure. Geotechnical Testing Journal. 2012;35(4):629–40.

12. Jiang H, Bian X, Chen Y, Han J. Impact of Water Level Rise on the Behaviors of Railway Track Structure and Substructure. Transportation Research Record: Journal of the Transportation Research Board. 2015 Oct 1;2476:15–22.
13. Esveld C. Modern railway track, 2nd editon. Delft university of Technology. 2001;
14. Brough M, Stirling A, Ghataora G, Madelin K. Evaluation of railway trackbed and formation: a case study. Ndt & E International. 2003;36(3):145–56.
15. Kostrzewski M, Melnik R. Condition monitoring of rail transport systems: A bibliometric performance analysis and systematic literature review. Sensors. 2021;21(14):4710.
16. Zohuruzzaman A, Wamai DP, Feng W, Khan S, Downey AR, Wei J, et al. Highway slope monitoring using 3D laser scanning at different seasons. In SPIE; 2024. p. 36–43.
17. Chambers JE, Kuras O, Meldrum PI, Ogilvy RD, Hollands J. Electrical resistivity tomography applied to geologic, hydrogeologic, and engineering investigations at a former waste-disposal site. Geophysics. 2006;71(6):B231–9.
18. Samouëlian A, Cousin I, Tabbagh A, Bruand A, Richard G. Electrical resistivity survey in soil science: a review. Soil and Tillage research. 2005;83(2):173–93.
19. Zohuruzzaman AQM, Zulfikar M, Khan S, Beasley T, Cisko A. Enhancing Geotechnical Monitoring and Asset Management through IoT-Enabled Sensor Integration. In 2025.
20. Ngamkhanong C, Kaewunruen S, Costa BJA. State-of-the-art review of railway track resilience monitoring. Infrastructures. 2018;3(1):3.
21. V. J. Hodge, S. O'Keefe, M. Weeks, A. Moulds. Wireless Sensor Networks for Condition Monitoring in the Railway Industry: A Survey. IEEE Transactions on Intelligent Transportation Systems. 2015 Jun;16(3):1088–106.
22. Rahman MA, Taheri H, Dababneh F, Karganroudi SS, Arhamnamazi S. A review of distributed acoustic sensing applications for railroad condition monitoring. Mechanical Systems and Signal Processing. 2024 Feb 15;208:110983.
23. Fumeo E, Oneto L, Anguita D. Condition Based Maintenance in Railway Transportation Systems Based on Big Data Streaming Analysis. Procedia Computer Science. 2015 Jan 1;53:437–46.
24. Ghofrani F, He Q, Goverde RMP, Liu X. Recent applications of big data analytics in railway transportation systems: A survey. Transportation Research Part C: Emerging Technologies. 2018 May 1;90:226–46.
25. Wilkinson PB, Uhlemann S, Meldrum PI, Chambers JE, Carrière S, Oxby LS, et al. Adaptive time-lapse optimized survey design for electrical resistivity tomography monitoring. Geophysical Journal International. 2015 Oct 1;203(1):755–66.
26. Wiwattanachang N, Giao P. Monitoring crack development in fiber concrete beam by using electrical resistivity imaging. Journal of Applied Geophysics. 2011;75(2):294–304.
27. Neyamadpour A, Wan Abdullah W, Taib S, Neyamadpour B. Comparison of Wenner and dipole–dipole arrays in the study of an underground three-dimensional cavity. Journal of Geophysics and Engineering. 2010;7(1):30–40.

A Case Study of Pile Load Tests on a Driven Steel H-Pile in Kansas Shale

Kam Ng, P.E., Ph.D., M. ASCE (Corresponding author)
Professor
Department of Civil and Architectural Engineering and Construction Management,
University of Wyoming
1000 E. University Avenue, Department 3295, Laramie, Wyoming, 82071
Email: kng1@uwyo.edu

Nafis Bin Masud
Ph.D. Candidate
Department of Civil and Architectural Engineering and Construction Management,
University of Wyoming
1000 E. University Avenue, Department 3295, Laramie, Wyoming, 82071
Email: nmasud@uwyo.edu

Shaun S. Wulff, PhD.
Professor
Department of Mathematics and Statistics,
University of Wyoming
1000 E. University Avenue, Department 3295, Laramie, Wyoming, 82071
Email: wulff@uwyo.edu

Prepared for the 73rd Highway Geology Symposium, September, 2024

Acknowledgements

The authors express their gratitude to the research supports from the Wyoming Department of Transportation as the lead agency, Colorado Department of Transportation, Iowa Department of Transportation, Kansas Department of Transportation, North Dakota Department of Transportation, Idaho Transportation Department and Montana Department of Transportation under the Grant RS05219. Additional support from Mountain Plains Consortium is also acknowledged.

ABSTRACT

Pile foundations are crucial to support transportation infrastructure such as bridges because of the bedrock strata in the Midwestern United States. Thus, driven piles rely on Intermediate GeoMaterials (IGM) or weak rock strata to meet structural load demands. The performance and acceptability of driven piles in IGMs are commonly assessed using dynamic load test (DLT) methods. Although dynamic approaches offer considerable technological and economic benefits, a static pile load test (SLT) is necessary to understand the geotechnical behavior of driven piles in IGMs. This paper presents a field load test program recently completed on a HP 10x42 pile driven in shale for the K-55 bridge over Arkansas River in Sumner County, Kansas. Static analysis methods are used to determine geotechnical resistances in overburden soil and shale. Vibrating wire strain gauges are mounted along the pile length on both web faces to measure load distribution and determine shaft resistance and end bearing of the test pile. In addition, DLTs are conducted using a pile driving analyzer, followed by signal-matching analysis utilizing the Case Pile Wave Analysis Program. Eight failure criteria are considered in determining the total pile resistances from the SLT: 80% Brinch-Hansen, Chin-Kodner, Mazurkiewicz, Tangent, Load at maximum curvature, Davisson's, De Beer yield load, and 5% pile size. The differences in pile resistances from the DLT and SLT based on the eight failure criteria vary from 4% to 42%.

INTRODUCTION

Because of the shallow underlying bedrock layers in the Midwest regions of the United States, pile foundations are frequently used to support transportation infrastructure, including bridges. Driven piles in these regions are dependent on the resistance offered by soft rock or Intermediate GeoMaterials (IGMs), which are geomaterials that act as a transition between soil and rock due to the shallow underlying layers. IGMs pose challenge to design and installation of driven piles because of their properties, which lie in the transition zone between hard soils and soft rocks. IGMs can be divided into two parts: soil-based (*1-3*) and rock-based (*4-7*). Rock-based IGMs includes siltstone, sandstone, mudstone, claystone, and shale. Among different types of rock based IGMs, shale has been considered a problematic material due to its propensity for shrinkage, swelling, and deterioration of shear strength (*8*). Shale is a fine-grained, classic sedimentary rock created over time by the deposition and compaction of silt and clay-sized mineral particles. Shale is regarded as a transitional material because of its wide range of strength qualities. The current foundation design guidance needs more precise instructions and techniques to predict the resistances of piles driven in shale. The American Association of State Highway and Transportation Officials (AASHTO) Load and Resistance Factor Design (LRFD) Bridge Design Specifications (*9*) recommends analyzing shale or IGMs as soil for determining geotechnical resistances of driven piles. The current practice to determine the resistance of driven piles is performed using the signal matching technique CAse Pile Wave Analysis Program (CAPWAP). The resistances obtained from CAPWAP differ from those anticipated using the current soil-specific static analysis (SA) methods and those SA methods suggested by AASHTO (*10-12*). Furthermore, a static load test is warranted to validate the results obtained from CAPWAP. A load-displacement response is required to ascertain the load transfer mechanism from a pile to shale. In the literature, few static load tests (SLT) have been conducted on piles driven into shales (*13-14*) to help understand the pile responses in shale. To overcome these challenges, this study presents a case study of full-scale SLT, including pile instrumentation, installation, and dynamic load tests (DLT) on a steel H-pile driven in shale in Sumner County, Kansas. Vibrating wire strain, gauges were installed along the pile length on both web faces to capture the load distribution along the pile length and to determine the total pile resistance. The findings of this study will help understand shale behavior for driven pile design and validate the findings from the dynamic load tests. Eight failure criteria are considered when determine the total pile resistances from the SLT: 80% Brinch-Hansen, Chin- Kodner Extrapolation, Mazurkiewicz's Method, Tangent Method, Load at Maximum Curvature, Davisson's, De Beer yield load, and 5% pile size. Based on the eight failure criteria, the difference among pile resistances from the DLT and SLT vary from 4% to 42%.

SITE DESCRIPTION AND SUBSURFACE CONDITION

The test site was situated in Sumner County, Kansas reconstruction of the Highway K-55 bridge over the Arkansas River, shown by a red circle (Figure 1). The bridge is a 24-foot wide, seven-span structure supporting two lanes of traffic. The subsurface exploration consisted of eight test borings, B-1 to B-8, shown in Figure 2. The HP10×42 test pile was located near test boring B-8 (Figure 2). Figure 3 is the subsurface profile of boring B-8 starting from elevation 1191.1 ft to the termination elevation of 1131.1 ft. The finished ground elevation where the test pile was installed was 1182.4 ft. The groundwater table was encountered at elevation of 1173.4 ft, which was 9 ft below the finished ground surface. The overburdened soil comprised 1 ft of silty sand, 2 ft of poorly graded sand, 6.7 ft of lean clay, 18.8 ft of poorly graded sand and 31.5 ft of Wellington Formation, primarily consisting of 6 ft of soft shale, 0.5 ft of slightly weathered shale and 25 ft of moderately hard shale below the finished ground surface when the test pile was installed. The total pile penetration was 51 ft with 31.2 ft into the shale layer. Figure 3 shows the geotechnical subsurface profile and N-values from a standard penetration test.

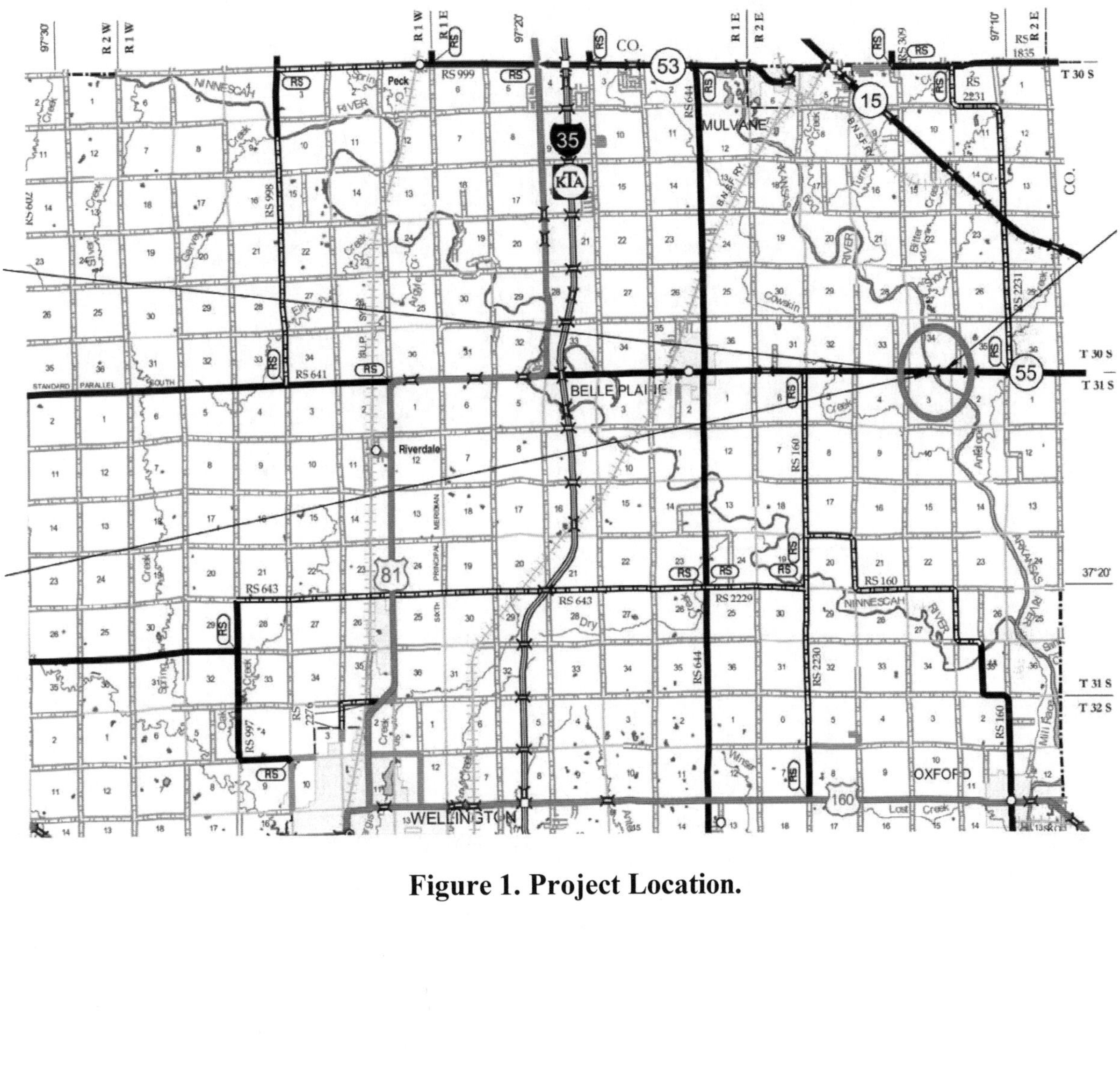

Figure 1. Project Location.

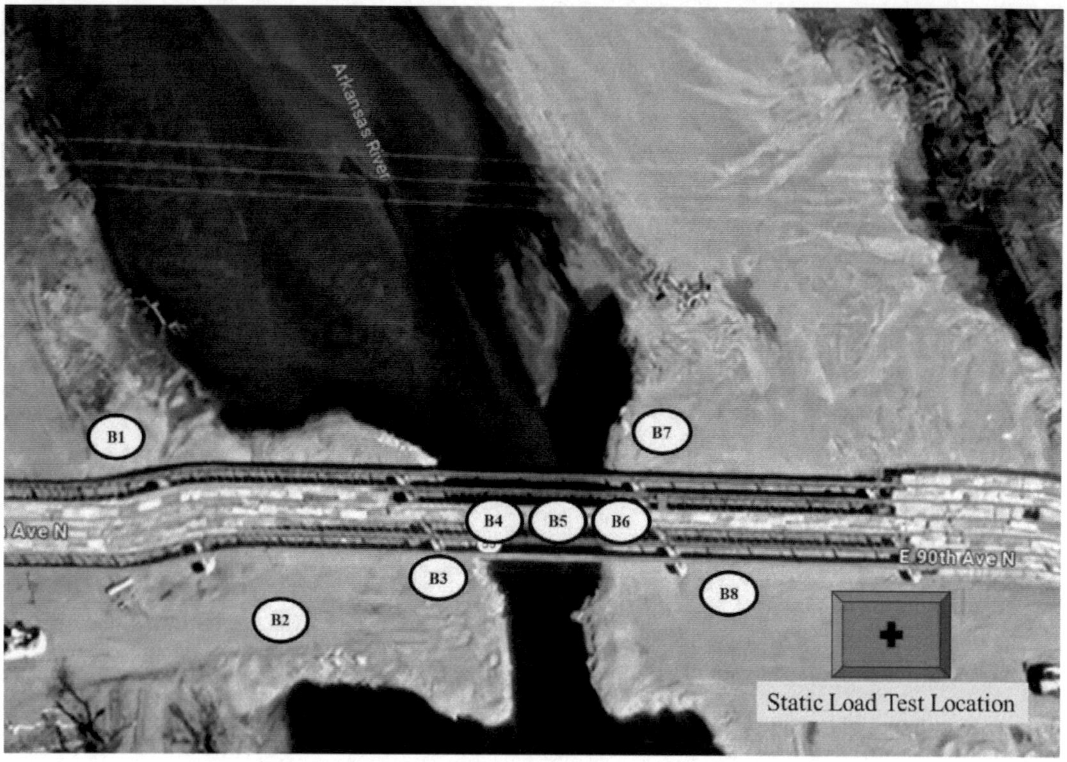

Figure 2. Borelog and Static Load Test Location.

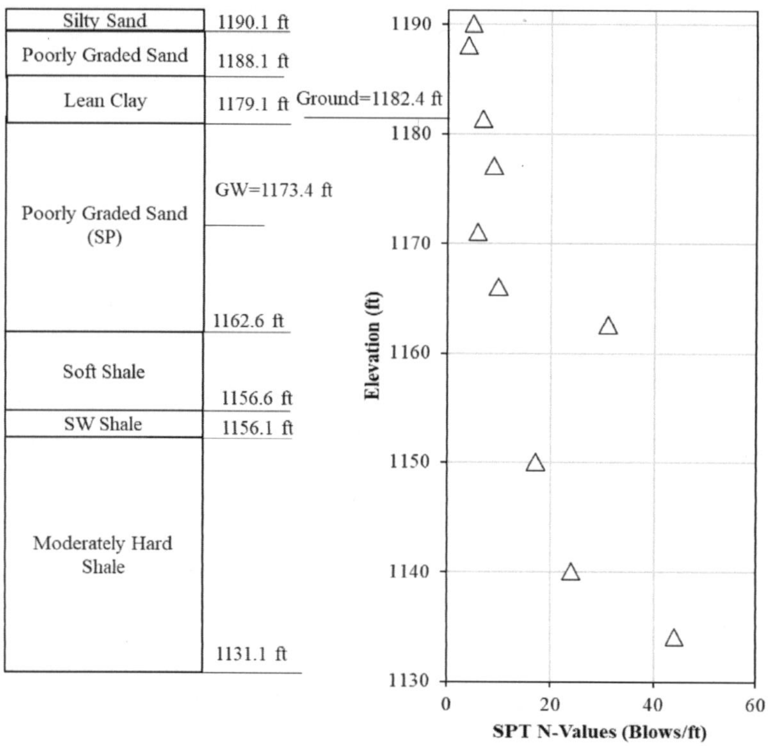

Figure 3. Subsurface Profile and SPT N-values for Borelog B-8.

TEST PILE PREPARATION

Strain Gauge Installation

To measure the load distribution, eight pairs of Geokon® vibrating-wire (VW) 4000A strain gauges were positioned throughout the length of the test pile on both web sides. The VW strain gauges were selected because of their long cable lengths, consistent signal transmission, and great resistance to water intrusion and lightning damage (Figure 4a). Every web face at the gauge site is cleaned with a grinder to eliminate dirt and rust before installing the mounting blocks. Subsequently, a spacing jig was used to establish the mounting blocks, and the single setscrew was kept in a higher position to weld the sides of the mounting blocks (Figure 4a). The central tube of each VW strain gauge was inserted through a hose clamp and secured to the two mounting blocks shown in Figure 4a. Finally, a coil was fastened to the hose clamp in the middle of the central tube (Figure 4a). The strain gauge measures the change in sensor reading (ΔR), and the micro strain (ε) can be calculated using Equation (1), where G is the gauge factor = 4.062, and B is the batch factor = 0.96.

$$\varepsilon = G \times B \times \Delta R \quad \text{(Eq. 1)}$$

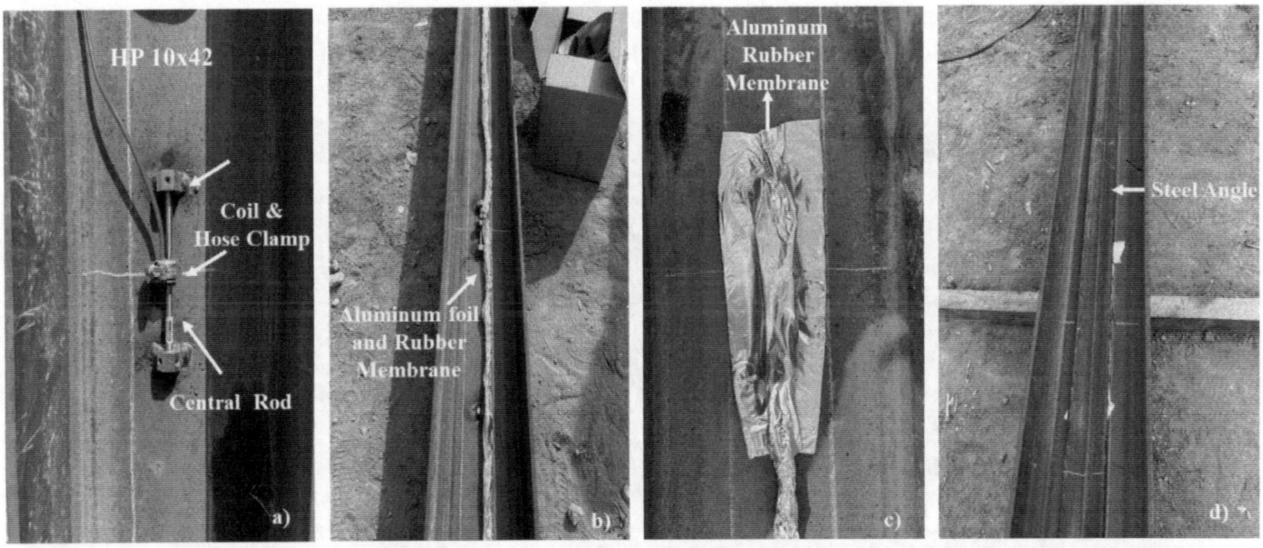

Figure 4. Test Pile Instrumentation: a) Strain Gauge Welded on the Web Face of the Pile, b) Protection of Strain Gauge Cables with Aluminum Foils, Aluminum Tapes and Membranes, c) Protection of Strain Gauge with Aluminum Rubber Membrane and d) Protection Using Tapered Steel Angles.

Strain Gauge Protection

The VW strain gauges and cables were covered with aluminum foil, aluminum tape, and aluminum-laminated rubber membrane to shield them from the heat and sparks generated during welding (Figure 4b, 4c). A 4-inch steel angle was welded to the test pile to cover the strain gauges and cables on each web face to avoid direct soil contact during pile installation. To securely attach the steel angles to the test pile, 24-inch intervals of continuous 6-inch fillet welds were utilized (Figure 4c). The steel angles were tapered at the pile tip to reduce the possibility of any geomaterial disturbance during pile driving. Equation (2) is used to compute pile stresses based on elasticity (E) of G50 steel, micro strains, and times before and after the end of drive (EOD), the beginning of restrike (BOR), and during the SLT of the steel test pile.

$$\sigma = \varepsilon E = G \times B \times \Delta R \times E \qquad \text{(Eq. 2)}$$

DATA ACQUISTION SYSTEM

A 16-channel Geokon Model 8032 multiplexer was used to collect the strain gauge readings (Figure 5a). LabVIEW software was used to give an ordered record and synchronize measurement for data processing. The circuit board's internal mechanical relay (Figure 5b) was utilized to set up the channels, enabling one sensor at a time to receive measurements. Every VW sensor is read by this multiplexer. The data logger system (Figures 5a and 5c) was used to convert the recorded resonant frequency from the VW sensors into stresses using a Model 8020-59 VW-analog converter (Figure 5d). This converter energized the vibrating wire, generated an output voltage according to the sensor reading, and measured the frequency 500 times for each measurement. For each sensor, this operation took one to two seconds. The data collection system (DAS) also included a hard drive, power supply, and power switch. The PC (Figure 5f) was utilized to record and monitor all DAS measurements in the USB flash drive (Figure 5e). The PC received data from the six displacement sensors and the load cell.

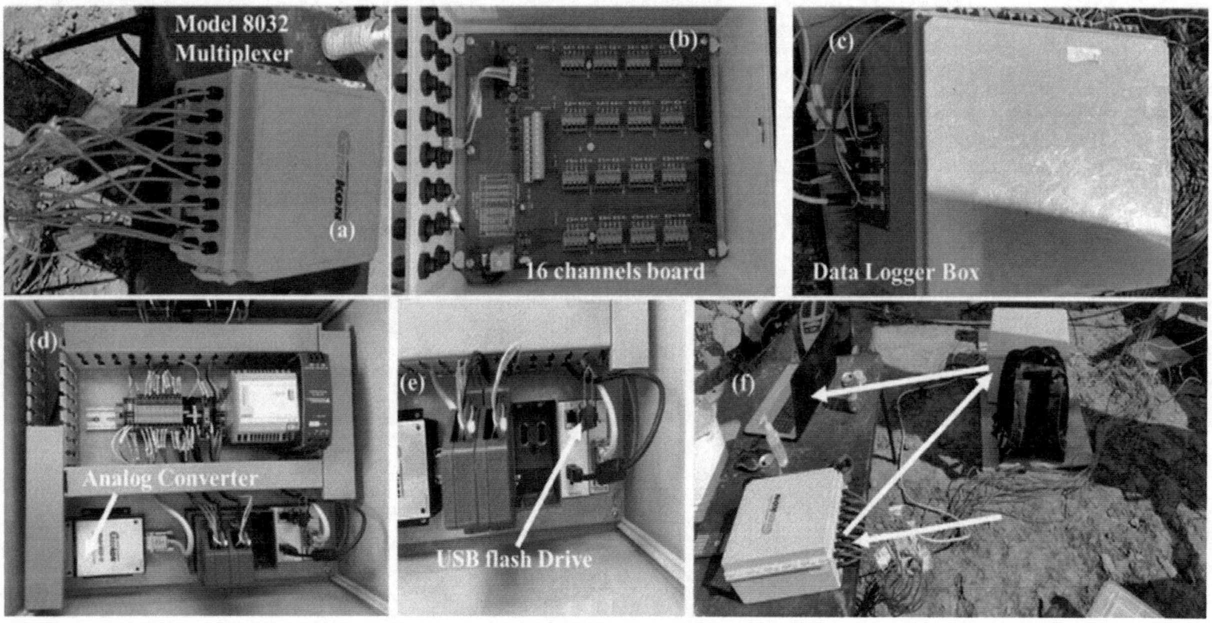

Figure 5. Data Acquisition System: a) Multiplexer, b) 16-Channels in the Multiplexer, c) Data Logger Box, d) VW Analog Converter, e) USB Flash Drive, and f) Computer System to Collect the Readings During Testing.

STATIC LOAD TEST SETUP

In this case study, the framing system, test pile, and reaction piles are not a part of the central bridge foundation system since an independent pile load system depicted in Figure 6 was used. Four HP14×89 reaction heaps (R-1 to R-4) and a test pile at the middle of the frame were installed. Eight feet was the minimum clear space ASTM (2007) advised between the test and reaction piles. For TP1, using the shorter clear distance was therefore appropriate. The SLT made use of a 10-in-stroke, 355-ton Power Team hydraulic jack. An electric double-acting hydraulic pump with a 10,000-psi maximum pressure output, driven by Power Team, applied the pressure to the jack. A digital pressure gauge (DG9042) from PowerTeam™ was used to track the pressure. A 500-ton Model MPB Honeywell load cell measured the applied load on the test pile. The load cell readout was calibrated against the hydraulic pump's pressure before the SLT. Four SPD-12-3 electronic string potentiometers from Measurement Specialties, Inc., with a complete stroke length of 12.5 inches, were used to measure the pile head vertical displacement. Away from the test pile, on each of the two wooden reference beams held up by two ladders, two potentiometers were installed (see Figure 6). The vertical displacement of the production or reaction piles was measured using two more JX-P420 series linear position transducers from UniMeasure, Inc., with a complete stroke length of 15 inches. The quick load test (SLT) complied with ASTM (2007) protocol A. Every load was applied with a 5% compression step and maintained for ten minutes. The test was carried out continuously until the final geotechnical or structural capacity was attained. 10% of the pile was removed at a time, with a 5-minute interval between each removal. A 16-channel Geokon® Model 8032 multiplexer was used to gather the strain gauge values. VW strain gauge readings from the multiplexer were transformed into analog readings with the help of a Geokon® Model 8020-59 converter. This converter, a key component within a data logger system (DAS), played a crucial role in the testing process. The DAS, equipped with a power supply, hard drive, and on/off switch, was used to gather data from the load cell and six displacement sensors. All DAS readings were transmitted and viewed on a computer system via a USB connection.

Figure 6. Static Load Test Setup.

DYNAMIC LOAD TEST RESUTLS

The Pile Driving Analyzer (PDA) was used to monitor the test pile during driving and the ensuing restrikes. CAPWAP was then used to analyze the PDA observations in order to predict pile shaft resistance and end bearing. The authors ran the dynamic test on TP1. The estimated pile resistances from CAPWAP at the EOD and BOR are summarized in Table 1. The CAPWAP analysis on the test pile at the EOD predicts the shaft resistance of 141.3 kips, end bearing of 59.3 kips, and a total pile resistance of 200.6 kips. At the 24-hour of restrike, there is a 35% increase in total pile resistance from 200.6 kips (CAWAP-EOD) to 311.3 kips (CAPWAP BOR). Both the end bearing and shaft resistance are responsible for the increase in the total resistance. Increase in pile resistance with time (known as pile setup) was also observed in Kansas shale by the study conducted by Islam et al (*6*).

Table 1 – Dynamic Load Test Results

Event	Total Resistance (kips)	Shaft (kips)	End (kips)
End of Driving (EOD)	200.6	141.3	59.3
24-Hour Restrike (BOR)	311.3	190	121.3

STATIC LOAD TEST RESUTLS

A static load test was performed one day after the restrike. The total pile length was 54 ft, and the embedded pile length was 51 ft. Figure 7a shows the load-displacement response for the test pile obtained from the SLT. The applied axial load increases linearly until the displacement reached about 0.33 in. After that, the load-displacement curve increases non-linearly with a significant increase in axial pile top displacement. Figure 7b shows the axial load distributions along the length of the pile. The axial load is calculated by multiplying the axial stress from Equation (2) by the cross-sectional area of the steel pile. The total pile resistance was determined based on eight different criteria: 5% of pile diameter or width (D) (*15*), Davisson Criterion (*16*), De Beer Yield Load (YL) (*17*), Tangent Method (*18*), Chin-Kondner Extrapolation (*19-20*), 80% Brinch Hansen Method (*21*), Mazurkiewicz's Method (*22*), and Load at Maximum Curvature (*18*). The highest total pile resistance of 400 kips is obtained from Chin- Kondner extrapolation, and the lowest total pile resistance of 180 kips is obtained from the 5%D criterion. Three criteria: De Beer YL, Tangent Method, and Load at Maximum Curvature ended with the same total ultimate resistance of 271 kips. On average based on the eight criteria, the shaft resistance and end bearing contribute 90% and 10% of total pile resistance, respectively. The increase in axial load at about 20 ft (a boundary from loose graded sand to shale) could be due to the settlement of the loose sand resulting from the axial deformation of test pile. The settlement of the sand creates a negative unit shaft resistance along the pile segment at about 20 ft and results in an increase in axial load shown in Figure 7b.

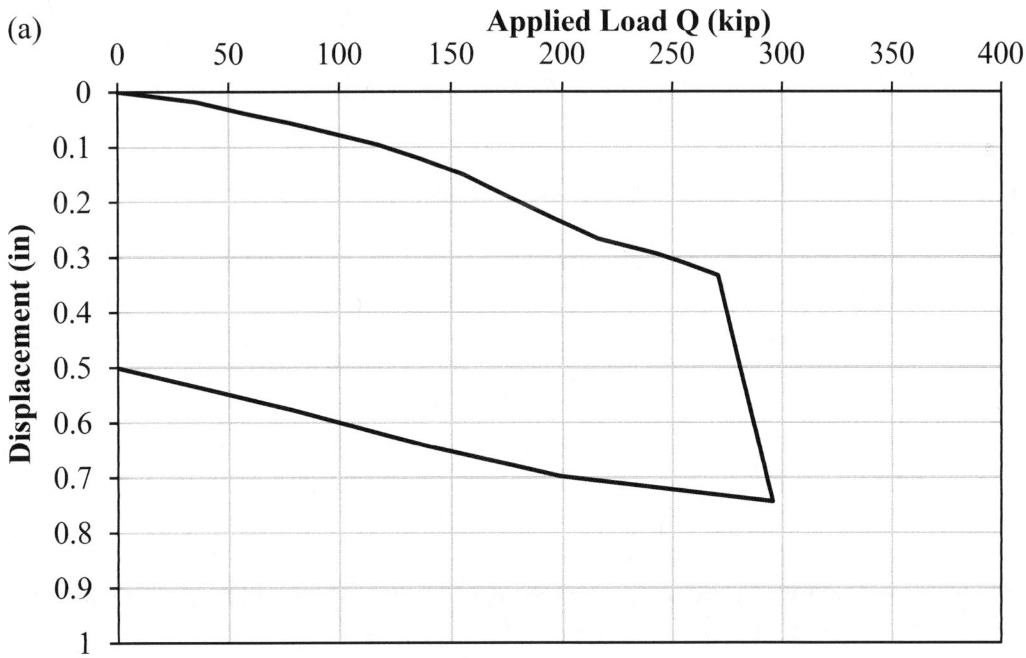

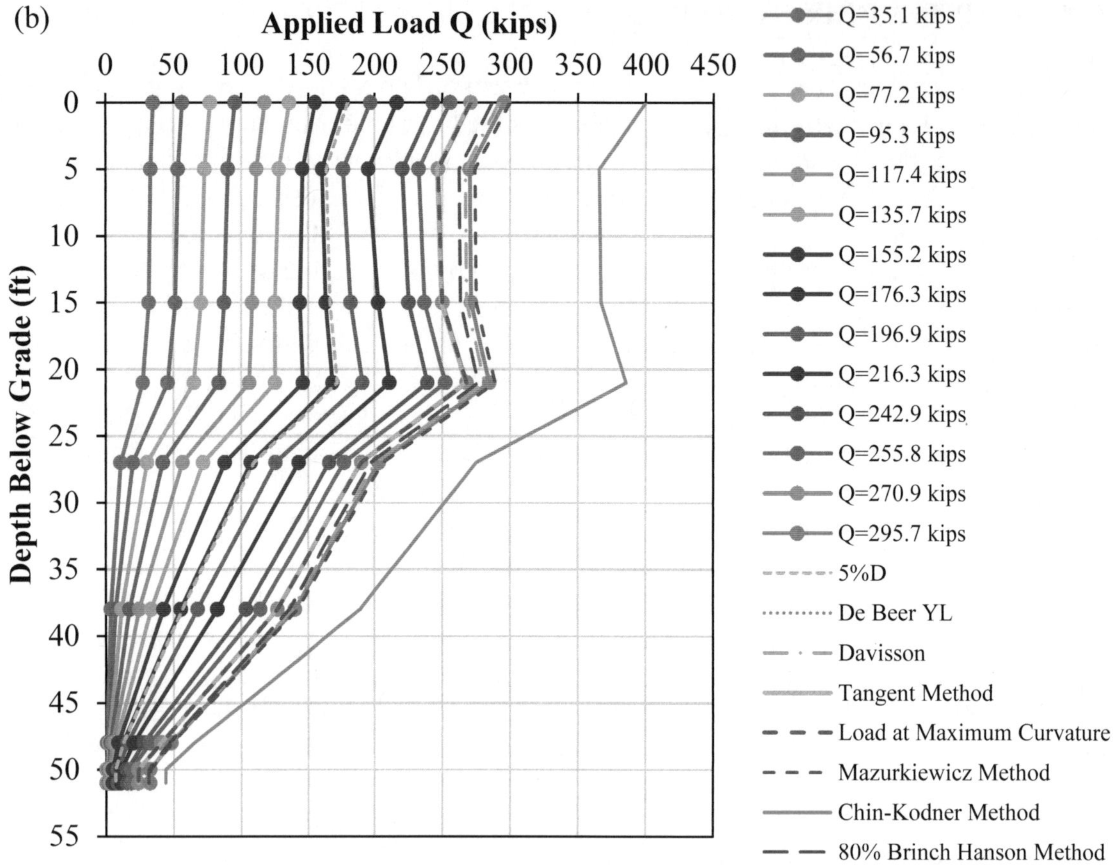

Figure 7. a) Load Displacement Curve, and b) Load Distribution Curve.

COMPARISON OF STATIC AND DYNAMIC LOAD TEST RESULTS

Figure 8 compares the total resistances determined for the dynamic and static load tests. The PDA data is input into CAPWAP for a signal-matching analysis to estimate pile shaft resistance and end bearing. The estimated pile resistances of the test pile from CAPWAP at the EOD and BOR are presented in Figure 8. A 35% increase in total pile resistance from 201 kips (CAWAP-EOD) to 311 kips (CAPWAP BOR) is observed within 24 hours (Figure 8). Compared to the total pile resistance from the CAPWAP BOR, seven criteria underpredict the total resistance except Chin-Kondner Extrapolation method. The total pile resistance for Chin- Kondner is 29% higher than that from CAPWAP BOR. The total pile resistance determined from CAPWAP-BOR was about 8%, 4%, 13%, 13%, 6%, 13%, and 42% higher than those based on 80% Brinch-Hansen, Mazurkiewicz Method, Load at Maximum Curvature, Tangent Method, Davisson Criteria, De Beer YL and 5%D criteria, respectively.

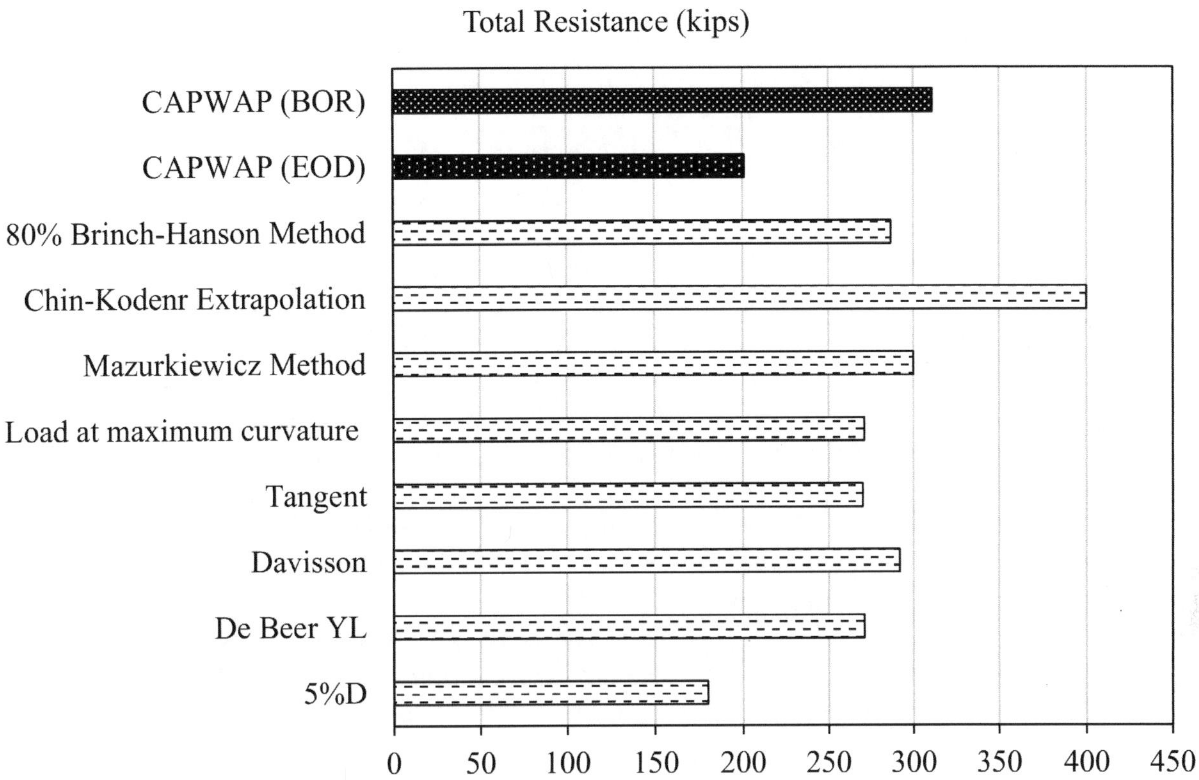

Figure 8. Comparison of Total Resistances Determined from Dynamic and Static Load Test Results.

SUMMARY AND CONCLUSION

An independent static pile load test and dynamic analysis using PDA/CAPWAP were conducted on a HP 10x42 steel H pile driven into shale in Sumner County, Kansas. This case study yields the following conclusions.

- The independent static pile load system required a loading and reaction frame and four reaction piles for conducting the SLT on a test pile. The test pile was loaded with the hydraulic jack and load cell according to the ASTM loading and unloading procedures. This independent SLT system requires additional cost and labor, but it will not be part of the bridge structures avoiding construction interruption and ultimate pile resistance can be attained.
- Dynamic load test results revealed that within 24 hours there is a 35% increase in total resistance from 201 kips (CAWAP-EOD) to 311 kips (CAPWAP BOR) for the test pile driven into shale. Both the end bearing and shaft resistance contribute to the increase in total pile resistance at 24 hours.
- Eight different failure criteria were used to predict total resistances using the static load test data. Comparison of static and dynamic load test results revealed that the total pile resistance from CAPWAP-BOR is higher than the total pile resistances determined based on 80% Brinch- Hansen, Mazurkiewicz Method, Load at Maximum Curvature, Tangent Method, Davisson Criteria, De Beer YL and 5%D criteria by around 8%, 4%, 13%, 13%,

6%, 13%, and 42%, respectively.

REFERENCES

1. Masud, N.B., Ng, K.W., Wulff, S.S. and Johnson, T., 2022. Driven Piles in Fine-Grained Soil-Based Intermediate Geomaterials. *Journal of Bridge Engineering*, *27*(6), p.04022037.
2. Masud, N.B., Ng, K.W. and Wulff, S.S., 2023. Resistance responses and design recommendations for driven piles in coarse-grained soil-based intermediate geomaterials. *Soils and Foundations*, *63*(6), p.101381.
3. Kalauni, H.K., Masud, N.B., Ng, K. and Wulff, S.S., 2024. Improved Wave Equation Analysis for Piles in Soil-Based Intermediate Geomaterials with LRFD Recommendations and Economic Impact Assessment. *Geotechnics*, *4*(2), pp.362-381.
4. Masud, N.B., Ng, K.W., Kalauni, H.K. and Wulff, S.S., 2023. Reliability-based design improvement and prediction of steel driven pile resistances in rock-based intermediate geomaterials. Acta Geotechnica, pp.1-23.
5. Masud, N.B., Ng, K. and Wulff, S.S., 2022b. New Static Analysis Methods for the Prediction of Driven Pile Resistance in Siltstone. In Geo-Congress 2022 (pp. 93-103).
6. Islam, M.S., Ng, K.W. and Wulff, S.S., 2022. Prediction of driven pile resistances in shales considering weathering and time effects. Canadian Geotechnical Journal, 59(11), pp.1851-1871.
7. Islam, M.S., Ng, K. and Wulff, S.S., 2022. Improved wave equation analysis of steel H- piles in shales considering LRFD and Economic Impact Studies. Journal of Bridge Engineering, 27(6), p.04022039.
8. Yagiz, S., 2001. Overview of classification and engineering properties of shales for design considerations. In Construction and Materials Issues 2001 (pp. 156-165).
9. AASHTO (American Association of State Highway and Transportation Officials). 2020. AASHTO LRFD Bridge Design Specifications 8th Ed. U.S. Custom-Ary Units. Washington, D.C.
10. Ng, K. and Sullivan, T., 2017. Case studies to demonstrate challenges of driven piles on rock. Geotechnical Research, 4(2), pp.82-93.
11. Ng, K. and Sullivan, T., 2017. Challenges and recommendations for steel H-piles driven in soft rock. Geotech. Eng, 48(3), pp.1-10.
12. Ng, K.W., Yasrobi, S.Y. and Sullivan, T.A., 2015. Current limitations and challenges of driven piles in rock as demonstrated using three case studies in Wyoming. In IFCEE 2015 (pp. 500-517).
13. Masud, N.B., Ng, K., Oluwatuyi, O.E., Islam, M.S., Kalauni, H.K. and Wulff, S.S., 2023. Evaluation of Static Load Test Systems for Driven Piles in Intermediate GeoMaterials. Transportation Research Record, p.03611981231163827.
14. Masud, N.B., Ng, K., Islam, M.S. and Wulff, S.S., 2022 Static and Dynamic Pile Load Tests on Steel H-Piles in Intermediate Geomaterials. In Geo-Congress 2022 (pp. 104- 112).
15. Seo, H., Yildirim, I.Z. and Prezzi, M., 2009. Assessment of the axial load response of an H pile driven in multilayered soil. Journal of Geotechnical and Geoenvironmental

Engineering, 135(12), pp.1789-1804.
16. Davisson, M.T., 1972. High capacity piles. Proc. Innovations in Found. Const., 52.
17. De Beer, E.E., 1968. Proefondervindelijke bijdrage tot de studie van het grensdraagvermogen van zand onder funderingen op staal (Deel 2–3). Annales destravaux publics de Belgique (2nd series), 69(1), pp.44-88
18. Fellenius, B.H., 2001. What capacity value to choose from the results a static loading test. Deep Foundation Institute, Fulcrum, pp.19-22.
19. Chin, F.K., 1970. Estimation of the ultimate load of piles from tests not carried to failure. In Proc. 2nd Southeast Asian Conference on Soil Engineering, Singapore, 1970.
20. Kondner, R.L., 1963. Hyperbolic stress-strain response: cohesive soils. Journal of the Soil Mechanics and Foundations Division, 89(1), pp.115-143.
21. Hansen, J.B., 1963. Discussion of "Hyperbolic stress-strain response: cohesive soils". Journal of the Soil Mechanics and Foundations Division, 89(4), pp.241-242.
22. Abdelrahman, G.E., Shaarawi, E.M. and Abouzaid, K.S., 2003, November. Interpretation of axial pile load test results for continuous flight auger piles. In Proc. of the 9 Arab Structural Engineering Conf., Abu Dhabi, UAE.

Top "Mistakes" Associated with Pile Design and Construction for Bridge Projects

William C. Jones, P.E., P.G
Foundation Testing and Consulting, LLC
Overland Park, KS 66221
(913) 626-8499
Email: cj@ftandc.com

Ronan Jones
Foundation Testing & Consulting LLC
16500 Lucille St.
Overland Park, KS 66062
913-263-2336
Email: rjones@ftandc.com

Prepared for the 72nd Highway Geology Symposium, September 2024

Disclaimer

Statements and views presented in this paper are strictly those of the author(s) and do not necessarily reflect positions held by their affiliations, the Highway Geology Symposium (HGS), or others acknowledged above. The mention of trade names for commercial products does not imply the approval or endorsement by HGS.

Copyright Notice

ABSTRACT

Many bridges are supported by driven pile foundations. Installation requirements for piling vary widely in practice among the various state DOTs based upon our extensive experience performing construction phase testing services at hundreds of bridge projects.

Based upon our experience, it is evident that there are often "mistakes" made relative to pile design and installation requirements that result in one or more of the following issues: (1) pile damage, (2) pile sections larger than could be otherwise used, (3) specifying piles that are either much longer or much shorter than what will ultimately be installed at the bridge project, (4) specifying the wrong type of pile and (5) not optimizing pile installation requirements to account for time-dependent effects.

In this paper, we will present specific examples (no specific DOTs or consultants will be identified) illustrating such issues and suggest solutions that will reduce the time and costs associated with the installation of pile foundations for bridge projects and ensure a higher quality and more reliable foundation.

INTRODUCTION

Examples and observations about what we generally refer to as "mistakes" in pile design and installation come from our experience testing hundreds of piles for bridge foundations among multiple states during the last 16 years and root cause analysis of piling issues from our internal database of Foundation Testing and Consulting, LLC (FTC) projects. This paper will illustrate what we mean by "mistakes" in that these are pile design and construction issues resulting in pile damage, project delays, and/or extra costs that could otherwise have been avoided by implementing better practices.

In 2007, AASHTO implemented load and resistance factor design standards for pile-supported bridges. These new standards have presented both challenges and opportunities related to designing and installing piling for bridge foundations. Many of these challenges stemmed from the different options in pile capacity verification requirements. The type of pile capacity verification to be used at a given bridge project has implications for both pile design and pile installation.

Aside from optimizing the selection of the type of pile capacity verification to be implemented at a given bridge project, other factors come into play that involve selecting the appropriate pile section type (H-pile or pipe pile) and diameter, the appropriate planned pile lengths, the appropriate pile end condition (open-ended or closed-ended for pipe piles and driving tip or no driving tip for H-piles). Other factors that influence the decision to select a given pile section type, size, length, and end-condition include required pile capacity, characteristics of subsurface soil and rock, and the historic pile driving capability of local or regional contractors (what size cranes, pile hammers and pile leads do they typically have available).

Research Justification

The need to analyze the major causes of piling issues in bridge design presented itself after FTC noticed recurrent piling design and construction issues over many years and over a wide geographical distribution. Throughout FTC's consulting practice, we obtained a clear understanding of these piling design and construction mistakes. While within FTC, institutional knowledge of these piling mistakes progressed, FTC noted the prevalence of these mistakes persisted. To understand the extent of these piling mistakes, FTC in 2022 developed an internal database of PDA records from our project files. This dataset has grown to include 703 distinct PDA tests spanning 150 bridges across the United States. FTC chose to incorporate bridge projects based on the quality of PDA test results data, pile installation records and availability of bridge plans and geotechnical records. The goal of this database is to analyze piling design and installation practices.

Statistical Relevance of Piling Mistakes

As previously defined, these "mistakes" are piling design and construction practices that result in pile damage, project delays, and/or extra costs that could otherwise have been avoided by implementing better practices. When interpreting our datasets from 150 pile supported bridge projects, it became apparent that pile length deviation from plan was the single most common indicator that a piling mistake had occurred. The mean deviation from pile plan length across the 150 bridge dataset was 14.4 percent. The deviations from plan pile lengths are often the result of one or more of the issues that will be presented.

In instances of overrun (where installed pile lengths were greater than plan length), the mean deviation from plan length was 17.6 percent with overruns accounting for 61 bridges out of the 150 bridge dataset. In instances of underrun, the mean deviation from plan length was 13.1 percent representing 89 of the 150 bridges in the dataset.

FTC performed root cause analysis of pile installation in cases where the installed pile length deviated 15 percent or more from the plan pile length. Out of the 150 bridges in the dataset, 91 projects met this criterion. Using this criterion as a filter of the dataset meant that the 59 projects were likely to have both design phase and construction phase issues. Through both FTC's extensive consulting experience and the data analytics used on this dataset, FTC has distilled the five most significant piling mistakes in bridge design and construction.

The data analysis suggested specific regional challenges, with some regions consistently facing significant pile length overrun and other regions having consistent underrun on pile lengths. With this, the location of a bridge site could make it particularly susceptible to specific piling mistakes or vastly amplify the consequences of both design phase and construction phase mistakes. Through this data-driven methodology, this paper has identified five piling errors that occur in both the design phase and construction phase of a bridge project. In this paper, we chose to select this list of sample project piling issues due to their historically high rate of occurrence within the FTC dataset and to illustrate the impact that improved design and construction practice may have on pile-supported projects.

Selection of Pile Verification Method

The minimum nominal axial compressive resistance for a pile is computed using the factored nominal axial pile compressive resistance and the driving resistance factor (phi). The value of the selected phi factor is based on the selected pile verification method as shown in the table below.

Verification Method	Resistance Factor (phi)
FHWA-modified Gates Dynamic Pile Formula (End of Drive condition only)	0.40
Wave Equation Analysis (WEAP)	0.50
Dynamic Testing (PDA) on 1 to 10% piles	0.65
Other methods	Refer to LRFD Table 10.5.5.2.3-1

In our experience, the selection of the pile verification method to be used at a given DOT bridge project is based on historical practice and the resulting computed minimum nominal axial compressive resistance for a pile. The direct costs for implementing a given pile verification method typically increase when going from Dynamic Driving Formula to WEAP to Dynamic Testing (PDA). We have seen apparent instances where the selected pile installation verification method appeared to be based mostly on the anticipated costs of the associated field verification method. As a result, indirect costs associated with lower-level verification methods can greatly exceed the apparent cost and/or schedule savings that had been anticipated by implementing lower-level verification methods as illustrated by one of our past projects.

Project Example 1 – Poor Pile Capacity Verification Method Selection

Pile Type	HP14X73
Plan Pile Length (feet)	25
Specified Pile Verification Method	WEAP
Pile Resistance Factor (phi)	0.5
Factored nominal axial pile compressive resistance (kips)	293
Minimum nominal axial compressive resistance (kips)	586
Overburden Soil Type	Silty Clay (soft)
Depth to bedrock contact (feet)	15
Total Number of Piles	14
Total Number of Pile Supported Substructures	2

From the above table, it can be seen that this is a relatively high required pile capacity and a relatively short pile length for a pile that would bear on weathered shale bedrock. This particular DOT requires submittal of a pre-construction WEAP analysis that accounts for the pile requirements, anticipated subsurface conditions, and contractor-selected pile-driving hammer. The allowable driving stress for this pile was 45 ksi (90 percent of the 50 ksi pile yield strength). The DOT's specifications for WEAP submittals require that the pile be driven to achieve minimum nominal axial pile resistance. Such driving shall also be within allowable driving stresses, with ending equivalent blow counts in the range of 24 to 120 blows per foot (2 to 10 blows per inch). We found a mutually exclusive set of results in our WEAP analysis such that if the pile were to achieve the minimum nominal axial pile compressive resistance (hereinafter referred to as required pile capacity), allowable driving stresses would be exceeded. If a smaller hammer were to be used to reduce the driving stresses to acceptable values, then the required pile capacity would not be met.

Our findings and recommendations for Project 1 were that PDA testing would have been a better pile verification method. This would have resulted in a required pile capacity of 455 kips which could have readily been handled by a smaller pile section such as an HP12X53. Also, the pile driving stresses could have been measured using the PDA to avoid pile damage. Many people who are not familiar with PDA test results do not understand that pile driving stresses can be twice as high near the pile tip compared to those at the top of the pile when driving short piles to hard rock and high required capacities. Instead, the use of WEAP verification resulted in the DOT having to base driving on "refusal" criteria which resulted in a considerable risk of pile driving damage and required the use of a larger pile section to accommodate the higher required pile capacity. The term "refusal" is problematic in that it refers to a pile set that is often not adequately defined. To some, “refusal” is a pile-driving set of 1 inch in 10 blows. To others, “refusal” is a pile driving a set of 1 inch in 20 blows. Another issue with specifying a set value is that hammer size or fuel setting is not mentioned in these DOT specifications. We recommend that the term “refusal” not be applied to pile driving installation verification requirements.

Project Example 2 – Poor Selection of Pile Type, Size and Length

Pile Type	36-inch Diameter Pipe (closed-ended)
Plan Pile Length (feet)	45
Specified Pile Verification Method	PDA
Pile Resistance Factor (phi)	0.65
Factored nominal axial pile compressive resistance (kips)	408
Minimum nominal axial compressive resistance (kips)	627
Overburden Soil Type	Sand (medium dense)
Depth to bedrock contact (feet)	>100 feet
Total Number of Piles	22
Total Number of Pile Supported Substructures	4

As we previously had performed PDA testing of piles at many bridge projects in this area, we were familiar with typical pile installation depths and the pile capacities that were achieved. Historically, pipe piles in that area would range from 14 inches to 24 inches in diameter with a closed-ended configuration for friction pile and open-ended for end bearing pile. What we immediately noticed was that at 36 inches in diameter, this was the largest diameter, closed-ended pipe pile that had ever been driven to date in that state. Also, we thought that not only would it be highly unlikely that the pile would achieve the required capacity within saturated, medium-dense sand (Navg=15) at plan embedment depths of 27 feet, but we also realized that due to the closed-end pile condition and relatively large diameter, pile driving would be very difficult due to generation of excess pore water pressures. As a result, pile driving would have to be done with several stops and waiting periods.

Our findings and recommendations for Project 2 were that a more constructible pile design would have involved a smaller diameter pile in line with historical usage and as a result, more and/or longer pile should have been used. As we had predicted, the piling had to be driven much deeper than planned (an extra 30 feet per pile) and driving was very time-consuming due to the high blow counts and multiple waiting periods that had to be used to allow for dissipation of excess pore water pressures. This resulted in significant cost overruns and project delays. Another apparent issue was that neither the DOT nor their design consultant appeared to have realized that the specified combination of pile size and capacity requirements had never been used before in that state. We have found that many DOTs and their design consultants simply do not have a source of historical pile data that is readily accessible and searchable for use during the design of a new pile-supported bridge project, and/or there is insufficient field experience on the part of some designers to have developed an understanding of typical pile installation capacities and driving requirement.

Project Example 3 – Disregarding PDA Results Because of Historical Pile Installation Practices

Pile Type	HP14X89
Plan Pile Length (feet)	100
Specified Pile Verification Method	PDA
Pile Resistance Factor (phi)	0.65
Factored nominal axial pile compressive resistance (kips)	214
Minimum nominal axial compressive resistance (kips)	329
Overburden Soil Type	Clayey Sand (dense)
Depth to bedrock contact (feet)	83
Total Number of Piles	16
Total Number of Pile Supported Substructures	2

As can be seen from the piling summary for Project 3 listed above, we had anticipated that the pile could readily be driven to the required capacity near the plan tip elevation in the weathered shale bedrock and that pile driving could be terminated when the required capacity had been met per the results of the PDA testing. Unfortunately, this DOT has a project specification that required that all piles were to be driven to a pile set value of 1 inch in 20 blows of driving (equivalent to ¼ inch pile penetration in 5 blows). We concluded our testing at a pile set value of 1.875 inches in 5 blows at a computed axial capacity of over 1,100 kips when the pile started reaching the allowable limits of driving stresses. Therefore, the pile was driven to over 3.3 times the required axial capacity in an attempt to reach a pile set of ¼ inch in 5 blows as directed by the owner's onsite representative. To make matters worse, as we refused to continue the pile driving and testing once allowable stresses had been met, the owner's onsite representative (consultant working for DOT) directed that pile driving continue without PDA monitoring until the ¼ inch 5 blow set criteria had been met even though we had informed him that limiting pile driving stresses and excess pile capacity had been met at the conclusion of our testing.

Our findings and recommendations for Project 3 were that pile driving should have been terminated when the PDA demonstrated that the pile had met the required capacity at or below minimum pile tip elevations. Given that the owner's onsite representative directed that pile driving continues with capacity well above the required value and without the ability to monitor driving stresses in the pile, there was a very high risk of pile damage for no other benefit other than trying to reach the set value listed in the standard DOT specifications.

Project Example 4 – Not Taking Advantage of Time-Dependent Pile Capacity Increases

Pile Type	20-inch Diameter, closed-ended
Plan Pile Length (feet)	85
Specified Pile Verification Method	PDA
Pile Resistance Factor (phi)	0.65
Factored nominal axial pile compressive resistance (kips)	254
Minimum nominal axial compressive resistance (kips)	390
Overburden Soil Type	Clay (dense)
Depth to bedrock contact (feet)	63
*Total Number of Piles	16
Total Number of Pile Supported Substructures	2

Our findings and recommendations for Project 4 were that the pile had to be driven to the bedrock contact to reach the required capacity at end-of-initial drive (EOID). This involved over 15 feet of additional length per pile at each abutment. There was no provision for optimizing pile lengths by incorporating PDA restrike testing. That is, the required pile capacity had to be met at EOID. While onsite, we noticed that there was considerable set-up (pile capacity increases following a waiting period) based on our observations of increased capacity following the completion of the splice. It was apparent that if waiting periods of 24 hours had been used to perform restrike testing at this project, much higher pile capacities would have been achieved relative to EOID conditions. Such restrike testing would have resulted in significantly shorter piles with lower cost, and faster installation during construction. This situation points to an underlying design issue. Typically, designers can compute static pile capacity based on the site soil profile and pile configuration. This static capacity represents the upper bound of geotechnical resistance on the pile. What is more difficult, and something that pile designers are typically not accounting for, is to determine the percentage of the static pile capacity that is likely to be achieved at EOID. This is another situation where having a compilation or understanding of historical pile data that includes both EOID and restrike PDA data that is readily accessible and searchable for use during the design of a new pile-supported bridge project would be of tremendous value in optimizing the pile design and improving constructability.

Project Example 5 – Specifying Unrealistic Length of Pile Penetration Below Bedrock Contact

Pile Type	HP14X117
Plan Pile Length (feet)	83
Specified Pile Verification Method	PDA
Pile Resistance Factor (phi)	0.65
Factored nominal axial pile compressive resistance (kips)	490
Minimum nominal axial compressive resistance (kips)	754
Overburden Soil Type	Sand (medium dense)
Depth to bedrock contact (feet)	28
*Total Number of Piles	92
Total Number of Pile Supported Substructures	4

The issue with this project is that the plan pile tip elevation corresponded to 52 feet of penetration below the top of a weathered shale bedrock contact. There was no pre-boring required for these piles. Plan pile length was based solely on the anticipation of how much pile penetration would be required to achieve the required axial capacity.

Our findings and recommendations for Project 5 were that based upon a review of the project plans, and our previous pile driving experience in the area, we advised that we did not anticipate that more than 20 to 25 feet of pile penetration into the shale bedrock would be necessary to develop required capacity. Further, it was our opinion that it was simply impossible to drive the piling 52 feet below the shale bedrock contact. The contractor followed our guidance and ordered 2,700 feet less than the planned quantity of piling. Our PDA testing confirmed that the required pile capacity was met at these shallower penetration depths. It was apparent that the pile designer for this project was not familiar with typical pile penetration depths into the weathered shale contact in the area.

CONCLUSION

It is critically important for those who are responsible for the design and installation requirements for pile supported bridge foundations to have a good understanding of historical pile installation and PDA test results. Such an understanding can facilitate the optimization of pile design (section size, pile diameter, pile length and pile end condition), and reduce the potential for pile damage and project delays. DOT's and their design consultants should consider bringing in pile testing consultants during the design stage for pile supported bridges. These pile testing consultants, if sufficiently experienced, can provide great insights into the viability of a given set of pile design and installation requirements based upon their knowledge of results for similar projects in the past.

A Comparison of the UNR and Classical Bearing Capacity Equations

Gary Norris
Emeritus Professor
University of Nevada, Reno
1664 N. Virginia St, MS258
Reno, NV 89557
norris@unr.edu

Sherif Elfass
Associate Research Professor
University of Nevada, Reno
1664 N. Virginia St, MS258
Reno, NV 89557

Horng-Jyh Yang
Professor
West Virginia University Institute of Technology
410 Neville St.
Beckley, WV 25801

Hugh Ezzell
Principal Engineer,
Soils Engineering, L.L.C .,
1000 Road Runner Road
Reno NV 89510

Prepared for the 73rd Highway Geology Symposium, September, 2024

Disclaimer

Copyright Notice

ABSTRACT

The classical shallow foundation bearing capacity equations of Meyerhof, Hanen and Vesic (MHV), as presented for instance in the Army Corps of Engineers EM 1110-1-1915 publication, is compared with the UNR (University of Nevada, Reno) equation. The UNR model's equation is very simple, $q_{net} = P_0*(\tan^6 \alpha_f - 1)$, the basis for which is presented herein. The UNR equation does not require foundation shape (s) and depth of embedment (d) correction factors as do the MHV equations. The UNR model is based on the characterization of the failure mass of a square/round foundation as composed of three zones of triaxial test assessed soil strength: a free-field zone I, a radial shear zone II and a zone III immediately below the foundation. The comparison provided is of the individual MHV bearing capacity factors (N_c, N_q and N_γ) along with their shape (s_c, s_q and s_γ) and depth (d_c, d_q and d_γ) correction factors with the equivalent UNR factors (with no s or d corrections), as well as typical resulting bearing capacity values.

INTRODUCTION

The University of Nevada, Reno (UNR) model for shallow foundation load-settlement-bearing capacity analysis was presented at the 71st Highway Geology Symposium (2022) in conjunction with field assessed shear wave velocity as input for preliminary foundation analysis during the exploratory phase of a project. The pressure-settlement response of five foundation tests (Briaud and Gibbens 1994 and Briaud 2007) performed at Texas A&M were analyzed with the UNR model and presented at GeoDenver (Elfass et. al 2007) as well as elsewhere (Norris et. al 2011, Elsayed et. al 2011 and Nimeri et al. 2017). Lastly, the bearing capacities of the Milovic and Muhs full scale foundation tests carried to failure, assessed by Bowles' 1996 (5th Ed) using the Terzaghi, Meyerhof, Hansen and Vesic bearing capacity equations, were compared with the UNR equation values as presented in Elfass et. al 2007 and Norris et. al 2011.

It is the purpose here to more closely compare the UNR equation with the classical shallow foundation bearing capacity equations of Meyerhof, Hanen and Vesic (MHV), as presented for instance in the Army Corps of Engineers EM 1110-1-1915 (1992) publication. The UNR equation does not require foundation shape (s) and depth of embedment (d) correction factors as do the MHV equations. The MHV equations are based on a model of plane strain failure of an infinitely long foundation, to which lab scale assessed correction factors are applied. The comparison provided is of the individual MHV bearing capacity factors (Nc, Nq and N_γ) along with their shape (s_c, s_q and s_γ) and depth (d_c, d_q and d_γ) correction factors with the equivalent UNR factors (that require no s or d corrections), as well as typical resulting bearing capacity values. Therefore, for example, it is intended to compare the N_c, s_c, and d_c factors of Meyerhof, Hansen and Vesic with UNR's N_c for different width (B), length to width shape (L/B) and depth to width (D/B) ratios for a given soil unit weight (γ) and cohesion (c), over the range in soil friction angle (ϕ) from 0 to 50 degrees. Besides comparing the bearing capacity factors with their attached shape and depth correction factors with UNR's factors, their combination yielding the ultimate bearing capacities (q_{ult}) is compared with the equivalent UNR bearing capacities for several different cases.

THE CLASSICAL BEARING CAPACITY EQUATIONS

The classical bearing capacity equation after MHV is

$$q_{ult} = 0.5B\gamma N_c(s_\gamma d_\gamma) + D\gamma N_q(s_q d_q) + cN_c(s_c d_c) \quad (1)$$

While MHV use the same dimensionless

$$N_q = e^{\pi \tan \emptyset} N_\emptyset \qquad N_\emptyset = tan^2\left(45° + {}^{\emptyset}/_{2}\right) \quad (2)$$

and

$$N_c = {}^{(N_q - 1)}\big/_{\tan \emptyset} \quad (3)$$

bearing capacity factors, they each have different N_γ equations. They are

$$N_\gamma = (N_q - 1)j \qquad j = \tan(1.4\emptyset) \qquad \text{Meyerhof} \quad (4a)$$

$$j = 1.5 \tan \emptyset \qquad \text{Hansen} \quad (4b)$$

$$N_\gamma = (N_q + 1)j \qquad j = 2 \tan \emptyset \qquad \text{Vesic} \quad (4c)$$

Table 1 from the Army Corps of Engineers EM 1110-1-1915 (1992) provides the values of these different dimensionless bearing factors versus friction angle (ϕ). Table 2 provides the shape (s) and depth of embedment (d) factor equations simplified from the EM 1110-1-1915 (1992) Tables 4-4, 4-5 and 4-6. as shown, for instance in Murphy (2003) page 505.

Table 1. Bearing Capacity Factors

				N_γ		
ϕ	N_ϕ	N_c	N_q	Meyerhof	Hansen	Vesic
0	1.00	5.14	1.00	0.00	0.00	0.00
2	1.07	5.63	1.20	0.01	0.01	0.15
4	1.15	6.18	1.43	0.04	0.05	0.34
6	1.23	6.81	1.72	0.11	0.11	0.57
8	1.32	7.53	2.06	0.21	0.22	0.86
10	1.42	8.34	2.47	0.37	0.39	1.22
12	1.52	9.28	2.97	0.60	0.63	1.69
14	1.64	10.37	3.59	0.92	0.97	2.29
16	1.76	11.63	4.34	1.37	1.43	3.06
18	1.89	13.10	5.26	2.00	2.08	4.07
20	2.04	14.83	6.40	2.87	2.95	5.39
22	2.20	16.88	7.82	4.07	4.13	7.13
24	2.37	19.32	9.60	5.72	5.75	9.44
26	2.56	22.25	11.85	8.00	7.94	12.54
28	2.77	25.80	14.72	11.19	10.94	16.72
30	3.00	30.14	18.40	15.67	15.07	22.40
32	3.25	35.49	23.18	22.02	20.79	30.21
34	3.54	42.16	29.44	31.15	28.77	41.06
36	3.85	50.59	37.75	44.43	40.05	56.31
38	4.20	61.35	48.93	64.07	56.17	78.02
40	4.60	75.31	64.19	93.69	79.54	109.41
42	5.04	93.71	85.37	139.32	113.95	155.54
44	5.55	118.37	115.31	211.41	165.58	224.63
46	6.13	152.10	158.50	328.73	244.64	330.33
48	6.79	199.26	222.30	526.44	368.88	495.99
50	7.55	266.88	319.05	873.84	568.56	762.85

Table 2. Shape (s) and Depth of Embedment (d) Factor Equations

Factors	Meyerhof	Hansen	Vesic
s_c	$1+0.2N_\phi \frac{B}{L}$	$1+\frac{N_q}{N_c}\frac{B}{L}$	The shape and depth factors of Vesic are the same as those of Hansen.
s_q	$1+0.1N_\phi \frac{B}{L}$ for $\phi > 10°$	$1+\frac{B}{L}\tan\phi$	
s_γ	$s_\gamma = s_q$ for $\phi > 10°$ $s_\gamma = s_q = 1$ for $\phi = 0$	$1-0.4\frac{B}{L}$	
d_c	$1+0.2\sqrt{N_\phi}\frac{D_f}{B}$	$1+0.4\frac{D_f}{B}$	
d_q	$1+0.1\sqrt{N_\phi}\frac{D_f}{B}$ for $\phi > 10°$	$1+2\tan\phi(1-\sin\phi)^2\frac{D_f}{B}$	
d_γ	$d_\gamma = d_q$ for $\phi > 10°$ $d_\gamma = d_q = 1$ for $\phi = 0$	1 for all ϕ Note; Vesic's *s* and *d* factors = Hansen's *s* and *d* factors	

THE UNR MODEL AND EQUATIONS

On the other hand, the very simple UNR bearing capacity equation is

$$q_{ult} = q_{net} + D\gamma_x \quad (5a)$$

$$q_{net} = P_0{}^*(tan^6\alpha_f - 1) \qquad \alpha_f = 45° + {}^{\emptyset}/_2 \quad (5b)$$

where

$$P_0{}^* = {}^{c}/_{\tan\emptyset} + P_0 \qquad P_0 = D\gamma_x + {}^1/_2\, Bj\gamma_y \qquad j = 1.5\tan\emptyset \text{ after Hansen} \quad (5c)$$

Figures 1 and 2 provide the background for the different components of the UNR equation. The ultimate bearing capacity pressure (q_{ult}) applied to the foundation base shown in Fig. 1 is broken up into two parts, the net ultimate bearing capacity (q_{net}) and the effective overburden pressure ($D\gamma_x$) from the ground surface to depth D. The vertical effective pressure (P_0) on an average element at depth 0.5jB in free-field zone I is ($D\gamma_x$ + ½ B j γ_y) where γ_x and γ_y are the effective unit weights of the soil above and below the base of the foundation. At failure, the major principal horizontal stress (P_h) of zone I becomes the minor principal stress at the left side of radial shear zone II. A 90-degree rotation of principal stresses through the radial shear zone yields a major principal stress at the right side of zone II that becomes the minor principal stress of zone III located immediately below foundation base at the same ½ B j depth as the average element of zone I. Note the average elements of zones I and III are characterized as if triaxial test specimens. The stresses at failure in the three zones of the failure mass shown in Fig. 2 are tangent to the Mohr-Coulomb failure envelope of strength parameters c (cohesion) and ϕ (friction angle) relative to origin O. However, rather than deal with a c-ϕ envelope and associated strength characterization, a new origin at O' at an offset distance of c /tan ϕ employed in the

UNR analysis yields a purely frictional (ϕ) characterization of the soil. Accordingly, the vertical effective stress applied to the average element of zone I (as if the confining pressure of a triaxial test specimen so envisioned) becomes P_0^* (= $P_0 + c / \tan \phi$) relative to origin O'. Based on P_0^* as the minor principal stress (relative to O') at the lower end of the Mohr circle of zone I, P_h of Fig. 2 becomes $P_0^* \tan^2 (45 + \phi/2)$, the minor principal stress of zone II. The major principal stress of zone II, equal to the minor principal stress of zone III, becomes $P_h \tan^4 (45 + \phi/2)$. The major principal stress of the zone III circle becomes $\{[P_h \tan^4 (45 + \phi/2)]\times[\tan^2 (45 + \phi/2)]\} = P_0^* \tan^6 (45 + \phi/2)$ relative to origin O'. The horizontal distance or difference between the upper end of circle III ($q_{ult} + \frac{1}{2} B j \gamma_y$ in Fig. 1) and the lower end of circle I ($D\gamma_x + \frac{1}{2} B j \gamma_y$ in Fig. 1) is

$$\left(q_{ult} + {}^{1}/_{2}\, Bj\gamma_y\right) - \left(D\gamma_x + {}^{1}/_{2}\, Bj\gamma_y\right) \qquad \text{relative to O}$$

or

$$\left[\left(q_{ult} + {}^{1}/_{2}\, Bj\gamma_y\right) + {}^{c}/_{\tan \emptyset}\right] - \left[\left(D\gamma_x + {}^{1}/_{2}\, Bj\gamma_y\right) + {}^{c}/_{\tan \emptyset}\right] \qquad \text{relative to O'}$$

which is $q_{ult} - D\gamma_x$ (where $q_{ult} = q_{net} + D\gamma_x$)

i.e. $q_{net} + D\gamma_x - D\gamma_x$ or just q_{net}.

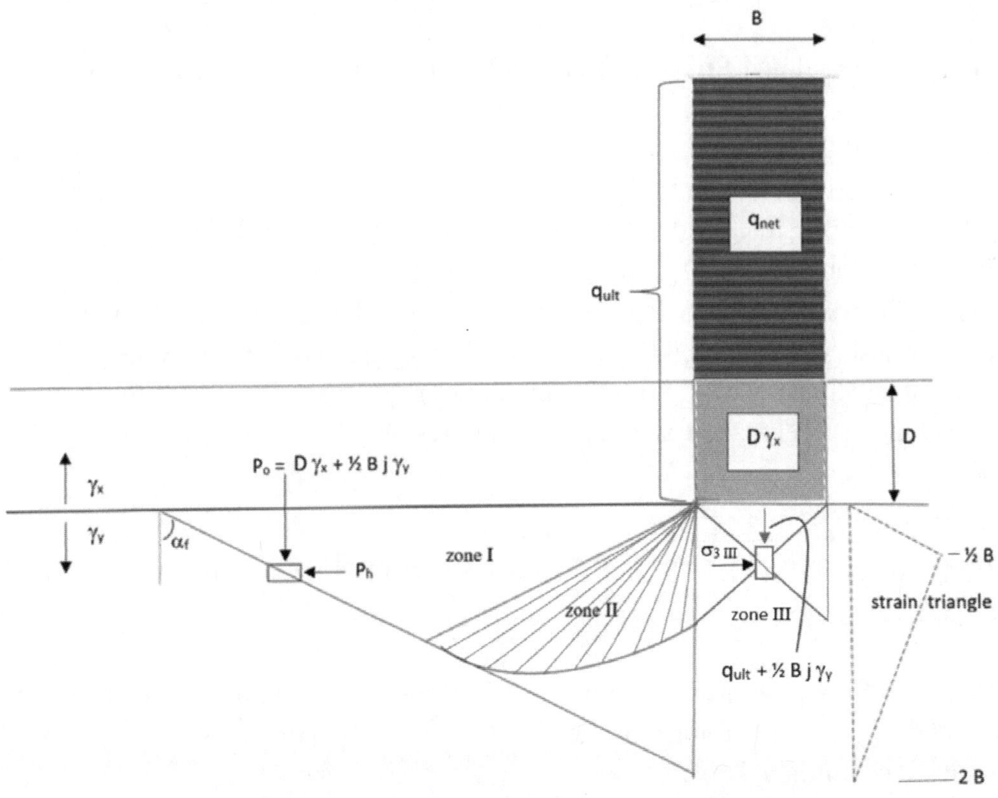

Figure 1. The Envisioned Three-zone Failure Mass.

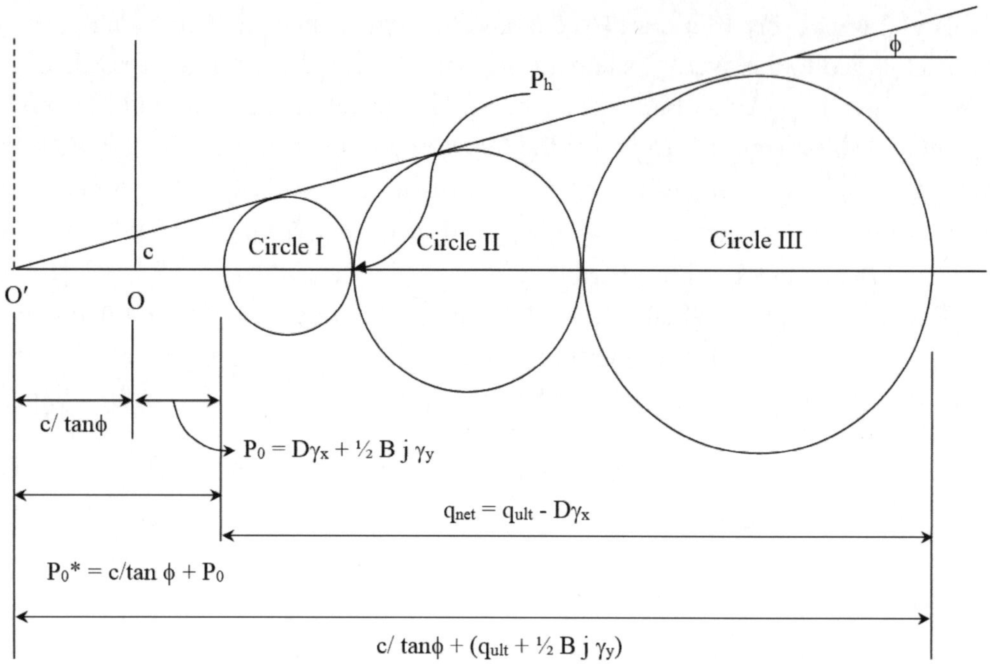

Fig 2. Interrelated Stresses of Zones I through III of the Failure Mass.

But this same distance/difference, $P_0{}^*\tan^6(45 + \phi/2) - P_0{}^*$ (relative to O'), is given in Eq. 5b. This pictured horizontal distance in Fig. 2 is a normal stress that represents the increase in vertical pressure above the free-field value (of Fig. 1) that causes bearing capacity failure of a square/round foundation. Adding $D\,\gamma_x$ to this q_{net} yields q_{ult}.

As noted, the triaxial test c and ϕ are implicit to the UNR analysis. However, the c and ϕ of the direct shear test (where width is equal to length) are equally relevant to the UNR analysis, as well as field correlations related to triaxial and direct shear tests. Finally, in the case where cohesion, c, is conservatively taken as zero, $P_0{}^*$ becomes just P_0.

THE UNR EQUATION EXPRESSED IN TERMS OF THE CLASSICAL BEARING CAPACITY EQUATION

While one can add $D\gamma_x$ to q_{net} (as assessed directly from Eq. 5) to obtain q_{ult}, it is the intent here to express the UNR equation in the form of the classical Eq. 1 to compare MHV and UNR bearing capacity factors. Accordingly,

$$q_{ult} = q_{net} + D\gamma_x = P_0{}^*(tan^6\alpha_f - 1) + D\gamma_x$$

$$= \left({}^{c}/_{\tan\emptyset} + D\gamma_x + {}^{1}/_{2}\,Bj\gamma_y\right)(tan^6\alpha_f - 1) + D\gamma_x$$

i.e. $$q_{ult} = {}^{1}/_{2}\,Bj\gamma_y(tan^6\alpha_f - 1) + D\gamma_x(tan^6\alpha_f - 1) + D\gamma_x + {}^{c}/_{\tan\emptyset}(tan^6\alpha_f - 1)$$

whereby

$$q_{ult} = {}^{1}/_{2}\,Bj\gamma_y(tan^6\alpha_f - 1) + D\gamma_x(tan^6\alpha_f) + {}^{c}/_{\tan\emptyset}(tan^6\alpha_f - 1) \qquad (6)$$

UNR's equivalent to Eq. 1, with s and d factors taken to be unity, and γ_x and γ_y substituted to distinguish the unit weight of the soil above verses that below the foundation base, is

$$q_{ult} = 0.5B\gamma_y N_\gamma + D\gamma_x N_q + cN_c \quad (7)$$

Comparing Eqs. 6 and 7 term by term, starting with the second term, the UNR versus MHV bearing capacity factors are these

UNR	MHV	
$N_q = tan^6\alpha_f$	$N_q = e^{\pi \tan\emptyset} N_\emptyset$	$N_\emptyset = tan^2\left(45° + \emptyset/2\right)$
or $N_q = tan^6\left(45° + \emptyset/2\right)$	$N_q = e^{\pi \tan\emptyset} tan^2\left(45° + \emptyset/2\right)$	(8)
$N_c = (N_q - 1)/\tan\emptyset$	$N_c = (N_q - 1)/\tan\emptyset$	(9)
$N_\gamma = (N_q - 1)j$	$N_\gamma = (N_q - 1)j$	Meyerhof/Hansen (10)
$j = 1.5\tan\emptyset$	$j = 1.5\tan\emptyset$	Hansen (10a)
	$j = \tan(1.4\emptyset)$	Meyerhof (10b)
	$N_\gamma = (N_q + 1)j$	Vesic (11)
	$j = 2\tan\emptyset$	Vesic (11a)

The one difference that affects N_q and therefore everything upon which N_q depends is the $e^{\pi\tan\phi}$ (of an infinitely long foundation) of the MHV N_q as opposed to $\tan^4(45+\phi/2)$ (for a square/round foundation) in the UNR N_q Eq. 8.

COMPARISION OF BEARING CAPACITY FACTORS

While one could compare the UNR and MHV factors across from each other in Eqs. 8-11, in use it is

UNR	MHV
N_q	$N_q(s_q d_q)$
N_c	$N_c(s_c d_c)$
N_γ	$N_\gamma(s_\gamma d_\gamma)$

that should be compared. As a result, the plot of the UNR versus the separate M, H and V bearing capacity factors with increasing friction angle, ϕ, will change with a change in the B/L and/or D/B value.

It should be noted that the UNR method uses a modified value of the given triaxial test friction angle based on foundation shape (B/L). In different additions of his textbook, Bowles (1996) computes q_{ult} using the Terzahi, Meryerhof, Hansen and Vesic equations to compare with eight full-scale field tests by Milovic and Muhs supported by carefully assed triaxial test c and ϕ values for the pressure range of the footing tests. While Bowles applied the shape and depth correction factors to the respective MHV equations, he found improved calculated results using a plane strain ϕ for the three non-square Muhs tests (B/L = 0.25). In the earlier edition of his textbook (Bowles 1988), the plane strain ϕ he used was $\phi_{ps} = 1.1\ \phi_{triax}$ (attributable to Meyerhof), with ϕ_{triax} being the triaxial test friction angle. In the later edition (Bowles 1996), he used $\phi_{ps} = 1.5\ \phi_{triax} - 17^o$ (attributable to Lade and Lee for ϕ_{triax} values ≥ 34°). In the paper by Elfass et. al (2007), the same eight tests were analyzed with the UNR equation. For the three rectangular footings, an intermediate $\phi_{B/L} = (1.1 - 0.1\ B/L)\ \phi_{triax}$ was employed yielding the closest to the recorded result in all three cases. Today, the senior author would prefer using a modified version of the Lade and Lee plane strain equation, i.e. $\phi_{B/L} = \phi_{triax} + (1 - B/L)(0.5\ \phi_{triax} - 17)$. Table 3 provides a comparison of the UNR equation results for the three tests employing the plane strain and the B/L modified Meyerhof and Lade and Lee friction angles against the Terzaghi, Meyerhof, Hansen and Vesic results calculated by Bowles based on the Lade and Lee plane strain ϕ.

Table 3. Composition of Calculated q_{ult} Values of Tests undertaken by Muhs

Muhs Tests	L/B = 4	Recorded	Calculated Based on UNR Equation for Different ϕ				Calculated by Bowles Using $\phi_{PS} = (1.5*\phi_{triax} - 17)$			
Test	D/B	q_{ult} (kg/cm^2)	A	B	C	D	Terzaghi	Meyerhof	Hansen	Vesic
1	0	10.8	14.62	12.71	10.50	9.93	9.4	8.2	7.2	8.1
2	1	12.2	16.20	14.26	10.88	10.60	9.2	10.3	9.8	10.4
3	1	24.2	33.13	28.69	26.11	24.05	22.9	26.4	23.7	25.1

Closest to the recorded value

A Meyerhof	$\phi_{PS} = 1.1\ \phi_{triax}$
B Modified Myerhof	$\phi_{B/L} = (1.1 - 0.1\ B/L)\ \phi_{tri}$)
C Lada and Lee	$\phi_{PS} = 1.5\ \phi_{triax} - 17$
D Modified Lada and Lee	$\phi_{B/L} = \phi_{triax} + (1 - B/L)(0.5\ \phi_{triax} - 17)$

Figure 3 is a plot of the different ϕ vs ϕ_{triax} variations for $\phi_{triax} \geq 34^o$ and L/B = 4 (i.e. B/L = 0.25). What is interesting is that plots of the UNR and MHV N_q, N_c and N_γ (to be shown) established based on $\phi_{B/L}$, versus $\phi_{B/L}$, overlie the same curves as the plots of N_q, N_c and N_γ established using ϕ_{triax}, plotted versus ϕ_{triax}. However, the variation of q_{ult} established using $\phi_{B/L}$ plotted versus ϕ_{triax} will be different depending on B/L and D/B.

The first item to consider is the difference in N_q vs ϕ (°) for MHV versus UNR. The UNR curve lies above the MHV curve since the UNR N_q equation employs $\tan^4(45 + \phi/2)$ in place of $e^{\pi \tan\phi}$ of the MHV expression (Eq. 8). See Fig. 4. However, the UNR N_q curve should not be compared directly to the MHV N_q but rather to ($N_q\ s_q\ d_q$). For instance, if one sets B/L and D/B both to unity, the Hansen/Vesic ($N_q\ s_q\ d_q$) curve separates from the Meyerhof ($N_q\ s_q\ d_q$) curve.

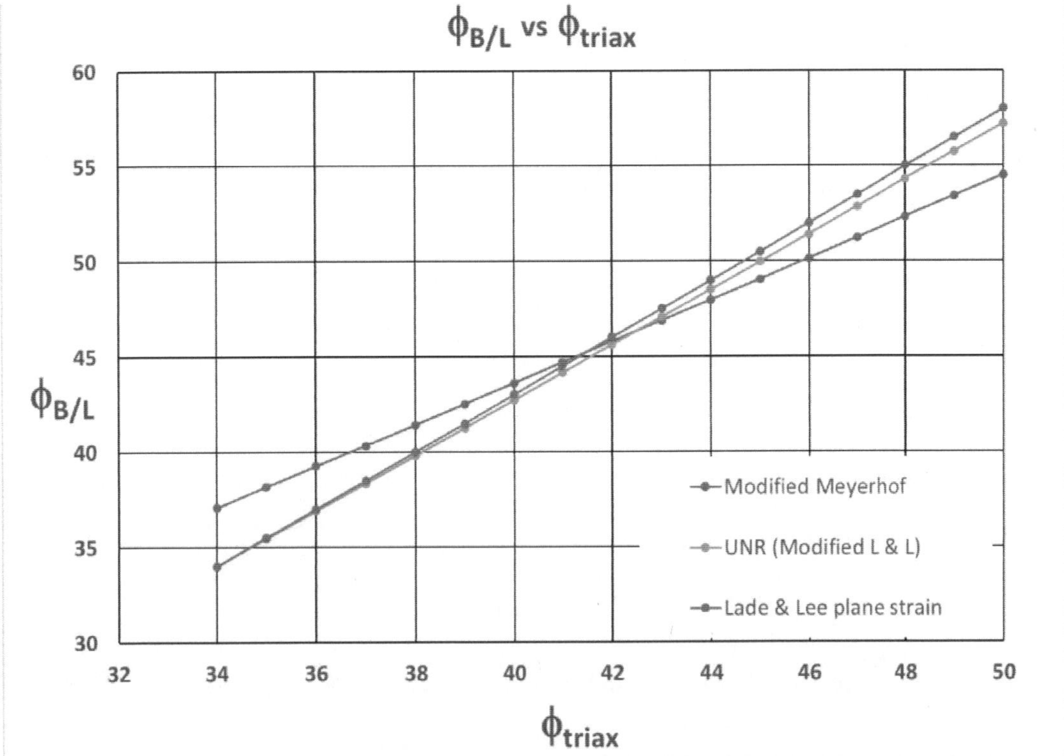

Figure 3. Different $\phi_{B/L}$ vs ϕ_{triax} Variations.

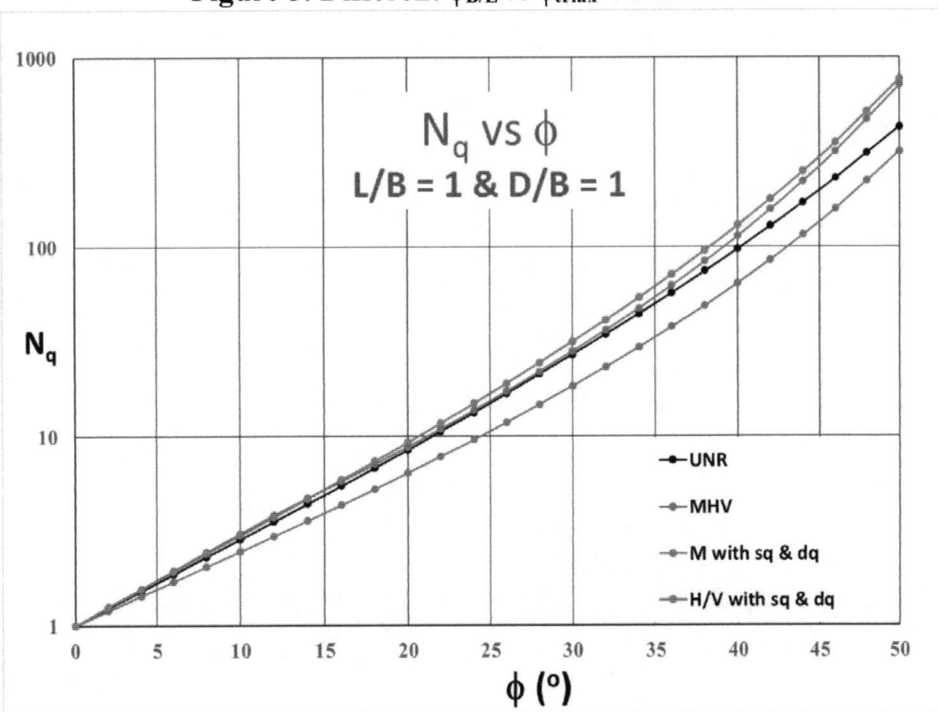

Figure 4. UNR (N_q), MHV (N_q), M (N_q s_q d_q) and H/V (N_q s_q d_q) vs ϕ (°) for L/B =1 and D/B =1.

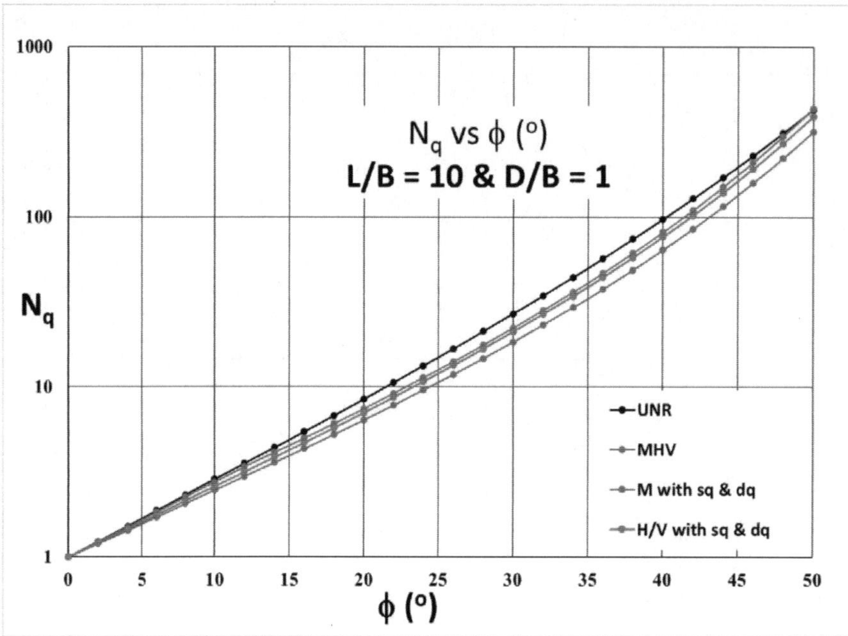

Figure 5. UNR (N_q), MHV (N_q), M (N_q s_q d_q) and H/V (N_q s_q d_q) vs ϕ (°) for L/B =10 and D/B =1.

Both the H/V and M curves fall slightly above the UNR curve in Fig. 4. If one changes L/B to 10 (D/B = 1) the curves shift as shown in Fig. 5, i.e. the H/V (s_q d_q N_q) and M (s_q d_q N_q) curves now

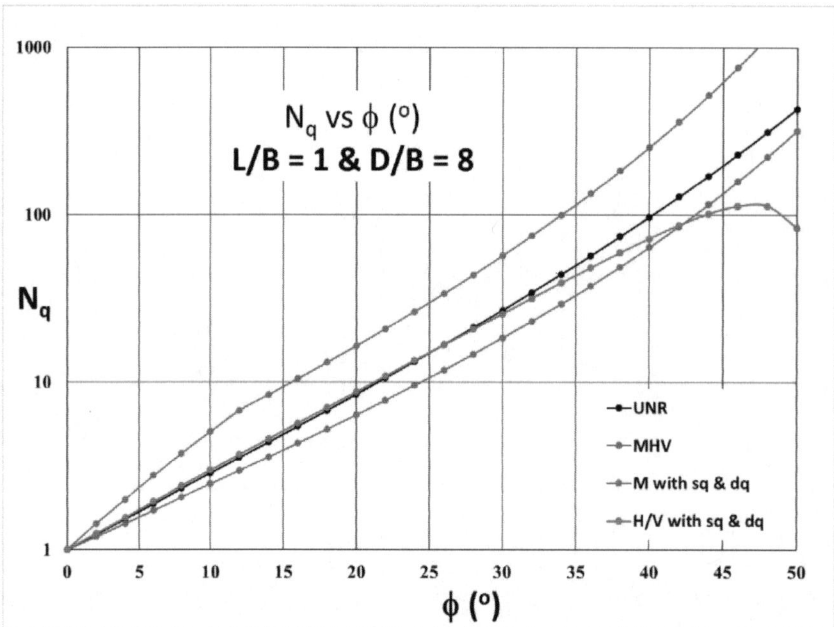

Figure 6. UNR (N_q), MHV (N_q), H/V (N_q) with s_q d_q and M (N_q) with s_q d_q vs ϕ (°) for L/B =1 and D/B =8.

fall below the UNR curve. If, on the other hand, L/B =1 but D/B is set to 8 (corresponding to the depth at which the footing reaches a limiting pressure as if considered as the tip of a pile), the curves of Fig. 6 result.

Realize that while the UNR N_q curve remains fixed, its use in assessing bearing capacity, q_{ult}, as L/B increases above unity, affects the $\phi_{B/L}$ value employed. Using $\phi_{B/L}$ is the way the UNR method approaches the shape effect. The MHV equations rely on the triaxial test or field correlated ϕ, unless, like Bowles, one considers changing the ϕ value. Recall the effect of $\phi_{B/L}$ on the results of Table 1 for full size footing tests. Furthermore, if, like Bowles, one uses a plane strain ϕ in the selection of N_q (thus affecting N_c and N_γ as per equations 9-11), does one use this ϕ in the calculation of the shape factors (see Table 2)?

While there are great number of N_q, N_c and N_γ comparisons that might be made, consider the comparing the $N_q\, s_q\, d_q$, $N_c\, s_c\, d_c$ and $N_\gamma\, s_\gamma\, d_\gamma$ of Meyerhof, Hansen and Vesic with the UNR's N_q, N_c and N_γ for just L/B = 2 and D/B =1. Furthermore, rather than have a figure with the traditional vertical axis in log scale form, consider a closer look with a linear vertical axis and a restricted ϕ_{triax} range from 28° to 42°. Figures 7 through 9 show the bearing capacity factor combinations mentioned versus ϕ_{triax}.

Note that the M, H/V and UNR N_q curves of Fig. 7 are almost one and the same. In Fig. 8, the M and H/V N_c curves are together while the UNR curve plots lower. The reason for this is that UNR's N_c derives from P_o* which, because c/tanϕ is an extension of the ϕ envelope, treats the cohesion (c) effect in proper proportion with the frictional (ϕ) effect. The authors suggest that, as a consequence, the c/tanϕ term is in correct proportion with the $D\gamma_x$ term (and ½ $Bj\gamma_y$

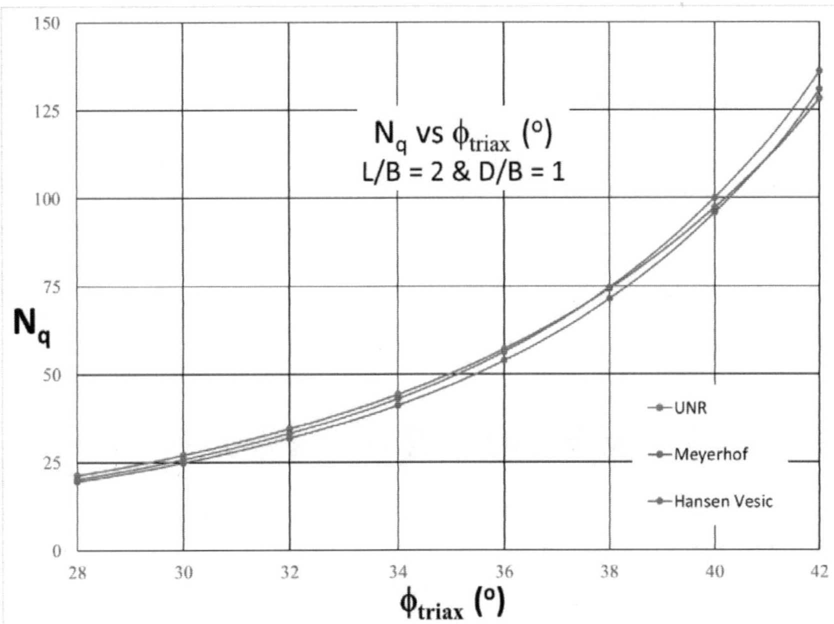

Figure 7. $N_q\, s_q\, d_q$ of M and H/V vs and N_q of UNR for L/B = 2 and D/B =1.

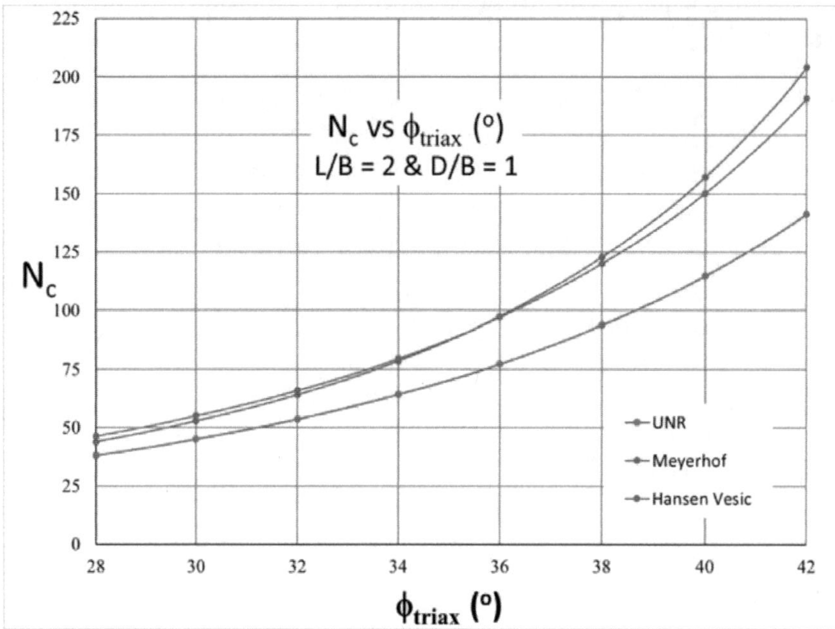

Figure 8. N_c s_c d_c of M and H/V vs N_c of UNR for L/B = 2 and D/B =1.

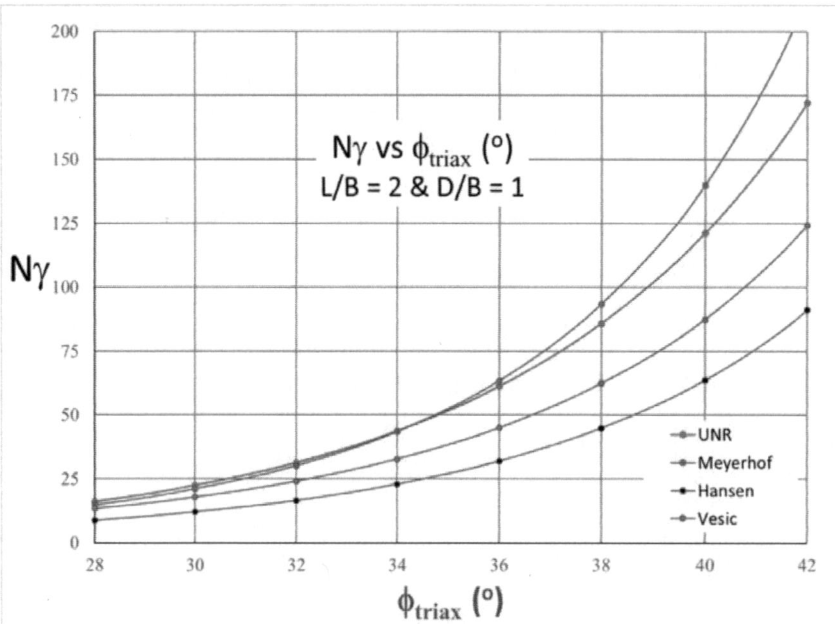

Figure 9. N_γ s_γ d_γ of M, H and V vs N_γ of UNR for L/B = 2 and D/B =1.

term) of the q_{ult} equation. However, it is in Fig. 9 that differences of all four arise. It is the UNR curve that more closely follows the Meyerhof curve.

COMPARISON OF q_{ult} VALUES

While it is one thing to compare bearing capacity factors, it is their combination in assessing q_{ult} that is of greater importance. Realize, however, the bearing capacity factors of the UNR approach are really not needed to assess q_{ult}, just the use of Eq. 5b. Putting aside differences that B, γ_x and γ_y might cause, it is c and B/L (or L/B) and D/B that are of interest here. Therefore, setting B = 2 ft, γ_x and γ_y = 120 lb/ft^2, consider the following six cases:

L/B = 1, D/B = 1   c = 100 and 500 lb/ft^2.

L/B = 10, D/B = 1   c = 100 and 500 lb/ft^2

L/B = 1, D/B = 8   c = 100 and 500 lb/ft^2

Figures 10-15 present the results of q_{ult} vs ϕ calculated using the M, H, V and UNR equations. In the case of the UNR results, the modified Lade and Lee correction of ϕ above 34° is used where L/B is greater than one. It is up to the reader, if desired, to recalculate MHV q_{ult} based on a modified friction angle (different from ϕ_{triax}) and whether or not to use it in the calculation of the shape and depth factors.

Figures 10 and 11 are for the square footing embedded at a depth equal to the foundation width, a commonly considered 2 ft depth. Note that for c = 100 psf in Fig. 10, the UNR variation falls below that of Meyerhof, Vesic and Hansen. For c = 500 psf in Fig. 11, the UNR variation moves further away (lower) from that of M, V and H. Therefore, the UNR variation is decidedly conservative in regard to the cohesive component of strength. But the UNR variation was meant specifically for a square foundation (L/B = 1) for which the triaxial test strength applies. The MHV variations are for a plane strain situation (and realistically, a plane strain friction angle) for which lab scale shape and depth correction factors (a function of a triaxial friction angle) are meant to bring it closer to the axisymmetric (L/B = 1) situation at hand.

Figures 12 and 13 are for an L/B of a long (i.e. strip) footing corresponding closely to a plane strain situation of infinite length. The UNR variation is shown splitting starting at 34° above which the modified Lade and Lee plane strain approximation applies. The lower UNR curve for ϕ_{triax} > 34° is the UNR variation using a triaxial ϕ, while the upper curve employs the modified Lade and Lee ϕ. Note that the curves are plotted versus ϕ_{triax}. Relative to the difference between Figs. 12 and 13 in regard to c = 100 versus 500 psf, note that as with Figs. 10 and 11, the greater c causes the UNR curve to move down relative to the MHV curves.

In regard to the M, V and H variations of Figs. 12 and 13, they were evaluated relative to a triaxial ϕ to which shape and depth correction factors are applied to try to bring them closer to the near plane strain situation that now exists. If, as Bowles found out, a plane strain ϕ were used, it would result in an upward shift of their curves. However, that begs the question of whether the shape and depth correction factors should be used in addition to the plane strain ϕ, and if the plane strain ϕ should be used in assessing the correction factors. Note that at best the practicing engineer has available is the triaxial (or direct sear) ϕ or a field correlation yielding a triaxial ϕ.

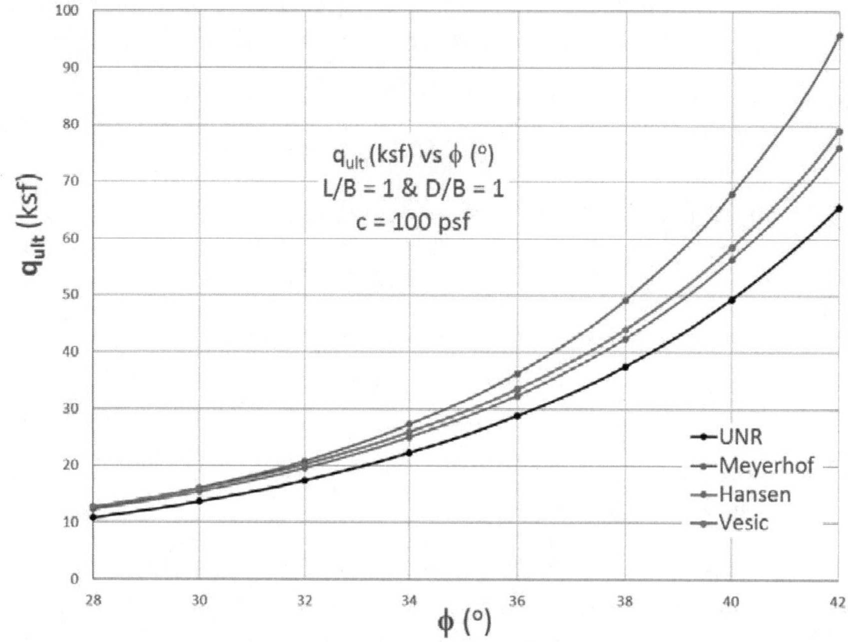

Figure 10. q_{ult} for c = 100 psf, L/B = 1, D/B = 1 (B = 2 ft, γ_x and γ_y = 120 lb/ft²)

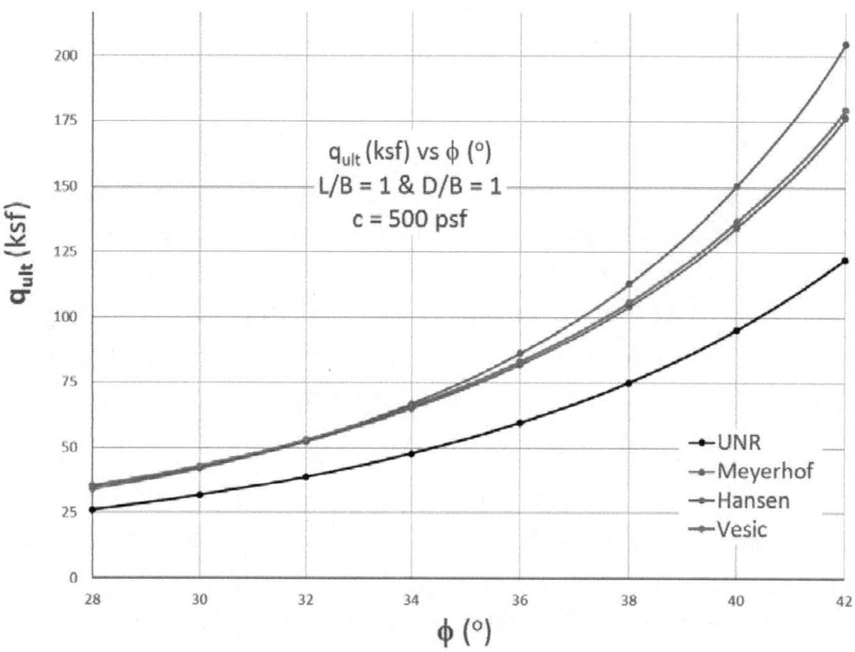

Figure 11. q_{ult} for c = 500 psf, L/B = 1, D/B = 1 (B = 2 ft, γ_x and γ_y = 120 lb/ft²)

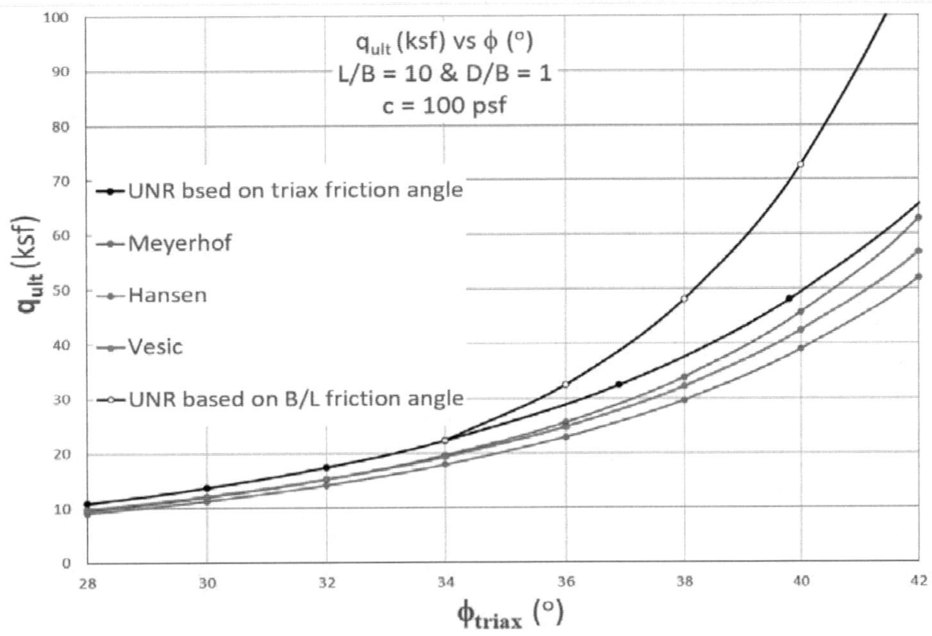

Figure 12. q_{ult} **for c = 100 psf, L/B = 10, D/B = 1 (B = 2 ft,** γ_x **and** γ_y **= 120 lb/ft²)**

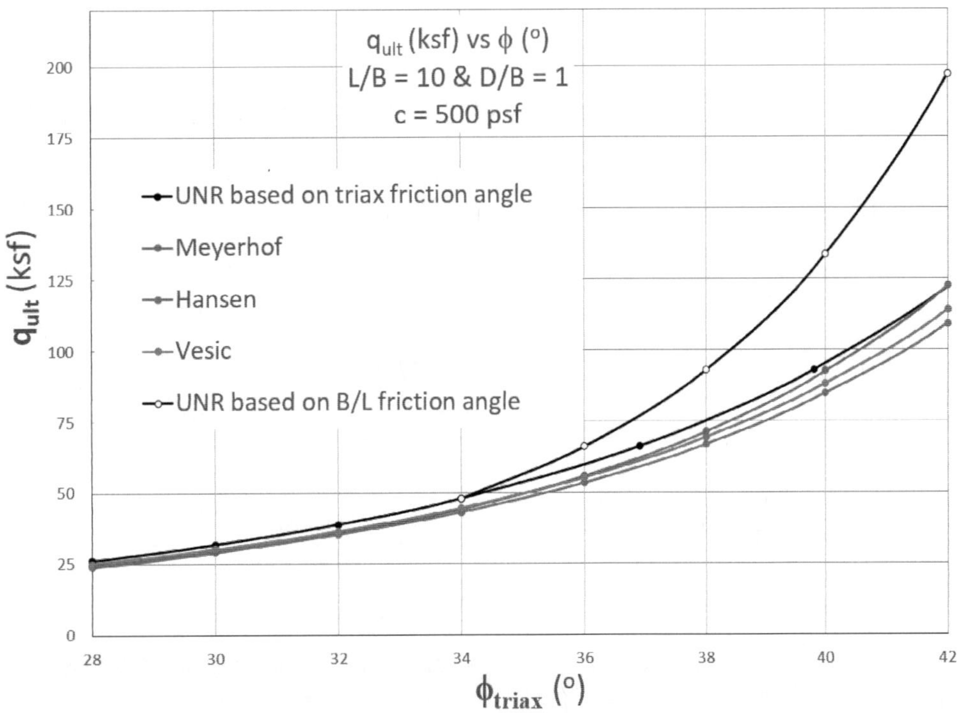

Figure 13. q_{ult} **for c = 500 psf, L/B = 10, D/B = 1 (B = 2 ft,** γ_x **and** γ_y **= 120 lb/ft²)**

Finally, in regard to the square footing at depth (D/B = 8), note that the UNR variation of Fig. 14 is extremely conservative and grows more so with increasing the cohesion of Fig.15. It

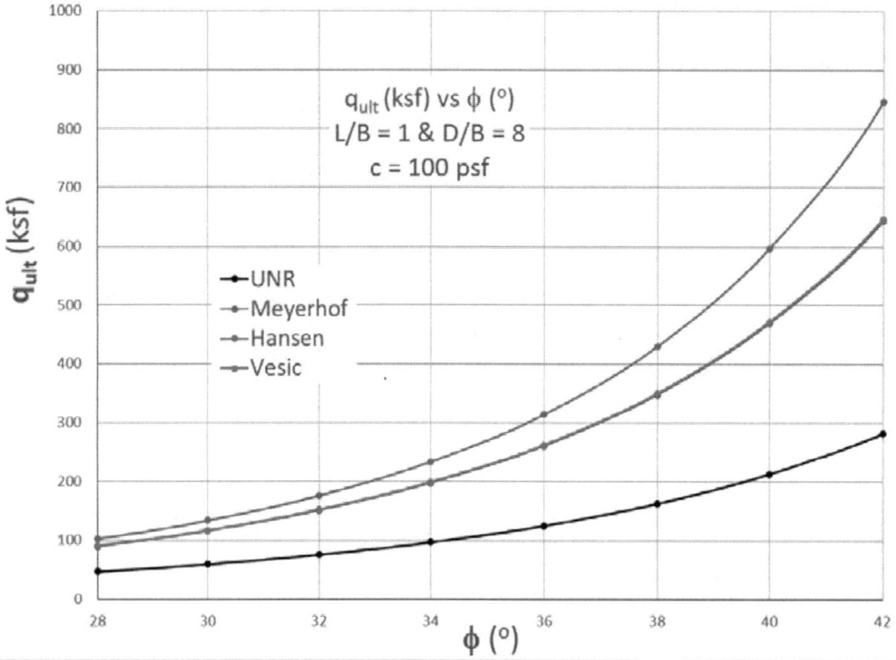

Figure 14. q_{ult} for c = 100 psf, L/B = 1, D/B = 8 (B = 2 ft, γ_x and γ_y = 120 lb/ft^2)

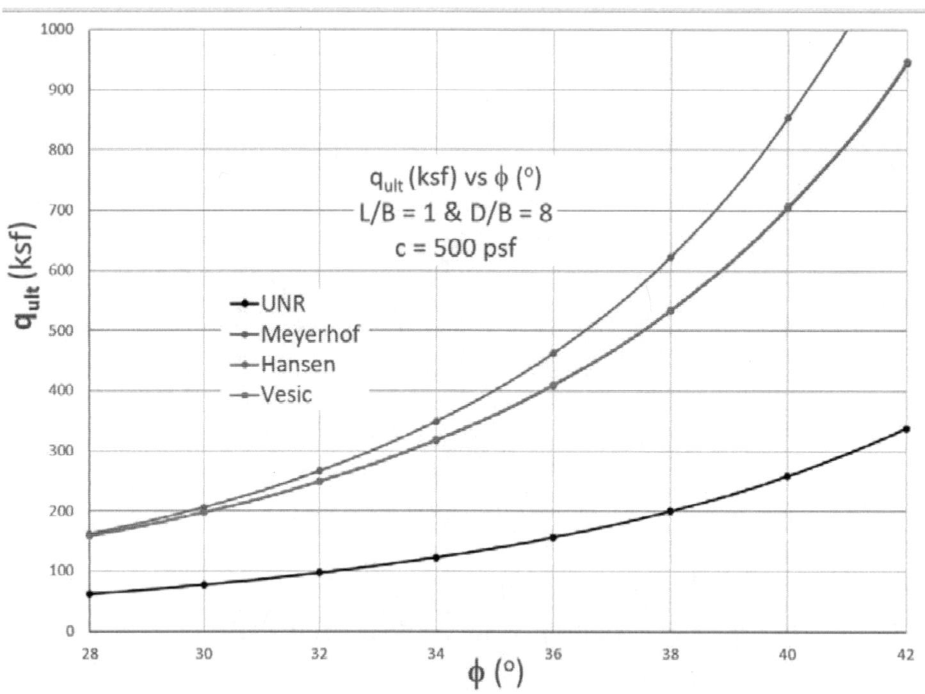

Figure 15. q_{ult} for c = 500 psf, L/B = 1, D/B = 8 (B = 2 ft, γ_x and γ_y = 120 lb/ft^2)

should be noted that The UNR equation ignores the strength of the soil along any shear surface above the foundation base. If this were added, the UNR curve would shift up closer to the M and overlapping H/V curves.

THE ISSUE OF COMPARESSIBILITY

In recent time, it has been suggested that another correction be considered, the compressibility of cohesionless soil. If the engineer only has the ϕ at lower confining pressure (e.g. in DM 7.1 p. 398 from correlation with dry unit weight and relative density), the Mohr-Coulomb envelope over the pressure range of q_{net} (i.e. P_o to $P_o + q_{net}$ of Fig.2) may be curved, due to compressibility. What is needed for classical equation analysis is the best-fit secant c-ϕ to that curved envelope. The authors suggest a ϕ such as that of DM 7.1 be attributed to a confining pressure of say 1 ton/ft^2, as is the correction from the standard penetration test blow count, N, to a value for 1 ton/ft^2 (tsf) overburden pressure, $N_{1,60}$. Then the secant ϕ at pressure P_o (or P_o* if cohesion is to be assumed) can be assessed from the value at 1 ton/ft^2 using the log scale change in ϕ with increasing confining pressure, $\Delta\phi$, obtained via a modified form of Bolton's equation (Elfass and Norris 2012). The ϕ at P_o* is then $\phi_{Po^*} = \phi_{1\ tsf} - \Delta\phi \log (P_o^*/1\ tsf)$. As shown in Fig. 16, from P_o* with ϕ_I at pressure P_o*, the pressure P_h becomes $P_o^* \tan^2 (45 + \phi_I/2)$. For confining pressure, P_h, a ϕ_{II} is obtained using $\Delta\phi$ and the confining pressure for zone III is assessed as $P_h = \tan^2 (45 + \phi_{II} /2)$. With confining pressure, σ_{3III}, ϕ_{III} is obtained and then σ_{1III}, is assessed as $\sigma_{3\ III} \tan^2 (45 + \phi\ \sigma_{3\ III} /2)$. The value of q_{net} is then $\sigma_{1\ III} - P_o^*$, after which $q_{ult} = q_{net} + D\gamma_x$. In this UNR approach,

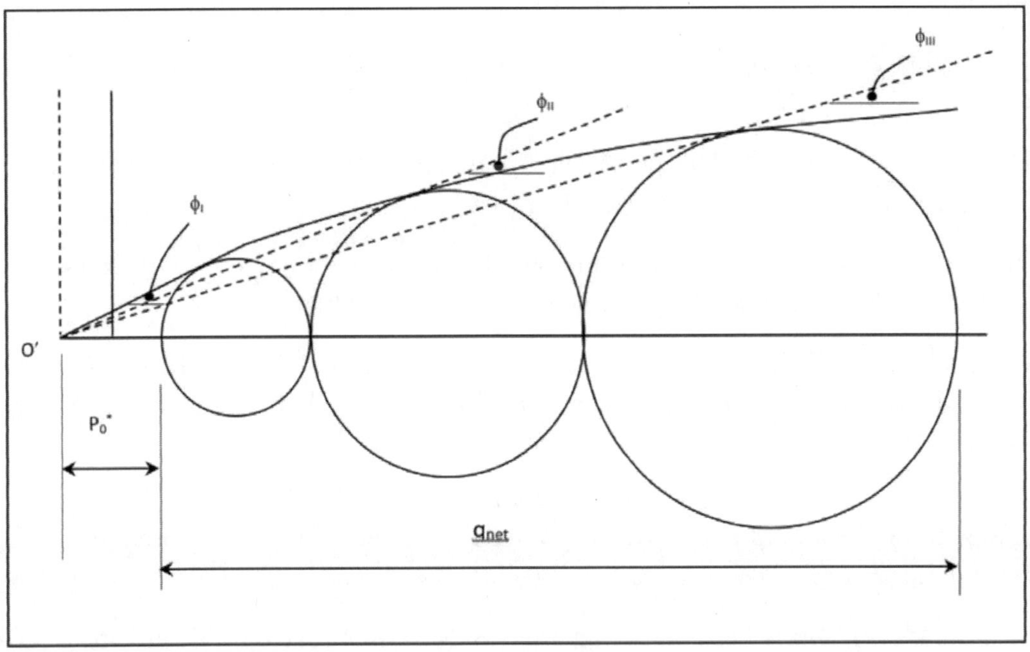

a

Figure 16. Use of the Modified Bolton Equation $\Delta\phi$ to Establish q_{net} Relative to a Curved Envelope.

a best fit secant c-ϕ to the curved envelope so constructed, required for classical equation assessment, is not needed (because q_{net} obtained in the curved envelope construction yields $q_{ult} = q_{net} + D\gamma_x$).

DISCUSSION AND CONCLUSION

This presentation has introduced the UNR method of assessing bearing capacity. A comparison to the classical Meyerhof, Hansen and Vesic equations with their shape and depth correction factors has been provided. This includes the bearing capacity factors individually and the assessed bearing capacity. In addition, the UNR concept can be used to assess net and then ultimate bearing capacity relative to a curved Mohr-Coulomb envelope associated with soil compressibility. What is noteworthy is that, as shown in Fig. 2 and Fig. 16, the net ultimate bearing capacity can actually be pictured in graphical form. Realize that the UNR equation was derived for an envisioned square/round triaxial test situation, while the MHV equations are based on a plane strain situation requiring correction factors. Lastly, while not a part of this paper, it is noted here that the UNR method can be used to assess the load-settlement response up to net bearing capacity failure. Such evaluation is presented in the other UNR papers mentioned in the references.

REFERENCES

Army Corps of Engineers. *Bearing Capacity of Soils*. Engineering and Design Manual EM 1110-1-1915. 1992.

Briaud RL and R.M. Predicted and Measured Behavior of Five Spread Footings on Sand. ASCE GSP 171, 1994, pp.

Briaud, J.L. Spread Footings in Sand: Load Settlement Curve Approach. *Journal of Geotechnical and Geoenvironmental Engineering*. vol. 133 (8), 2007, pp.: 905-920.

Elfass S., G. Norris and P. Vimalaraj. A Simple Bearing Capacity Equation. *Proc. GeoDenver Conference - New Peaks in Geotechnics: Advances in Shallow Foundations, Denver, CO,* ASCE, GSP 171, 2007, pp. 1-10.

Norris, G., S, Elfass, P. Vimalaraj and A. Lofty. A Simple Method for Assessing Load vs. Settlement of a Shallow Foundation up to Bearing Capacity Failure. *Proc. 1st Congress on Transportation Infrastructure, Costa Ric*a, 2011, 79 slides

Elsayed, A., S. Elfass, G. Norris and P. Vimalaraj. A New Approach for Non-Linear Load-Settlement Assessment of Shallow Foundations. *Proc. Geo-Frontiers Congress 2011 - Advances in Geotechnical Engineering, Dallas, TX*, ASCE GSP 211, 2011, vol. 1, 2011, pp. 372-381.

Nimeri, M., S. Elfass and G. Norris. G. Load–Settlement Response of Shallow Foundations Resting on Granular Soil. *GeoMEast2017 International Conference: Innovative Infrastructure Solutions, Cairo, Egypt.* Springer International Publishing 2 (33), 2017, 11 pp.

Bowles, J. E.,), *Foundation Analysis and Design, Fifth Edition*, McGraw-Hill Companies, Inc. New York, 1996.

Murphy, V.N.S. *Geotechnical Engineering: Principles and Practices of Soil Mechanics and Foundation Engineering*, Marcel Decker Inc., New York, 2003, Table a2.3, p 505.

Bowles, J. E. *Foundation Analysis and Design, Fourth Edition*. McGraw-Hill Companies, Inc. New York, 1988.

United Facilities Criteria. *Soil Mechanics (DM 7.1)*, Report UFC 3-220-10, UFC, 2022, 546 pp.

Elfass. S. and G. Norris. A Modified Bolton Equation and Its Application. *Proc. 44th Symposium on Engineering Geology and Geotechnical Engineering: New Approaches to Traditional Problems, Reno,* NV, Google Books, 2012, pp. 76-95.

Innovative Ground Improvement Solutions for Pennsylvania State Route 420 (Wanamaker Avenue) Bridge Replacement over Darby Creek

Sarah McInnes, P.E.
Principal Assistant Construction Engineer
Pennsylvania Department of Transportation
Engineering District 6-0
7000 Geerdes Boulevard
King of Prussia, PA 19406
610-205-6544
smcinnes@pa.gov

Abhijit Sheth, P.E.
Principal Geotechnical Engineer
Gannett Fleming, Inc.
1010 Adams Avenue
Audubon, PA 19403
610-783-3853
asheth@gfnet.com

Brian Teles, P.E.
Vice President
Gannett Fleming, Inc.
1010 Adams Avenue
Audubon, PA 19403
610-783-3866
bteles@gfnet.com

Prepared for the 73rd Highway Geology Symposium, September, 2024

Acknowledgements

The author(s) would like to thank the individuals/entities for their contributions in the work described:

Pennsylvania Department of Transportation, Engineering District 6
Richard E. Pierson Construction Company, Inc.
GeoConstructors, Inc.
The Collin Group
Schnabel Engineering

Disclaimer

Statements and views presented in this paper are strictly those of the author(s), and do not necessarily reflect positions held by their affiliations, the Highway Geology Symposium (HGS), or others acknowledged above. The mention of trade names for commercial products does not imply the approval or endorsement by HGS.

ABSTRACT

This paper presents a case study of implementation of a combination of column supported embankment (CSE) and lightweight fill/load balancing for the south approach embankment of the Pennsylvania State Route 420 Bridge (Wanamaker Avenue) over Darby Creek in Delaware County, Pennsylvania. The south approach is being widened and raised up to 7.5 feet. The site consists of poor subsurface conditions along with utility, right-of-way and environmental constraints. The subsurface profile includes up to 15 feet of highly compressible peat and organic soils which presented design challenges including mitigating large settlements, maintaining embankment stability, and minimizing construction costs. The CSE ground improvement technique addresses these challenges by utilizing a grid of stiff vertical columns to transfer embankment loads through soft soils to a firm bearing layer. A flexible load transfer platform (LTP) is also recommended to help transfer embankment loads to the columns. Additionally, load balancing using ultra lightweight foamed glass aggregate was recommended for portions of the lower height embankment to minimize CSE construction. Several aspects of design will be discussed in the paper including defining the types and limits of necessary mitigation, the load-displacement compatibility (LDC) method used for analysis and design of the CSEs, recommendations for critical embankment height, and constructability of CSEs during staged construction. The performance of the CSEs is assessed via sacrificial testing of columns and a robust instrumentation and monitoring program including piezometers, inclinometers, and settlement plates. Finally, conclusions are presented regarding the construction aspects of CSEs along with importance of contract specifications.

INTRODUCTION

The overall project, SR 420 (Wanamaker Avenue) over Darby Creek, involves the replacement of two two-lane bridges carrying northbound and southbound SR 420 over Darby Creek, in Tinicum Township and Prospect Park Borough in Delaware County, Pennsylvania. On the south side of the bridge, the existing Wanamaker Avenue approach embankment will be widened to accommodate the proposed roadway widening and a shared use path. At this location, PENNDOT's right-of-way is bordered on the southeast and southwest by United States Government right-of-way, associated with the John Heinz National Wildlife Refuge (part of US Fish and Wildlife's National Wildlife Refuge System). On the northwest, the roadway is bordered by the historic Morton Homestead which is part of the Morton Homestead State Park. Thus, limiting right -of way impacts is critical to preserve important adjacent properties.

To prevent encroachment on adjacent properties, four (4) new retaining walls will be constructed on the south side of the bridge to support fill associated with the reconfigured approach roadway and the addition of the new shared use path. The project also involves profile modifications along SR 420 approaching the bridge so the shared use path is continuous under the bridge. To accomplish this, the northbound and southbound portions of the shared use path are connected in a U-shape by a "duck-under" passing beneath the proposed replacement bridge. The duck-under will pass directly in front of the southern abutment for each bridge. At the duck-under location, the proposed shared use path grade will be above the Darby Creek high tide elevation with an overhead clearance of 10 feet that will be maintained between the trail and the underside of the bridges. The project will also replace both SB and NB bridges in a way that minimizes impacts to traffic and the surrounding natural and cultural resources while not over restricting the contractor's access. The proposed bridge will be dual structures separated by a 1" longitudinal open joint. The site location is shown in Figure 1. A general site layout for the south approach is shown in Figure 2.

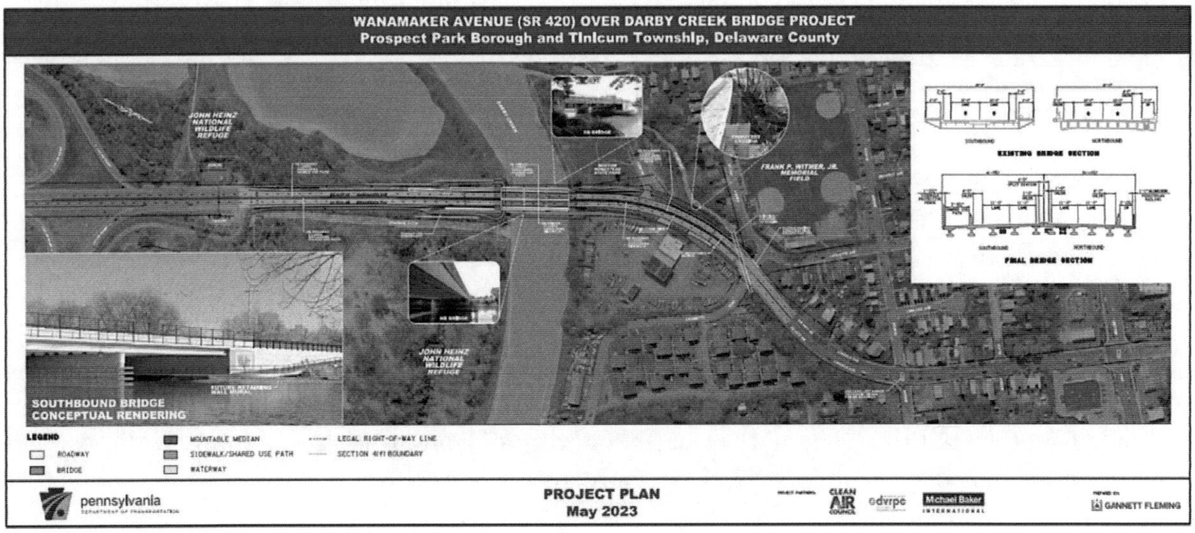

Figure 1 –Site Location Map

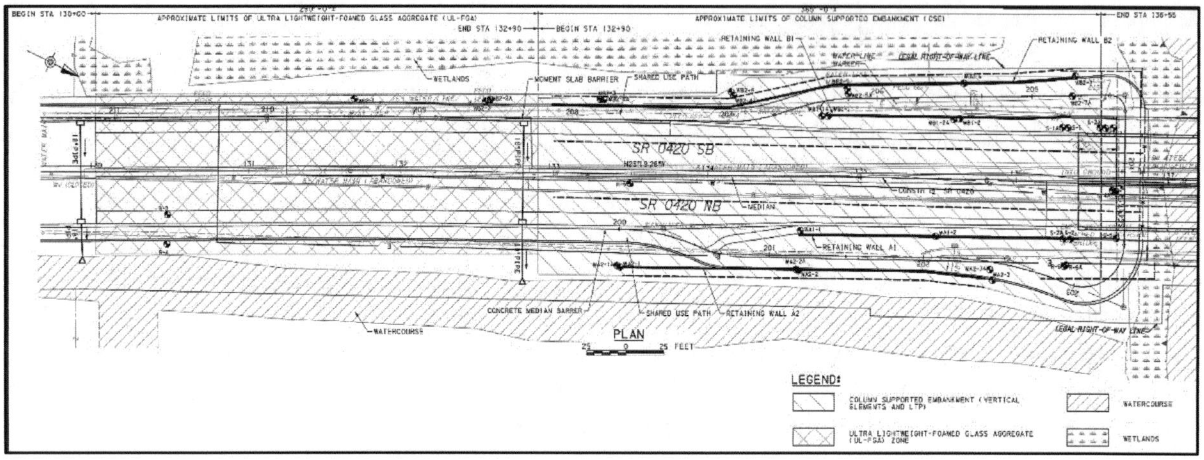

Figure 2 – South Approach General Site Layout

The proposed roadway grade approaching South Abutment (Abutment 1) will be approximately 7.5 feet higher than existing grade to maintain the required under clearance for the shared use path. On the north side of the bridge, slight profile adjustments will be made to tie the proposed structure to the existing roadway profile while minimizing impacts to the Morton Homestead State Park. The limits of embankment construction and ground improvement are provided in Table 1 below.

Table 1 – Summary of Embankment Construction

Approx. SR 420 Station (ft)		Approx. Length (ft)	Approx. New Fill Height* (ft)	Proposed Embankment Widening Support	
Begin	End			Side Slope	Retaining Wall
130+00	133+00	300	0 to 3.5	2:1	N/A
133+00	136+55	355	3.5 to 7.5	N/A	Walls A1, A2, B1, B2
136+55	139+50	295	Bridge Replacement		
139+50	146+00	650	0 to 2.5	2:1	N/A

*Note that fill height varies both longitudinally and transverse to the proposed roadway alignment. The fill height shown is at the centerline of the roadway.

The proposed construction was originally planned in three stages. However, the contractor, R.E. Pierson, developed a plan to cantilever a sidewalk off the existing northbound bridge to allow the bridge to be constructed in two stages. After the cantilever sidewalk, shoulder reconstructions, and utility relocations are completed, the first stage commenced. In the first stage, the southbound structure and Retaining Walls B1 and B2 will be constructed along the southbound roadway. To minimize wall lengths in this area, moment slab barriers with toe walls will also be constructed as the roadway transitions to existing grade. In the second stage, traffic is diverted to the newly constructed southbound roadway and the northbound bridge along with Retaining Walls A1 and A2 will be constructed along the northbound roadway.

PROJECT CONSTRAINTS AND EVOLUTION OF DESIGN

Project Constraints

The site is located within the boundaries of John Heinz National Wildlife Refuge on the south side and at the edge of Morton Homestead at the NW quadrant, both are Section 4(f) resources. In addition, materials within the right-of-way are within a known EPA Superfund site. Thus, it is imperative that the proposed construction adhere to strict testing and management of fill policies to ensure all earth moving is properly handled as well as both workers and the environment are safe during construction. Also, the tidal Darby Creek is considered navigable by the US Coast Guard due to its proximity to the Delaware River. Thus, all construction and final conditions (clearances, signage, navigation lighting, etc.) need to be in accordance with the US Coast Guard Bridge Permit obtained for the project.

As described in more detail in the subsequent sections, the area of the proposed profile raise, i.e., on the south side of the bridge, contains a relatively thick compressible soil layer with a history of settlement since the northbound bridge was constructed in 1967. Conversations with PennDOT personnel who were involved with the 1967 design and subsequent maintenance of the structure confirmed that settlement was an issue during construction and continued to be over the life of the facility. This coincided with the results of the subsurface investigation for the current project and validated that a careful evaluation and solution was critical when constructing the raised profile.

Evolution of Design

Given the site's complex subsurface conditions and a history of settlement, PennDOT and the design consultant collaboratively explored multiple design alternatives. The primary objectives were to mitigate short-term and long-term settlement, minimize construction and anticipated maintenance costs, and reduce potential risks. These items were discussed in detail and scrutinized with the goal of finding a solution that balances long term risk, future maintenance, and cost (both initial and maintenance). Keeping these factors in mind, it was agreed to take a slightly more conservative approach with the design and performance of a ground improvement solution, even though the solution was not the least expensive to build.

The original approach consisted of prefabricated vertical drains (PVDs) with surcharge to accelerate the time required for primary consolidation settlement of the underlying soft compressible soils. The use of PVDs would provide a shorter and more rapid drainage path to relieve excess pore water pressure and consequently help achieve 90% of the primary consolidation settlement within the anticipated Stage 1 (center portion of the embankment, approximately 40 feet wide) construction duration. Due to the site constraints discussed previously, constructing PVDs did not appear to be feasible at the wall locations. Therefore, to mitigate settlement concerns at the walls, lightweight aggregate backfill (expanded shale aggregate or similar) was recommended as wall backfill. Furthermore, to accommodate any post-construction differential settlement, slip joints were recommended for the proposed precast modular walls. This approach would minimize initial construction costs; however, there were concerns regarding long-term serviceability and maintenance of the roadway. There were also concerns that surcharging would need to be in place for extended construction durations which could impact the rideability of the adjacent roadway and serviceability of underground utilities.

Consequently, it was decided to consider a more robust solution, in this case column-supported embankments (CSE), to achieve the project goals but in a way which further minimizes potential future maintenance concerns. This approach was the preferred alternative and is discussed in more detail in subsequent sections.

SUBSURFACE CONDITIONS

The subsurface investigation program for this section of the project consisted of ten (10) roadway borings and fourteen (14) retaining wall borings. The boring locations are shown in Figure 2. Based on the borings, the soil encountered at the site was mainly divided into two strata: fill and alluvium. Each stratum is described in detail below. The borings indicated that the subsurface conditions (layer depths and properties) varied considerably across the site, especially within the underlying highly compressible organic soil layer. Generalized subsurface profile is shown in Figures 3, 4 and 5.

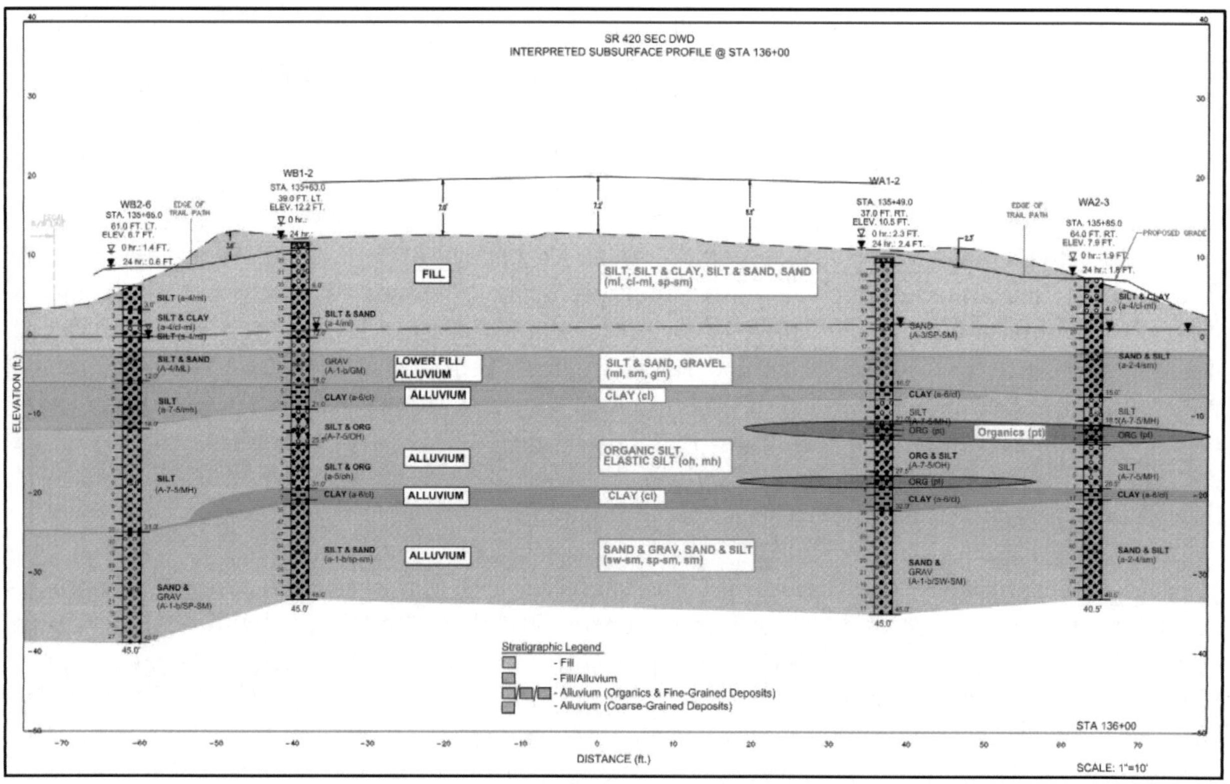

Figure 3 – Generalized Subsurface Profile @ STA 136+00

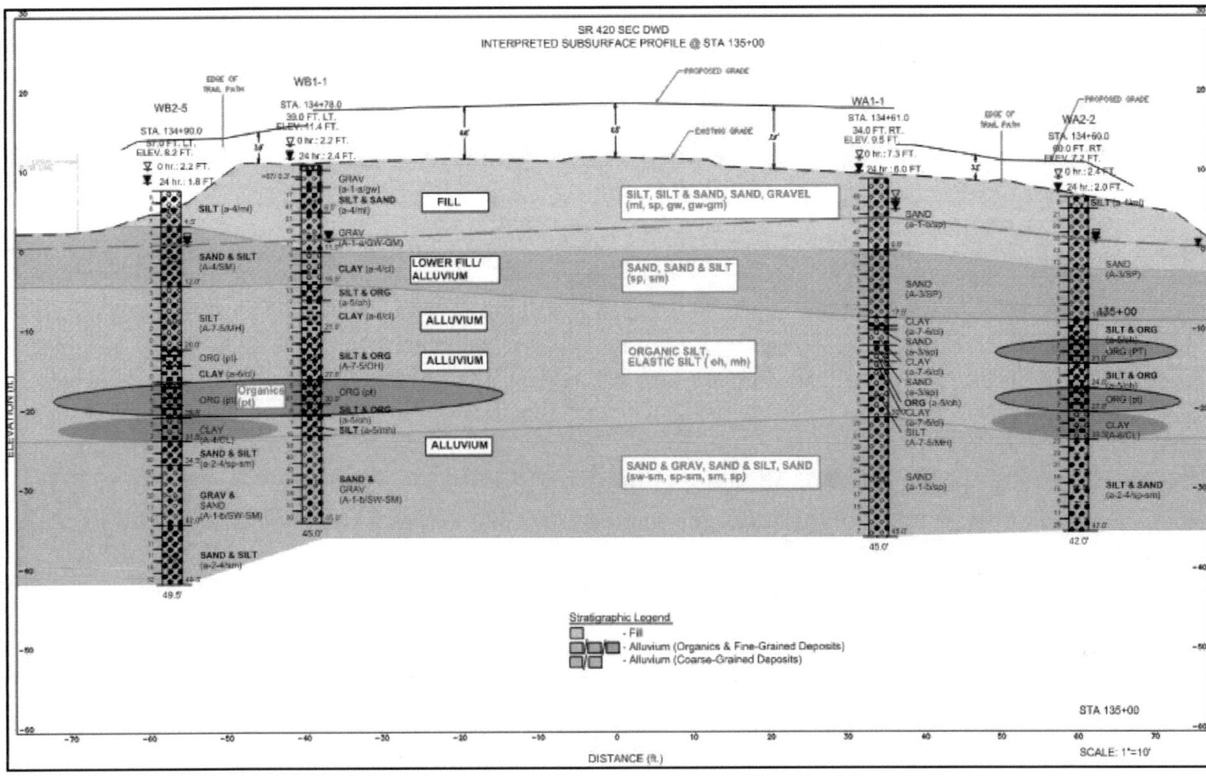

Figure 4 - Generalized Subsurface Profile @ STA 135+00

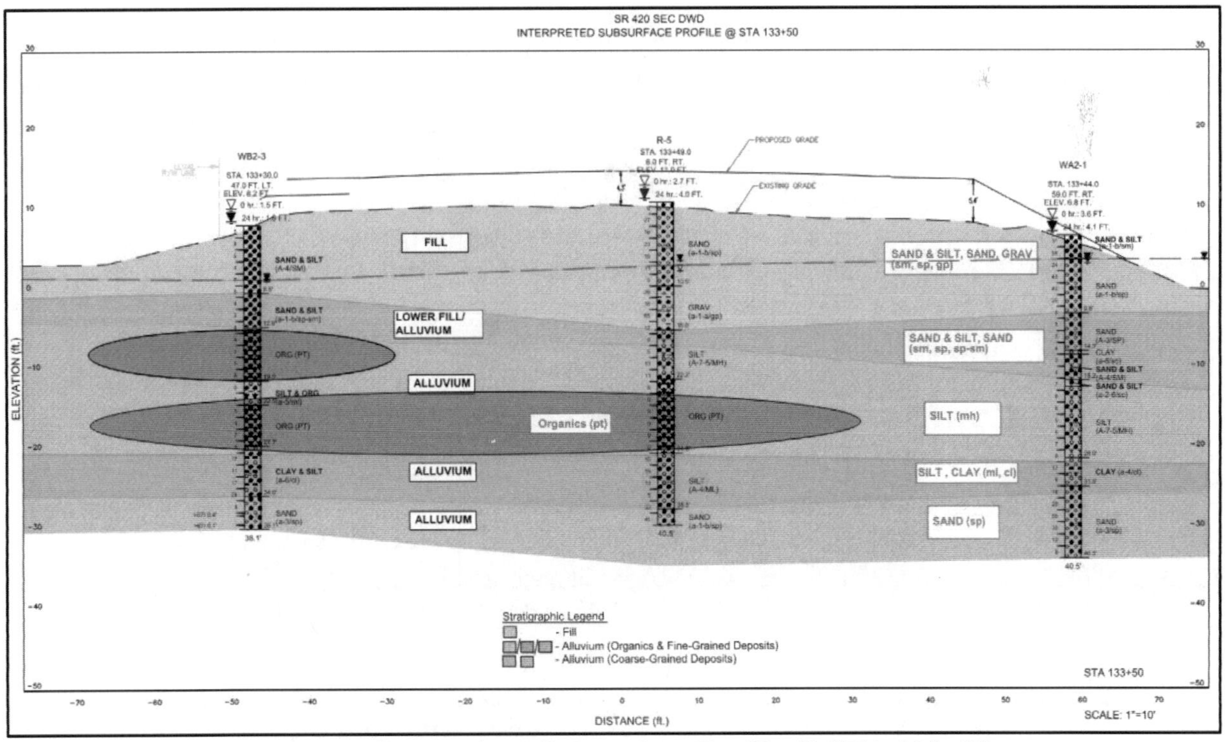

Figure 5 - Generalized Subsurface Profile @ STA 133+00

A summary of the subsurface exploration data is provided in Table 2.

Table 2 – Summary of Subsurface Exploration Data			
Stratum (in order of increasing depth)	Soil Type	Relative Density / Consistency	Estimated Layer Thickness
Fill	SP, SM, ML, GW	very loose to very dense	3.5 ft to 15 ft
Alluvium (Coarse-Grained soils)	SP, SP-SM, SC	very loose to dense	6 ft to 12.5 ft
Alluvium (Organics/Clayey Soils)	PT, OH, MH, CL, ML	very soft to medium	12.5 ft to 20 ft
Alluvium (Coarse-Grained Soils)	SP-SM, SW-SM, SP, SM	Medium to very dense	10 ft to 14.5 ft

All borings were terminated within the coarse-grained alluvium layer at depths of 38 feet to 49.5 feet below ground surface. The groundwater ranged from EL 0.6 to EL 6.0 feet. The 100-yr flood elevation for the area is at EL 10 feet.

DESIGN SOIL PARAMETERS

The embankment fill heights, retaining wall heights and configurations, along with the subsurface conditions vary significantly across the site. Therefore, to optimize the design, representative sections were selected, and design parameters established for each section. To aid design, an extensive laboratory testing program was undertaken to establish design soil parameters for the representative sections. Because the compressible layer consisted of alternating layers of peat, organic silt, clay, and silty clay, it was important to isolate these individual layers and establish design soil parameters and depths/thicknesses for each compressible sub-layers ensuring settlement was calculated correctly. Regarding the consolidation testing data, laboratory testing indicated considerable spatial variability; therefore, consolidation parameters for individual soil layers were estimated based on scatter plots of the laboratory consolidation test data. To estimate secondary compression parameters, a preconsolidation pressure of 2 tsf was conservatively assumed. This represented the anticipated upper bound of the vertical effective stress in the compressible layer beneath the raised embankment section. The design soil parameters are provided in Table 3A and Table 3B.

Table 3A – Design Soil Consolidation Parameters							
Layer/Stratum (in order of increasing depth)	Approx. Layer Thickness (feet)	Saturated Unit Weight, γ_S (pcf)	E_s (ksf)	Consolidation Parameters			
				C_c	C_r	C_v (ft^2/day)*	C_α*
Upper Fill	10 to 14	125 to 130	300	-	-	-	-
Lower Fill/Alluvium	4 to 5	110	75	-	-	-	-
Compressible Sub-Layers within Alluvium							
PT	6.5	70	-	1.1	0.23	0.10	0.0250
MH	2.0	96	-	0.55	0.09	0.08	0.0170
PT	7.5	70	-	2.7	0.42	0.68	0.0360

Table 3A – Design Soil Consolidation Parameters							
Layer/Stratum (in order of increasing depth)	Approx. Layer Thickness (feet)	Saturated Unit Weight, γ_S (pcf)	E_s (ksf)	Consolidation Parameters			
				C_c	C_r	C_v (ft^2/day)*	C_α*
CL/ML	6.0	125	-	0.29	0.04	0.30	0.0110
Alluvium	9 to 12.5	130	400	-	-	-	-

Table 3B – Design Soil Strength Parameters						
Layer/Stratum (in order of increasing depth)	Approx. Layer Thk. (feet)	Moist Unit Weight, γ_m (pcf)	Saturated Unit Weight, γ_S (pcf)	Friction Angle, φ' (°)	Cohesion (psf)	E_s (ksf)
Upper Fill	10 to 14	115 to 120	125 to 130	33	0	300
Lower Fill/Alluvium	4 to 5	100	110	28	0	75
Alluvium (Organics and clayey soils)	15 to 22	NA	80	14 (undrained) / 25 (drained)	324 (undrained) / 0 (drained)	NA
Alluvium	9 to 12.5	120	130	35	0	400

ROADWAY SETTLEMENT ANALYSIS AND SIGNIFICANT ISSUES

The presence of highly compressible organic/clayey soils poses a significant challenge to the stability of the roadway embankment and retaining structures. Therefore, it is important to understand the settlement characteristics of these soils prior to evaluation of ground improvement alternatives. A baseline settlement analysis was performed to evaluate the maximum settlement magnitude at the site. The analyses considered elastic settlement of the granular soils, time-dependent consolidation settlement of the soft, cohesive soils, and secondary time-dependent compression (creep) of the cohesive and organic soils. To capture the critical settlement case for analysis, an embankment height of 6.5 feet and a 22-foot-thick compressible soil stratum with sub-layers indicated in Table 3A, was assumed.

Settlement analysis was performed using the Settle3 software by Rocscience. Settle3 is a 3-dimensional program that utilizes a Boussinesq stress distribution model to compute vertical displacements resulting from the placement of the fill load. Construction staging was incorporated into the Settle3 model, and the settlement duration considered construction staging (approximately 34 months), roadway maintenance (typically every 20 years), and service life of the structure (100 years). A graphic showing the Settle3 input is provided in Figure 6 and a summary of the estimated settlement at the center of the proposed roadway embankment without ground improvement is provided in Table 4.

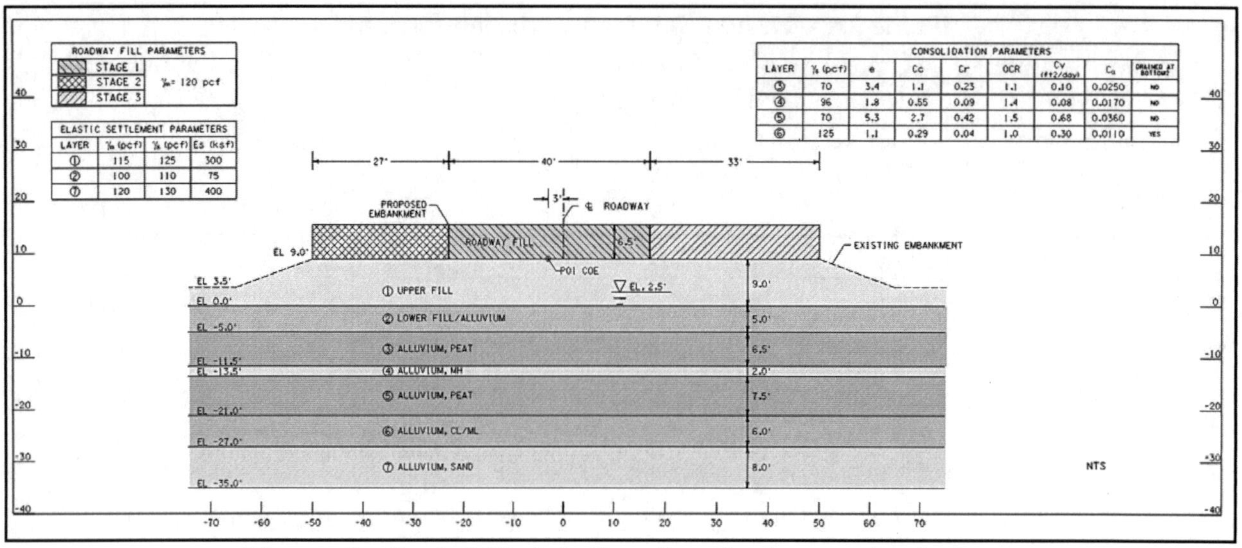

Figure 6 – Settle3 Model - Baseline Settlement Analysis

Table 4 – Baseline Settlement Summary (without Ground Improvement)

Settlement Type	At End of Construction (34 months)	At 20 years	At 100 years
Elastic	1.0	1.0	1.0
Primary Consolidation	6.9	7.6	7.6
Secondary Compression	0.3	1.6	2.7
Total	8.2	10.2	11.3

*Total Settlement = Elastic + Consolidation + Secondary

The estimated total settlement at the end of construction is more than 8 inches. The post-construction settlement is more than 2 inches, which could cause substantial pavement distress and cracking and was not considered tolerable. Furthermore, these settlements could induce large differential settlement (longitudinal and transverse) of the proposed retaining structures, which could affect the long-term serviceability of these structures. Because the south abutment foundations are on piles, these settlements could also induce downdrag on the piles. Consequently, to reduce the total settlements and post-construction settlements to tolerable limits (i.e., post-construction settlement of final pavement surface ≤ 1 inch in 20 years), ground improvement was needed.

GROUND IMPROVEMENT ALTERNATIVES

Many ground improvement alternatives were evaluated to minimize settlement, increase bearing resistance, and transfer embankment loads to a firm bearing stratum including: overexcavation and replacement of highly compressible soils with lightweight fill, preloading with PVDs, load balancing with lightweight fill, and rigid inclusion (RI) column supported embankments (CSEs). For the RIs, augured piles, controlled modulus columns (CMCs), vibro concrete columns (VCCs), and driven piles were considered as potential alternates to transfer the proposed load to a firm bearing stratum. An extensive summary of the four ground improvement options utilized for the project with their advantages and disadvantages can be found in Siddiqui et al. (2017). The following factors influenced ground improvement alternative evaluation:

- Embankment fill heights
- Cost effectiveness and feasibility
- Limits of ground improvement
- Uniformity in ground improvement system
- Impact on existing and proposed utilities
- Constructability and construction staging

Traditionally, when embankments are constructed on soft compressible soils, surcharging (with or without PVDs) helps accelerates the natural consolidation process, squeezing out excess water from the soil matrix to create a firm base. The height of surcharge depends upon the depth of the soft compressible soils below the existing roadway, target consolidation level, and construction staging. After the target consolidation is achieved, the surcharge is removed and the construction of the roadway proceeds.

It should be noted that effectiveness of surcharging is often dependent on the consolidation properties of the compressible soils. The soft compressible soils encountered on this project have variable consolidation rates which could increase the duration of the surcharge placement thus impacting the overall construction schedule. Conversely, CSEs do not require waiting periods and can be installed with minimal impact to the construction schedule. While surcharging offers a solution for soft soil construction, CSEs provide a means to transfer the roadway embankment load directly to firmer bearing strata. The load is either distributed by a flexible or rigid load transfer platform, which typically range from 2 to 4-feet-thick. Additional discussion regarding the load transfer platforms is provided later in this paper.

From a cost standpoint, CSEs are more expensive compared with the other ground improvement alternatives indicated above. However, the advantages of the CSEs outweigh its cost impact including: reduced construction time, reduced impact on existing utilities, mitigation of long-term settlement, no downdrag on the adjacent bridge foundations, and reduced long-term maintenance requirements for the retaining walls. When considering these advantages, CSEs offered a value-added ground improvement approach for this project; and therefore, were determined to be the most feasible ground improvement method for embankment fill heights greater than 5 ft. For embankments with a fill height of 5 ft or less, load balancing with lightweight aggregate was the most viable option.

RIGID INCLUSION (RI) COLUMN SUPPORTED EMBANKMENTS (CSE)

RIs are small-diameter concrete columns designed to transfer embankment load through soft compressible soil layer(s) to a firm foundation, mitigating settlement concerns and increasing bearing capacity of the subsurface soils. Selection of the type of column used for the CSE typically depends on project specific needs, design loads, constructability of the columns, cost, etc. The CSE system is designed to effectively transfer the embankment load to the columns and prevent punching of the columns through the embankment fill causing differential settlement at the surface of the embankment. Therefore, center-to-center spacing of the columns is a critical aspect of design, if the columns are placed close enough together, soil arching will occur, and the full embankment load will be transferred to the columns.

For this project, unreinforced concrete columns were selected. This type of column is typically formed using a hollow displacement tool rotated and drilled into the ground at a controlled penetration rate to a predetermined tip elevation. At the tip elevation, the drill string is withdrawn from the ground at a constant speed and during withdrawal concrete is pumped under constant pressure through the hollow stem of the drill string to fill the void left from the drill. Typical sections showing various elements of the CSE system and the effect of soil arching are shown in Figures7 and 8, respectively.

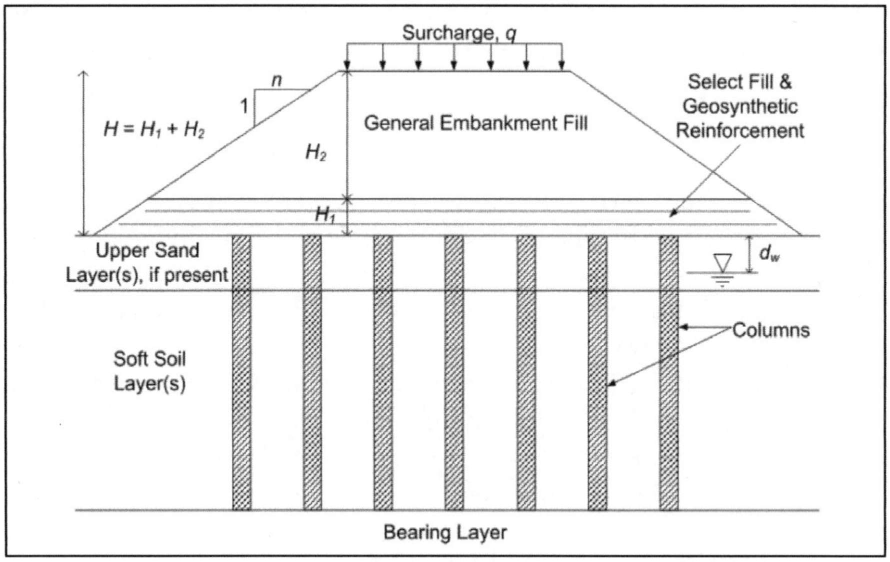

Figure 7 – CSE with Flexible Load Transfer Platform
Image Reference – FHWA - NH-16-028 GEC 013 – Volume II

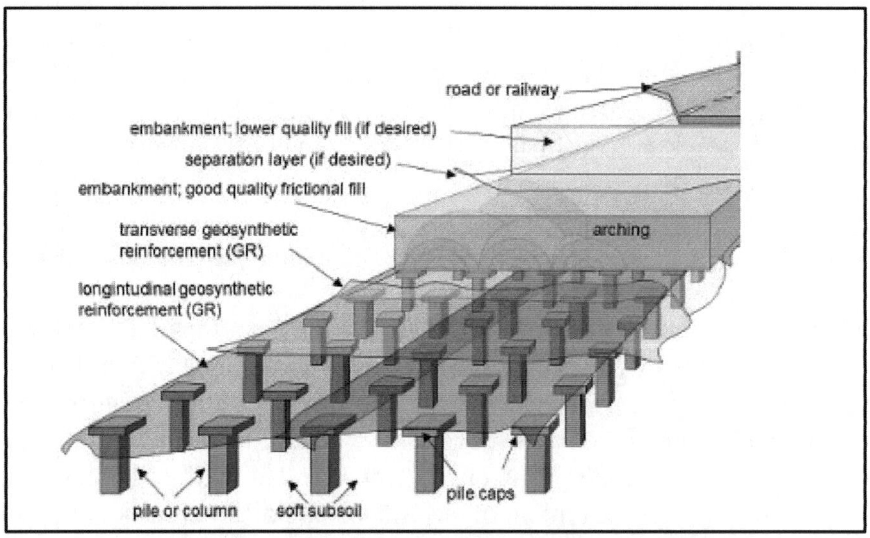

Figure 8 – Soil Arching Schematic for CSE
Image Reference – https://publicwiki.deltares.nl/display/PE/Piled+embankments

Performance Criteria

Performance approach specifications allows for contractor flexibility when selecting feasible and cost-effective ground improvement methods which satisfy specified performance criteria by the Project Geotechnical Engineer. In addition, this approach allocates most of the risk to the Contractor. Below are the main performance criteria for the CSE system for this project categorized as general requirements, settlement requirement, and lateral clearance requirements.

General Requirements

- Axial resistance factor for RIs = 0.45
- Resistance factor for global stability of embankments and PM walls supported on RIs = 0.75
- Minimum depth of RIs = 3 times the vertical element diameter or 4 feet, whichever is greater, into the underlying dense to very dense alluvium stratum
- Minimum diameter of RIs = 18 inches
- Maximum center-to-center spacing of RIs = 7 feet
- Minimum thickness of load transfer platform = 3 feet
- Minimum layers of evenly spaced reinforcement in the load transfer platform = 3 (biaxial)
- $H_{embankment} > H_{critical\ embankment\ height}$

Settlement Requirements

- Total settlement at the base of the embankments or PM walls ≤ 2.5 inch
- Post-construction settlement ≤ 1 inch in 20 years
- Differential settlement per 100 feet of the pavement subgrade surface ≤ 1.0 in. at completion of the full load condition
- Embankments and PM walls supported on rigid inclusions cannot cause any additional loading on existing/proposed foundations or utilities
- CSEs cannot cause dimpling of the final pavement

Lateral Clearance Requirements

- Extend the edge of LTP a minimum of 1 foot beyond the outer edge of RIs supporting the LTP
- Maintain a minimum clear distance of 3 feet between the vertical elements and existing or proposed substructures, piles or drilled shafts

DESIGN METHODOLOGY

RI Configuration

The design of the RIs is associated with the tributary area of soil surrounding each column. The load from the embankment and retaining walls is assumed to be carried in its entirety by the columns.

The key to RI design lies in finding an optimal balance between the soil's load bearing capacity and the RI's contribution, which involves finding the right combination of column diameter, center-to-center spacing, or area replacement ration (ARR). The typical range for ARR is approximately 3.5 to 10%, however it could be as low as 2.5% depending on project requirements. The design of RIs requires a rigorous analysis that accounts for both, settlement and lateral movements of the structure(s). RI designers typically use numerical methods, i.e., finite element or finite difference to perform RI design. They often incorporate relatively high factors of safety in their models to prevent bearing capacity failures and overall stability issues, for both short and long-term conditions.

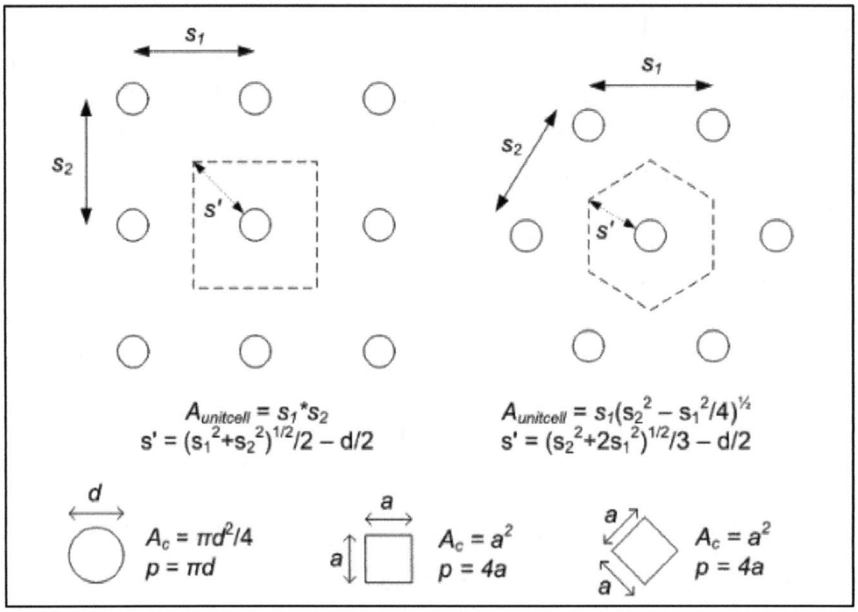

Figure 9 – Column Layout
Image Reference – FHWA - NH-16-028 GEC 013 – Volume II (2017)

A general industry standard to establish RI configuration is shown in Figure 9. For a square column pattern, the effective diameter (D_e) is equal to 1.13 times the center-to-center column spacing. For a triangular column pattern, the effective diameter is equal to 1.05 times the center-to-center column spacing, with typical center to center spacing ranging from 5 to 10 feet).

Load Transfer Platform (LTP)

Unlike traditional piling, which relies solely on deep foundation elements to transfer the entire embankment load axially to a firm bearing layer, column supported embankments employ a more integrated approach. In this system, a geosynthetically reinforced platform (typically 3 feet thick), called the LTP, acts as a stress distribution element, transferring most of the load from the embankment onto the RIs by soil arching within the embankment, tension developed in the LTP, and negative skin friction between the settling soil and the columns.

For this site, the CSE designer used a minimum 3-foot-thick load transfer platform with 3 layers of biaxial geogrids, Stratabase SGB30, each with an allowable tensile strength of 822 lb/ft. This results in a total allowable tensile capacity of 2,466 lb/ft. A schematic showing the LTP is shown in Figure 10.

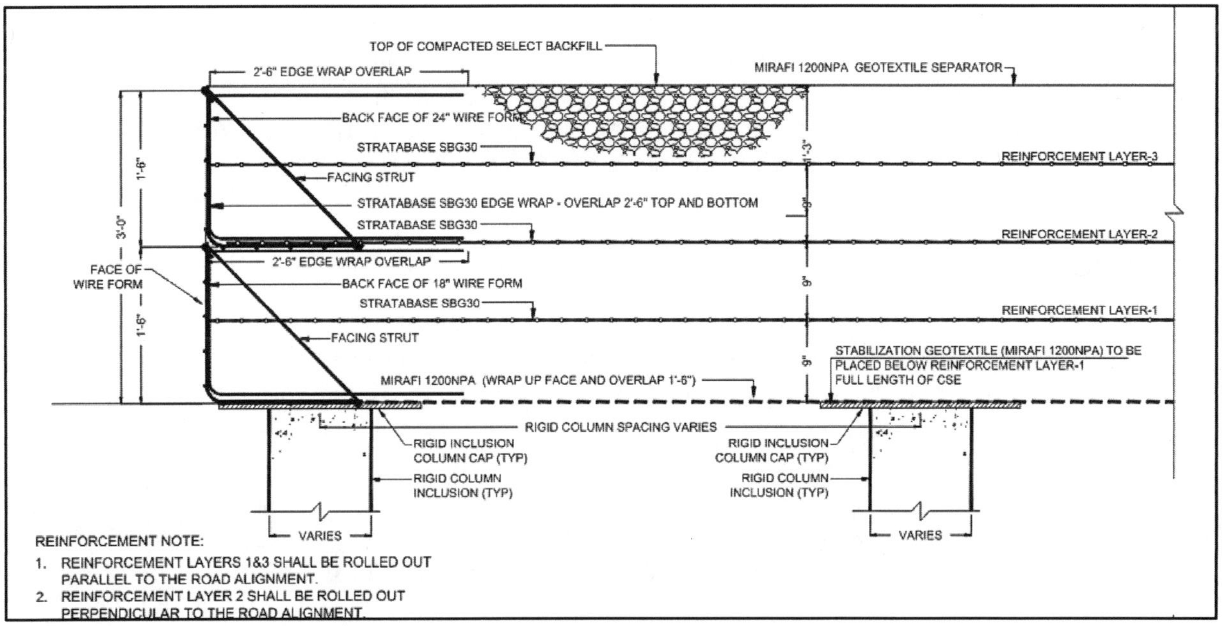

Figure 10 – Load Transfer Platform Section
Image Reference – D6-0 SR 420 Wanamaker Avenue over Darby Creek CSE Design, The Collin Group (2024)

Design of CSEs

The CSE design was performed by the ground improvement contractor (GeoConstructors, Inc.) and its design engineer (The Collin Group, Inc.) to meet the requirements of the performance criteria.

During the design phase, a preliminary design was performed by the design consultant using the FHWA methodology (Chapter 6, Publication No. NHI-16-028, to establish preliminary vertical element center-to-center spacing, critical embankment height, and design vertical load on the vertical elements. For analysis, maximum and minimum fill heights (accounting for depth of load transfer platforms), 9 and 6.5 feet, respectively, were considered. Since LTPs were recommended for load transfer, the design fill heights considered the depth of the LTP (i.e. 3ft). The 9-foot fill section was analyzed using 18-inch column diameter with center-to-center spacing of 7 feet and the 6.5-foot fill section was analyzed using 12-inch column diameter with center-to-center spacing of 5 feet used. These analyses required some iterations of column diameter and spacing to satisfy the performance requirements. These calculations provided a conceptual-level design of the CSE system to validate the feasibility of using a CSE system.

FHWA (Publication No. NHI-16-028, Chapter 6) provides a design methodology to establish preliminary vertical element center-to-center spacing, critical embankment height, and design vertical load on the vertical elements. Specialty geotechnical contractors have differing approaches to the design and construction of RI-column supported embankments and each will typically utilize their own proprietary finite element methods or other numerical analysis techniques to design the RI-column supported embankments. Typically, these methods have been verified and implemented over time through contract specifications, testing and successful project experience.

For this project, the design engineer used an in-house spreadsheet called GeoGridBridge 3 (GGB3). GGB3 is a Microsoft Excel spreadsheet that applies the Load-Displacement Compatibility (LDC) method for analyzing and designing RI-column supported embankments. GGB3 solves simultaneous nonlinear equations to satisfy vertical equilibrium within the embankment for the long-term, drained, condition. The results from the GGB3 analysis typically include total settlement, critical embankment height check, and maximum column load. The geogrid reinforcement in the LTP is also evaluated within the GGB3 spreadsheet.

One critical aspect of this project was that to limit dimpling of the final pavement surface, which is generally controlled by the critical embankment height, i.e. H_{crit}. Per FHWA, H_{crit}, is defined as "*the embankment height above which differential settlements at the base of the CSE do not produce measurable differential settlement at the embankment surface.*" The critical height is a function of the column spacing and the diameter of the column/cap (McGuire et al. 2012). The effect of dimpling tends to get magnified for RIs spaced too far apart with shallow embankment heights and relative thin LTPs. Therefore, it is critical to consider these aspects during design to ensure dimpling does not impact the rideability and long-term performance of the roadway.

The design process involved several iterations of the RI configuration to support the elevated roadway embankment and the two-tier retaining walls. The design process also involved several review meetings to ensure their expectations were met and the design satisfied the required performance criteria.

Performance criteria for this project required minimum 18-inch diameter columns with maximum 7-foot center-to-center spacing. During the initial review of the CSE design, the Project Geotechnical Engineer expressed concerns related to lateral spread/squeeze (discussed further in the paper) on utilities and adjacent structure foundations. To mitigate the concerns related to lateral squeeze, the CSE designer provided a solution involving a slightly smaller diameter column size, i.e. 16-inch, and a relatively tight grid spacing ranging between 4.5-ft^2 and 6-ft^2. In addition, the CSE designer recommended that a test section of production columns be installed under Wall B2 where the columns would be spaced 4.5 feet apart (i.e., closest spacing for settlement) and install inclinometer and piezometer to assess lateral squeeze and the buildup of pore pressure from the column installation. Meeting was held to discuss this approach and to further understand the testing and documentation requirements for the test area. During the meeting, the CSE designer came up with another alternative solution to mitigate the concerns related to lateral squeeze. This alternative solution involved further reducing the column diameter, i.e. 12 inches, and increase the column spacing to maximum 7.5 feet. In addition, the CSE designer proposed a pile cap (option provided in the performance specification) to increase the effective diameter of the column and help distribute the load from the embankment fill more evenly across the underlying piles. The pile cap grouped with the LTP would provide a stable anchor point for the geosynthetic reinforcement to interact with and distribute the embankment load more effectively across the entire column system. Based on the various alternatives provided by the CSE design, it was agreed upon to use 12-inch diameter columns with an effective square grid spacing ranging between 6-ft^2 and 7.5-ft^2.

The CSE designer completed the final design in July 2024. Based on the final design, the total load on the columns ranges between 32 kips and 135 kips. In general, the geotechnical factor of safety for the RIs ranges between 1.5 and 7.1, and the structural factor of safety for the RIs ranges between 1.6 and 5.4. The cap on the column consists of either a 27-inch x 27-inch x 1-inch thick or a 30-

inch x 30-inch x 1-inch-thick galvanized steel plate. The steep plate will be connected to the column via a galvanized steel bar, i.e. #5 bar (0.625 inches nominal diameter) with an embedment of 36 inches. A typical section of column supported embankment is provided in Figure 11 and RI details are provided in Figure 12 .

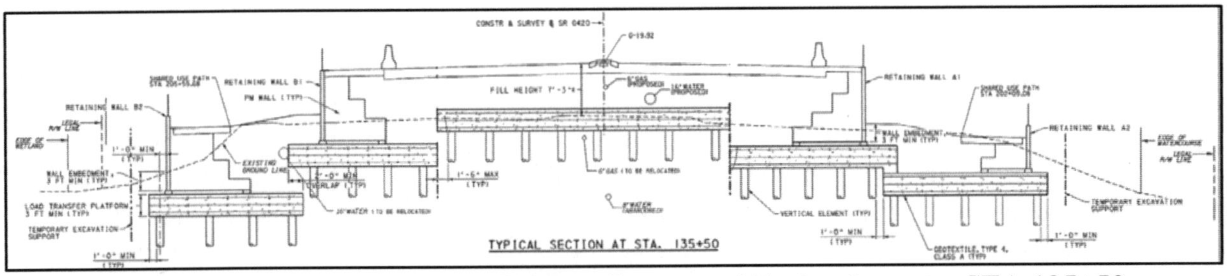

Figure 11 – Typical Section – Column Supported Embankment – STA 135+50 (Max Fill Height Section)

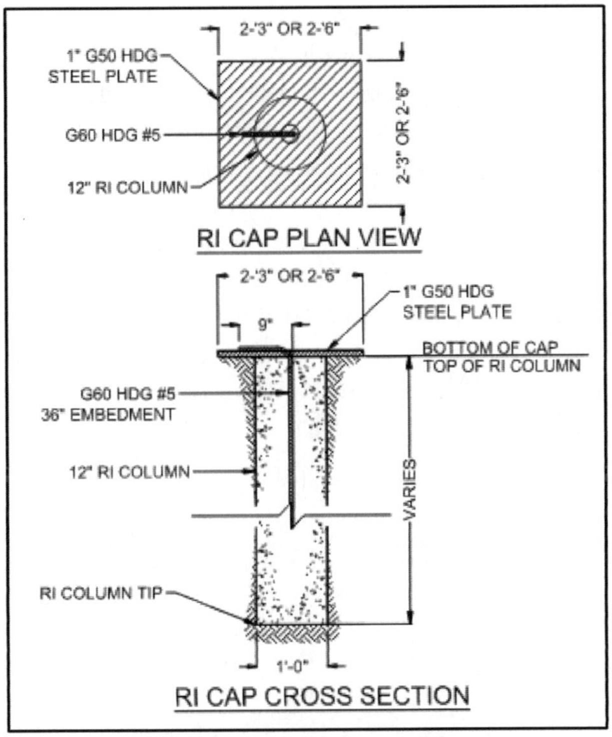

Figure 12 – RI Typical Details
Image Reference – D6-0 SR 420 Wanamaker Avenue over Darby Creek CSE Design, The Collin Group (2024)

Lateral Spreading and Global Stability

Embankments built on soft soils can be susceptible to lateral spreading, where the embankment squeezes laterally due to internal horizontal stress and ground movement below the LTP, often magnified if soft soils are present at the surface. A typical schematic of lateral spreading behavior is provided in Figure 13. This site does not have this condition; therefore, lateral spreading is not a real concern for this site. Global stability analysis is another mode of failure that needs to be evaluated for RI-column supported embankments.

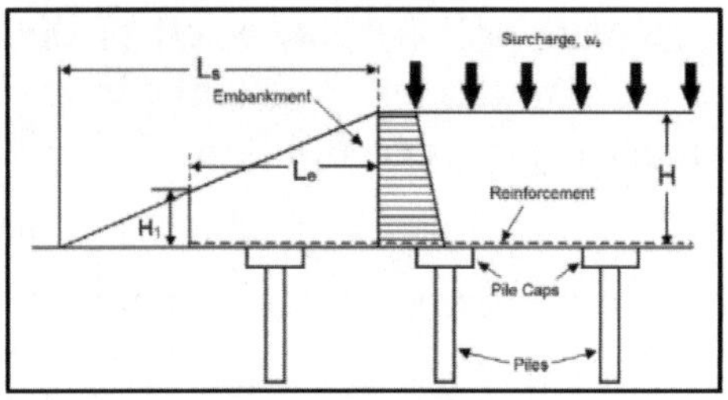

Figure 13 – Typical Schematic of Lateral Spreading Behavior
Image Reference – FHWA - NH-16-028 GEC 013 – Volume II (2017)

Limit equilibrium analysis is not well suited to analyze a complex system like RI-column supported embankments because it does not consider strain (i.e., deformation) in the components of the system but rather the ultimate strength at different magnitudes of strain. However, 3D numerical modeling (i.e., FLAC 3D, PLAXIS, etc.) considers both the stress and strain in each component of the system which gives a much better indication of performance. To capture the modes of failure, the RI designer for this project used FLAC 3D analysis to consider both stresses and strains in each component of the RI-column supported embankment system.

For comprehensive validation of the design, it is necessary to evaluate the global stability of the system. FLAC 3D analysis provides magnitudes of lateral deformations; however, does not explicitly provide the factor of safety (FOS) of the system. During the design phase, a global stability analysis was performed for the critical condition, i.e., maximum fill height condition with two-tier retaining walls, using the Slide2 computer program by Rocscience, Inc. The analyses assumed a traffic live-load surcharge of 360 psf at the roadway level. Both the simplified Bishop method and a non-circular failure analysis were evaluated for drained and undrained soil conditions. Because the project is located with the 100-year floodplain, a rapid drawdown condition from a 100-year flood was also evaluated. The analysis indicated a FOS greater than 1.5 (required FOS = 1.3) for the static case and a FOS greater than 1.3 (required FOS = 1.1) for the rapid drawdown condition. The Slide output for the critical design case for static and rapid drawdown condition is shown in Figure 14 and 15.

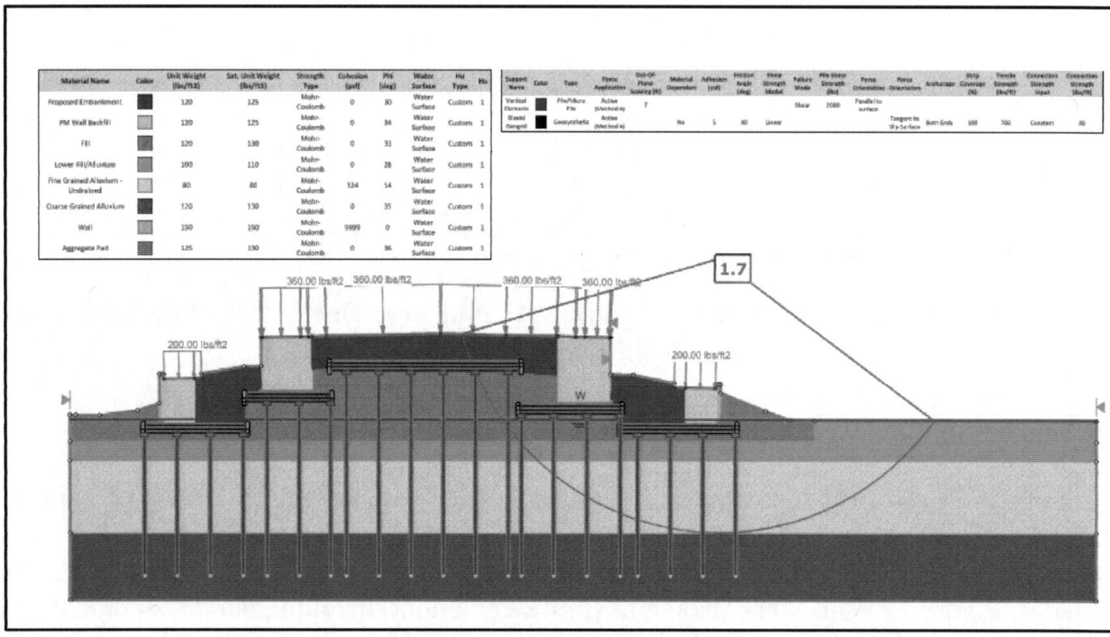

Figure 13 – Global Stability STA 135+50 – Static Case

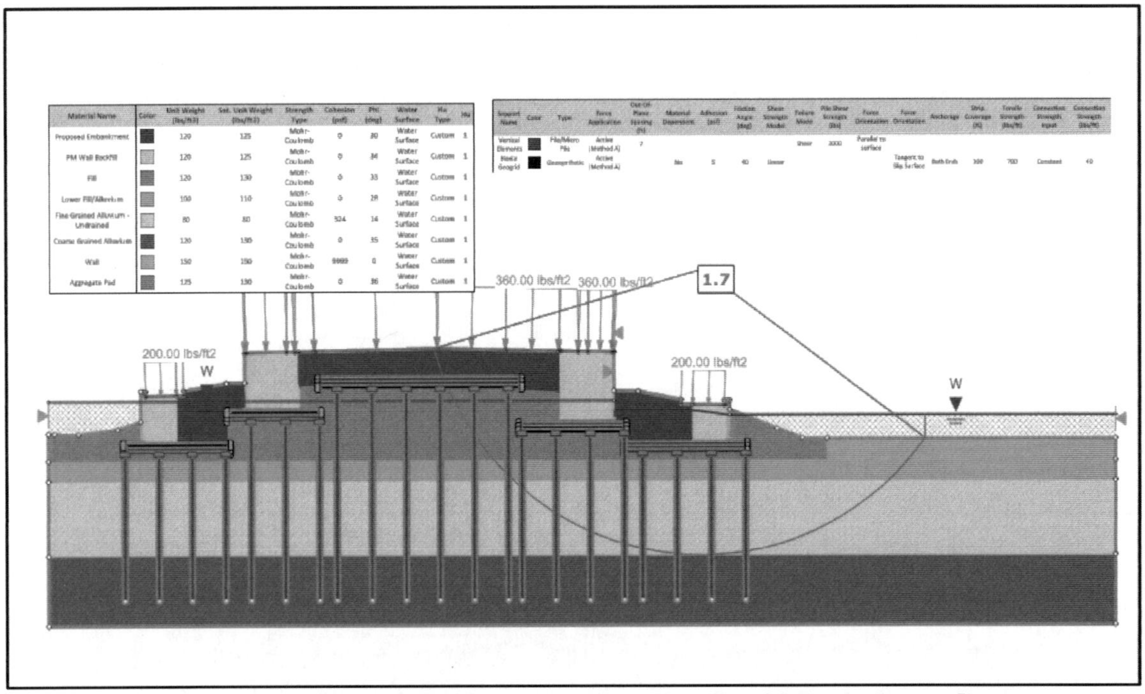

Figure 14 – Global Stability STA 135+50 – Rapid Drawdown Case

The CSE designer also used the computer program Slide to evaluate global stability; however, the effect of the RIs was ignored in their model. Their model showed FOS > 1.3 for the static case, without considering RIs.

LOAD BALANCING

As indicated previously, a critical aspect for RI design on this project was to limit dimpling of the final pavement surface, which is generally controlled by critical the embankment height design criteria (H_{crit}). To mitigate dimpling, the proposed embankment height must be greater than the critical embankment height. Therefore, in areas with less than 5 ft. of new embankment, load balancing options, including lightweight fill, were considered.

To minimize settlements and decrease the net load applied to the underlying highly compressible soils, three types of lightweight fill were considered: low density cellular concrete (LDCC), lightweight expanded shale aggregate, and ultra-lightweight foamed glass aggregate (UL-FGA®). LDCC consists of neat cement slurry mixed with a synthetic foaming agent, resulting in a unit weight as low as 20 pcf and up to a maximum of 120 pcf, with compressive strengths ranging from 20 to 3000 psi, depending on the unit weight. Lightweight aggregate, such as expanded shale aggregate, has an in-place unit weight ranging from about 50 to 60 pcf. Ultra-lightweight foamed glass aggregate has an in-place unit weight ranging from about 18 to 23.5 pcf.

One significant advantage of UL-FGA® is the low unit weight of the material compared with the other lightweight fill options. Based on the proposed embankment height, i.e., less than equal to 4 feet, a zero-net increase in the effective stress could be achieved with use of UL-FGA®. However, one disadvantage of this material is its buoyancy, the density of this material is less than water and therefore it can cause uplift forces in a submerged condition. The site is located within 100-year flood plain; and therefore, buoyancy had to be accounted for in the design. During the design phase, the design consultant reached out to the UL-FGA® manufacturer to understand how buoyancy effects could be negated. Per the manufacturer's recommendations, a minimum 18-inch soil cover was used to weigh down the UL-FGA® and provide a sufficient factor of safety. A typical section showing the UL-FGA® section along with transition details is provided in Figure 15 and 16.

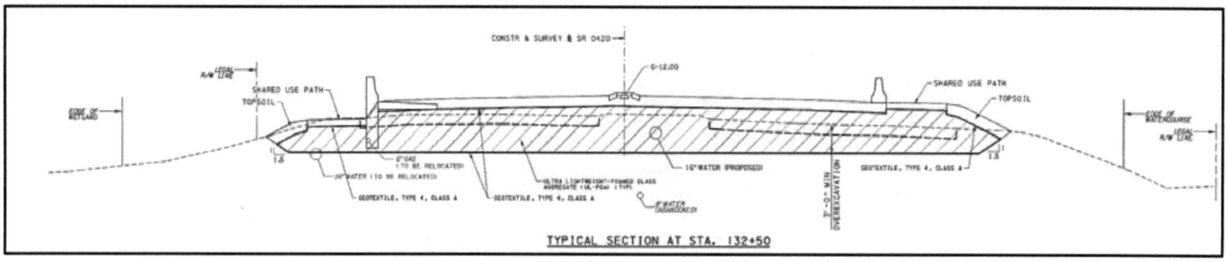

Figure 15 – Typical UL-FGA Section

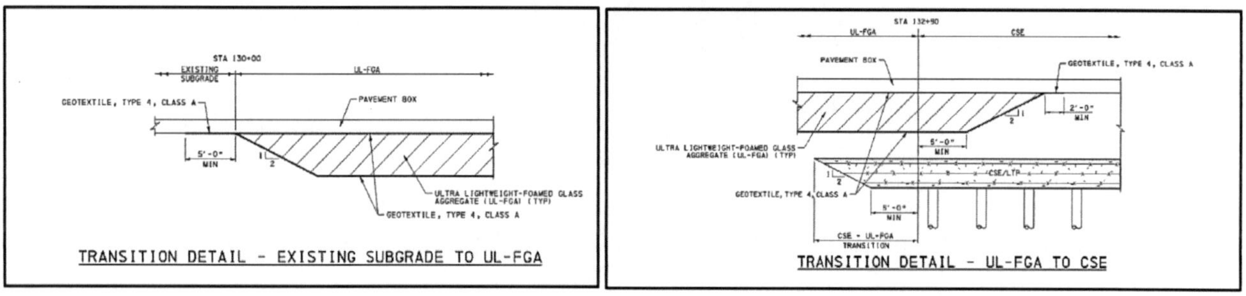

Figure 16 – Typical UL-FGA Section – Transition Detail

INSTRUMENTATION AND MONITORING

An important aspect of the QA/QC for CSEs is a requirement for the contractor to demonstrate the effectiveness of the system in accordance with the project performance criteria. Generally, a field instrumentation and monitoring program is required to measure changes in pore water pressures in the foundation soils to evaluate the predicted settlement magnitudes during and post-construction. For this project, an extensive instrumentation and monitoring program including standpipe piezometers, inclinometers, settlement monitoring points, and settlement platforms was proposed. Figure 17 shows the proposed instrumentation types and locations required to validate the performance of the CSEs. To monitor performance of CSEs during embankment fill placement, a total of 12 pairs of piezometer/ settlement plates were recommended. In addition, 4 inclinometers were recommended to monitor the lateral movements during fill placement. To monitor the performance of UL-FGA® fill placement, a total of 4 settlement plates were recommended for this project.

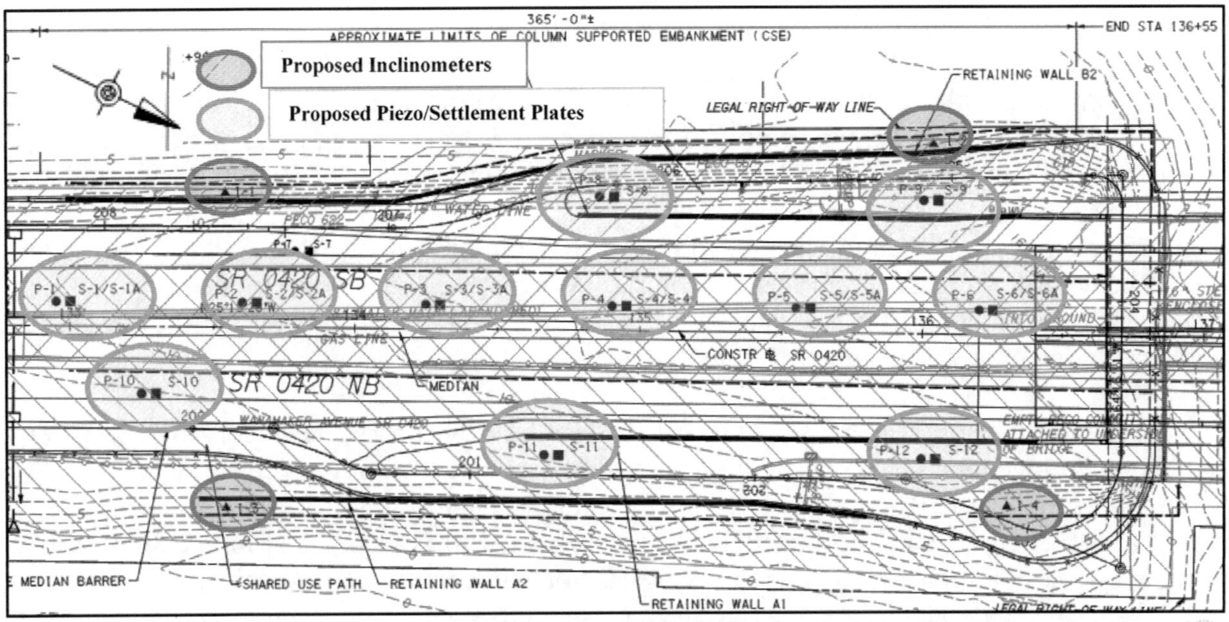

Figure 17 – Proposed Instrumentation within CSE footprint – South Approach STA 132+90 to STA 136+55

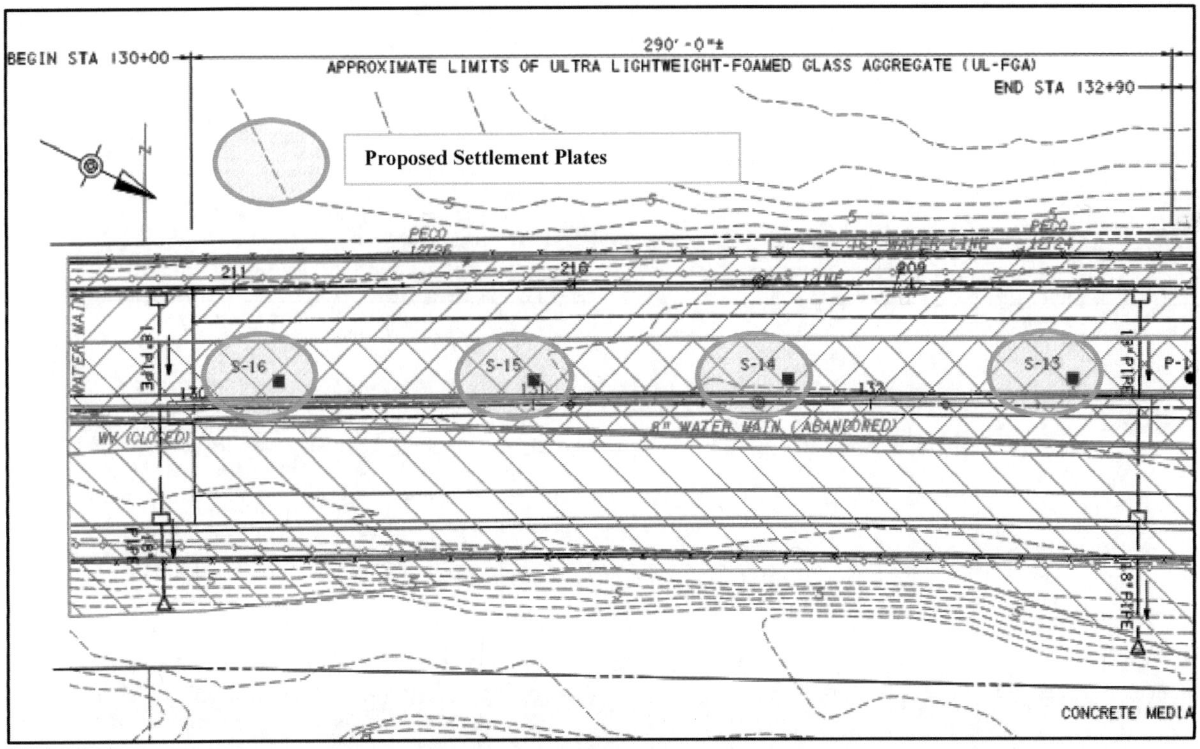

Figure 18 – Proposed Instrumentation within UL-FGA® Footprint - STA 130+00 to STA 132+90

CONCLUSIONS AND NEXT STEPS

This project involved the construction of a roadway embankment with two-tier retaining walls on soft, fine-grained cohesive soils and organic deposits. Various ground improvement techniques were considered, however, column supported embankment with rigid inclusions provided an efficient and innovative solution to improve the subsurface conditions and provide the desired balance between initial construction costs, risk, and future maintenance. Load balancing with ultra-lightweight foamed glass aggregate proved to be more cost-effective in areas with minimal fill heights and without two-tier walls. Performance specifications, developed collaboratively with the owner and engineer, further enhanced project value and efficiency. The proposed settlement monitoring and instrumentation program will allow for monitoring settlements during construction and validate the project's performance criteria. A strong collaboration between the design consultant, general contractor, specialty contractor, and the owner is crucial for the successful implementation of any ground improvement system.

The project was Let on March 23, 2023and is currently in the first stage of construction where southbound bridge foundation construction is underway. Once the bridge is constructed, the CSE work will begin in fall 2024 and the UL-FGA® installation will begin in the spring of 2025. The first stage of CSE and UL-FGA® installation is anticipated to be completed in late spring/early summer of 2025. Then the second stage of construction will commence where the remaining CSE and UL-FGA® will be constructed on the northbound side. Overall project completion is anticipated to be summer 2026.

REFERENCES

Cacciola, Daniel V. 2018. "Rigid Inclusion Supported Embankments for New Jersey Turnpike Interchange 14A: A Case Study." Paper presented at the *International Foundations Congress and Equipment Expo, Orlando, FL, March 5-10, 2018.*

Collin, J.G. (2004). " Column Supported Embankment Design Considerations." Proc. University of Minnesota 52nd Annual geotechnical Engineering Conference, Minneapolis, Minnesota.

Collin, J.G.; Han, J. and J. Huang (2005), "Geosynthetic-reinforced column-support embankment design guidelines." Proceedings, the North America Geosynthetic Society Conference.

Elias, V., Lukas, R., Bruce, D., Collin, J.G., and Berg, R.R. (2004). "Ground Improvement Methods Reference Manual". Federal Highway Administration FHWA NHI-04-001.

Gannett Fleming, Inc. (2020), *S.R. 0420, Section DWD over Darby Creek, Phase II and II Environmental Site Assessment (ESA) Findings Report.*

Gannett Fleming, Inc. (2018), *S.R. 0420, Section DWD over Darby Creek, Subsurface Exploration and Planning Submission.*

Gannett Fleming, Inc. (2022), *S.R. 0420, Section DWD over Darby Creek, Structure Foundation Report, Retaining Walls A1, A2, B1 and B2.*

U.S. Department of Transportation Federal Highway Administration, Publication No. FHWA NHI-16-028, (2017), NHI Course No. 132034 *Ground Modification Methods Reference Manual – Volume II.*

Climate Change Effects on Slope Stability in the Northeast United States

David J. Scarpato, P.E.
Scarptec, Inc.
P.O. Box 326
Monument Beach, MA 02553
(603) 361-0397
dave@scarptec.com

Jay R. Smerekanicz, P.G., C.P.G.
Scarptec, Inc.
P.O. Box 326
Monument Beach, MA 02553
(603) 493-6274
jay@scarptec.com

Peter C. Ingraham, P.E.
Scarptec, Inc.
P.O. Box 326
Monument Beach, MA 02553
(603) 785-0262
pete@scarptec.com

Ethan J. Thomas
AOT Geologist III | Highway Division | Construction & Materials Bureau
Geotechnical Engineering Section | Geology Unit
Vermont Agency of Transportation
2178 Airport Road | Dill Building Unit B
Berlin, VT 05641-8641-8628
(802) 595-6752
Ethan.Thomas@vermont.gov

Prepared for the 73rd Highway Geology Symposium, September 9-12, 2024

Acknowledgements

The author(s) would like to thank the individuals/entities for their contributions in the work described:

Krystle Pelham, NHDOT

Disclaimer

Statements and views presented in this paper are strictly those of the author(s), and do not necessarily reflect positions held by their affiliations, the Highway Geology Symposium (HGS), or others acknowledged above. The mention of trade names for commercial products does not imply the approval or endorsement by HGS.

ABSTRACT

While climate change may be a debatable topic among politicians, changing atmospheric river courses, rainfall concentration and intensity, longer freeze-thaw cycles, and attendant landslides, flooding and rock slope failures are issues we as engineering geologists must address. The effects of the increasing intensity of storms on transportation infrastructure in Glenwood Canyon, British Columbia, and the Olympic Peninsula have been the subjects of past HGS papers. New England has recently faced adverse flooding and temperature fluctuation conditions in Vermont and New Hampshire along with record storm surges along the New England coastline. This changing weather pattern causes stress on capital spending and asset management programs, interrupting regularly scheduled projects as resources need to be reallocated for emergency response. Storm data for design and asset management programs must be updated for these new conditions to keep transportation infrastructure adequately resilient. We highlight some of these recent challenges transportation agencies in New England face regarding rainfall intensity changes (e.g., five 100+ year storm events in two months), runoff drainage structures that are undersized for current conditions, longer and earlier freeze-thaw seasons affecting rock slope maintenance/repair programs, and the loss of roadway segments and bridge structures, all causing major impacts to transportation.

INTRODUCTION

The past two decades have seen many improvements in asset management in our industry. Monitoring of rock and soil slopes has improved from inventory development to high-tech change detection (e.g., LiDAR), remote sensing and even satellite tracking of slope movements (e.g., InSAR). As we have become more adept at tracking changes in slopes and embankments, we have also been able to observe changes in key design factors such as storm intensity, duration and recurrence intervals. These changes, combined with aging infrastructure, ongoing weathering, degradation of infrastructure, required maintenance, and replacement projects, are sorely taxing department of transportation budgets. A single slope failure along a critical transportation corridor can shift priority from planned maintenance or capital improvement to emergency repairs, delaying planned infrastructure investment, and requiring reliance and increasing pressure on existing assets identified as needing replacement/renewal.

It is frequently said that "water is the root of all evil in geotechnics", and perhaps the most obvious change in our weather patterns in the northeast has been the changes in intensity and concentration of rainfall. In the late 1990's we observed five 100-year rainfall events in the span of six weeks. The events were not widespread, but concentrated and caused a few washouts locally and the failure of a landfill cap in Vermont. Nonetheless, it begged the question – were they really 100-year storm events or a shorter interval event? In any case, we were observing a paradigm shift, and Hurricane Irene in 2011 sealed the deal when the eye of the hurricane (later downgraded to a tropical storm) did not track out to sea south of New York City, but instead tracked due north through Connecticut, western Massachusetts and into Vermont. The toll in Vermont was the loss of nearly 200 miles of roadway and over 100 bridges.

Similarly, in July 2023 a 3-day intense rainfall event with a rainfall signature very much like the Irene storm (Figure 1), caused surprising similar extensive flooding in Vermont. The rainfall triggered debris flows, rockfalls, eroded roadway and railway embankments, damaged transportation and other structures, and caused flooding in the state capital of Montpelier, along with other central Vermont cities and towns. By all accounts thus far, 2024 also looks to be a very wet year.

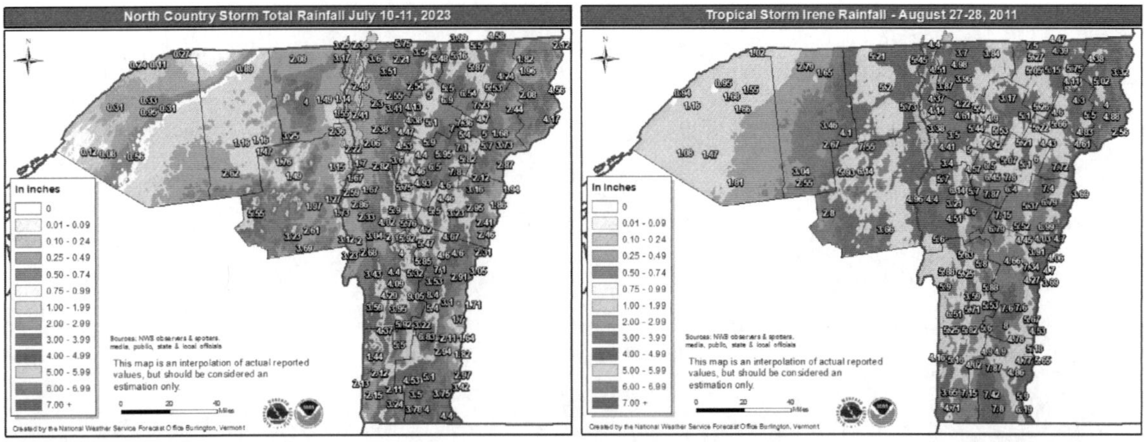

Figure 1. Comparison of 10-11 July 2023 rainfall event with Tropical Storm Irene 27-28 August 2011 (Banacos, 2023).

Aging infrastructure certainly plays a part in the susceptibility of our transportation corridors to intense weather events. With many of our interstate corridors beginning construction in the late 1950's, considering rock slopes as "assets" within our highway systems has become inevitable. When asked how much longer we will need to track rock slopes, Carter and Miller (1996) provide some guidance. Ultimately the answer is "forever, depending on the geology", but Carter and Miller's plot (Figure 2) of abandoned mine opening stability shows a plot of failures with poorly designed openings failing in the first 30 years and lithology-controlled degradation failures following in a 40- to 70-year timeframe.

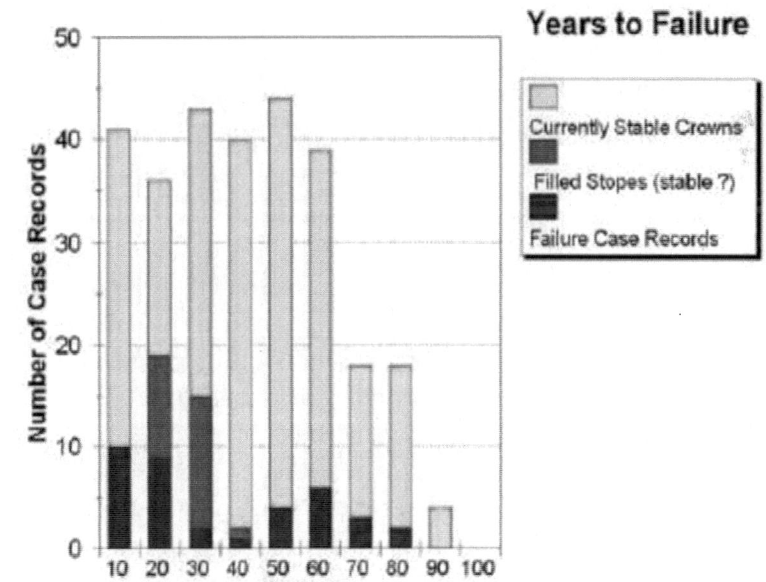

Figure 2 – Time dependency of failure based on over 400 case records and nearly 50 failures in Golder-Canmet Crown Pillar database (Carter & Miller 1996).

CASE NO. 1 – WARREN, NEW HAMPSHIRE FLOODING, FALL 2017

After an already wet fall, Tropical Storm Philippe dropped approx. 2- to locally over 5-inches of rain in northern New England in October 2017, resulting in numerous landslides across the region. On 30 October 2017, the Baker River gauge station in Rumney, NH recorded a peak stream discharge of 18,300 cubic feet per second and gage height of almost 15 feet. Approximately 15-miles upstream of Rumney, the small Town of Warren, NH suffered a 300-foot long by 20-foot-high section of soil slope fail due to repeated scour action. The scour was focused along the lower half of the coarse-grained soil slope (Figure 3), resulting in an over-steepened upper section, which subsequently failed.

Figure 3. Distribution of exposed slope materials subject to failure along NH S.R. 25 undermining roadway stability in fall of 2017.

The upper reaches of the failure scarp impacted the edge of State Route (S.R.) 118 which connects S.R. 25 with U.S. Route 93 near Lincoln in the White Mountains Region of New Hampshire (Figure 4). A private residence was also impacted by the failure event, resulting in demolition of the house structure. Multiple smaller failures had taken place along this stretch of slope over the last two decades; however, the October 2017 event was considered significant due to its impact on the stability of S.R. 118 and its importance as a regional roadway.

Figure 4. Persistent flooding damage to NH S.R. 25 undermining roadway stability in fall of 2017.

In lieu of more traditional scour mitigation alternatives, The New Hampshire Department of Transportation (NHDOT) considered an anchored riprap wedge concept for enhanced scour resistance along the exposed slope face. The selected general contractor, slope stabilization subcontractor and their subconsultant geotechnical designer (three of the authors) modified the anchored riprap wedge concept by phasing the fall 2018 work in lifts, wrapping each lift in G65/4 mm Tecco Mesh and by installing predrilled vertical and inclined Titan hollow bar anchors that were terminated into bedrock (Figures 5 and 6). The anchored riprap wedge concept was a novel approach for slope stabilization in scour and ice floe prone streambank environments.

Figure 5. Construction of anchored riprap lifts for streambank stabilization fall of 2018.

Figure 6. Final constructed anchored riprap cells in spring of 2019.

CASE NO. 2 – JULY 9 – 11, 2023 FLOODING, NORTHERN NEW ENGLAND

In mid-July 2023, catastrophic flash flooding and river flooding occurred in Vermont and northern New England, heavily affecting areas of mountainous terrain by causing rockslides, mudslides and debris flows. Total rainfall amounts from the 3-day storm ranged from 3- to over 9-inches, with the highest amounts along the spine of the Green Mountains. Damage estimates from just this one storm, including those to Vermont's infrastructure system, exceed \$300M. Extensive flooding wreaked havoc to towns such as Montpelier, Barre and Ludlow, cutting residences and business off. The damage from this storm rivaled, and in some areas, even exceeded the rainfall amounts generated by Tropical Storm Irene in 2011, as shown by the comparison of total rainfall between these two storm systems (Figure 1; Banacos, 2023).

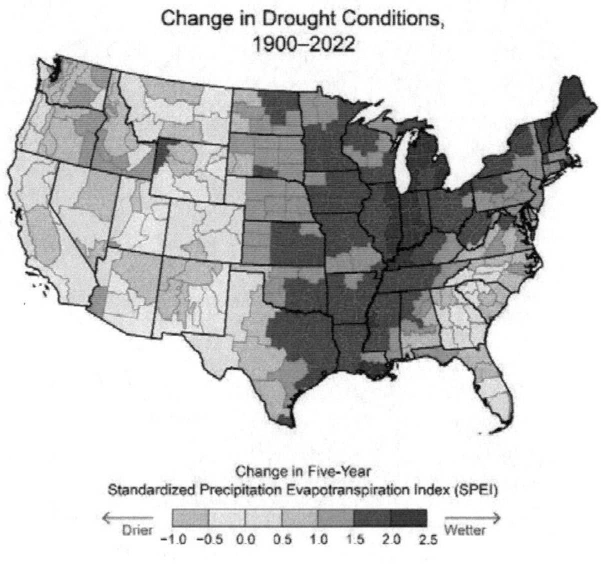

Figure 7. Wetter conditions occurring in central and northeast U.S. (Crimmins et al., 2023).

A change toward wetter conditions in the Northeast from drought studies of 1900 to 2022 indicate that more precipitation is occurring (Figure 7). As more water accumulates, pore pressures (i.e., groundwater levels) increase and negatively affect slope stability. The July 2023 storm was preceded by higher-than-average rainfall, which lead to reduced capacity for absorption and dissipation of precipitation, and exacerbated stability conditions. Not only is the amount of precipitation increasing, but the intensity of that precipitation is also increasing. Days with 2+, 3+, 4+ and 5+ inches/day appear to have increased by 49%, 62%, 84% and 103%, respectively, since 1958 (Figure 8).

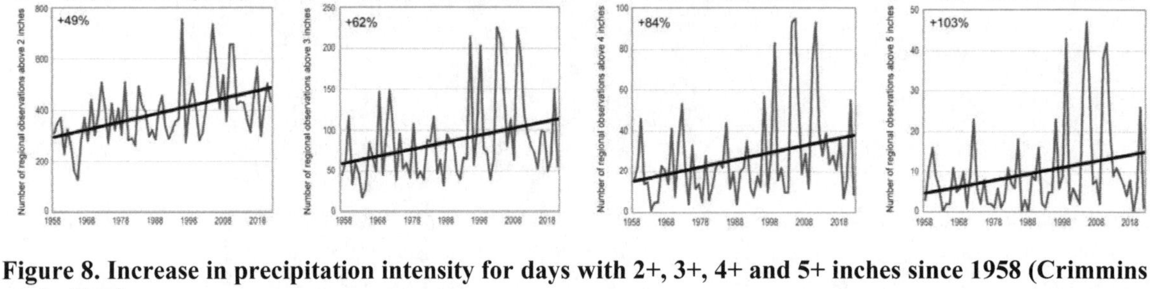

Figure 8. Increase in precipitation intensity for days with 2+, 3+, 4+ and 5+ inches since 1958 (Crimmins et al., 2023).

Representative flooding effects from the July 2023 storm are shown in Figures 9 through 12.

Figure 9. Flooding surrounding the Vermont state capitol building in Montpelier, 11 July 2023 (courtesy of USGS).

Figure 11. Flooding damage to US Rt. 4 in Killington, VT, 10 July 2023.

Figures 12A & 12B. Post-Flooding debris at dam rockfall mitigation site in Quechee Gorge, July 2023.

CASE NO. 3 – FAIRLEE AND HARTLAND, VERMONT ROCK SLOPES

Interstate 91 in Vermont was constructed in 1968 (Eliassen and Ingraham, 2000). The roadway was aligned, for the most part, parallel to U.S. Route 5 and the Connecticut River. The roadway comprises two northbound and two southbound lanes which run from the Massachusetts-Vermont border to the Canadian border. The geology consists predominantly of Cambrian and Ordovician phyllites and schists, along with minor metavolcanic rocks and younger Devonian, rare, igneous plutons that strike sub parallel to the course of the Connecticut River. The northern reaches of the interstate roadway north of White River Junction were the subject of an Interstate Safety Project in the late 1990's.

A large escarpment consisting of quartz monzonite in Fairlee, Vermont, adjacent to the southbound lanes, was draped in 1996 and is roughly 300 feet high (Eliassen, 1997; Westerman et al., 2003). The drape consisted of double-twist mesh and was installed to control small rocks generated by slope weathering. Because the slope had been addressed in 1996, the Interstate Safety Project did not rate the slope or identify issues in 1998.

In late November 2018 the north end of Fairlee slope that was lower in elevation and not draped, experienced a minor toppling rockfall which led to a minor traffic accident. The rockfall occurred during a freeze/thaw event and caused a vehicle strike that disabled a box truck. The northern portion of the slope was repaired in 2019 and 2020, with mitigation consisting of scaling, rock dowels and grouting of open joints. The southern draped escarpment shed some large rocks in

2022, with rocks again reaching the interstate roadway. These failures occurred in the lower 100 feet of the rock cut and caused tearing and ripping of existing netting. The Vermont Agency of Transportation (VTrans) repaired the drape in 2023 with a double twist patch laced into the original mesh. At this point the original slope mesh was nearly 27 years old, and the slope had aged roughly 55 years since interstate construction when VTrans removed its talus slope, conducted bench blasting, installed spot rock dowels, and conducted some rudimentary high scaling. Although the slope performed well for decades, long term weathering, joint dilation and soil infilling as well as ice riving during freeze-thaw seasons allowed larger and larger blocks to fall from the slope. More recently, a large toppling rockslide in late February 2024 tore the drape and buried the southbound lanes with about 200 CY of debris. Although no significant injuries or accidents were reported (Figure 13), this event indicated that the drape could no longer function as an effective rockfall mitigation element against the size and quantity of rocks likely to fall. Due to this most recent slide, VTrans has elected to remove the drape, conduct extensive scaling, and install spot rock dowels as part of an interim measure to secure the slope until a more robust, permanent rock slope mitigation can be designed and constructed.

Figure 13. Recent rockfall on SB I-91 in Fairlee, VT. Note rockfall debris in ditch and roadway, along with start of removal of damaged mesh, 6 March 2023 (courtesy VTrans).

Figure 14. Wedge rockslide on SB I-91 in Harland, VT, 25 November 2023.

A smaller rockslide of about 100 CY of rockfall debris impacted the southbound lanes of I-91 in Hartland, Vermont about 30 miles south of Fairlee on the evening of Thanksgiving Day in 2023 (Figure 14). At least 15 vehicles suffered tire damage from the debris, but no major accidents or injuries occurred. The slide emanated from a wedge feature in thinly bedded but highly jointed metavolcanics. Emergency mitigation took place during winter conditions and consisted of scaling of remaining hanging rock masses and installation of pattern and spot rock dowels.

Both slide events were likely triggered by more extreme swings in temperatures compounded by excessive precipitation, leading to more freeze-thaw cycles. Figure 15 shows that for both slopes, temperatures reached nearly 60°F with rainfall a few days before each slide, followed by sudden freezing events. We suspect that the warming climate is leading to more and longer duration freeze-thaw cycles. These cycles accelerate rock slope degradation as ice jacking expands open joint systems, adhesion freeze (ad-freeze) temporarily binds rock blocks, and sudden thaws associated with warm rainfall melting the adfreeze and providing hydrostatic uplift, leading to rockfalls. While these effects are normal for the "shoulder" seasons (i.e., late fall and early spring), we are now seeing longer shoulder seasons with more rockfalls closer to the middle of the usually subfreezing winter season (Smerekanicz et al., 2024).

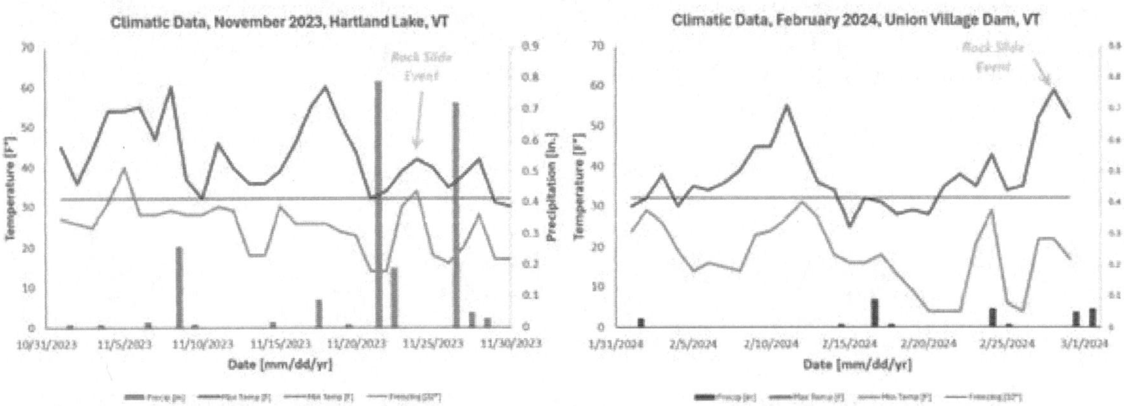

Figure 15. Comparison of daily precipitation, and maximum and minimum temperatures for the Hartland (left) and Fairlee (right) rock slopes (National Regional Climatic Center, 2024).

CASE NO. 4 – PROCTORSVILLE GULF, VERMONT

One of the many mass wasting events of the 9 to 11 July 2023 rainstorm was a debris flow that blocked a portion of Vermont Route 103 in Proctorsville Gulf (a "gulf" is a local term for a narrow, deeply incised Northern Appalachian valley). The debris flow originated in an upper steep portion (~40°) of the valley with thin colluvial cover, with exposed bedrock above (Figures 16 and 17). We postulated that an intense amount of rainfall, likely exceeding 9-inches from the meteorological models, led to sheet flow over upper bedrock slopes onto already saturated thin colluvium/ablation till, leading to slope failure. The debris consisted of soil, rock and trees, flowing over and completely blocking the three-lane highway. Fortunately, the event occurred during the evening, and no accidents were reported. Geologic mapping revealed relict similar scars indicating this is a common mass wasting method in such steep terrain. We postulate that as longer duration precipitation events become more common, we will continue to see a corresponding increase in the frequency and magnitude of mass wasting events.

This debris flow was one of several that impacted Vermont. Mitigation for this debris flow consisted only of removal of sloughed soil and tree debris and cleanout of the slope toe ditch as the debris flow itself scoured much of the thin soil cover, exposing bedrock along its path. The slope failure area lies well beyond VTrans' right-of-way in state forest owned land, thereby making future mitigation efforts less likely; however, the increasing occurrence of these events requires emergency response allocation of resources to clean up, repair and mitigate the damage with funding that would otherwise be allocated for maintenance and infrastructure improvement.

Figure 16. Proctorsville Gulf debris flow scar. Note thin colluvial cover scoured to bedrock surface by sliding debris, 12 July 2023.

CONCLUSIONS

Aging infrastructure, adverse daylighting joints, weathering slopes and intense weather events played a part in these case histories. Changing weather due to shifts in atmospheric rivers, stalled storm systems and longer freeze-thaw seasons has "moved our cheese" and our rapidly aging infrastructure require that we adapt our asset management programs to weather newly recognized design storm events. This will undoubtedly necessitate moving up the schedule for many assets to be renewed/upgraded/repaired to preserve the integrity and resilience of our transportation corridors. Additionally, this is putting more stress on the limited pool of engineering geologists, geological engineers and geotechnical engineers tasked with conducting emergency response and mitigation design – we are now used to being on alert when drastic weather conditions are forecast.

Figure 17. Proctorsville Gulf debris partially blocking VT Rt 103, 12 July 2023.

Finally, a short plug for the HGS Proceedings – There are many answers in our past proceedings. Last year, in Tacoma, a paper by Robert Humphries outlined methods to assess culvert adequacy, address scour and upgrade fish passages – all the tools are in the paper – and many more tools from over 70 years of HGS papers are available online for free. Check it out and avoid the learning curve.

REFERENCES:

Banacos, P., August 5, 2023. The Great Vermont Flood of 10-11 July 2023: Preliminary Meteorological Summary. National Weather Service, National Oceanic and Atmospheric Administration, https://www.weather.gov/btv/The-Great-Vermont-Flood-of-10-11-July-2023-Preliminary-Meteorological-Summary#:~:text=The%20most%20widespread%20and%20significant,2).

Bounty, M. and Cheesungworn, A., April 15, 2024. July 2023 Flood Economic Injury Report – Central Vermont. Central Vermont Economic Development Corporation, 8 p.

Carter, T.G. and Miller, R.I., 1996. Some Observations on the Time Dependency of Collapse of Surface Crown Pillars, Proceedings, 2nd North American Rock Mechanics Symposium, Montreal, Canada, Vol. 1, p. 285-294.

Crimmins, A.R., Avery, A.R., Easterling, D.R., Kunkel, K.E., Stewart, B.C. and Maycock, T.K., 2023. Fifth National Climate Assessment, U.S. Global Change Research Program, Washington, DC, USA. https://doi.org/10.7930/NCA5.2003.RiB

Eliassen, T.D., 1997. Wire Mesh Netting of the Palisades, Interstate 91, Fairlee, Vermont. Northeastern States Geotechnical Engineering (NESGE) September/October 1997 Meeting, 12 p.

Eliassen, T.D. and Ingraham, P.E. 2000. Review, Rockfall Hazard Rating and Repair of Aging Interstate Rock Cuts in Northeastern Vermont. Proceedings, 51st Annual Highway Geology Symposium, Seattle, WA, August 29-September 1, 2000, p. 233-244.

Scarpato, D.J., Innovate Streambank Slope Stabilization – Use of Anchored High-Strength Rockfall Netting and Wrapped Stone Along S.R.118 in Warren, New Hampshire; 2 December 2020 Webinar with Geobrugg, N.A.

Smerekanicz, J.R., Thomas, E.J., and Sophis, J., 2024. Engineering Geology's Role in Reopening Vermont's Transportation Infrastructure in Response to Climate Change. Geological Society of America Northeastern Section Meeting, March 17 – 19, 2024, abstracts with program, Manchester, NH.

Westerman, D.S., Eliassen, T.D, Finucane, R. and Wright, S.F., 2003. Highway Geology Symposium Field Trip Guide for Central Vermont, 54th Highway Geology Symposium, Burlington, Vermont, September 25, 2003, p. 3-7.

Recent Applicable UNR Geotechnical Research

Gary Norris
Emeritus Professor
University of Nevada Reno
1664 N. Virginia St, MS258
Reno, NV 89557
(775) 331-5582
norris@unr.edu

Prepared for the 73rd Highway Geology Symposium, September, 2024

Acknowledgements

The following former students and current colleagues that undertook the cited (and earlier supporting) investigations with the author are:

Sherif Elfass

Horng-Jyh Yang

Mohamed Ashour

Patrick Pilling

Tung Nguyen

Jeyasuthan Pooranampillai

Jeffery Palmer

Mike Alps

Panchalingam Vimalaraj

Ahmed Elsayed

Mohamed Nimeri

Emilio Sanchez

Hugh Ezzell.

It is only through their efforts that the work discussed here was completed.

ABSTRACT

It is from collaboration with past students of the University of Nevada, Reno (UNR) that the following research has been produced:

A Nonlinear Stress Strain Equation for Soil and Rock – as employed in analysis of laterally loaded pile deflection and shallow foundation settlement

Load-Settlement-Bearing Capacity Assessment– with graphical Mohr circle representation of the net mobilized and ultimate bearing capacities

The Characteristic Friction Angle: Its Determination and Use – to assess the drained volume change response under triaxial test loading and its application to undrained response evaluation

The Relationships of Peak, Constant Volume, Characteristic and Dilatant Friction Angles

Resulting accepted/proposed/papers from the above studies undertake to compare the

UNR bearing capacity equation with the classical Meyerhof, Hansen and Vesic equations

UNR load–settlement assessment with a modified (nonlinear) Schmertmann analysis

UNR nonlinear stress strain equation with the Vucetic-Dobry and Seed-Idriss small strain and the Duncan-Chang large strain variations

INTRODUCTION

Research by the University of Nevada, Reno (UNR) geotechnical group has recently yielded the following studies:

A Nonlinear Stress Strain Equation for Soil and Rock

Load-Settlement-Bearing Capacity Assessment

The Characteristic Friction Angle, Its Determination and Use

A Comparison of the Peak, Constant Volume, Characteristic and Dilatant Friction Angles

It is the aim of this presentation to briefly high point the basis and application of each of these studies.

1. A NONLINEAR EQUATION FOR SOIL AND ROCK

The axial strain, ε, in the drained frictionless cap and base triaxial test, conducted at constant confining pressure, as a function of stress level, SL, can be expressed in power equation form as

$$\varepsilon = SL\, e^{3.707SL}\, \varepsilon_{50} / \lambda \quad (1a)$$

where $$SL = \sigma_d / \sigma_{df} \quad (1b)$$

Stress level, SL, is the current deviator stress, σ_d, divided by the deviatoric stress at failure, σ_{df}, and parameter Lambda, λ, is itself a function of SL, i.e.

$$\lambda = 3.19 \qquad \text{for SL} \leq 0.5 \tag{1c}$$

$$\lambda = -7.1219\ \text{SL2} + 7.0592\ \text{SL} + 1.4403 \qquad \text{for SL} > 0.5 \tag{1d}$$

The strain, ε_{50}, is the strain at SL = 0.5, or $\sigma_d = \sigma_{d,\,50} = ½\sigma_{df}$.

The ε_{50} value in Eqs. 1a is a function of confining pressure,

$$\varepsilon_{50} = (\sigma_3 / \sigma_{3\ ref})^{n}\ \varepsilon_{50\ ref} \qquad n = 0.2 \tag{1e}$$

corresponding to $\varepsilon_{50\ ref}$ at a reference pressure, $\sigma_{3\ ref}$. (Figures 2 and 3 of Norris and Yang 2024 provide tentative $\varepsilon_{50\ ref}$ values for cohesionless and cohesive soils.) This power equation is applicable over the entire range of strain, 1 x 10-6 to failure, ε_f, or from SL = 0 to 1.

Alternatively, the power equation can be rewritten as

$$\varepsilon / \varepsilon_{50} = \text{SL}\ e^{3.707\text{SL}} / \lambda \tag{2}$$

i.e. normalized strain ($\varepsilon/\varepsilon_{50}$) as a function of normalized stress (SL) as shown in Fig. 1. The corresponding mobilized friction angle, ϕ_m, is

$$\phi_m = \sin^{-1} [\text{SL A} / (\text{SL A} + 2)] \tag{3a}$$

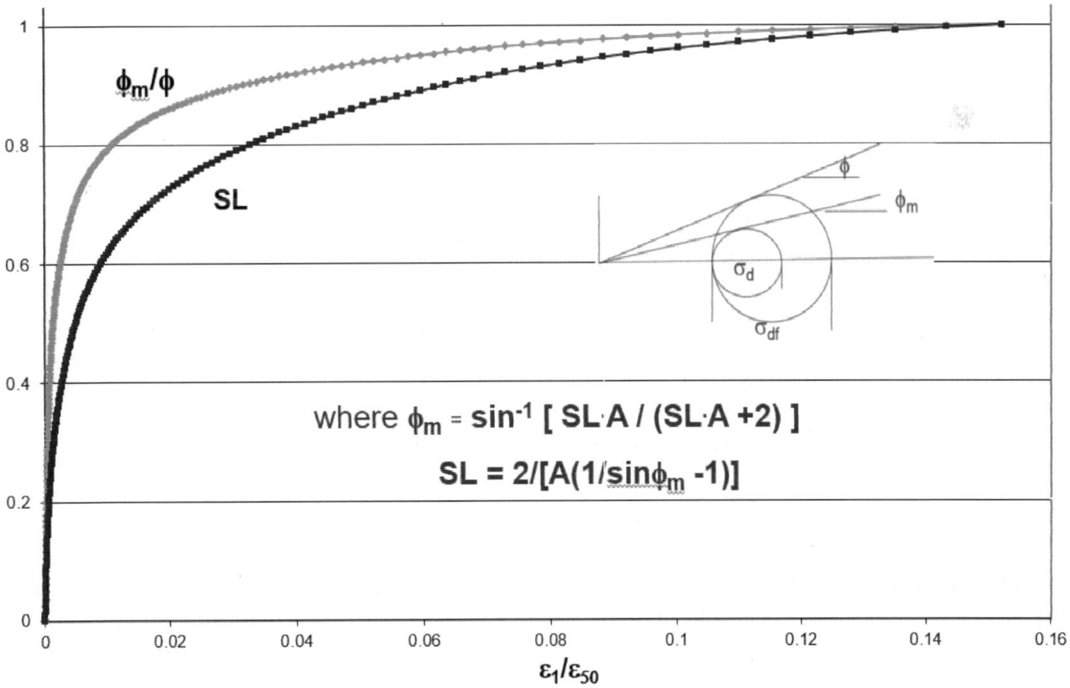

Figure 1 – SL and ϕ_m/ϕ (for $\phi = 37^o$) vs $\varepsilon/\varepsilon_{50}$.

where $\qquad A = 2\sin\phi/(1-\sin\phi) \qquad \phi$ = peak friction angle $\qquad$ (3b)

The mobilized friction angle divided by the peak value (ϕ_m/ϕ) versus normalized strain is also shown in Fig.1, specifically for $\phi = 37^o$.

By contrast, the commonly considered hyperbolic equation employs one version for small strain problems, as applicable to machine foundation vibrations and seismically induced strain, and another version applicable to larger finite element applications of deformation, e.g. slope, earth dam and foundation movement.

1a. Small Strain Characterization

Vucetic and Dobry (1991) provide shear modulus reduction curves for clay while Seed and Idriss (1970) have curves for sandy soils. It can be shown that UNR's power equation can generate curves that are close matches to these curves. If Eq. 1a is rewritten with σ_d/σ_{df} replacing the first SL

$$\varepsilon = \left(\frac{\sigma_d}{\sigma_{df}}\right) e^{3.707\, SL\, \frac{\varepsilon_{50}}{\lambda}} \tag{1a}$$

rearranging the equation yields the secant Young's modulus, E,

$$E = \frac{\sigma_d}{\varepsilon} = \sigma_{df}\, e^{-3.707\, SL\, \frac{\lambda}{\varepsilon_{50}}} \tag{4a}$$

the secant line to a point on the stress strain curve. Substituting that $\sigma_{df} = A\,\sigma_3$, where A is given in Eq. 3b, E becomes

$$E = A\,\sigma_3\, e^{-3.707\, SL\, \frac{\lambda}{\varepsilon_{50}}} \tag{4b}$$

At SL = 0, $$E = E_i = 3.19\, \frac{A\,\sigma_3}{\varepsilon_{50}} \tag{4c}$$

such that $$\frac{E}{E_i} = e^{-3.707\, SL} \left(\frac{\lambda}{3.19}\right) \tag{4d}$$

At SL = 0.5, $\lambda = 3.19$, $E = E_{50}$ whereby

$$E_i = 6.38\, E_{50} \tag{4e}$$

The secant shear modulus, G, corresponding to the Young's modulus is

$$G = \frac{E}{2(1+\nu)} \tag{5a}$$

for which the initial modulus at SL = 1 x 10^{-6}, G_o, becomes

$$G_0 = \frac{E_i}{2(1+\nu_i)} \tag{5b}$$

The modulus reduction, G/G_o is then

$$\frac{G}{G_0} = e^{-3.707SL} \left(\frac{\lambda}{3.19}\right) \frac{1+\nu}{1+\nu_i} \tag{5c}$$

in which the secant Poisson's ratios, ν and ν_i, are a function of the characteristic friction angle, ϕ_c, which will be discussed in a following section.

The corresponding shear strain, γ, is

$$\gamma = \varepsilon\,(1 + \nu) \tag{6}$$

Figure 2 shows both the Young's and shear modulus reduction curves generated based on the nonlinear power equation. Figure 3 is a plot of two G/G_o curves derived using the power equation for cohesionless material. One curve is for ϕ of 45° ($\phi_c = 22.5°$) and an ε_{50} of 0.22% while the other is for ϕ of 34° ($\phi_c = 33°$) and ε_{50} of 0.6%. Both are for a confining pressure (σ_3) of 100 kPa. The horizontal position of the power curves is greatly affected by the value of ε_{50}, which changes with confining pressure as given in Eq. 1e for a chosen reference $\sigma_{3\ ref}$ and $\varepsilon_{50\ ref}$. As shown, the power function curves bracket the range of the upper and lower limits of the superposed Seed and Idriss (1970) shear modulus reduction curves.

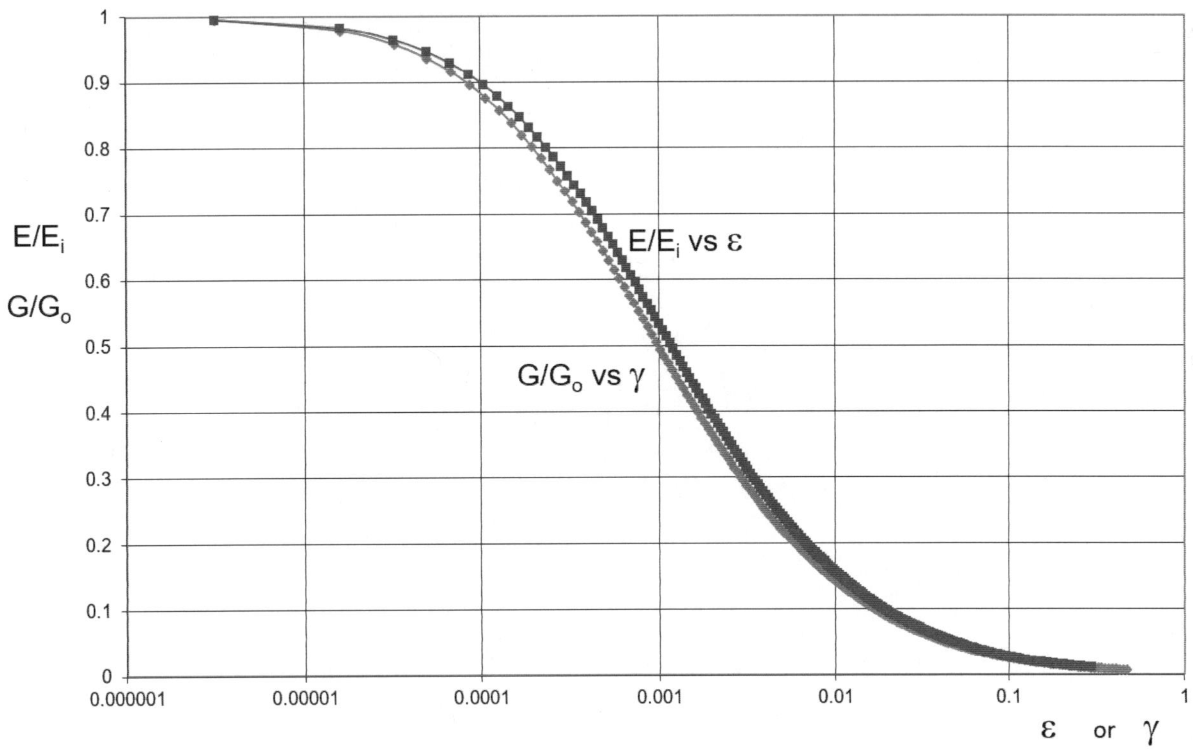

Figure 2 – Generated Young's and Shear Modulus Reduction Curves (σ_3 = 100 kPa, ε_{50} = 0.01, ϕ = 36° and ϕ_c = 28°).

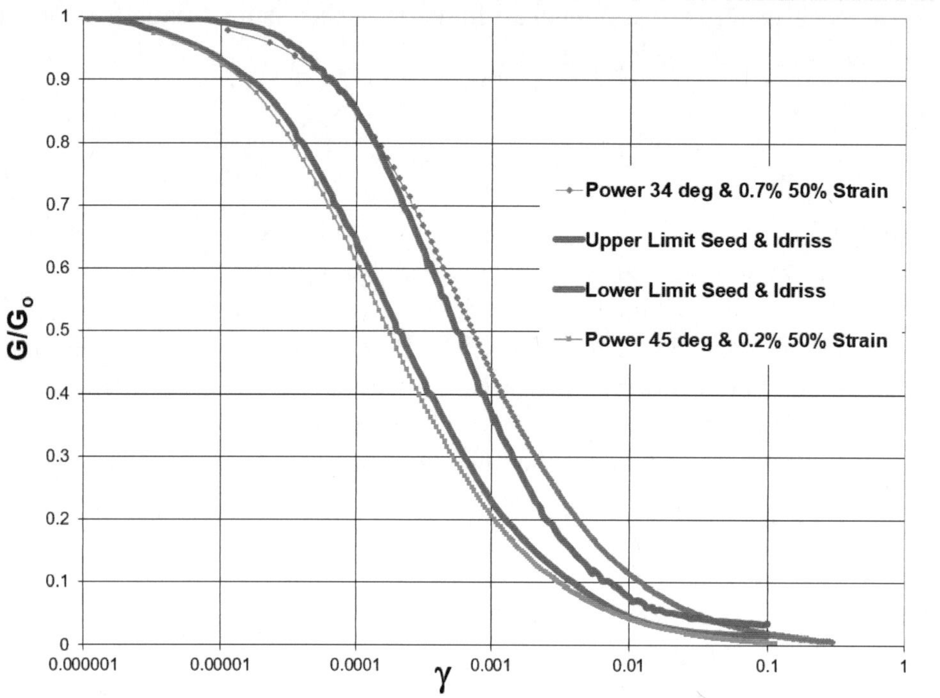

Figure 3 – Two Power Equation Curves versus the Seed and Idriss (1970) Sandy Soil Upper and Lower Limit Shear Modulus Reduction Curves.

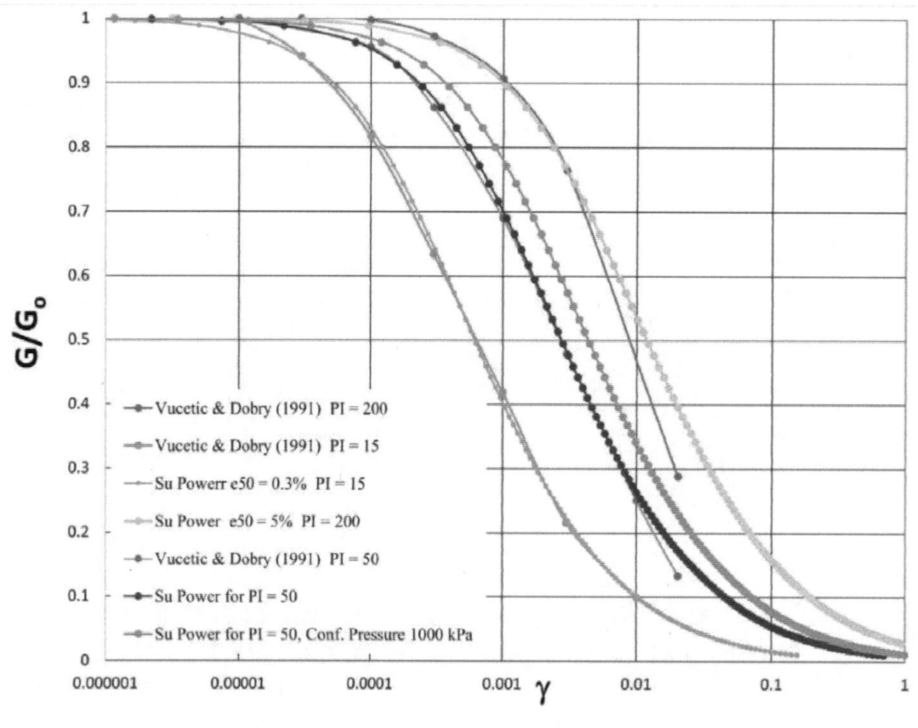

Figure 4 – Three Power Equation Curves versus the Vucetic and Dobry (1991) Clay Shear Modulus Reduction Curves for Plasticity Index, PI = 15, 50 and 200).

As shown in Fig. 4, the power equation can generate shear modulus reduction curves for clay that closely compare to the Vucetic and Dobry (1991) curves for $\sigma_3 = 100$ kPa (with one for 1000 kPa for comparison). The detail for such power equation match is presented in Norris and Yang (2024).

1b. Large Strain Characterization

The hyperbolic equation for normal stress-normal strain for soil, can be expressed as

$$\sigma_d = \frac{\varepsilon}{a+b\varepsilon} \quad a = \frac{1}{E_i} \quad b = \frac{1}{\sigma_{d,ult}} \quad \sigma_{d,ult} = \frac{\sigma_{df}}{R_f} \tag{7a}$$

after Duncan and Chang (1970). The peak of the hyperbolic stress-strain curve at the deviatoric stress of failure, σ_{df}, is taken as a fraction, R_f, of the ultimate asymptotic limit, $\sigma_{d,ult}$. Substituting for a and b in Eq. 7a, σ_d becomes

$$\sigma_d = \frac{\varepsilon}{\frac{1}{E_i}+\frac{R_f}{\sigma_{df}}\varepsilon} \tag{7b}$$

Rearranging Eq. 7b,

$$\frac{1}{E_i} + \frac{R_f}{\sigma_{df}}\varepsilon = \frac{\varepsilon}{\sigma_d} \tag{8a}$$

$$\frac{1}{E_i} = \varepsilon\left(\frac{1}{\sigma_d} - \frac{R_f}{\sigma_{df}}\right) \qquad \sigma_d = SL\,\sigma_{df} \tag{8b}$$

$$\frac{1}{E_i} = \frac{\varepsilon}{\sigma_{df}}\left(\frac{1}{SL} - R_f\right) \tag{8c}$$

$$\varepsilon = \frac{\sigma_{df}/E_i}{\left(1/SL - R_f\right)} \tag{9}$$

$$E_i = \frac{\sigma_{df}}{\varepsilon}\,\frac{1}{1/SL - R_f} \tag{10a}$$

At SL = 0.5, $\sigma_d = \frac{1}{2}\sigma_{df}$ (or $\sigma_{df} = 2\,\sigma_{d,50}$) and $\varepsilon = \varepsilon_{50}$, Eq. 10a becomes

$$E_i = 2E_{50} \,/ \left(\frac{1}{0.5} - R_f\right) = \frac{2E_{50}}{2-R_f} \tag{10b}$$

Typically, R_f is taken to be 0.9, in which case $E_i = 1.818\,E_{50}$. This E_i is considerably smaller than $E_i = 6.38\,E_{50}$ as given by the proposed power relationship. This demonstrates that this large strain hyperbolic relationship doesn't reliably extend to the small strain range. However, the Duncan-Chang equation is really meant for the moderate to large strain range ($\varepsilon = 0.001$ to failure).

Given that the majority of triaxial test stress strain curves presented in the literature are for the standard test with end restraint of the soil at the cap and base, a modified power equation is proposed for curve fit that is not for the ideal cylindrical deformation of the frictionless cap and base test. Accordingly, instead of Eq. 1a, a modified power version is given as

$$\varepsilon = SL\, e^{SL\, Coef}\, X\, \frac{\varepsilon_{50}}{\lambda} \quad (11a)$$

where

$$X = \frac{6.38}{e^{0.5\, Coef}} \quad (11b)$$

The Coef is a value that will be less than 3.707 of Eq. 1a because end restraint reduces the recorded axial, ε. (Smaller overall lateral strain reduces overall axal strain via Poisson's ratio.) The associated dependent value of *X*, as given in Eq. 11b, forces curve fit through the recorded ε_{50} at SL = 0.5. (If the Coef is chosen to be 3.707, the resulting *X* equals 1.)

To fit a recorded triaxial test curve where the strength, σ_{df}, and ε_{50} are known, $E_{50} = (\frac{1}{2}\sigma_{df})/\varepsilon_{50}$, such that E_i of Eq. 10b of the Duncan-Chang equation becomes a function of the specified R_f value. The σ_d can then be assessed for increasing values of ε from Eq. 7b. Accordingly, the R_f value is varied to yield the best fit Duncan-Chang curve to the test curve. Alternatively, the SL of the test curve points is σ_d/σ_{df} and the associated modified power equation ε is assessed via Eqs. 11a and b based on the chosen value of Coef. The value of Coef is varied to yield the corresponding best fit modified power curve to the test curve. In this fashion, the best fit Duncan-Chang R_f and the best fit modified power Coef curves can be compared to the recorded triaxial test stress strain curve. Such side by side comparisons have been made relative to 23 tests from the literature: five isotopically consolidated, undrained (CIU) tests of soft Bangkok clay by Likitlersuang et. al (2013) and nine drained tests each of loose and dense Sacramento River sand by Lee and Seed (1970).

It is only possible to present a few modified power and Duncan-Chang comparisons in the space available here. Of the five CIU tests of Bangkok clay by Likitlersuang et. al (2013), Fig. 5 is typical of the best fit modified power curve and the best fit Duncan-Chang curve plotted against the test curve. Even if, in the Duncan-Chang assessment, both E_i and R_f are varied independently of each other, the modified power curve still compares better than the Duncan-Chang curve in four of the five cases.

Figure 6 is one of nine Lee and Seed loose Sacramento River sand tests. As can be seen, the best fit modified power (Coef = 1.35) and Duncan-Chang (R_f = 0.88) curves are almost identical and are above the test curve between the SL of 50% and 90%. Above 90% until failure they are below. Below 50% SL the original power curve is a better match to the test curve. This leads to the suggestion that a better overall fit is to use the power curve to 50% SL and then the modified power or Duncan-Chang curve above 50%. For tests below the confining pressure of 2 MPa (i.e. 100, 200, 300, 1300 kPa) the best fit modified power curves are a better match to the test curves, while above 2 MPa (i.e. 2.9, 3.9, 7.8, and 13.7 MPa), the Duncan-Chang offers the closer match to the test curves (albeit for associated R_f values uncommonly decreasing from 0.82 to 0.4).

Figure 7 is one of nine Lee and Seed dense Sacramento River sand test curves (confining pressures of 100, 300, 600 kPa, 1, 2, 2.9, 3.9, 7.8 and 13.7 MPa). It is at this 7.8 MPa (and 13.7 MPa) confining pressure that the Duncan-Chang best fit curve is a slightly better match to the test curve than the modified power curve at SL greater than 50%. Below 50%, the original power curve is a decidedly better fit to the test curve.

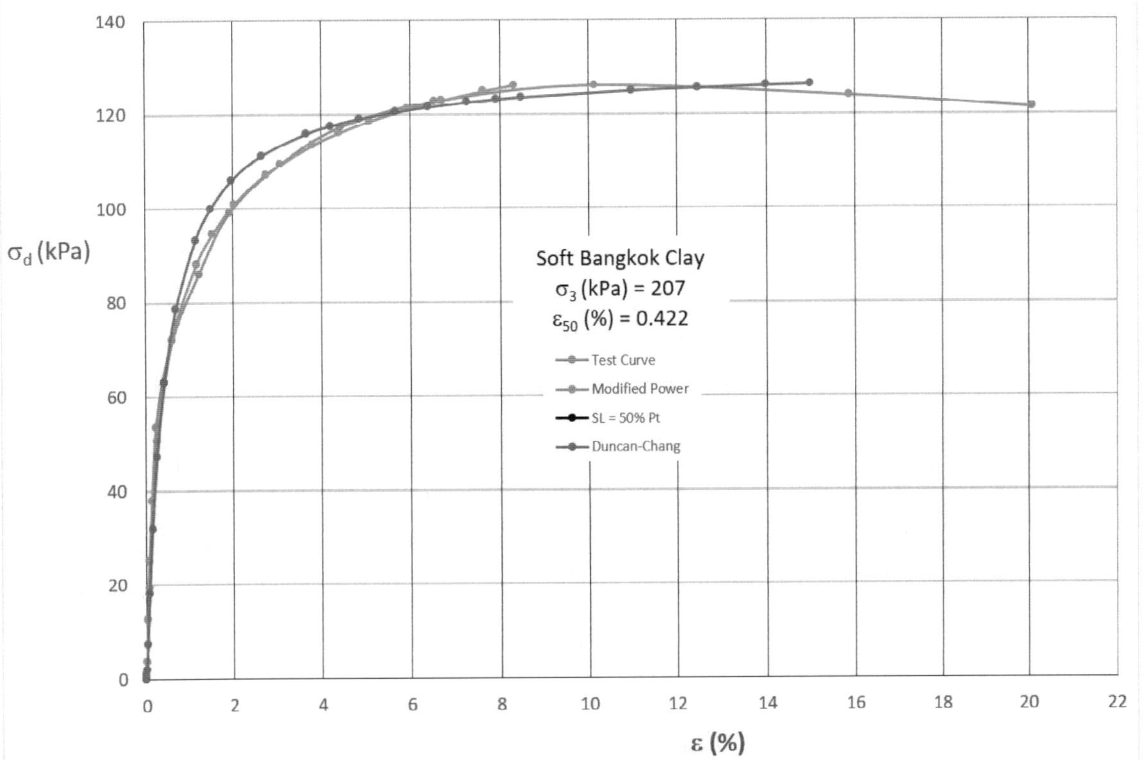

Figure 5 – CIU Soft Bangkok Clay Test at 207 kPa Confining Pressure. (Coef = 2.9, R_f = 0.97)

In total, 38 triaxial test stress strain curves from the literature have been fitted with a modified power curve. The power curve up to 50% stress level has been added as well, but it is not always distinguishable to the naked eye from the modified curve for SL $\leq$ 0.5. However, at very high confining pressures, there is a visible difference, and the power curve is a much better fit than the modified power curve in the lower SL/strain range. It is intended to present the modified power and power curve matches to the 38 test curves from the literature as well as the 23 Duncan-Chang matches in a future report, one of the CCEER reports from the Civil and Environmental Engineering Department of the University of Nevada, Reno. The paper by Norris and Yang (2024), while not the full coverage of the proposed CCEER report, should be consulted for additional background of the coverage presented so far.

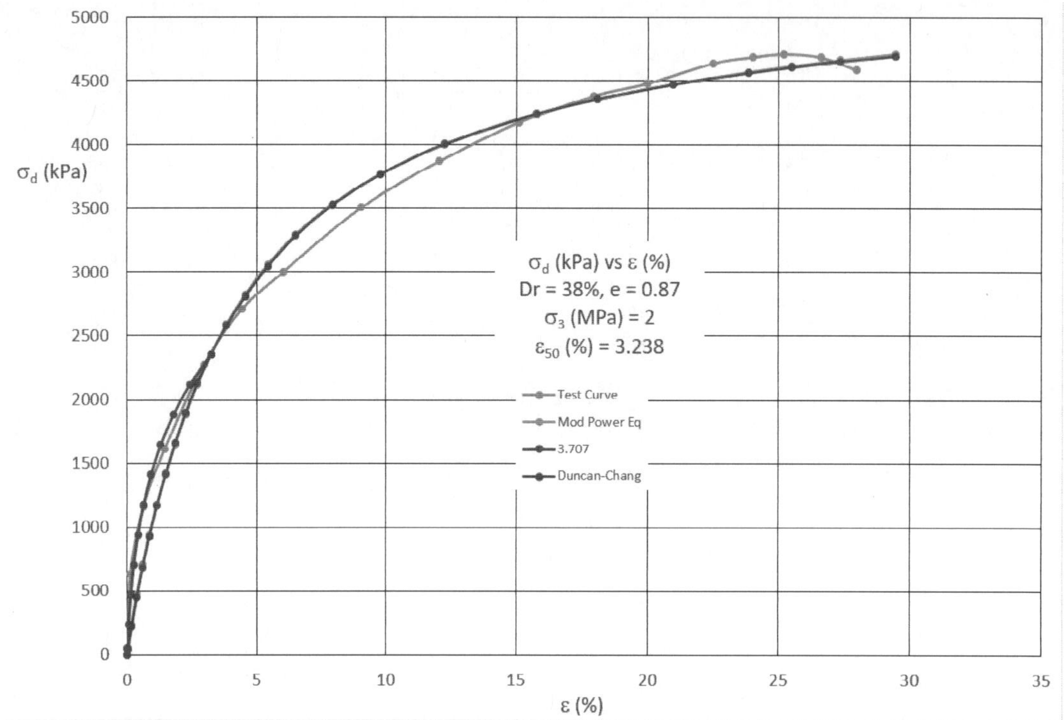

Figure 6 – Loose Sacramento River Sand Test at 2 MPa Confining Pressure. (Coef = 1.35, R_f = 0.88)

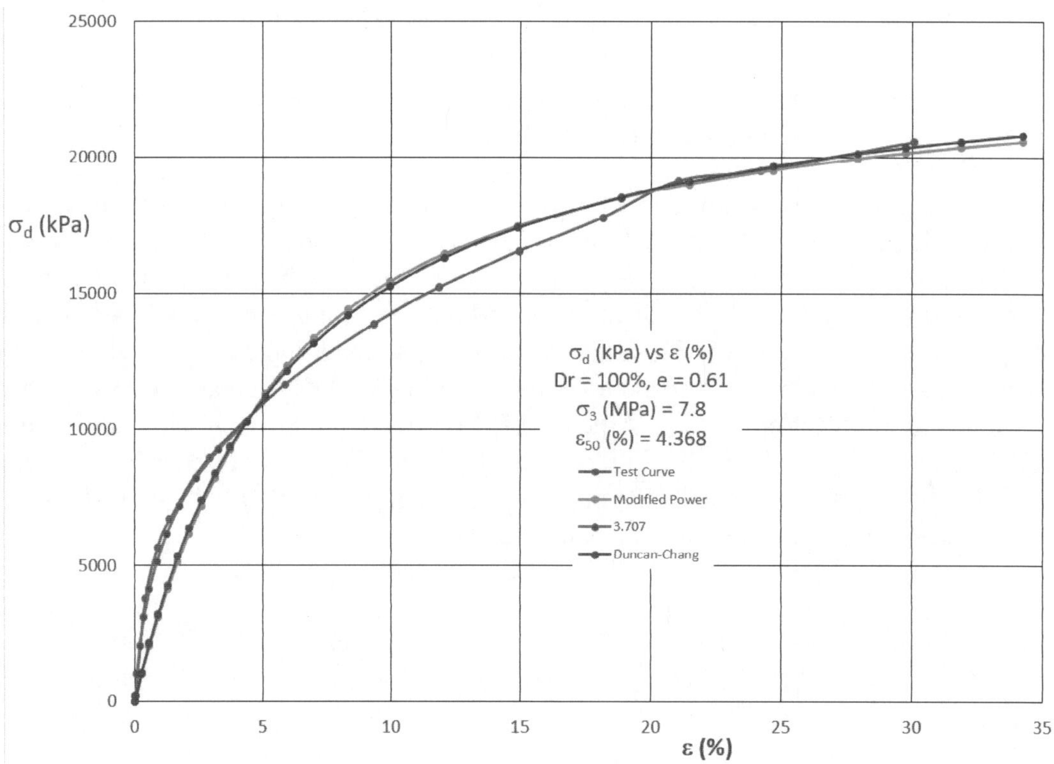

Figure 7 – Dense Sacramento River Sand Test at 7.8 MPa Confining Pressure. (Coef = 1.5, R_f = 0.84)

1c. Post Peak Curve Fit

All the modified power and the Duncan-Chang fits to the standard triaxial test curves from the literature are up to the peak strength, σ_{df}. However, there is a means (an equation) for characterizing the post peak portion of the test curves. Figures 8 and 9 are plots showing the

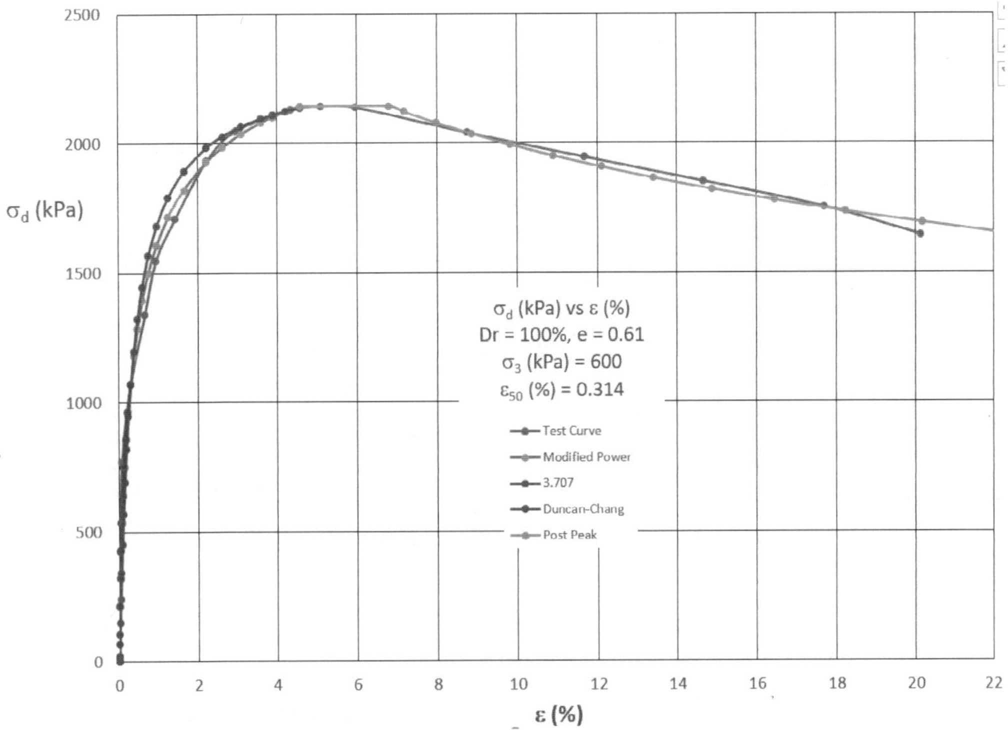

Figure 8 – Dense Sacramento River Sand Test at 600 kPa Confining Pressure. Coef = 1.5, R_f = 0.84; B = 9 & r = 0.45 for the Post Peak Curve Attached to the Modified Power Curve.

result of such capability. Using Eq. 11a, the variation in λ with decreasing SL beyond the peak can be represented by the expression,

$$\lambda = De^{B\,SL} \tag{12a}$$

where D is a function of B and r,

$$D = 3.19\ e^{-(B+r)} \tag{12b}$$

B primarily yields the slope of the post peak curve, while parameter r shifts this sloped portion of the curve horizontally, i.e. to a value of strain of the first point of say, SL = 0.99, past the peak. Generally, one should choose the value of B to match the slope of the recorded test curve, then r is used for fine tuning the horizontal position of the assessed curve.

1d. UNR Equation for Rock-Like Material

Separate from the 38 triaxial tests of soils from the literature, an additional 51 unconfined compressive strength test curves of cored samples from blocks of Las Vegas area caliche presented in an NDOT (2018) report were similarly fitted. The unconfined compressive strength, q_u (or σ_{df}), of these samples varied from 37 to 20,000 psi. Figure 10 is an example of one such

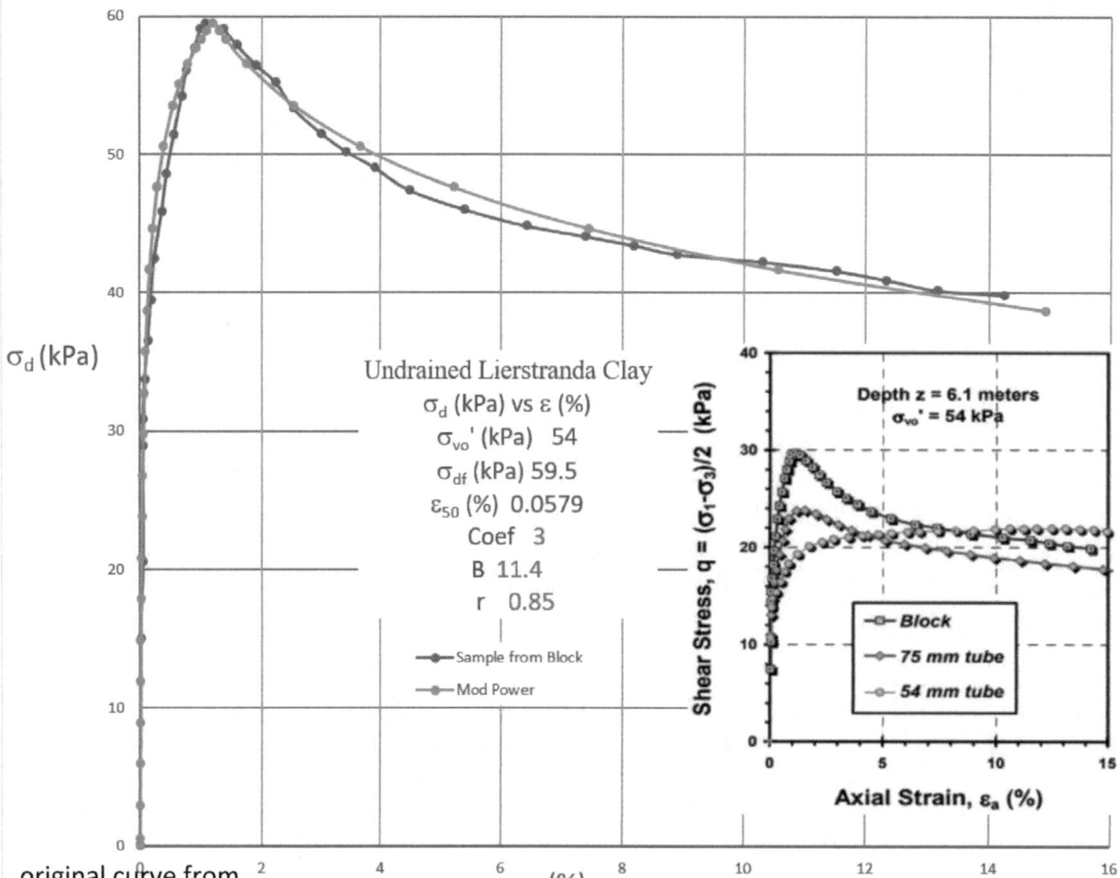

original curve from https://geosystems.ce.gatech.edu/Faculty/Mayne/papers/SOA1%20ICSMGE%202009%20Alexandria

Figure 9 – Test with Post Peak Fit Requiring B and r Values in Addition to the Pre Peak Coef.

test curve. Unlike soil characterization, the total strain at any SL for such cemented soil/rock is given as

$$\varepsilon_{total} = \varepsilon + \Delta\varepsilon \quad (13)$$

Category	q_u (psi)	Δε (%)
Strongly Cemented	> 10,000	0.015
Moderately Cemented	3,000 – 10,000	0.035
Weakly Cemented	< 3,000	0.056

where Δε is the strain of joint/fracture closure (a function of unconfined compressive strength). and ε, evaluated with Equation 1a, is for SL > 0.5, from the end of the linear portion ($E_{sec} = (q_u/2)/\ \varepsilon_{50}$) established from the post closure origin. In Eq. 1a, λ is taken to vary with q_u in psi and SL (Norris and Yang 2024), i.e.

$$\lambda = [10.42\ LN(q_u) - 55.23]\ SL^2 - [12.16\ LN(q_u) - 79.79]\ SL + [3.53\ LN(q_u) - 23.32]\ (32)$$

Such variation with q_u and SL is shown in Figure 11.

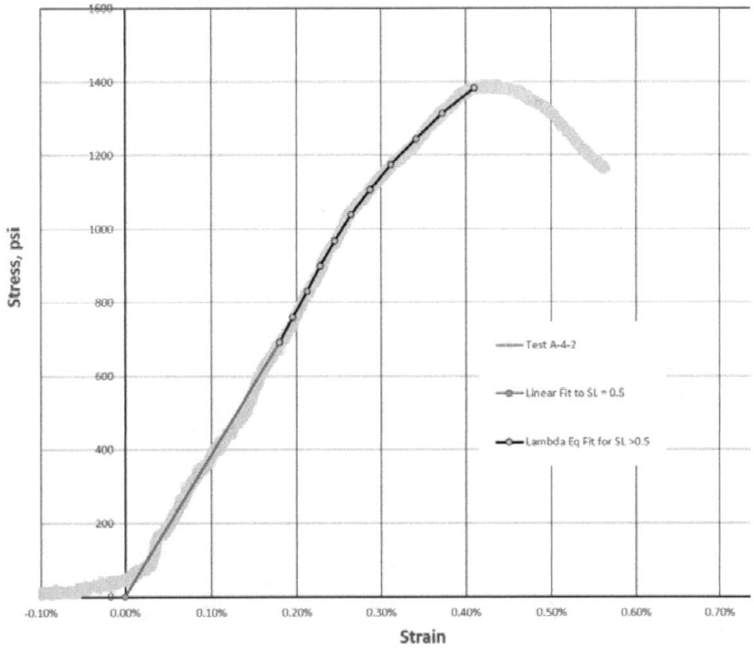

Figure 10 – Las Vegas Area Caliche Unconfined Compressive Strength Test Curve Fitted with the UNR Equation for SL > 0.5.

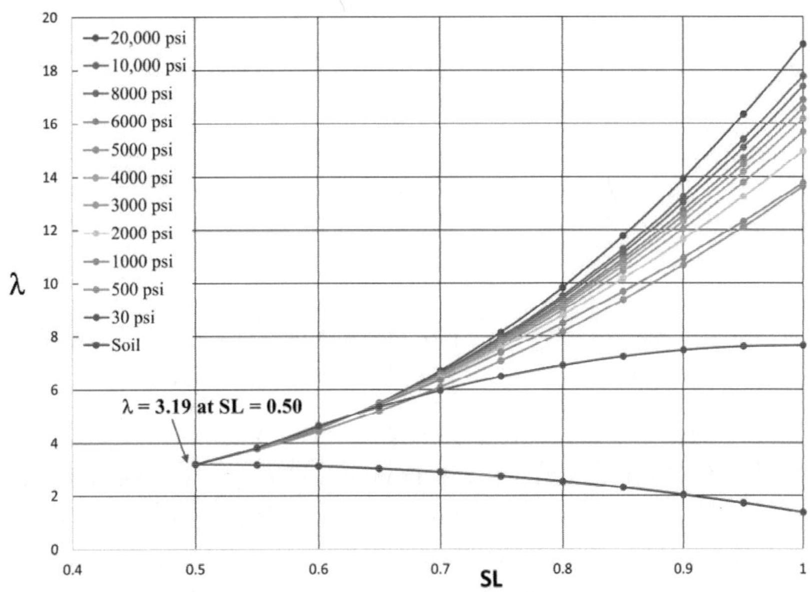

Figure 11 – Lambda (λ) Variation to be Used in Eq. 1a for Cemented Soil/Rock.

2. NONLINEAR LOAD-SETTLEMENT AND BEARING CAPACITY ASSESSMENT

Figure 12 shows the UNR vision of the three zones of soil in the failure mass at bearing capacity failure of a shallow foundation. As shown, the ultimate (gross) bearing capacity, q_{ult}, can be broken up into the stress equivalent to backfilling soil around the foundation up to the ground surface and the net ultimate bearing capacity, q_{net}, which is available to support the structural load at failure. Figure 13 shows the Mohr circles of the interrelated stresses of the three zones at failure, tangent to the Mohr-Coulomb c-ϕ (cohesion-friction) strength envelope. The vertical effective free-field stress at the representative depth in zone I, representing the minor principal stress of circle I, is $P_o = D\gamma_x + ½ B j \gamma_y$ relative to the origin at O. D is the embedment depth, B is the foundation

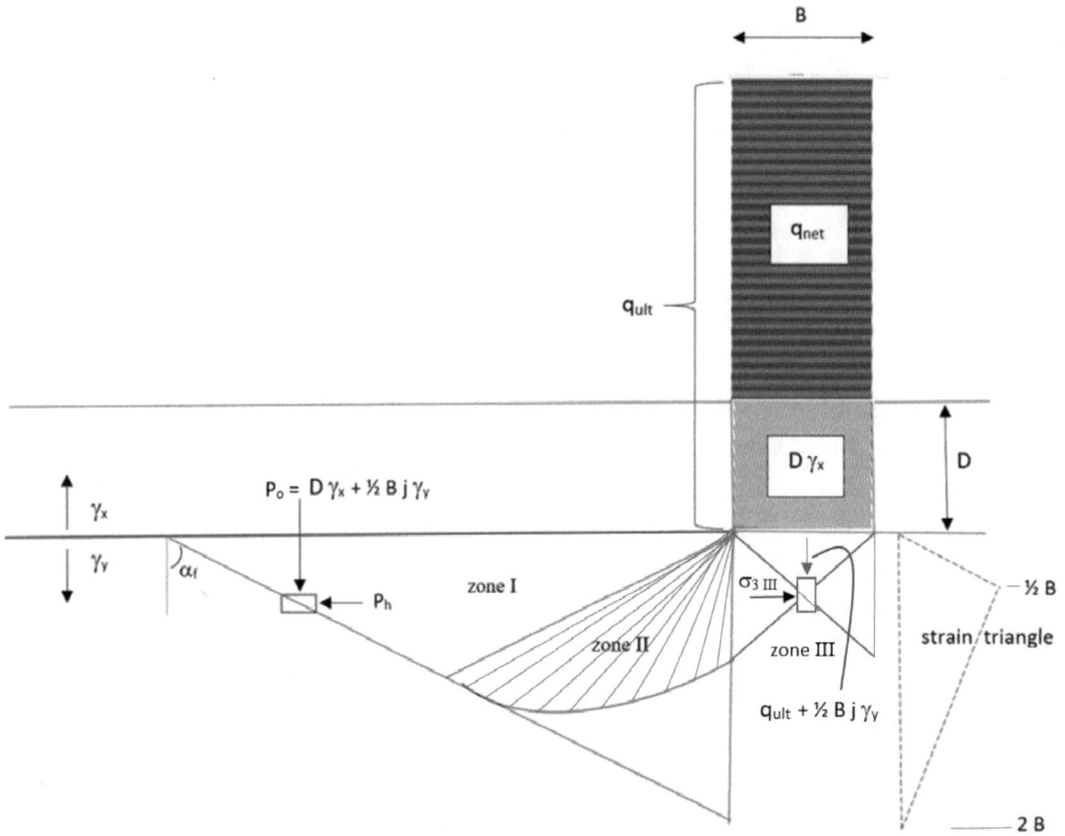

Figure 12 – The Envisioned Three Zone Failure Mass of a Square/Round Foundation and Associated Vertical Strain Triangle for Zone III.

width and unit weights. γ_x and γ_y are the effective soil unit weights above and below foundation base. The j in $½ B j \gamma_y$ is 1.5 tan ϕ (after Hansen). Instead of proceeding with a c-ϕ strength envelope relative to the origin at O, an origin shifted horizontally by c /tan ϕ, to O' yields a purely frictional ϕ envelope to consider. Accordingly, the lower end of circle I is $P_o^* = P_o + c/\tan\phi$, from origin O'. As shown in Fig. 13, the net ultimate bearing capacity becomes

$$q_{net} = P_o^*(\tan^6\alpha_f - 1) \quad (14)$$

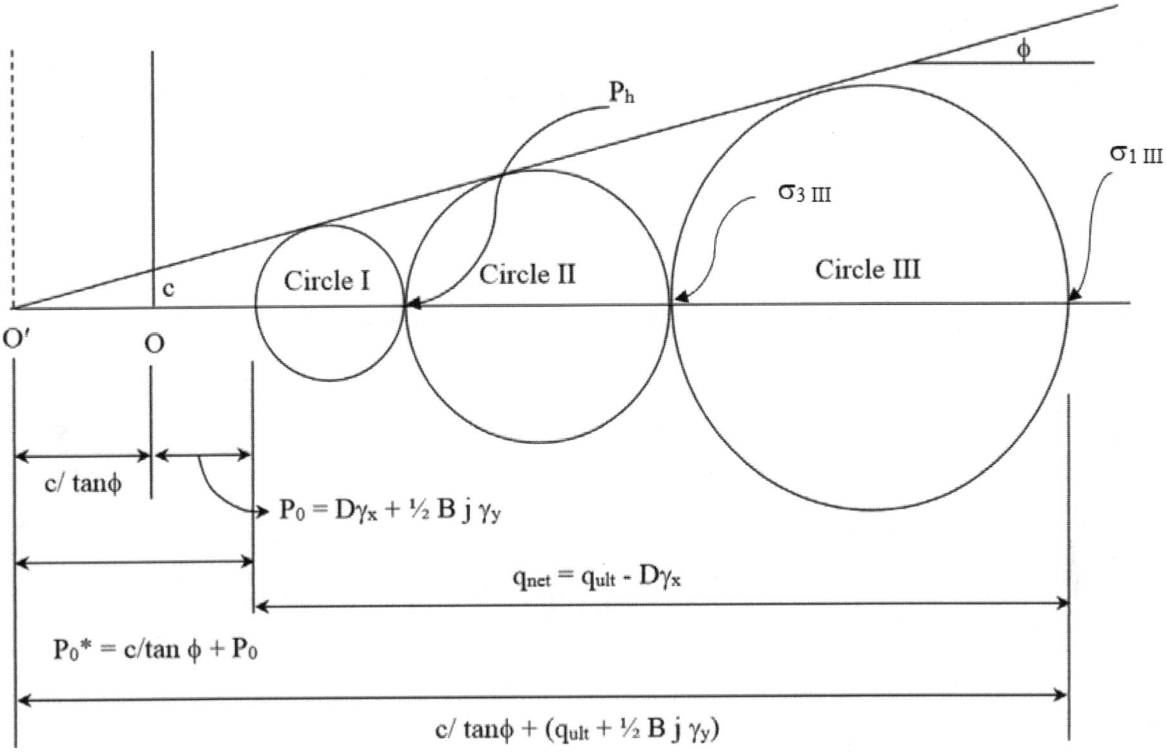

The distance from O' to $\sigma_{1\,III}$ is $P_o{}^*\tan^6\alpha_f$ where $\alpha_f = 45 + \phi/2$

The distance from the lower end of circle I from O' to $\sigma_{1\,III}$ is

$$= c/\tan\phi + (q_{ult} + \tfrac{1}{2} B\, j\, \gamma_y) - P_o{}^*$$

$$= c/\tan\phi + (q_{ult} + \tfrac{1}{2} B\, j\, \gamma_y) - [c/\tan\phi + D\gamma_x + \tfrac{1}{2} B\, j\, \gamma_y]$$

$$= q_{ult} - D\gamma_x = q_{net}$$

Distance from lower end of circle I to $\sigma_{1\,III}$ is also

$$= P_o{}^* \tan^6\alpha_f - P_o{}^*$$

so that

$$q_{net} = P_o{}^*(\tan^6\alpha_f - 1)$$

Figure 13 – Interrelated Stresses of Zones I through III of the Failure Mass.

What is most interesting is that from Fig. 13, the net ultimate bearing capacity is a graphically visible normal stress range. It represents the increase in normal pressure above the free-field pressure on zone I to initiate failure.

By the same token, the net mobilized pressure, q, is

$$q = P_o{}^*(\tan^6\alpha_m - 1) \quad \text{where } \alpha_m = 45 + \phi_m/2 \qquad (15)$$

which can be pictured in Fig. 13 but with ϕ_m in place of ϕ yielding a $c_m = \tan \phi_m (c / \tan \phi)$ for the same origin at O'. For this q and therefore, ϕ_m, a rearranged Eq. 3a yields the stress level, SL, corresponding to the mobilized ϕ_m envelope to which all three mobilized Mohr circles are tangent. It is with this SL that the peak strain, ε form Eq. 1a, at ½ B depth (for the square/circular foundation) in zone III, immediately the beneath foundation, is calculated. (See the strain triangle in Fig. 12.) It is from the strain triangle with its peak corresponding to the SL (for the assumed q and thus ϕ_m), the immediate settlement, ρ_i, is calculated, i.e.

$$\rho_i = \varepsilon B \tag{16}$$

the area of the strain triangle. Systematic evaluation of ρ_i for the applied net pressure, q, up to $\phi_m = \phi$ yields the pressure/load-settlement curve. The one item to mention is the dependence of the strain at 50% SL, ε_{50}, in zone III, upon which the assessment of ε from Eq 1a depends, changes with confining pressure from the free-field value at the constant confining pressure, P_o*, to the current confining pressure of Mohr circle III equal to

$$\sigma_{3\ III} = P_o^* \tan^4 \alpha_m \tag{17}$$

For instance, in the load-settlement analysis, if ε_{50} of the free-field in zone I is known and taken to be $\varepsilon_{50\ ref}$ corresponding to the free-field $\sigma_{3\ ref}$ of P_o*, then ε_{50} of zone III to be used to assess ε and then ρ_i (= ε B) is

$$\varepsilon_{50\ III} = (\sigma_{3\ III} / P_o^*)^n\, \varepsilon_{50\ I} = (P_o^* \tan^4 \alpha_m / P_o^*)^{0.2}\, \varepsilon_{50\ I} = \varepsilon_{50\ I} (\tan^4 \alpha_m)^{0.2} \tag{18}$$

The order in which the calculations are undertaken is

SL ^{yields} ϕ_m (Eq 3a) ^{yields} $\alpha_{m\ \&\ q}$ (Eq 15) ^{yields} $\sigma_{3\ III}$ (Eq 17) ^{yields} $\varepsilon_{50\ III}$ (Eq 18) ^{yields} ε (Eq 1a) ^{yields} ρ_i (Eq 16).

The reader can consult the reference papers with co-authors, Elfass (2007), Elsayed (2011) and Nimeri (2017) to judge the viability of UNR's load-settlement assessment relative to different reported cases from the literature.

It follows that the UNR equivalent to the Schmertmann (1970) I_z profile can be assessed as a function of stress level, SL, as shown in Fig. 14. Note that σ_d in zone III is $\Delta\sigma_1 - \Delta\sigma_3$ relative to the free-field stress state, P_o*. The I_z at its peak for the square/circular foundation is a straightforward function of α_m, i.e.

$$I_z = (\tan^6 \alpha_m - \tan^4 \alpha_m)/(\tan^6 \alpha_m - 1) \tag{19}$$

3. THE CHARACTERISTIC FRICTION ANGLE: ITS DETERMINATION AND USE

The characteristic friction angle, ϕ_c, is the mobilized friction angle, ϕ_m, in the drained triaxial test where the volumetric strain, ε_v, reaches its maximum compressive value. See the middle and lower part of Fig. 14. The characteristic friction angle is an integral factor in linking volume change behavior to developing axial stress-axial strain, as expressed in terms of the mobilized friction angle, as given by the expression

$$d\varepsilon_v/d\varepsilon_1 = 1 - \phi_m/\phi_c \tag{20}$$

From Eq. 20, it follows that

$$\varepsilon_v = \int (d\varepsilon_v / d\varepsilon_1) d\varepsilon_1 \tag{21a}$$

The volumetric strain, ε_V, at any axial strain, ε_1, is the offset from a 1:1 line of ε_v versus ε_1, equal to the area under the ϕ_m versus ε_1 curve, divided by ϕ_c, up to the value of ε_1 in question.

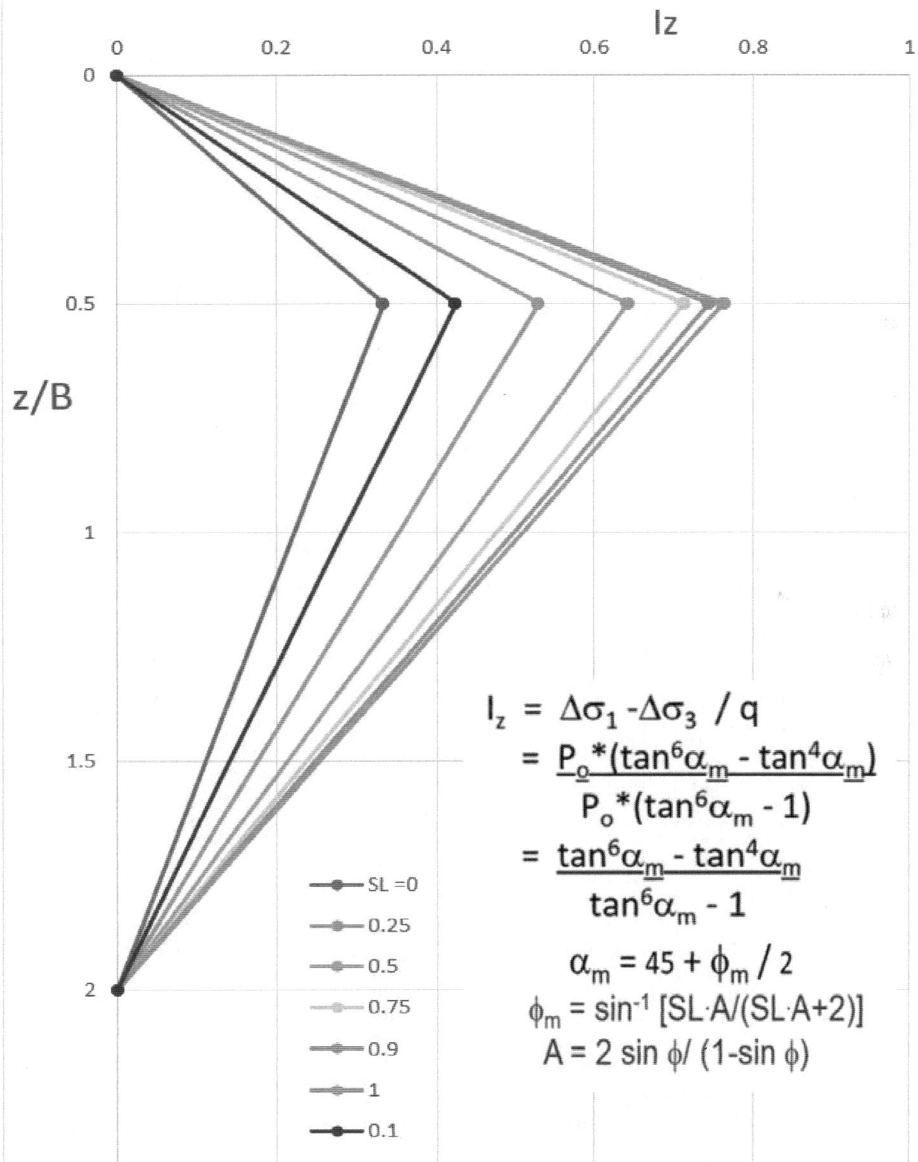

Figure 13 – The UNR Equivalent Iz Profile for a ϕ of 37°. (Iz = 1/3 at SL = 0.)

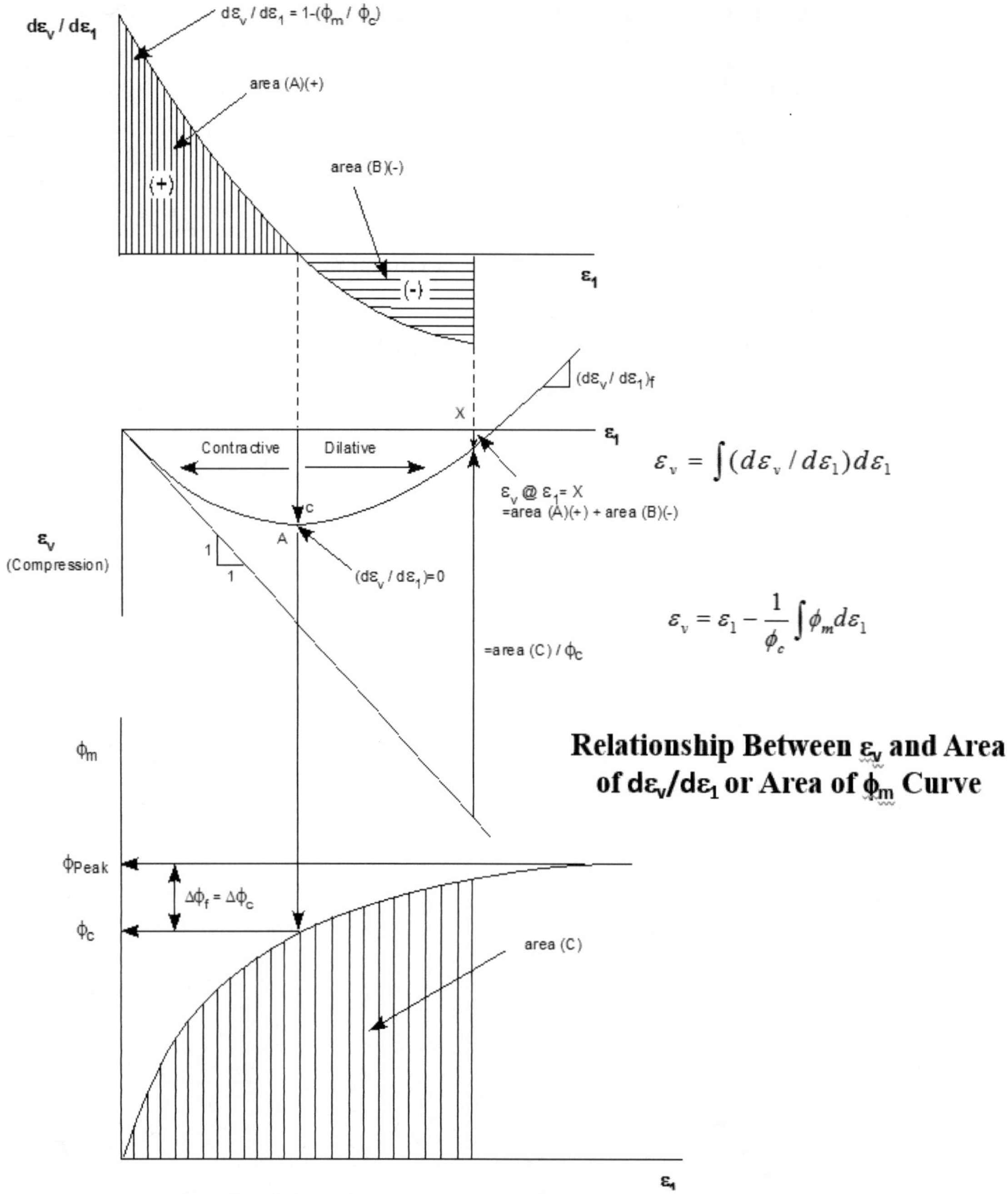

Figure 14 – Relationship Linking Stress-Strain and Volume Change Behavior: $d\varepsilon_v/d\varepsilon_1 = 1 - \phi_m/\phi_c$.

Figures 15 and 16 demonstrate how the practical evaluation of ϕ_c from a loose and a dense sand test is achieved. Given the numerical integration of the recorded ϕ_m versus ε_1 data in Eq. 21b, the resulting ε_v versus ε_1 curve is plotted for an assumed ϕ_c. Figures 15 and 16 show curve deviations from the best fit values of ϕ_c for ± 1 degree variations. An approximate relationship for the characteristic friction angle as a function of porosity, n (independent of particle shape, surface roughness and confining pressure) established from 144 frictionless cap and base triaxial tests on a variety of naturally occurring uniform size quartz sand fractions is

$$\phi_c\,(^o) = 0.60\ n\,(\%) \tag{22}$$

From the same tests, the peak friction angle normalized by the characteristic friction angle (ϕ/ϕ_c) as a function of void ratio minus minimum void ratio ($e - e_{min}$) is

$$(\phi/\phi_c) = 1.8688\ EXP[-1.612(e-e_{min})] \geq 1 \tag{23}$$

Equation 23 yields the possibility of establishing the characteristic friction angle from the peak friction angle for a triaxial test where volume change was not recorded, e.g. on a partially saturated soil. Such value of characteristic friction angle might then be compared with that based on knowledge of porosity (Eq. 22) for confirmation. If there is agreement, the value obtained might then be used to establish the entire drained volume change curve, and from drained response, the undrained behavior is possible (Norris et. al 1997). Consult the paper by Norris (2019) for additional coverage as well as the aspect of establishing the Poisson's ratio as a function of the characteristic friction angle. Assessment of the secant Poisson's ratios, ν and ν_i, was mentioned in conjunction with Eq 5c.

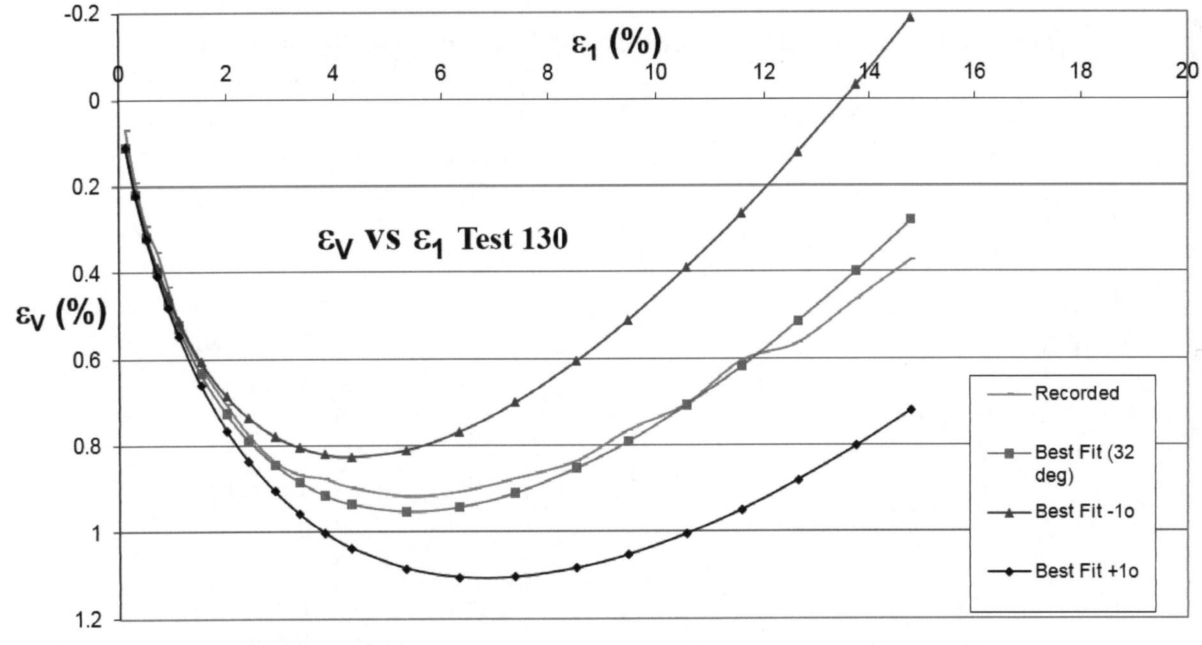

Figure 15 – Determination of Best Fit Value of ϕ_c of Loose Sand

4. PEAK, CHARACTERISTIC, CONSTANT VOLUME AND DILATANT FRICTION ANGLES

Figure 17 portrays the stress strain and volume change curves for a very loose and a dense sand tested at the same confining pressure. In the case of the dense sand test, note that the peak deviatoric stress, corresponding to the peak friction angle, occurs where the dilatant volume change slope, $-d\varepsilon_v/d\varepsilon_1$, is its greatest. Points x and u corresponding to slope, $-d\varepsilon_v/d\varepsilon_1 = 0$, are the deviatoric stresses associated with the characteristic friction and constant volume friction angles, respectively. They are not necessarily equal. In the case of the very loose sand, the peak, characteristic and constant volume friction angles are one and the same.

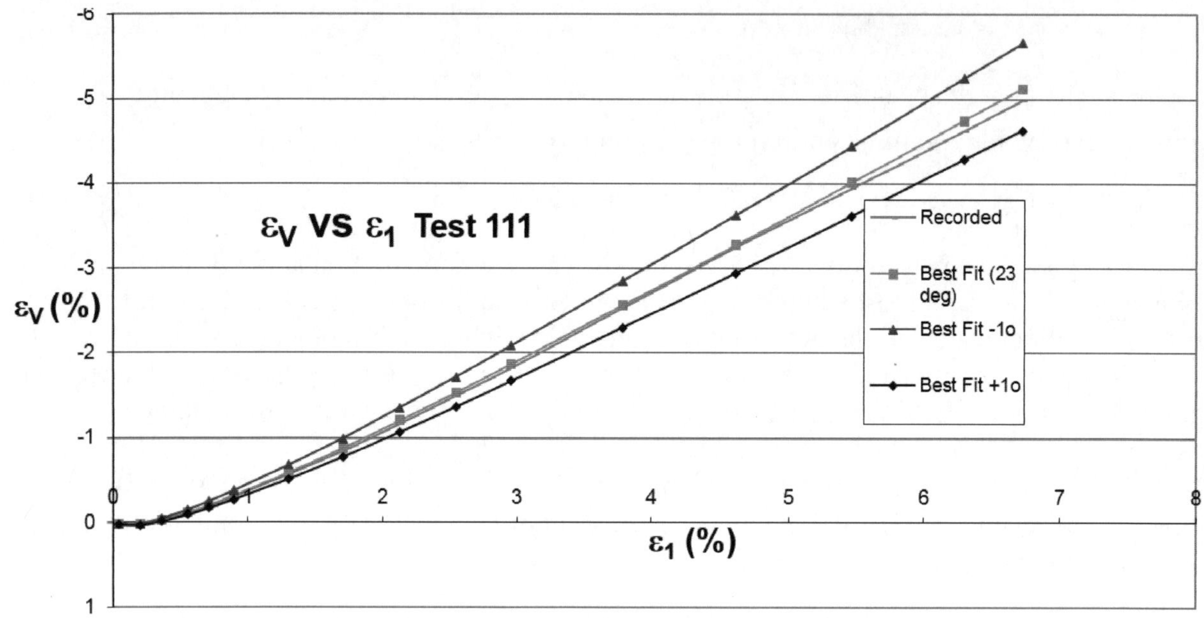

Figure 16 – Determination of Best Fit Value of ϕ_c of Dense Sand

It is unlikely that the relative density, Dr, of the very loose sand, where the constant volume and peak friction angles are the same, can be selected arbitrarily for testing. Alternatively, a dense sand test carried to very large strain often yields an incorrect constant volume friction angle due

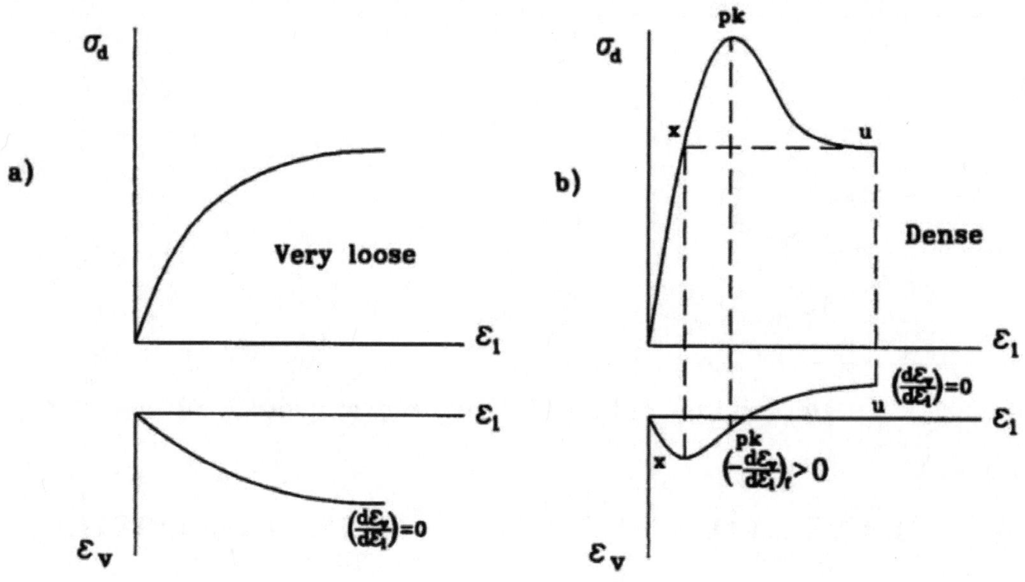

Figure 17 – Triaxial Test Stress Strain and Volume Change Curves for a) Very Loose and b) Dense Sand.

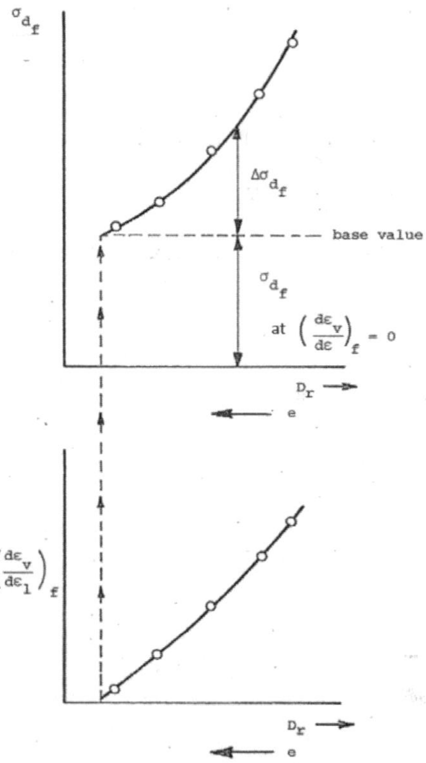

Figure 18 – Extrapolated Constant Volume Friction Angle

to failure along a single weak plane or a band in the test sample. Figure 18 pictures a means of assessing the constant volume friction angle from a series of tests of increasing void ratio, e, or decreasing Dr. From the associated plots of peak σ_{df} or ϕ and $d\varepsilon_V/d\varepsilon_1$ at failure, where $(d\varepsilon_V/d\varepsilon_1)_f$ diminishes to zero, the corresponding peak ϕ is the constant volume ϕ_{CV}. Figure 19 is just such a plot from frictionless cap and base tests of P45-50 sand. (See Norris 2019 for details of this and other materials tested.) Note the extrapolated $(d\varepsilon_V/d\varepsilon_1)_f$ curve and the different ϕ_{CV} 's for each of the three different confining pressures employed: 0.425, 1.20 and 3.40 kgf/cm^2, the lowermost curve. A common misconception is that, ϕ_{CV} is constant; it varies with confining pressure.

A different angle, one that relates incremental volumetric strain to the incremental maximum shear strain causing it, $d\varepsilon_V/d\gamma_{max}$, is the angle of dilatancy, ψ. While Bolton (1986) and others have concluded that

$$\sin\psi = -\,d\varepsilon_V/d\gamma_{max} \tag{24a}$$

Figure 20, on the other hand, shows the Mohr circle of incremental strain for the triaxial test, for both ψ less than (compressive) and greater than zero (dilative volume change). As shown, for the triaxial test, rather than sin ψ, it is tan ψ that is $-d\varepsilon_V/d\gamma_{max}$,

$$\tan\psi = -\,d\varepsilon_V/d\gamma_{max} \tag{24b}$$

In the triaxial test, the incremental volumetric strain is equal to the incremental vertical major principal strain, $d\varepsilon_1$, plus equal incremental horizontal minor principal stains, $d\varepsilon_3$, i.e.

$$d\varepsilon_V = d\varepsilon_1 + 2d\varepsilon_3 \tag{25}$$

Note that the (negative, expansive) $2d\varepsilon_3/2$ arrow of Fig. 20 is shown acting in the opposite direction of the (compressive) $d\varepsilon_1/2$ because of the Poisson, ν, effect ($\nu = -\,\varepsilon_3/\varepsilon_1$).

Using Eq. 25, it can be shown that

$$\tan\psi = -\,2\,d\varepsilon_V/d\varepsilon_1 \,/\, (3 - d\varepsilon_V/\, d\varepsilon_1) \tag{26}$$

and if one substitutes $d\varepsilon_v/d\varepsilon_1 = 1 - \phi_m/\phi_c$ from Eq. 20 into Eq. 26,

$$\tan\psi = -\,2\,(1 - \phi_m/\phi_c)/[3 - (1 - \phi_m/\phi_c)]$$

$$= -\,(2\phi_c - 2\phi_m) \,/\, (2\phi_c + \phi_m)$$

$$\tan\psi = (\phi_m - \phi_c) \,/\, (1/2\,\phi_m + \phi_c) \tag{27}$$

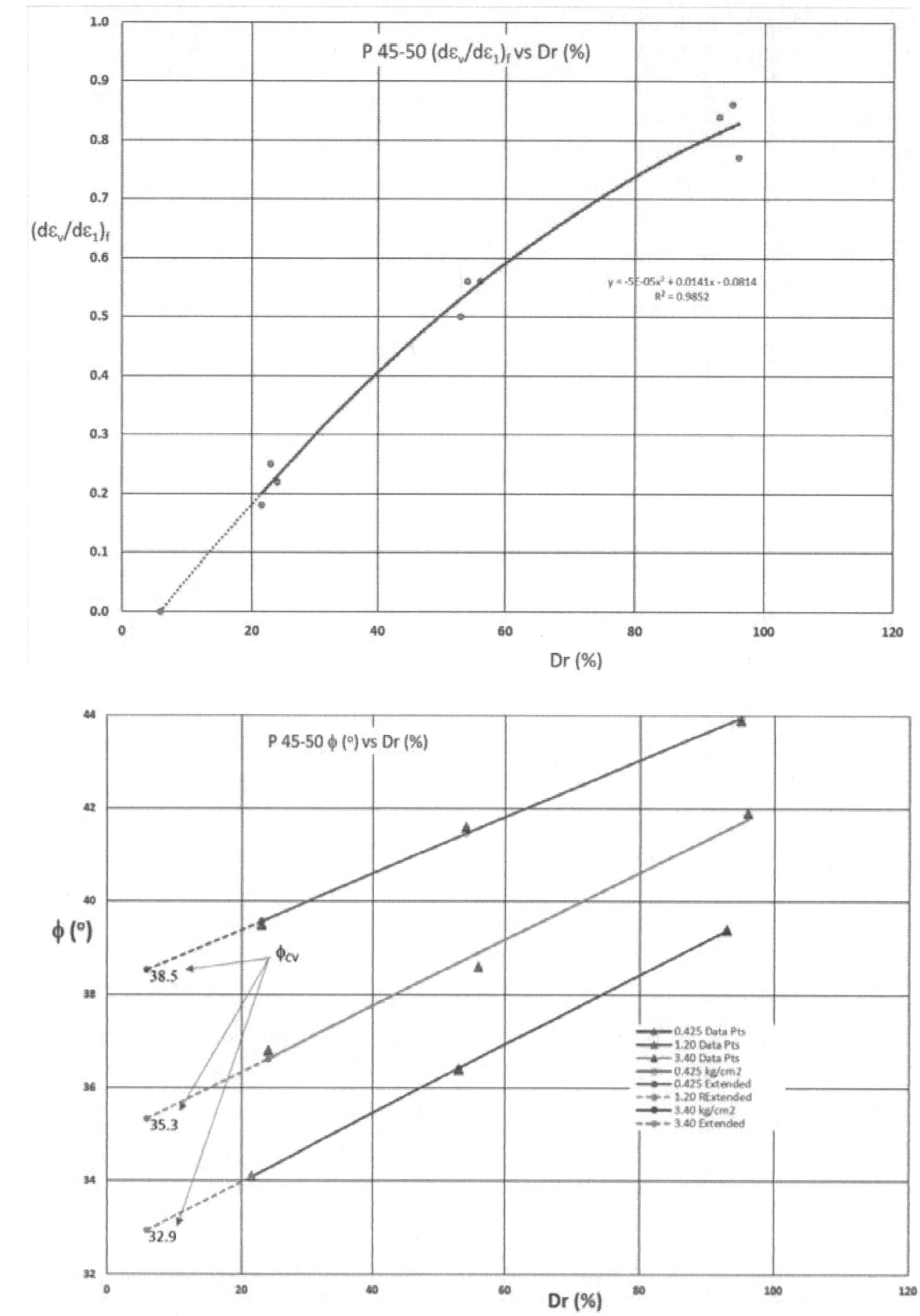

Figure19 – The Constant Volume Friction Angles of P 45-50 Sand for Confining Pressures of 0.425, 1.20 and 3.40 kgf/cm².

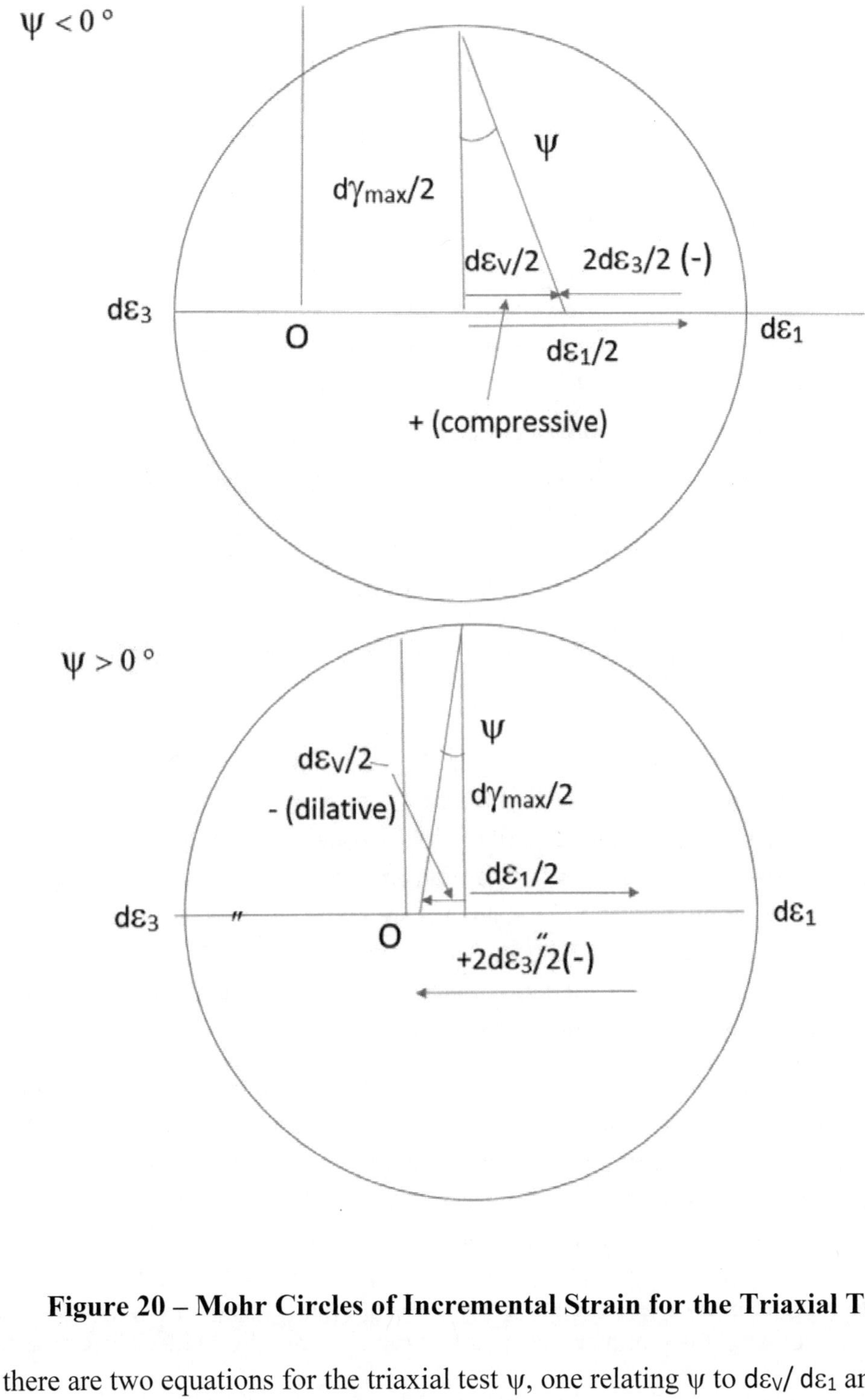

Figure 20 – Mohr Circles of Incremental Strain for the Triaxial Tests.

Therefore, there are two equations for the triaxial test ψ, one relating ψ to $d\varepsilon_V/d\varepsilon_1$ and another to ϕ_c. Furthermore, Eqs. 26 and 27 should be viewed as yielding a mobilized dilatancy angle. Equation 26 yields ψ for the current $d\varepsilon_V/d\varepsilon_1$ and Eq. 27 corresponds to the mobilized ϕ_m. At failure, $\phi_m = \phi$, i.e. the peak friction angle, and $d\varepsilon_V/d\varepsilon_1$ is $(d\varepsilon_V/d\varepsilon_1)_f$.

Shanz and Vermeer (1996) employ the relationship,

$$\sin \psi = -2\, d\varepsilon_V/d\varepsilon_1 / (2 - d\varepsilon_V/ d\varepsilon_1) \qquad (28)$$

for the triaxial test, while Bolton (1986) proposed the empirical relationship,

$$\phi = \phi_{CV} + 0.8\, \psi \qquad (29a)$$

or $$\psi = (\phi - \phi_{CV})/0.8 \qquad (29b)$$

which relates ψ to the difference between the peak and constant volume friction angles.

Figure 21 shows a spreadsheet exhibiting the evaluation of ψ for one of 144 frictionless cap and base tests of naturally occurring fractions of quartz sands. (See Norris 2019 for details.) Note that the recorded data appears in columns B, C and D. Older calculations undertaken prior to the evaluation of ψ occur in columns E through J. The recorded volume change curve (column C)

and calculated curve (column I), based on the visually determined best fit value of ϕ_C in cell F7, are plotted versus ε_1 in Fig. 22. Columns L through O of the newer calculations derive from the data of columns B and C numerically differentiated. On the other hand, column P is ψ evaluated from Eq. 26 based on $d\varepsilon_V/d\varepsilon_1$ values of column O. Note that ψ of columns N and P turn out to be identical because they rely on the same differentiated data.

One will note there is a column S of the mobilized ϕ_c values based on a rearranged Eq. 27, i.e.

$$\phi_c = \phi_m (1 - 1/2 \tan \psi)/(1 + \tan \psi) \qquad (30)$$

assessed from the values of ψ of column N. There is a plot below columns B through I of ϕ_c versus ε_1 showing a fairly stable mobilized value of ϕ_c of 27.5°, which matches the visually determined value of cell F7 determined as part of earlier calculations. Therefore, ϕ_c derived from a mobilized ψ is not a constant. In fact, at failure, where ψ is 6.12 ° (cell N34), ϕ_c is 29.46 ° (cell S34). This value is copied to cell G7 to be compared with the best fit ϕ_c of cell F7. If this failure value of ϕ_c is copied into cell F7, note what happens to the calculated volume change curve shown in Figure 23. The calculated curve moves away from the recorded curve but matches the recorded curve's slope at failure, the last two data points.

Figure 24 shows part of a table of 144 tests comparing ϕ_c, ϕ_{CV}, ϕ and ψ from the test and ψ assessed from Eqs. 26 through 29b. Note that the ψ values are at failure which yield a ϕ_c that differs from the best fit ϕ_c of column J as shown in the difference in Figs. 22 and 23. Column H is the ϕ_c that is varied such that ψ from Eq. 27 in column O matches the test ψ value at failure in column M. By contrast, the Bolton and Shanz and Vermeer values of ψ as evaluated from Eqs. 28 and 29b shown in columns N and R, differ sometimes significantly, from the test values. A more complete presentation of the foregoing material on dilatancy and its connection to the characteristic friction angle is planned for as part of the aforementioned CCEER report.

Test J62

	Dr (%)	23.5	σ_3	3.4	
	e	0.748	σ_{df}	8.90	0.32005
	e_{max}	0.894	ϕ	34.5	
	e_{min}	0.638	A	2.62	
Angularity	ρ	0.454	ϕ_{PT}	27.5	29.45
Surface Roughness	n	0.04	ε_{50}(%)	0.32005	

Newer Calculations

Coef		3.707
x		0.999668

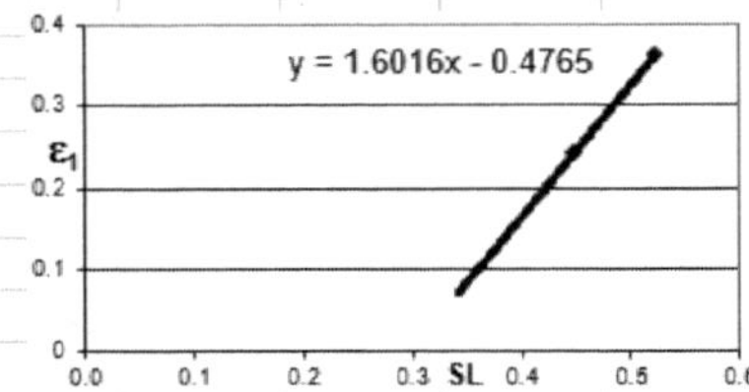

%	%	kso			(°)		from $\phi_{m,area}$ %							Calculated %			
ε_1	ε_V	σ_d	SL	σ_d/σ_3	ϕ_m	$\phi_{m,area}$	ε_V	λ	γ_{max}	tan ψ	ψ	$d\varepsilon_V/d\varepsilon_1$	ψ	ε_1	ϕ_C		
0	0.02954	0.35342	0.040	0.104	2.829	0.000	0.000	3.19	0					0.004613	2.828591	NG	
0.02008	0.02954	0.89647	0.101	0.264	6.682	0.095	0.017	3.19	0.03012					0.01467	6.681609	NG	
0.07027	0.07878	2.02849	0.228	0.597	13.271	0.596	0.049	3.19	0.050665	-0.97187	-44.1828	0.981072	-44.1828	0.053185	701.1039	NG	
		3.0446	0.342					3.19						0.121872			
0.24092	0.14771	3.98749	0.448	1.173	21.676	2.611	0.146	3.19	0.287525	-0.51373	-27.1909	0.613108	-27.1909	0.236369	56.0255		
0.36137	0.1871	4.65699	0.523	1.370	23.965	5.360	0.166	3.184169	0.16098	-0.24469	-13.7495	0.327024	-13.7495	0.365481	35.61061		
0.66252	0.23634	5.72019	0.643	1.682	27.166	13.059	0.188	3.035769	0.427105	-0.11529	-6.57647	0.163507	-6.57647	0.733112	32.47627		
1.03393	0.22649	6.42512	0.722	1.890	29.047	23.498	0.179	2.825368	0.56204	0.017525	1.004031	-0.02652	1.004031	1.186633	28.29616		
1.45553	0.19695	6.93853	0.779	2.041	30.314	36.011	0.146	2.615919	0.64717	0.045645	2.613445	-0.07007	2.613445	1.713958	28.32887		
1.89721	0.12802	7.28186	0.818	2.142	31.118	49.578	0.094	2.449422	0.696985	0.098897	5.648037	-0.15606	5.648037	2.21628	26.91726		
2.34893	0.05908	7.56299	0.850	2.224	31.753	63.778	0.030	2.297313	0.71205	0.096819	5.530086	-0.15262	5.530086	2.759051	27.54844		
2.83076	-0.01969	7.774	0.873	2.286	32.216	79.189	-0.049	2.173812	0.76213	0.103355	5.900857	-0.16348	5.900857	3.272419	27.6892		
3.23228	-0.09847	7.95717	0.894	2.340	32.609	92.203	-0.121	2.060116	0.64167	0.122773	6.999369	-0.1962	6.999369	3.814507	27.26035		Overall
4.05541	-0.28557	8.23157	0.925	2.421	33.183	119.281	-0.282	1.87851	1.328245	0.140863	8.018076	-0.2273	8.018076	4.851367	27.03733	27.47071	Best Fit
4.82835	-0.46283	8.43597	0.948	2.481	33.600	145.090	-0.448	1.734437	1.24804	0.142031	8.083693	-0.22933	8.083693	5.863215	27.33153		
5.66151	-0.65977	8.55425	0.961	2.516	33.836	173.182	-0.636	1.647636	1.34821	0.146075	8.310712	-0.23638	8.310712	6.57461	27.36737		
6.49468	-0.84687	8.65104	0.972	2.544	34.028	201.454	-0.831	1.574736	1.343305	0.139283	7.929334	-0.22456	7.929334	7.242923	27.78785		
7.25758	-1.02413	8.73359	0.981	2.569	34.190	227.475	-1.014	1.51123	1.23298	0.143766	8.181101	-0.23235	8.181101	7.885769	27.74363		
8.06063	-1.19153	8.7982	0.988	2.588	34.316	254.982	-1.211	1.460671	1.288275	0.129941	7.403599	-0.20846	7.403599	8.443208	28.39624		
8.86368	-1.37863	8.84642	0.994	2.602	34.409	282.577	-1.412	1.422449	1.298125	0.144131	8.201614	-0.23299	8.201614	8.894412	27.90691		
9.66673	-1.55588	8.86957	0.996	2.609	34.453	310.227	-1.614	1.40395	1.2932	0.137063	7.804507	-0.22072	7.804507	9.122707	28.22386		
10.46979	-1.72329	8.89143	0.999	2.615	34.495	337.912	-1.818	1.386394	1.288295	0.129947	7.403923	-0.20847	7.403923	9.345683	28.54487		
11.27284	-1.86115	8.90234	1.000	2.618	34.516	365.622	-2.022	1.3776	1.273505	0.108252	6.178348	-0.17167	6.178348	9.459762	29.45915	At peak ϕ	

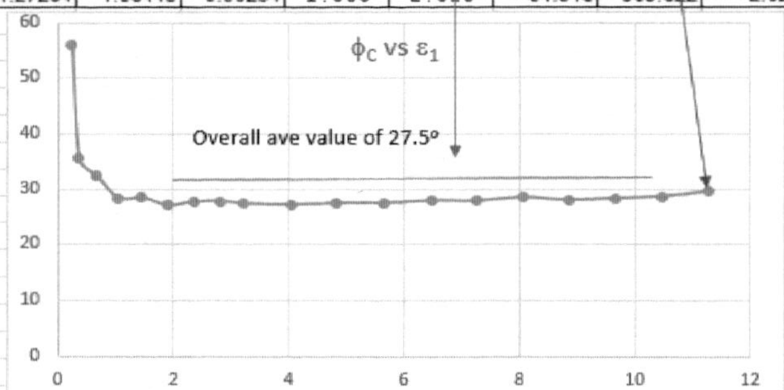

Figure 21 – Spreadsheet for Test 62 Data

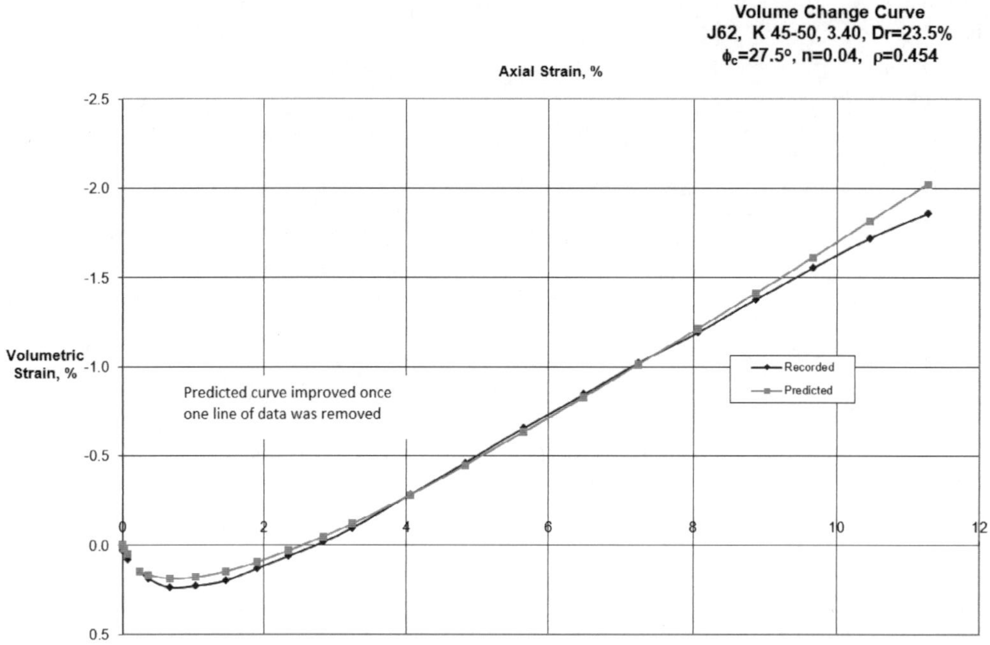

Figure 22 – Volume Change Curve of Test 62.

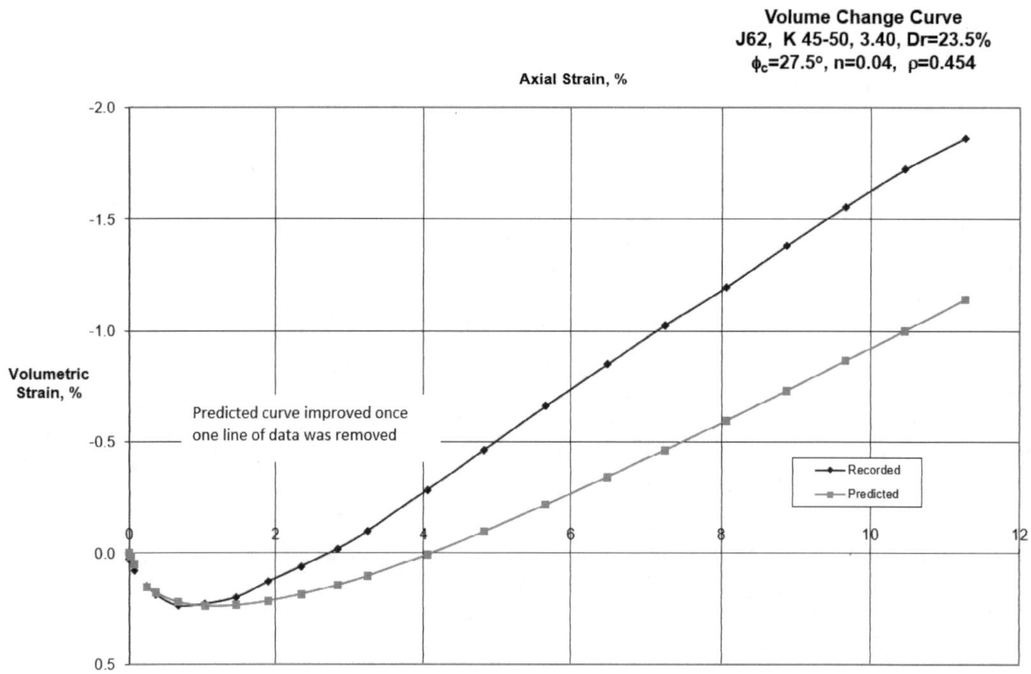

Figure 23 – Volume Change Curve of Test 62 but with ϕ_c of 29.46 ^{o} Substituted for the Best Fit Value to Produce the Calculated Curve.

B	C	D	E	F	G	H	J	K	L	M	N	O	P	Q	R	S	T	U
					Adjusted ϕ_c so Col O matches Cols M & Q		ϕ_c for Best Vol Curve Fit			Test	Bolton	From ϕ_c Equation		From our $d\varepsilon_V/d\varepsilon_1$ Eq	From S & V Eq	Difference in ϕ_c Col J vs Col H		
Test	Soil type	σ_3(kg/cm^2)	e_c	D_r(%)	ρ	ϕ_c (°)	ϕ_c (°)	ϕ_{CV} (°)	ϕ (°)	ψ_{max} (°)	ψ_{max} (°)	ψ_{max} (°)	$d\varepsilon_V/d\varepsilon_1$	ψ_{max} (°)	ψ_{max} (°)			
1	M 7-8	0.425	0.674	20	0.612	28.4	26	34.0	36.4	9.7	3.0	9.7	-0.28	9.7	7.1	-2.4	9.698378	-0.28031
2	M 7-8	0.425	0.544	76	0.612	22.25	23	34.0	42.3	24.8	10.4	24.8	-0.90	24.8	18.1	0.8	24.83967	-0.90347
3	M 7-8	1.20	0.678	18	0.612	28.15	28	32.6	34.7	8.2	2.6	8.2	-0.23	8.2	6.0	-0.1	8.230637	-0.23389
4	M 7-8	1.20	0.523	85	0.612	22.1	23	32.6	39.0	23.1	8.0	22.1	-0.81	23.1	16.8	0.9	23.11638	-0.81406
5	M 7-8	3.40	0.682	16.5	0.612	27.05	26.8	31.0	33.0	7.8	2.5	7.8	-0.22	7.8	5.7	-0.3	7.76803	-0.2196
6	M 7-8	3.40	0.548	74	0.612	21.8	22.4	31.0	36.5	20.2	6.9	20.2	-0.68	20.2	14.6	0.6	20.19058	-0.67589
7	F 7-8	0.425	0.748	26	0.496	30.85	29.5	35.7	37.4	7.5	2.1	7.5	-0.21	7.5	5.5	-1.4	7.464909	-0.21032
8	F 7-8	0.425	0.587	83	0.496	23.45	23	35.7	42.3	22.9	8.2	22.9	-0.80	22.9	16.6	-0.4	22.89556	-0.80307
9	F 7-8	1.20	0.752	24	0.496	28.1	28.5	33.7	35.4	9.1	2.1	9.1	-0.26	9.1	6.7	0.4	9.13413	-0.26226
10	F 7-8	1.20	0.574	88	0.496	22.95	22	33.7	40.6	22.2	8.6	22.2	-0.77	22.2	16.1	-0.9	22.21337	-0.76971
11	F 7-8	3.40	0.736	30	0.496	28.7	27	31.6	33.3	5.8	2.1	5.8	-0.16	5.8	4.3	-1.7	5.800996	-0.16055
12	F 7-8	3.40	0.574	88	0.496	23.75	23	31.6	36.7	17.1	6.4	17.1	-0.55	17.1	12.4	-0.8	17.12897	-0.54651
13	BR 7-8	0.425	0.770	24.5	0.451	29.75	29.5	36.2	37.7	9.3	1.9	9.3	-0.27	9.3	6.8	-0.3	9.32624	-0.26838
14	BR 7-8	0.425	0.578	88.5	0.451	23.9	23.5	36.2	42.5	22.4	7.9	22.4	-0.78	22.4	16.3	-0.4	22.36722	-0.77715
15	BR 7-8	1.20	0.772	24	0.451	29.7	28	33.3	35.7	7.2	3.0	7.2	-0.20	7.2	5.3	-1.7	7.248292	-0.20373
16	BR 7-8	1.20	0.583	87	0.451	22.9	23	33.3	40.1	21.8	8.5	21.8	-0.75	21.8	15.8	0.1	21.76952	-0.74849
17	BR 7-8	3.40	0.718	42	0.451	24.95	24.5	30.4	34.1	12.3	4.6	12.3	-0.37	12.3	8.9	-0.4	12.26526	-0.36587
18	BR 7-8	3.40	0.595	83	0.451	23.15	23.5	30.4	37.6	19.0	9.0	19.0	-0.62	19.0	13.7	0.4	18.99109	-0.62352
19	G 18-20	0.425	0.608	45.5	0.665	25.7	24	30.0	33.0	9.8	3.8	9.8	-0.28	9.8	7.2	-1.7	9.842464	-0.28496
20	G 18-20	0.425	0.499	89	0.665	22.1	20.5	30.0	38.5	21.6	10.6	21.6	-0.74	21.6	15.7	-1.6	21.61251	-0.74107
21	G 18-20	1.20	0.616	42.5	0.665	25.4	23.5	29.0	31.3	8.2	2.9	8.2	-0.23	8.2	6.0	-1.9	8.212824	-0.23333
22	G 18-20	1.20	0.571	60.5	0.665	21.15	20.5	29.0	34.0	18.6	6.3	18.6	-0.61	18.6	13.4	-0.6	18.56979	-0.60566
23	G 18-20	1.20	0.466	102	0.665	17.95	19	29.0	37.9	28.4	11.1	28.4	-1.11	28.4	20.9	1.1	28.37919	-1.11023
24	G 18-20	3.40	0.608	30.5	0.665	24.85	23.5	28.3	30.5	8.0	2.8	8.0	-0.23	8.0	5.8	-1.4	7.994901	-0.22659
25	G 18-20	3.40	0.481	96	0.665	20.2	20.5	28.3	34.9	21.3	8.3	21.3	-0.73	21.3	15.5	0.3	21.34382	-0.72848
26	M 18-20	0.425	0.713	36	0.462	25.8	25.5	38.1	40.6	17.8	3.1	17.8	-0.57	17.8	12.9	-0.3	17.7682	-0.57239
27	M 18-20	0.425	0.556	94	0.462	23.05	20	38.1	44.5	25.3	8.0	25.3	-0.93	25.3	18.5	-3.1	25.31174	-0.92914
28	M 18-20	1.20	0.679	48.5	0.462	26.75	25	34.0	37.4	13.2	4.3	13.2	-0.40	13.2	9.5	-1.8	13.17375	-0.39763
29	M 18-20	1.20	0.640	63	0.462	21.8	22	34.0	39.5	23.1	6.9	23.1	-0.81	23.1	16.8	0.2	23.09862	-0.81317
30	M 18-20	1.20	0.551	95.5	0.462	24.2	22	34.0	41.8	21.3	9.8	21.3	-0.73	21.3	15.5	-2.2	21.34208	-0.7284
31	M 18-20	3.40	0.734	28	0.462	25.65	25	31.4	34.8	12.0	4.3	12.0	-0.36	12.0	8.7	-0.6	11.96863	-0.35568
32	M 18-20	3.40	0.558	93	0.462	22.1	23	31.4	39.8	22.9	10.5	22.9	-0.80	22.9	16.7	0.9	22.90054	-0.80331
33	F 18-20	0.425	0.736	46	0.344	27.85	26	37.7	38.5	12.7	1.0	12.7	-0.38	12.7	9.2	-1.9	12.74218	-0.38244
34	F 18-20	0.425	0.597	89.5	0.344	21	20	37.7	43.6	27.8	3.4	27.8	-1.07	27.8	20.4	-1.0	27.79199	-1.07349
35	F 18-20	1.20	0.730	48	0.344	29	26.5	34.8	37.0	9.6	7.8	9.6	-0.28	9.6	7.0	-2.5	9.597993	-0.27708

Figure 24 – Portion of Table of 144 Tests of ϕ_c, ϕ_{CV}, ϕ and ψ Values.

DISCUSSION/CONCLUSIONS

Of particular note in the foregoing presentation is

that net ultimate bearing capacity can be visibly presented (Fig. 5), in terms of the Mohr circles of zones I through III, as the increase in pressure above the free-field value (P_o*) to cause failure, and

the validity and utility of the nonlinear UNR stress strain equation to represent soil response over both small (machine foundation and seismic) and large (shallow foundation and laterally loaded pile/shaft) strain range. The equation requires only the designation of the Coef (3.707 or other) as input.

Coverage of the different topics discussed appears in the papers cited in the references.
Future papers/report will:

compare the UNR bearing capacity equation with the classical equations of Meyerhof, Hansen and Vesic
compare a modified, nonlinear Schmertmann settlement analysis with UNR's analysis
show additional Duncan-Chang versus UNR triaxial test curve fits as well as UNR's post peak equations
Provide full coverage of the interrelationship of the characteristic friction angle and the angle of dilatancy

Participants interested in complimentary spreadsheet files of these and other material are welcome to email the author for copies.

REFERENCES

Norris, G. and H. Yang. A Nonlinear Stress-Strain Relationship for Soil and Rock. *ICE Geotechnical Engineering Journal* (Accepted manuscript doi: 10.1680/jgeen.22.00108) 2024.

Vucetic, M. and R. Dobry. Effect of Soil Plasticity on Cyclic Response. *Journal of Geotechnical Engineering* vol. 117 (1), 1993, pp. 89-107.

Seed, H. B. and I. M. Idriss, *Soil Moduli and Damping Factors for Dynamic Response Analysis.* Report EERC 70-10, Earthquake Engineering Research Center, Berkeley, CA, 1970.

Duncan, J. M. and C. Y. Chang. Non-linear Analysis of Stress and Strain in Soils. *Journal of Soil Mechanics and Foundations Division, ASCE.* vol. 96 (5), 1970, pp. 629-1653.

Likitlersuang, S., C. Surarak, A. Balasubramaniam, E. Oh, E, S. Ryull and D. Wanatowski. Duncan-Chang Parameters for Hyperbolic Stress Strain Behavior of Soft Bangkok Clay. In *18th International Conference on Soil Mechanics and Geotechnical Engineering, Paris,* 2013, pp. 381-384.

Lee, K. and H. Seed. Drained Strength Characteristics of Sands. *Journal of Soil Mechanics and Foundation Engineering Division, ASCE.* vol. 93 (6), 1967, pp. 117-141.

Elfass, S., E. Saint-Pierre, M. Ashour, R. Watters. and G. Norris. *Improving Strain Wedge Model Capabilities in Analyzing Large Diameter Drilled Shafts Subjected to Lateral Loading in Cemented Soils.* Report No. 667-15-803, NDOT, 2018.

Elfass S., G. Norris and P. Vimalaraj. A Simple Bearing Capacity Equation. *Proc. GeoDenver Conference - New Peaks in Geotechnics: Advances in Shallow Foundations, Denver, CO,* ASCE, GSP 171, 2007, pp. 1-10.

Elsayed, A., S. Elfass, G. Norris and P. Vimalaraj. A New Approach for Non-Linear Load-Settlement Assessment of Shallow Foundations. *Proc. Geo-Frontiers Congress 2011 - Advances in Geotechnical Engineering, Dallas, TX*, ASCE GSP 211, 2011, vol. 1, 2011, pp. 372-381.

Nimeri, M., S. Elfass and G. Norris. G. Load–Settlement Response of Shallow Foundations Resting on Granular Soil. *GeoMEast2017 International Conference: Innovative Infrastructure Solutions, Cairo, Egypt*. Springer International Publishing 2 (33), 2017, 11 pp.

Schmertmann, J.H. Static Cone to Compute Static Settlement Over Sand. *Journal of Geotechnical Engineering Division, ASCE*. vol.96 (3), 1970, pp. 1011-1043.

Norris, G., R. Siddharthan, Z. Zafir and R. Madhu. Liquefaction and Residual Strength of Sands from Drained Triaxial Tests. *Journal of Geotechnical Engineering, ASCE*. vol. 123 (3), 1997, pp. 220–228.

Norris, G. The Characteristic Friction Angle, Its Determination and Use. *Proc. 70th Highway Geology Symposium, Portland, Oregon*. Paper 14, 2019, 16 pp.

Bolton, M. D. The Strength and Dilatancy of Sands. *Geotechnique*. vol. 36 (1), 1986, pp. 55–78.

REMEDIATION OF ACID PRODUCING ROCK ON THE CSVT PROJECT

Jason M. Gardner, P.E.
Gannett Fleming, Inc.
202 Senate Avenue
Camp Hill, PA 17011
(717)-886-5429
jgardner@gfnet.com

Prepared for the 73rd Highway Geology Symposium, September 2024

Acknowledgements

The author would like to thank the following individuals/entities for their contributions in the work described:

Robert E. Johnson, P.E. – PENNDOT 3-0, District Geotechnical Engineer (retired)
Isaac R. Bragunier, P.E. – PENNDOT 3-0, District Geotechnical Engineer
Andy Smithmyer, P.G. – Gannett Fleming, Inc., Senior Geologist

Disclaimer

Statements and views presented in this paper are strictly those of the author(s), and do not necessarily reflect positions held by their affiliations, the Highway Geology Symposium (HGS), or others acknowledged above. The mention of trade names for commercial products does not imply the approval or endorsement by HGS.

ABSTRACT

The Central Susquehanna Valley Transportation (CSVT) Southern Section is approximately a six-mile, four lane, limited access highway located in central Pennsylvania that is currently under construction. The alignment traverses the Devonian age Hamilton Group and Trimmers Rock Formation consisting primarily of siltstone and shale. Pyrite was observed within a rock core sample collected on the project. An extensive subsurface exploration program was performed to delineate Acid Producing Rock (APR), and the borings and laboratory testing confirmed that much of the rock contained concentrations of acid producing sulfide minerals (%Sulfur $\geq$0.5%) that warranted treatment based on PennDOT standards.

The paper will concentrate primarily on the remediation of APR on the project. Remediation included adjusting the highway alignment to minimize APR; excavated APR was treated with alkaline material and encapsulated either within the highway embankments or in a waste area located within the project limits. The exposed APR slopes were temporarily covered with a copolymer product to reduce the APR's exposure to air and water to minimize the risk of the acid producing chemical reaction from occurring. The exposed APR rock slopes were permanently covered with 10 feet of low permeability soil. A Professional Geologist was on-site during earthwork activities to direct the Contractor on how to appropriately handle the APR during construction. The remediation also considered stormwater runoff from the APR areas, treatment of re-aligned stream channels in APR areas, as well as handling seeps in APR areas through an anoxic drain.

INTRODUCTION

The SR 15, Section 088 project, also known as Central Susquehanna Valley Transportation (CSVT) Southern Section, is a new 6.5 mile, four-lane, limited-access highway located in Monroe Township and Shamokin Dam Borough, Snyder County, Pennsylvania. The CSVT Southern Section begins at the SR 11/SR 15/SR 522 interchange located just north of Selinsgrove, PA and extends to the CSVT Northern Section, which starts west of SR 15, approximately ⅓ mile south of County Line Road (SR 1022). When completed, the CSVT Southern Section will reduce commercial vehicle traffic along the highly congested "Golden Mile" portion of SR 11/SR 15 between Selinsgrove Borough and Shamokin Dam Borough. A map showing the general location of the project is included as Figure 1 and a detailed site location map is presented as Figure 2.

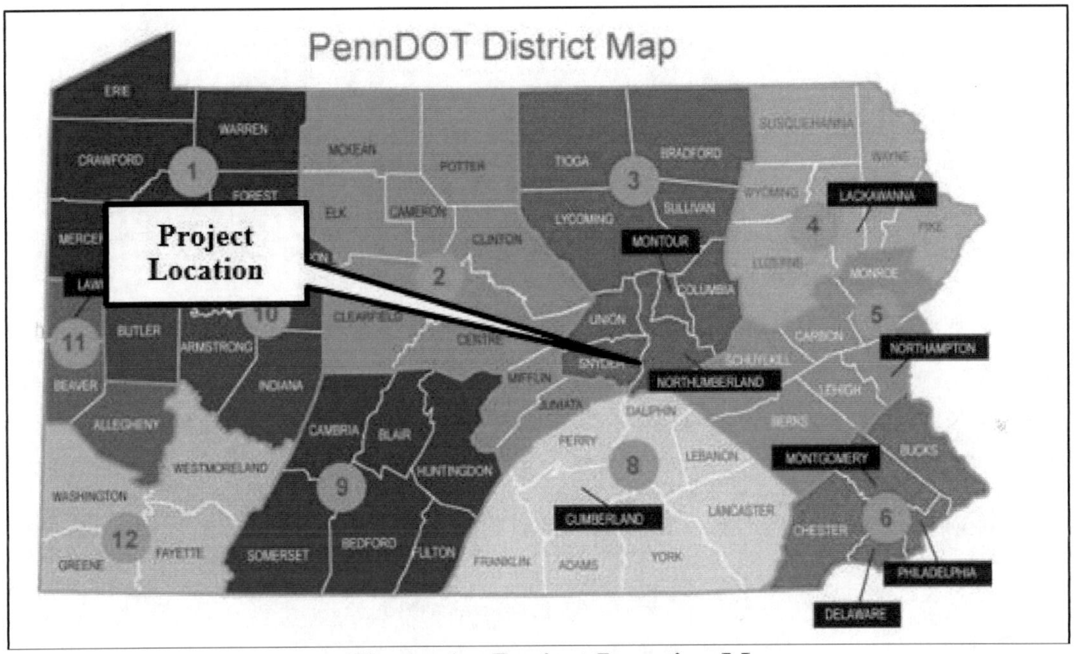

Figure 1 – Project Location Map

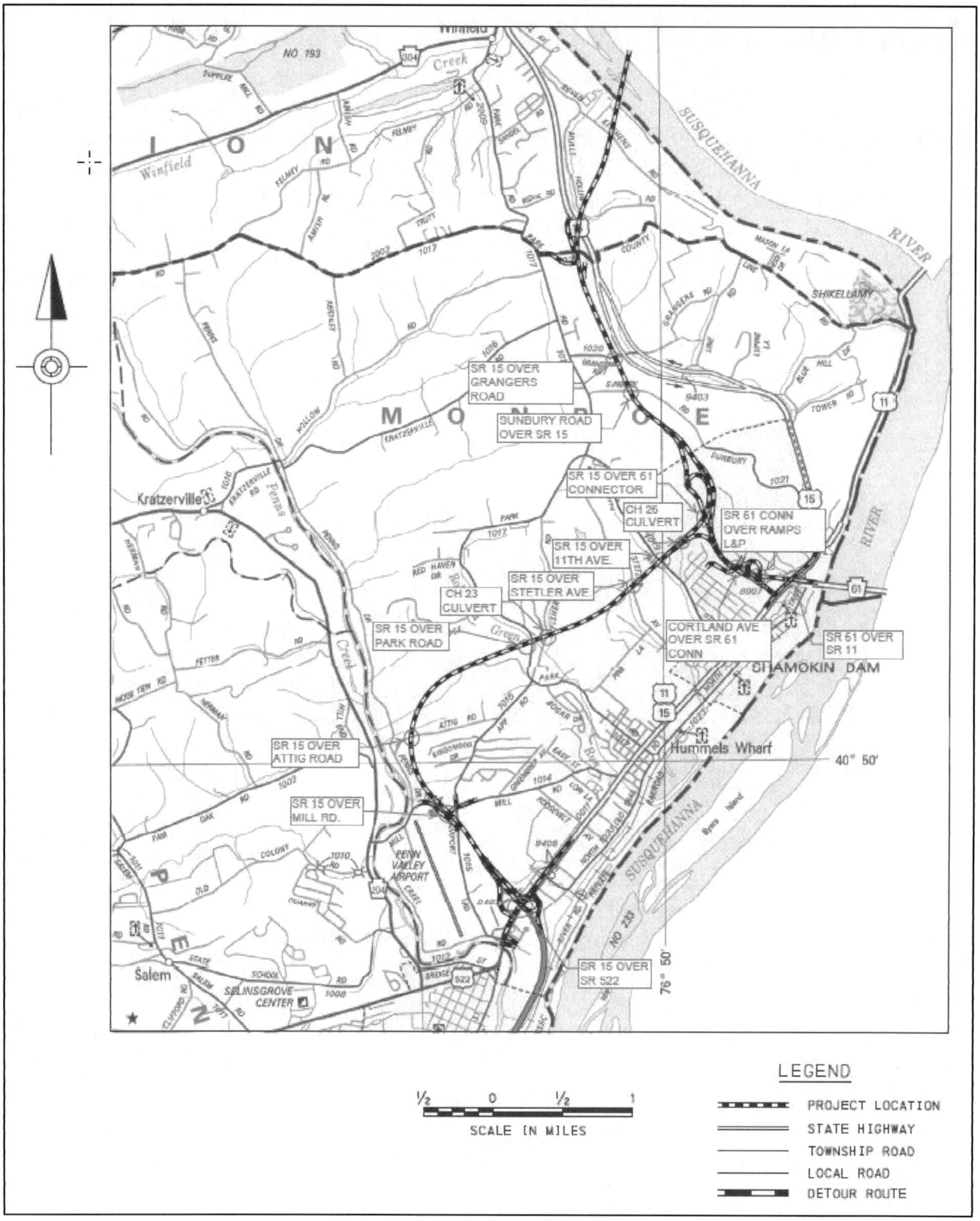

Figure 2 – Site Location Map

SITE GEOLOGY

A Geologic Map containing the CSVT South Section alignment and structure locations is included as Figure 3 below.

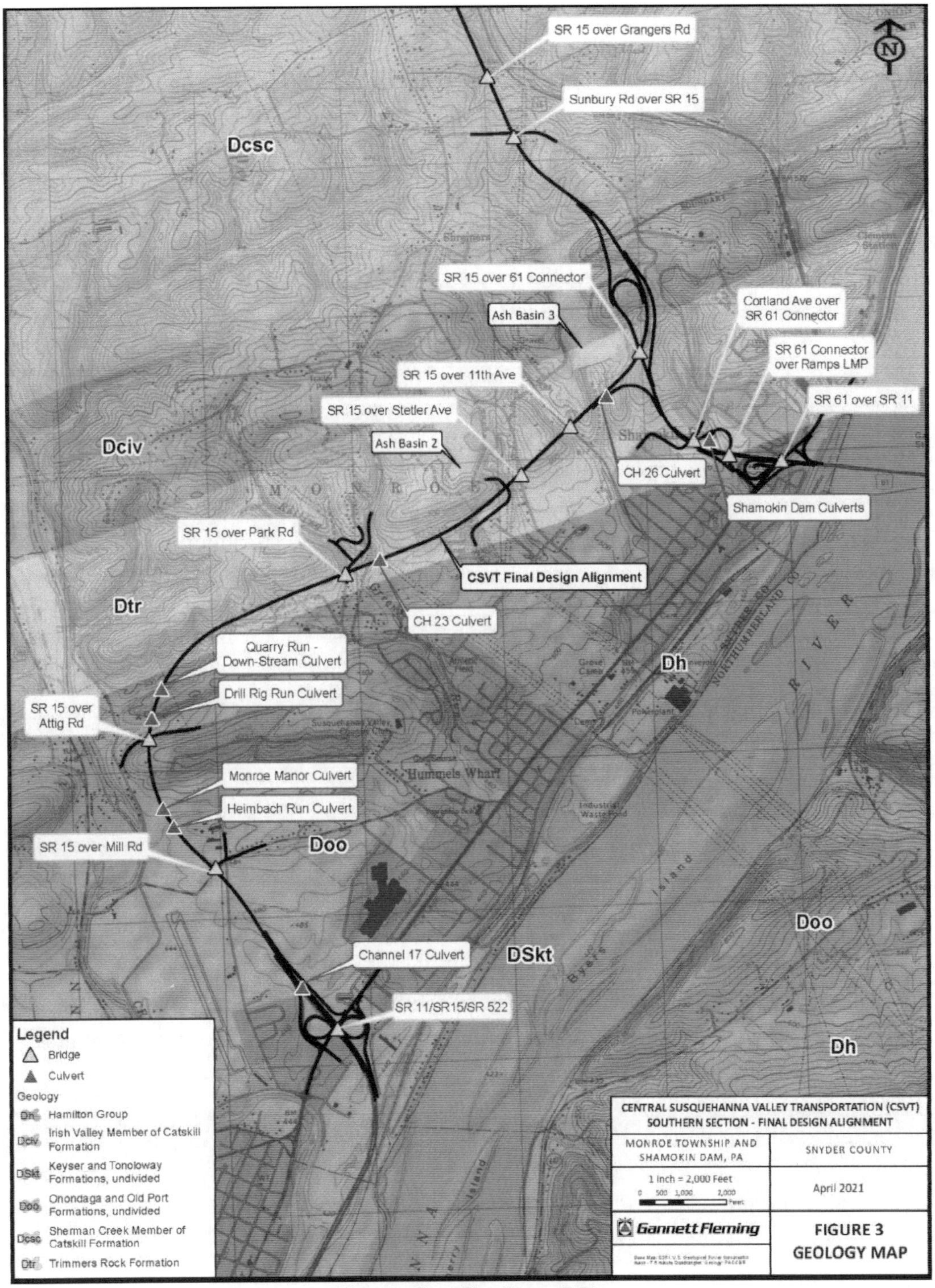

Figure 3 – Geology Map

The CSVT project lies within the Appalachian Mountain Section of the Ridge and Valley physiographic province. The surrounding topography has been developed through erosional processes of the regional drainage network and reflects the bedrock structural trends and relative northeast, oriented parallel to bedrock strike. The CSVT Southern Section alignment crosses six distinct geologic units, spanning from the Lower Devonian to the Upper Devonian, (408 to 360 million years ago) becoming younger to the north and represents an overall transition from shallow marine to deltaic depositional conditions.

The underlying rock units, described in ascending order within the project area include: the Keyser and Tonoloway Formations, undivided, the Onondaga and Old Port Formations undivided, the Hamilton Group (Marcellus Formation and Mahantango Formation), the Trimmers Rock Formation, and the Irish Valley and Sherman Creek Members of the Catskill Formation. Of these formations, the Hamilton Group and Trimmers Rock Formation were identified during the subsurface exploration program to contain sulfide bearing minerals. These two geologic units are described in more detail below.

The Hamilton Group consists of the Marcellus and Mahantango Formations which are described in ascending order. The Marcellus Formation is approximately 240 feet thick and consists of dark-gray to black, highly fissile, shales containing locally abundant pyrite and few fossils *(1)*.

The Marcellus Formation is documented as a potential acid producing rock unit *(2)*. Shale generally predominates but tends to become siltier and less fissile upward. A thin limestone unit occurs near the middle of the Marcellus and is commonly referred to as the Purcell Limestone, but has also been called the Upper Selinsgrove Limestone *(1)*.

The Mahantango Formation consists of the following four members:

- Fisher Ridge, consists of medium-gray to olive-gray, commonly laminated silty claystone, siltstone, and some very fine-grained sandstone.
- Montebello Sandstone, consists of light-olive- to medium-light-gray, very fine- to medium-grained sandstone, commonly having abundant marine fossils and containing interbeds of siltstone and silty claystone.
- Sherman Ridge, consists of light-olive- to medium-gray silty claystone with zones of siltstone and very fine-grained sandstone in upward coarsening sequences.
- Tully Limestone, a thin, but important marker bed at the top of the Mahantango Formation consisting of 10 inches of medium-dark-gray shale limestone and 2 feet of dark-gray calcareous silty shale, both abundantly fossiliferous.

Published geologic data indicates that the Trimmers Rock Formation is not a potential acid producing rock unit *(2)*. The Trimmers Rock Formation consists of olive gray and medium gray siltstone and silty shale with some very fine-grained sandstone in its upper part. Measured thickness is 2,000 feet at nearby Shamokin Dam *(1)*.

HISTORICAL BACKGROUND OF ACID PRODUCING ROCK

Acid Producing Rock (APR) is rock which contains iron sulfide minerals such as pyrite, marcasite, and pyrrhotite with pyrite being the most prevalent sulfide mineral in sedimentary rocks within Pennsylvania's Appalachian Plateaus and Ridge and Valley Provinces. These minerals may be altered during oxidation (exposed to both air and water), ferrous sulfate and sulfuric acid are produced, potentially resulting in the generation of acid rock drainage (ARD). The sulfuric acid can generate acidic runoff containing high levels of dissolved metals, resulting in water quality degradation by reducing the pH, lowering the dissolved oxygen available, and mobilizing heavy metals such as iron, aluminum, magnesium, and zinc. In general, acid generation can begin in as little as 2 to 3 weeks after exposure. Rock containing more than 0.5 percent by weight pyrite and having little to no alkaline content have the potential to be significant sources of ARD when exposed during construction *(3)*. The APR identified in the CSVT alignment is of syngenetic origin or formed entirely from the original sediments and the pyrite is disseminated throughout the rock mass.

The identification, handling, treatment and disposal of APR encountered during highway construction projects has its roots in the mining industry. The treatment of APR and prevention of ARD began with the Surface Mining and Conservation and Reclamation Act of 1971. The Act required a mining plan to provide a practical method of avoiding acid mine drainage (AMD) or other stream pollution. In the ensuing decades, regulatory requirements advanced laboratory testing methods, treatment techniques, and disposal of both ARD and APR.

In Pennsylvania, great attention to the identification, handling, treatment and disposal of APR began in 2003, after approximately 1 million tons of APR was exposed during the construction of Interstate 99 in Centre County Pennsylvania *(4)*. Project excavations exposed pyrite veins associated with zinc-lead deposit (epigenetic) that was unidentified during design. Shortly after placement of this material as embankment throughout the project corridor, acidic runoff was identified that impacted streams and groundwater and delayed construction, resulting in years of ARD treatment at a cost of over 100 million dollars to PennDOT. Treatment of ARD for this project is still ongoing. This prompted the Pennsylvania Department of Environmental Protection and PennDOT to develop detailed guidance documents to assist engineers and contractors on identification and mitigation of APR on earthwork projects.

In 2006, PennDOT authored a chapter dedicated to APR in its Publication 293- Geotechnical Engineering Manual *(3)*. The chapter provided guidelines, recommendations, and considerations for the investigation, testing, identification, prevention and treatment of potential APR and ARD in highway projects. The APR chapter in Publication 293 has undergone several revisions since its initial publication in 2006. In addition, several other State DOT's that have known Formations that can produce acidic runoff, have developed guidelines to handle APR on projects within their borders.

IDENTIFICATION AND DELINEATION OF ACID PRODUCING ROCK

During the final design boring program in 2016, pyrite was observed during examination of rock cores obtained within both the Hamilton Group and Trimmers Rock Formation. A typical occurrence of pyrite in recovered core is presented in Figure 4. After the initial discovery of pyrite in these 2 Formations, an additional 26 borings consisting of 1,450 lineal feet of rock core and 489 Acid Base Accounting (ABA) tests were completed within proposed excavations in the Hamilton Group and Trimmers Rock Formation to obtain an initial assessment of the presence of APR throughout the corridor. The data was also used to evaluate the extent of oxidized caprock *(3)*. The ABA testing was performed on all rock encountered within the borings, generally in three-foot sample lengths, and includes 1) The "Fizz" Rating, 2) Neutralization Potential (NP), and 3) Total Percent Sulfur. The Percent Sulfur and NP results are used to determine the following parameters:

- Maximum Potential Acidity (MPA), where MPA = (Total % Sulfur) x (31.25 ppt $CaCO_3$/1% Sulfur)
- Potential Ratio (PR), where PR = NP/MPA
- Net Neutralization Potential (NNP), where NNP = NP – MPA

For this project, based on concurrence from Pennsylvania Department of Environmental Protection (PADEP), material was considered potentially acidic if the ABA test results indicated Percent Sulfur > 0.5%, unless the NP of the sample was greater than 24 parts per thousand (ppt). Based on the established criteria, the preliminary ABA test results indicated that APR would be encountered along a nearly 1.5 mile stretch of the alignment, designated as the Acid Rock Focus Area, as shown in Figure 5.

Figure 4 – Observed pyrite in Marcellus Shale Formation

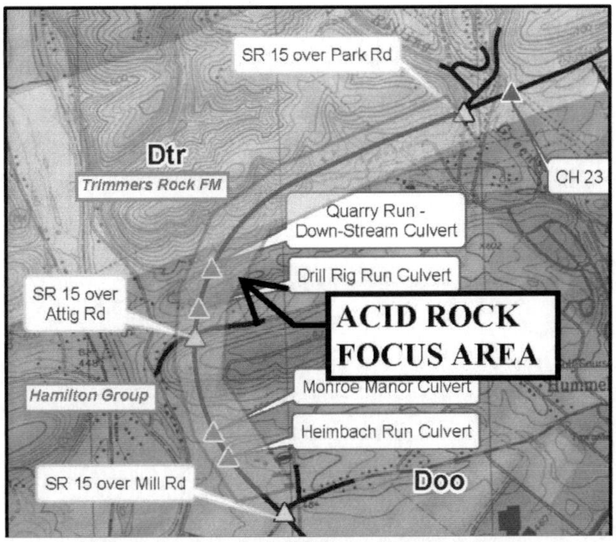

Figure 5 – Acid Rock Focus Area

Based on the preliminary ABA test results, the estimated quantity of APR within the Acid Rock Focus Area was over 1,000,000 cubic yards. After discussions with PADEP and PennDOT District 3-0, it was decided to move the alignment between 50 and 100 feet to the south of the original alignment to reduce the anticipated APR excavation quantity. The realigned roadway resulted in 4 distinct cut locations, designated at Cut 1 through Cut 4, within the Acid Rock Focus Area as shown in Figure 6.

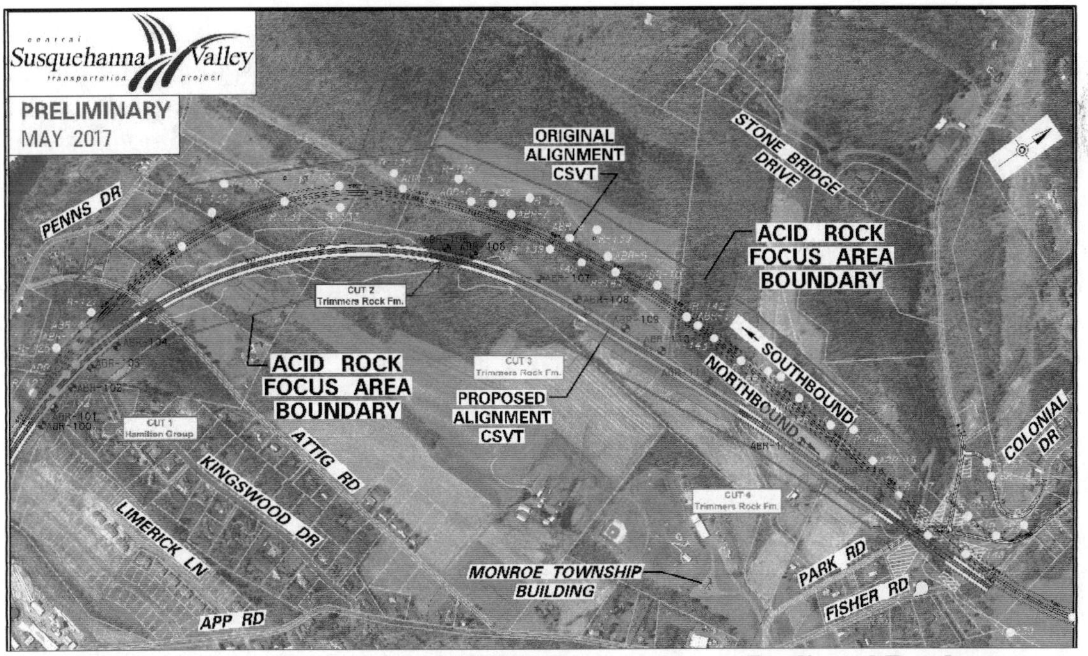

Figure 6 – Cut Slopes in Acid Rock Focus Area on Realigned Roadway

During final design, fifty-six (56) additional borings were completed with ABA laboratory testing along the new alignment within the Acid Rock Focus Area to assist with delineating APR. The borings were drilled vertically, and the locations considered the dip of the bedrock in order to provide complete coverage of all beds that would be encountered during excavation. Figure 7 illustrates the number and distribution of borings to characterize a portion of Cut 3. A total of 937 additional ABA tests (Cut 1, 484 tests, Cut 2, 13 tests, Cut 3, 405 tests, and Cut 4, 35 tests) were performed during final design and the tests provided full depth sampling of the rock units that would be excavated during construction. The ABA test results indicated that APR would be encountered in Cut 1 through Cut 3, but the rock encountered in Cut 4 would not be considered APR.

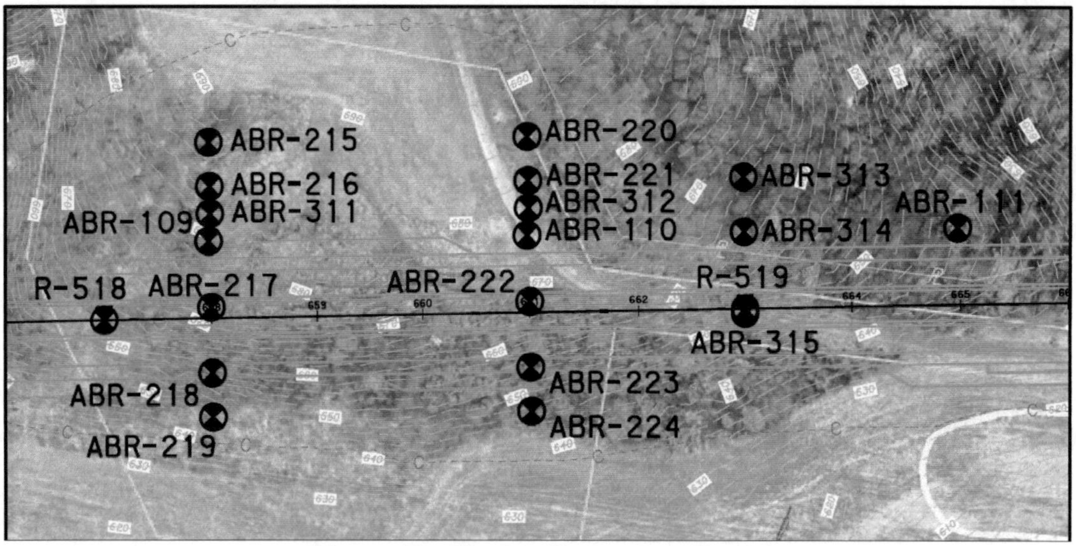

Figure 7 – Boring Layout at Cut 3

As final design of the project progressed, additional subsurface explorations, consisting of borings and ABA laboratory testing, were performed at any location where proposed excavations would be performed within the Hamilton Group or Trimmers Rock Formation. A total of 254 additional ABA test were performed. The results of these additional subsurface explorations indicated that APR would be encountered at 4 other locations of the project: 1) Park Road Bridge Pier, 2) Culvert 23, 3) Culvert 28A, and 4) Ramp LMP. Note that Culvert 28A and Ramp LMP are in close proximity to each other. Figure 8 shows the six project locations where APR was delineated and Table 1 provides a summary of APR locations, along with estimated APR excavation volume at each location. In total, and estimated 250,000 cubic yards of APR was anticipated on the project, which was a significant reduction from the original quantity of over 1,000,000 cubic yards of APR.

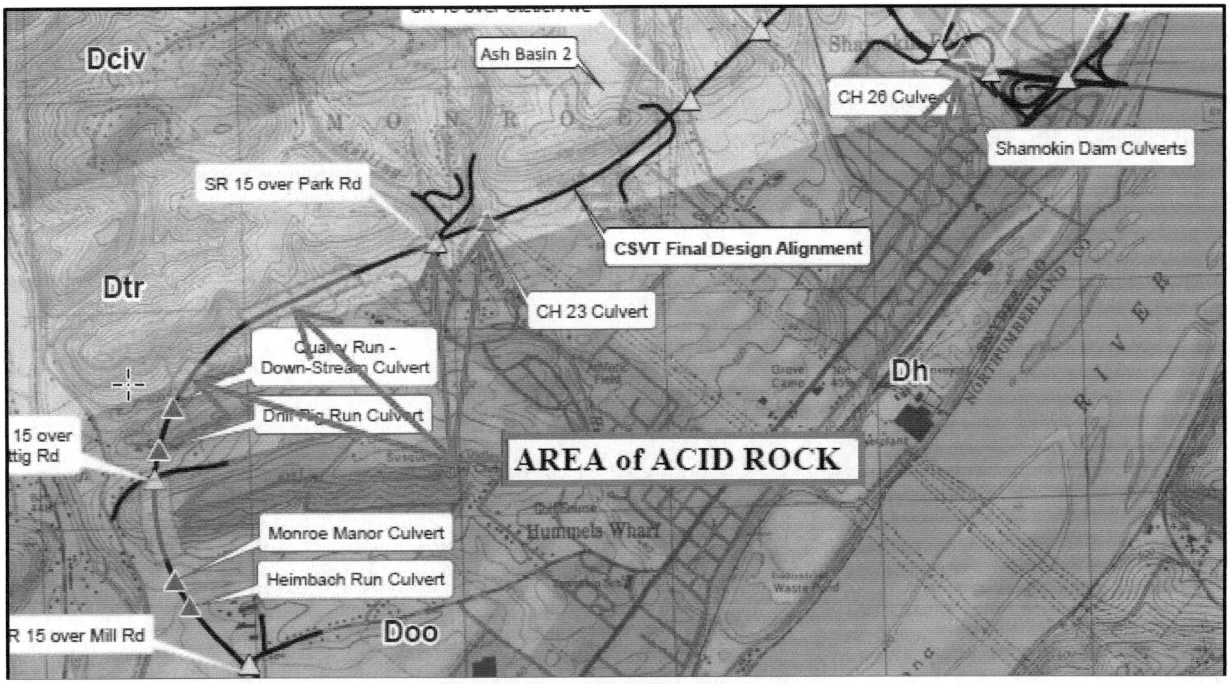

Figure 8 – Acid Rock Areas

Table 1 – Summary of APR Areas

Area	Formation	Number of Borings	Approx. Exc. Volume
Cut No. 1	Hamilton Group (Marcellus Sh. Fm., Purcell Mbr.) and Mahantango Fm.	21	92,000
Cut No. 2	Trimmers Rock Fm.	2	8,000
Cut No. 3	Trimmers Rock Fm.	28	133,000
Park Road Pier	Trimmers Rock Fm.	2	<1,000
Culvert 23	Trimmers Rock Fm.	5	<1,000
Channel 28/Ramp LMP*	Marcellus Sh. Fm., Trimmers Rock Fm.	15	15,000

* Included together due to close proximity of these areas to each other.

REMEDIATION OF APR

Many alternatives were considered to remediate both excavated APR material and exposed APR slopes on the project. Ultimately, it was determined that the preferred remediation was to minimize exposure of APR to air and water, which reduces the potential for oxidation of the APR, and the subsequent release of acid into the surrounding environment. Since acid generation can begin in as little as 2 to 3 weeks after exposure, it was recommended to cover any material that was delineated as APR within 7 days of exposure during construction to avoid oxidation of the material *(3)*. A detailed discussion of both temporary and permanent measures employed to cover the APR on the project is provided below.

The comprehensive subsurface exploration program was used to delineate the areas where APR would be encountered on the project. These areas, as well as the recommended remedial efforts, were clearly identified in the Acid Rock Handling Plan (ARHP) that was prepared for the project. The ARHP was reviewed and approved by PADEP, and the ARHP, along with the accompanying project special provisions became part of the construction contract documents. Furthermore, PennDOT provided an onsite licensed geologist during construction to review rock excavation to identify any additional APR on the project, as well as to monitor and document the excavation and handling of APR on the project was done in accordance with the ARHP *(5)*. The remediation efforts for each of the various aspects of APR are discussed in detail in the sections below.

Excavated APR

PennDOT Publication 293 provides guidance on treatment of excavated APR. The excavated APR should be treated immediately upon excavation by adding supplemental alkaline material (SAM) and encapsulating the treated APR with geotextile and a minimum of 3 feet of capping soil. Two encapsulation areas were provided within fill slopes for the project. The SAM added to the APR was determined based on the ABA testing such that calcium carbonate ($CaCO_3$) available in the treated APR results in a net neutralization (NNP) of at least 24 parts per thousand (ppt) $CaCO_3$, which should result in an alkaline condition with a factor of safety (FS) of 2. The bottom of treated APR should be placed at least 5 feet above the seasonal high-water table elevation and 100-year flood elevation. The typical APR encapsulation detail is shown in Figure 9, Figure 10 shows SAM being added to excavated APR, and Figure 11 presents a photo of the Contractor placing treated APR within one of the encapsulation areas during construction *(3)*.

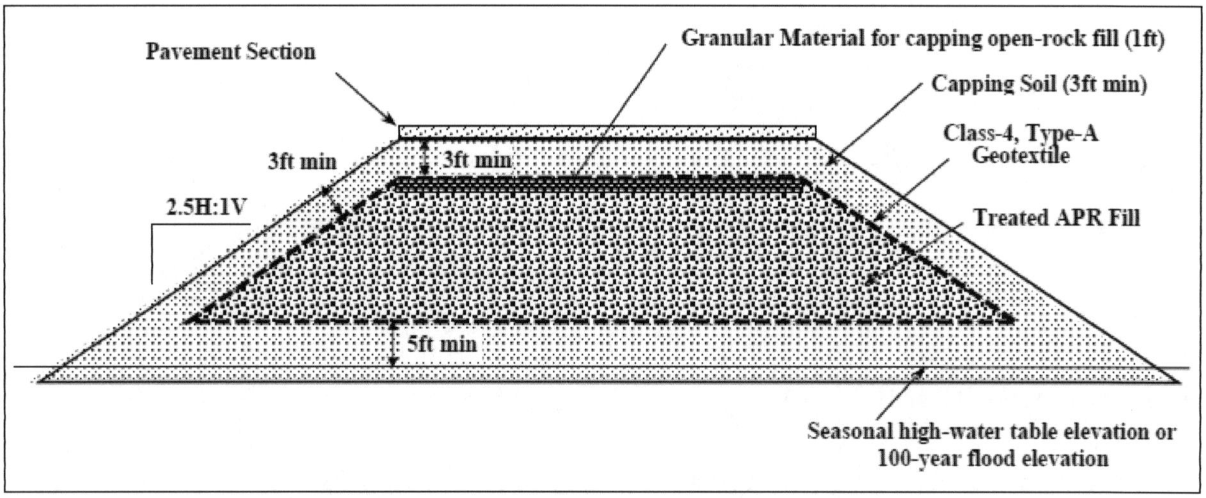

Figure 9 – Typical APR Encapsulation Detail

Figure 10 – SAM Addition to Excavated APR

Figure 11 – APR Encapsulation During Construction

Per Publication 293, the SAM treatment levels are based on the average NNP of the APR such that the amount of SAM added will result in the treated fill achieving a target NNP of 24 ppt $CaCO_3$ *(3)*. However, based on the amount of ABA testing that was performed on the project and performing statistical analyses, it was determined that an alternative treatment methodology of APR based on the lowest NNP value to a FS = 1.5 resulted in an average treated fill FS greater than 10. The revised methodology to determine SAM treatment levels was accepted by PennDOT and PADEP for this project and resulted in significantly less SAM required, which resulting in saving the Department approximately $422,000. The additional drilling and ABA testing required to obtain a comprehensive understanding of the APR cost approximately $130,000, resulting in a total cost savings to PennDOT of approximately $300,000. The SAM treatment rates for each APR area is provided in Table 2.

Table 2 – Summary of SAM Treatment Rates

APR Area	Station Limits (SR 15 BL or as noted)	SAM Treatment Rate (Tons of 96% CCE/445 CY of APR)
Cut No. 1	609+00 to 613+00	66.6
Cut No. 1	617+00 to 620+75	50.4
Cut No. 2	642+50 to 646+75, LT.	78.4
Cut No. 3	649+00 to 655+00, LT.	95.4
Cut No. 3	656+00 to 666+00	95.4
Park Road Pier	684+21	NA*
Culvert 23	689+50	83.7
Ramp LMP	10+50 to 13+25 (Ramp LMP)	81.6
Culvert 28A	536+77 (SR 61)	109.7

* CCE = Calcium Carbonate Equivalent of the SAM.

** Material from Park Road Pier to be disposed of off-site due to sequence of construction.

A portion of Cut No. 1, from Station 613+00 to Station 617+00, was determined to not contain APR based on the ABA test results. This section of Cut No. 1 consists of the Purcell Limestone Member of the Marcellus Formation and even though pyrite was observed in the matrix of the rock, and sulfur content was in excess of 0.5 percent, the ABA test results showed NNP values typically in excess of 30 ppt $CaCO_3$., which is above the target NNP of 24 ppt $CaCO_3$. Therefore, this material was delineated as clean fill for use in embankment construction on the project *(5)*.

Exposed APR in Excavations

Exposed APR slopes within excavations would have potential to produce acidity if air and water are permitted to contact the exposed slopes. Therefore, the remediation of exposed APR slopes was to restrict exposure of these slopes to air and water to the best extent possible upon excavation. Due to the height of many of these slopes, it would not be feasible to completely excavate these cuts from top to bottom within 7 days; therefore, both temporary and permanent methods of restricting air and water exposure was required.

Temporary coverage of the exposed APR surface using an emulsified resin or co-polymer product was required within 7 days of exposure of APR within proposed cut slopes. Temporary coverage of the exposed APR surface within a structure foundation or realigned stream channel using a concrete mud mat was required within 7 days of exposure of APR (i.e., Park Road Pier 1, Culvert 23 and Culvert 28A). A photo of the Contractor temporarily covering exposed APR slopes with a co-polymer product is provided as Figure 12, and APR coverage with a concrete mud mat is included as Figure 13.

Figure 12 – Temporary Coverage of Exposed APR Cut Slopes

Figure 13 – Temporary Coverage of Exposed APR Foundation

Permanent coverage of the exposed APR within foundation excavations is achieved through PennDOT's backfill requirements for their structures. Permanent coverage of exposed APR slopes was accomplished using a minimum of 10 feet of soil cover for slopes in excess of 20 feet in height and 4 feet of soil cover for slopes less than 20 feet in height. The requirements of the soil cover was that the material must meet the requirements of soil per Publication 408,

Section 206 *(6)*, with an additional requirement that the soil must have a plasticity index of at least 3 in order to minimize the risk of erosion during storm events prior to establishing vegetation. In addition, a turf reinforcement mat (TRM) was installed to minimize surficial erosion until vegetation is established. A photo of the Contractor permanently covering exposed APR with soil is provided as Figure 14.

Figure 14 – Permanent Coverage of Exposed APR Slopes with Soil

The bottom of cuts (i.e., the roadway location) in APR also required coverage in order to mitigate risk of acid production. In these subgrade areas, a 4-foot-thick soil cover was recommended to mitigate potential for acid production *(3),* and in order to ensure the material placed was adequate for pavement design purposes, the subgrade soil cover had specific requirements. The subgrade soil cover was required to exhibit a plasticity index greater than 6 (i.e., no clay permitted), and must be a coarse-grained material (i.e., less than 50 percent passing the No. 200 sieve), but also have at least 20 percent of the material passing the No. 200 sieve.

Groundwater Seeps from APR

Based on the proposed bottom of cut elevations and water levels encountered in the borings, it was expected that water would be encountered in several APR cuts. The design to handle seeps encountered in APR excavations included collection of the water from the seep via a seepage interceptor drain that would outlet into an anoxic drain at the base of the cut. Both the seepage interceptor and anoxic drains were to be constructed from aggregate with calcium carbonate equivalent (CCE) of 85 percent to ensure some buffering capacity in the event the drainage was acidic. The anoxic drain would outlet into APR designated stormwater management basins. Figure 15 shows the typical detail used for the seepage interceptor drain and Figure 16 provides the anoxic drain typical detail. Note that no seeps were encountered during construction of the APR cut slopes; however, the anoxic drain was included at the base of all

APR cut slopes to ensure any future water emanating from APR areas would be collected and outlet to APR designated stormwater management basins *(5)*.

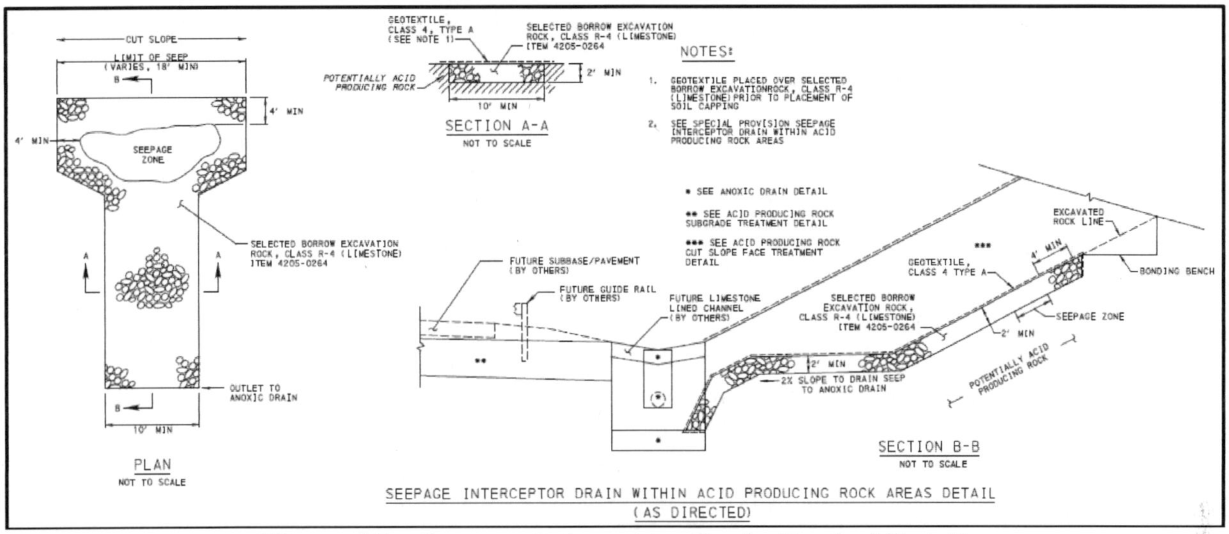

Figure 15 – Seepage Interceptor Drain Typical Detail

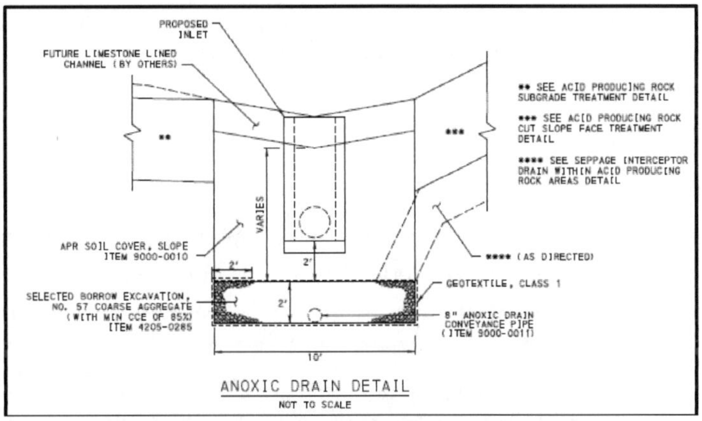

Figure 16 – Anoxic Drain Typical Detail

Stormwater Management and Monitoring Program

Construction methods and sequencing within designated APR areas on the project were intended to reduce groundwater and surface water runoff. Stormwater runoff and seeps encountered during construction in APR areas were conveyed and retained in an appropriately sized and lined retention basins. A total of five dual cell retention basins were provided to store water from APR areas on the project. These basins were lined with limestone rock as a measure to provide alkaline material in the unlikely event that acidic runoff was generated on the project. Runoff captured in the basins after each precipitation event exceeding 0.10 inch was screened for pH, temperature and specific conductance, and the results were provided to PennDOT and PADEP. If pH readings were less than 5.5, PADEP would require regulatory permits to neutralize the acidic runoff. To date, no acidic conditions were encountered in the APR basins during construction.

Groundwater and Surface Water Monitoring Program

A comprehensive water monitoring program, consisting of both groundwater and surface water, was performed pre-construction, during construction and will continue post-construction. The pre-construction phase of testing was performed quarterly for at least 1 year prior to commencing construction activities. Quarterly sampling has been performed throughout the duration of construction and sampling will continue quarterly for a period of 1-year post-construction. If at any point the conditions/groundwater quality trends warrant additional testing, more frequent testing will be considered.

To monitor groundwater, a network of monitoring wells installed by PennDOT and residential well sampling was performed within ¼ miles of identified APR areas. After extensive coordination with PADEP, groundwater samples were tested for the following analytical parameters:

Total acidity	Total alkalinity	Chloride
Nitrate/Nitrite-N	Sulfate	Specific Conductance
Total suspended solids	Total Dissolved solids	pH
Turbidity	Hardness	Aluminum, total
Arsenic, total and dissolved	Iron, total and dissolved	Lead total and dissolved
Manganese, total and dissolved	Zinc, total	Magnesium, total
Sodium, total	Calcium, total	E. Coli
Total Coliform		

The surface water monitoring program was intended to record the water quality conditions of receiving water bodies located near construction activities associated with the excavation, treatment, and placement/encapsulation of APR in the project area. A total of 12 surface water bodies were tested both upstream and downstream of construction activities to verify if the project area was contributing to the surface water quality in the project area. After extensive coordination with PADEP, surface water samples were tested for the following analytic parameters:

Total acidity	Total alkalinity	Sulfate
Sulfate	Total suspended solids	pH
Total Aluminum	Total Arsenic	Total Calcium
Total Copper	Total Iron	Total Lead
Total Magnesium	Total Manganese	Total Nickel
Total Zinc	Specific Conductance	

Conclusions

To date, there have been no known releases of acid drainage on the project. In addition, the water monitoring program did not indicate any significant changes to the surface or groundwater in the area. It is believed that as long as the presence of APR on a project is known, it can be treated effectively. Proper handling and treatment can be attributed to performing a comprehensive sampling and testing program, which permits delineation of what is and what is not considered APR on the project. Close coordination with the Owner (PennDOT) and Regulatory Agencies (PADEP) on the presence of APR on the project, collaborating on proper handling techniques, and developing an ARHP and Contract Documents, as well as having an on-site Geologist to assist the Contractor on handling and treating APR are considered crucial elements to the success of this project.

REFERENCES

(1) MacLachlan, D. B., Hoskins, D. M., and Payne, D. F., 1995, Bedrock geology of the Freeburg 7-1/2‘ quadrangle: Pennsylvania Geological Survey, 4th ser., Open File Report 95-04.

(2) Pennsylvania Geological Survey, (2004), Geologic units containing potentially significant acid-producing sulfide materials: Pennsylvania Geological Survey 4th ser. Open File Report OFMI 05-01.1, 1 map, scale 1:500,000.

(3) Pennsylvania Department of Transportation, Publication 293, *Geotechnical Engineering Manual 4,* 2016

(4) Hammarstrom, J. M., Brady, Keith, and Cravotta, III, C. A., 2005a, Acid-rock drainage at Skytop, Centre County, Pennsylvania, 2004: U.S. Geological Survey, Open-File Report 2005-1148, 45 p

(5) Gannett Fleming, Inc., *Pyritic Material Handling Plan for Project Excavations*, prepared for Pennsylvania Department of Transportation, April 2021

(6) Pennsylvania Department of Transportation, Publication 408, *Specifications,* 2020.

Application of Geological Mapping to Evaluate Karstic Units Affecting Critical Infrastructure in East Tennessee

Case Study: John Sevier Dam

David Hannam, P.G.
WSP USA Inc.
5170 Peachtree Road, Building 100, Suite 300
Atlanta, Georgia 30341
347-502-0267
david.hannam@wsp.com

Randy L. Kath, Ph.D., P.G.
Petrologic Solutions, Inc.
Professor of Geology
University of West Georgia
Callaway Science Building
125 East Roberts Drive, Carrollton, GA 30117
678-313-4146 / 678-839-4063
rlkath@comcast.net / rkath@westga.edu

Matthew T. Huebner, Ph.D, P.G.
Geologist – Dam Safety Geotechnics
1101 Market Street, Chattanooga, TN 37402
865-414-3218
mthuebner@tva.gov

Prepared for the 73rd Highway Geology Symposium, September, 2024

Disclaimer

Statements and views presented in this paper are strictly those of the author(s), and do not necessarily reflect positions held by their affiliations, the Highway Geology Symposium (HGS), or others acknowledged above. The mention of trade names for commercial products does not imply the approval or endorsement by HGS. Any views or opinions expressed or implied are that of the author and do not necessarily represent the opinions of the Tennessee Valley Authority (TVA) or the U.S. government. TVA does not intend or imply an endorsement of any kind.

ABSTRACT

Detailed geological mapping and lineament analysis were conducted as part of a geotechnical investigation for improvements to the Right Embankment for the John Sevier (JSF) dam, a run-of-river dam on the Holston River in Hawkins County, east Tennessee.

Karstic foundations had been treated near the center of the overfall spillway of the dam during construction in the early 1950s. Previous studies interpreted that the Right Embankment is underlain by the Newala Formation. The Newala Formation, an Ordovician biostratigraphic unit of which the Mascot Dolomite is the lithographic equivalent, forms the uppermost part of the Knox Group. The unit is known to be susceptible to karst and there is geomorphological evidence of karst features in the local area.

Geological mapping showed the Right Embankment is founded on the southeastern limb of a plunging, overturned anticline with an axis of Mascot Dolomite trending northeast-southwest. Geological mapping demonstrated that the Right Embankment is underlain by the upper member of the Lenoir Limestone, an interbedded shale and limestone that is relatively less susceptible to karst dissolution. The immediate upper slope and hilltop are underlain by the Mosheim Member, a pure, massive homogenous limestone that is more susceptible to dissolution and forming a karst topography.

This revised geological interpretation explains the conspicuous geomorphic evidence for sinkholes in the nearby area, yet why only minor evidence of karst dissolution was encountered during the geotechnical investigation. The geological mapping supports interpretations of drilling and geophysics and has refined the understanding of perceived risk related to karst in the Right Embankment foundation.

INTRODUCTION

JSF is a run-of-river dam on a reach of the Holston River in Hawkins County, Tennessee. Originally, the Tennessee Valley Authority (TVA) constructed the John Sevier (JSF) dam in 1955 to provide cooling water for the John Sevier Fossil Plant, which has since been decommissioned. The dam now provides a reservoir of water for use at the John Sevier Combined Cycle Power Plant on the south side of the river. JSF also raises water elevation upstream facilitating local boating and fishing. The dam includes a gravity overflow section serving as the main spillway, a gated section, and earthen embankments on both the left (south) and right (north) sides of the Holston River (**Figure 1**).

Figure 1: Aerial View of John Sevier Dam (Google Earth, 2019) with location map of east Tennessee inset.

The Right Embankment has a comparatively low crest height (design elevation of 1,085 feet) and is susceptible to overtopping. Previous information indicated a potential for karst features to affect the Right Embankment. This initiated a geotechnical investigation to better understand karst risk and support future design.

PHYSIOGRAPHY

The site is within the Valley and Ridge physiographic province, which is characterized by a series of long, narrow northeast/southwest-trending ridges separated by wide parallel valleys. These Paleozoic-age sedimentary rocks have been folded into a series of anticlines and synclines broken by numerous thrust (reverse) faults and lateral strike-slip faults. The ridges are primarily held up by layers of sedimentary rock more resistant to weathering, such as sandstone, whereas the valleys are underlain by shale and limestone that weathers more deeply than the sandstones.

The site resides at the base of a northeast-trending ridge north of the Holston River. Topographic relief in the general area is over 200 feet. Topographic lows are about elevation 1,080 feet (North American Vertical Datum 1988) at the Holston River. Topographic highs are up to elevation 1,360 ft. on the ridge to the southeast and elevation 1,200 to the northwest of the Holston River. The Holston River flows to the southwest within a broad flat valley.

REGIONAL GEOLOGY

Based on regional geologic mapping (Rodgers, 1953; Bultman, 2005), rocks exposed near the project site occur within the lower Ordovician section of the Knox Group and middle Ordovician Chickamauga Group stratigraphy (**Figure 2**).

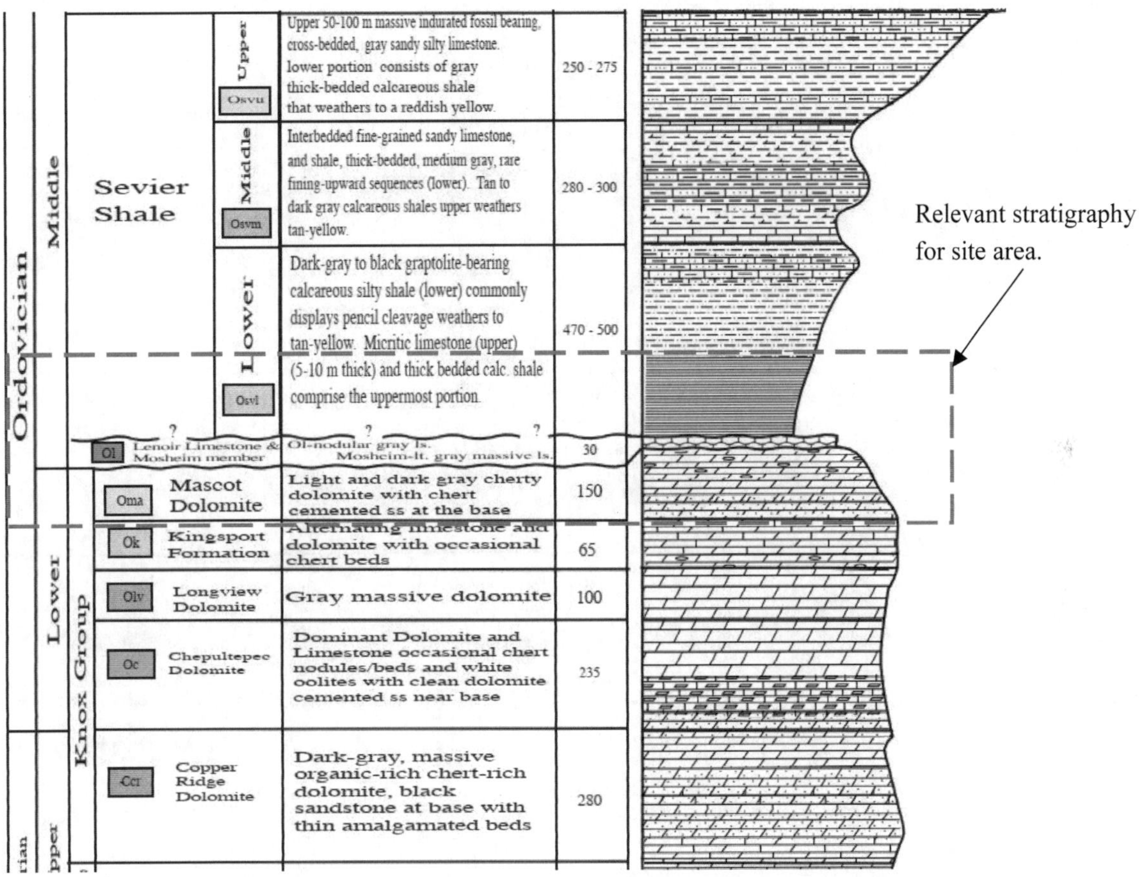

Figure 2: Generalized stratigraphic column for the Knox Group and Chickamauga Group, from Bultman (2005).

The Knox Group (mapped as Newala Limestone at the site by Rogers (1953) and Mascot Dolomite by Bultman (2005)) variably comprises cherty dolostone, dolomitic limestone and limestone. As these rocks are calcareous they are prone to chemical as well as physical weathering and typically develop karstic terrain.

Stratigraphically above the Knox Group, the Chickamauga Group is a sequence of mixed clastic and carbonate rocks. These consist of, from bottom to top, the Lenoir Limestone, Sevier Shale, and Bays Formation. The Knox Group and Chickamauga Group are separated by an erosional unconformity. Based on published geologic mapping in the region, a single thrust fault is shown to occur near the right abutment of the JSF between the Sevier Shale and Newala Limestone (**Figure 3**).

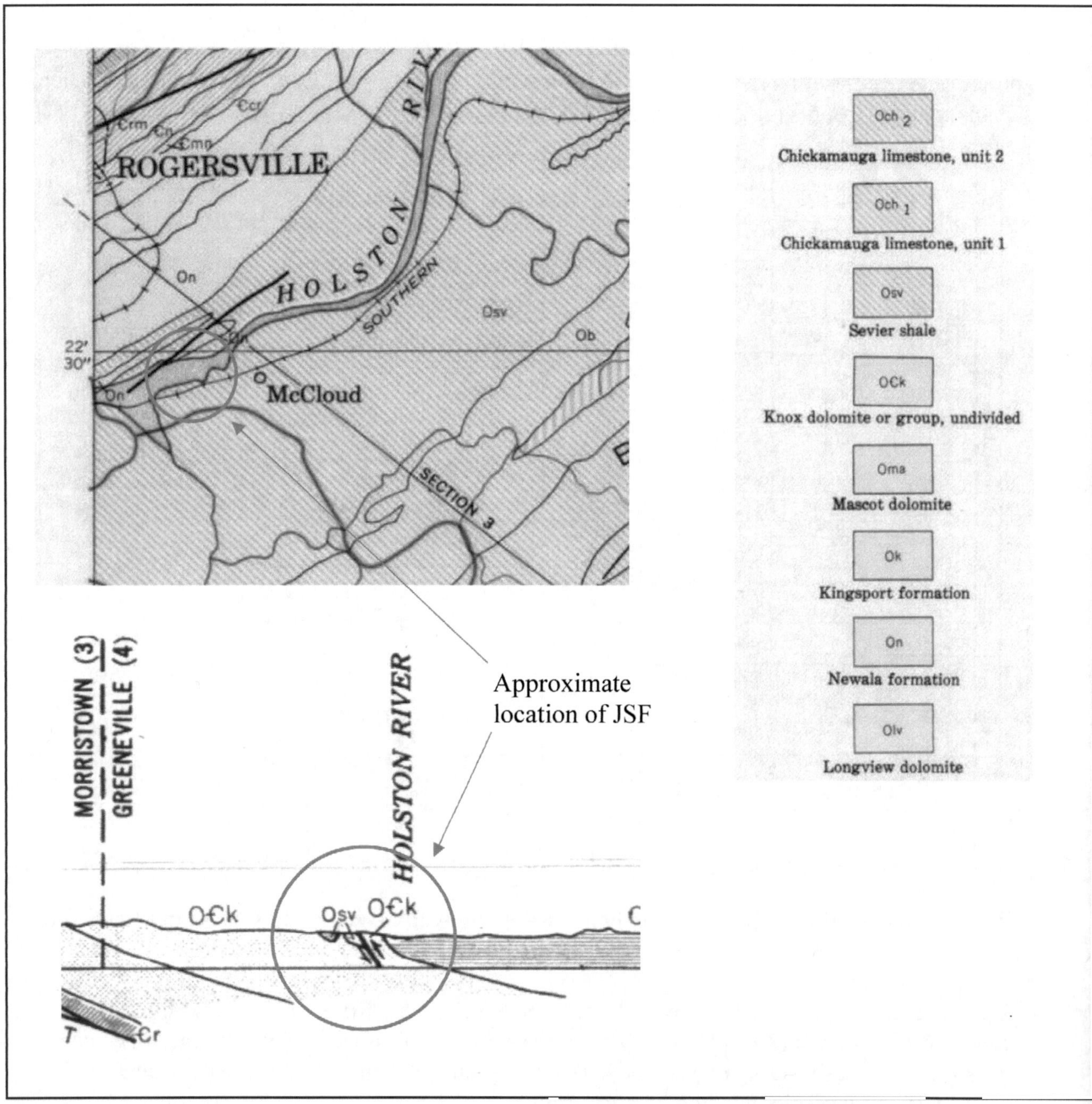

Figure 3: Extract of geological map and cross section from Rogers 1953. The site is mapped as a contact between the Newala Formation and the Sevier Shale.

RIGHT EMBANKMENT CONSTRUCTION

The Right Embankment was designed to be approximately 340 feet in length with a crest width of 10 feet at elevation 1,085.0 feet and 2H (horizontal) to 1V (vertical) side slopes. Impervious rolled fill was placed over the cutoff trench and remaining alluvial soils to construct the embankment. A continuous precast concrete cutoff wall was constructed on the upstream crest of the embankment. A natural aggregate filter composed of first a fine filter and then a coarse filter was placed over the rolled fill to address potential piping of the soils on the downstream side. The entire embankment was armored with rip rap which was subsequently grouted in 1956. Additional grouted rip rap was added in 2020.

TVA Drawing 822K1209 (TVA, 1953) shows the geological conditions encountered during construction (**Figure 4**). The drawing shows limestone underlying the Right Embankment, and shale underlying the Left Embankment. The contact to a shale is around the center of the channel underlying the spillway.

The drawing shows open cavities near the shale – limestone contact. Solutioning of the bedrock was encountered during construction in the form of open cavities and downstream air and water boils. As a result, a cut-off trench and a grout curtain were installed. The grout curtain was installed around 2 feet upstream of the base line, extending 20 feet below top of rock.

The construction drawings show variable grout takes during installation of the grout curtain. The highest grout takes were towards the center of the river, at the shale – limestone contact between spillway monoliths 5 – 7. Documents indicate a total of nearly 3,000 bags of grout were used in the grout curtain and most, around 85%, were used at this lithological contact.

Near the training wall (a structural feature separating the embankment from the spillway) of the Right Embankment, 104 bags of grout were recorded. This is a relatively high number compared to the 0 to 3 bags of grout recorded for other sections of the Right Embankment (**Figure 4**). The geological section in this area also shows faulting near the end of the Right Embankment.

Original Geological Interpretation

Prior to site specific geological mapping, available published geological map for the area (Rogers, 1953), show a thrust fault contact between the Newala Formation and Sevier Shale near the center of the Hoston River. In addition, the cross section generated during the construction shows a limestone – shale fault contact near the center of the river. Limited geotechnical drilling conducted in 2015 (TVA, 2019) confirmed the presence of limestone underlying the Right Embankment and shale underlying the Left Embankment

Based on this information, the original ground model concluded that the Newala Formation underlies the Right Embankment and the Sevier Shale underlies the Left Embankment and the contact in the center of the river channel was a thrust fault (**Figure 6**).

The Newala Formation/Mascot Dolomite is a geological unit locally known to be highly susceptible to karst. A digital elevation model was generated for the area and showed clear geomorphic evidence of karst in the vicinity of the dam (TVA, 2019) (**Figure 5**). Surface geophysics conducted in 2020 indicated several geophysical anomalies in the area of the Right Embankment. A combination of 1) the interpretation underlying Newala Formation/Mascot

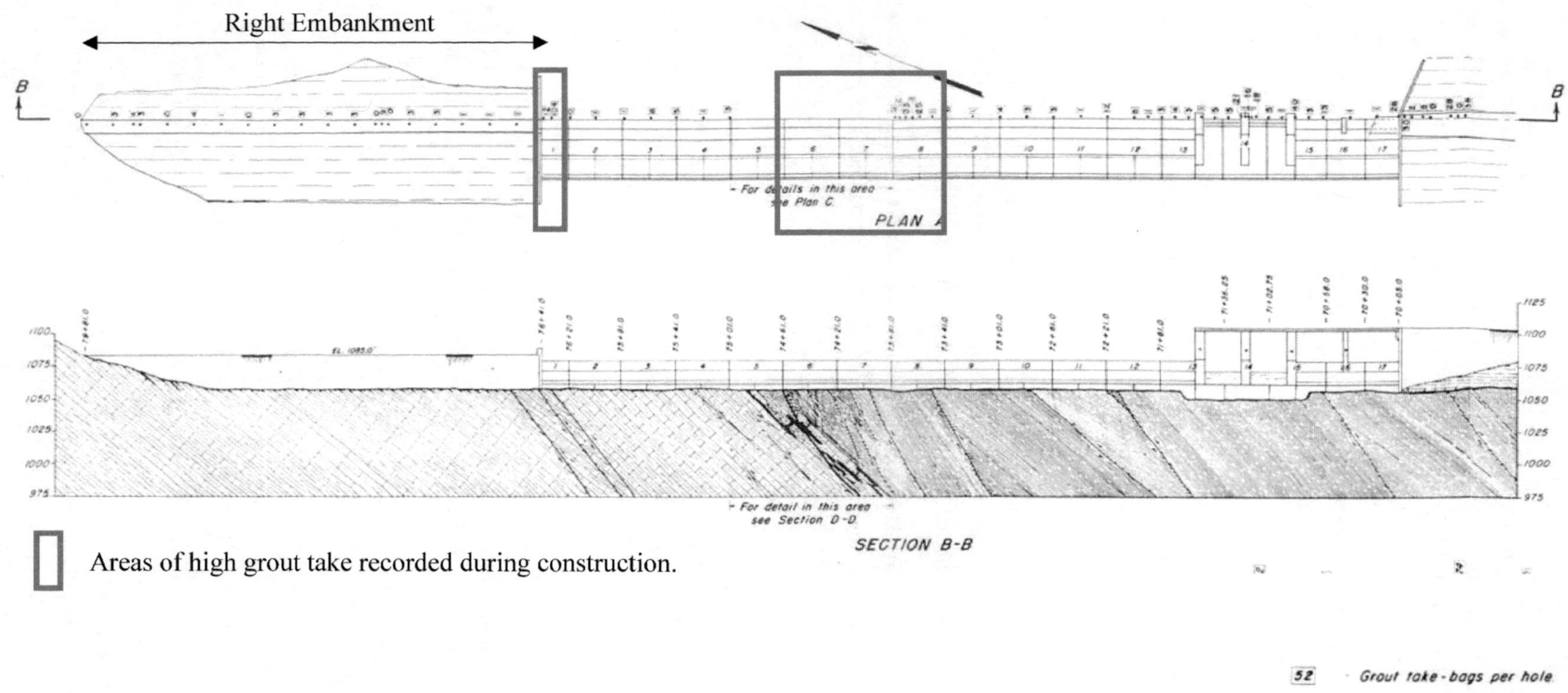

Figure 4: TVA, 1954 construction drawing of geologic section through dam showing limestone shale contact, encountered open cavities and grout takes during cutoff wall construction.

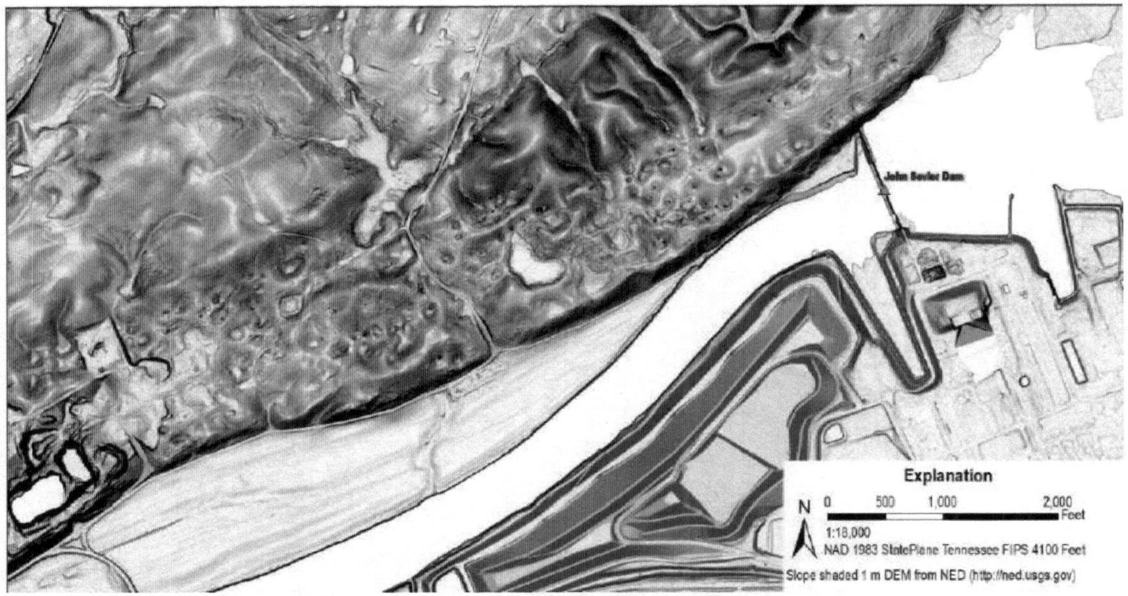

Figure 5: Digital Elevation Model showing karst terrain in the vicinity of JSF (TVA, 2019)

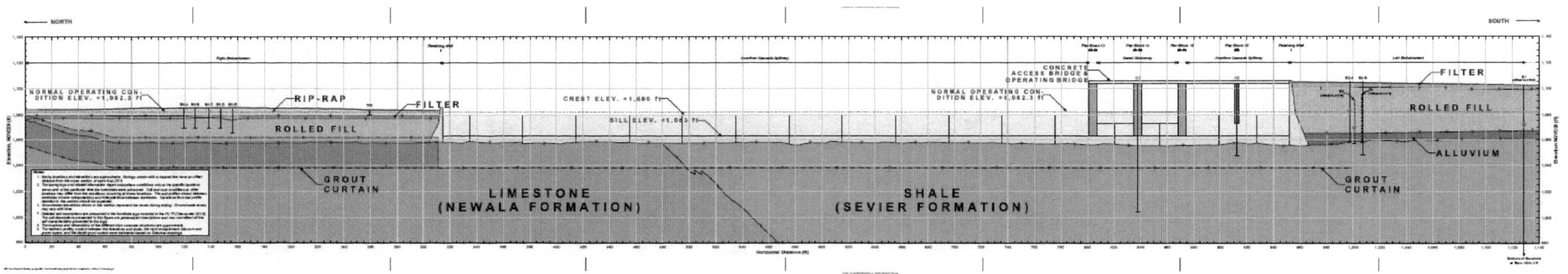

Figure 6: Section through dam showing original geological interpretation (TVA, 2015))

Dolomite, 2) local geomorphic evidence of karst, 3) geophysical anomalies and 4) karst encountered during construction, determined that a high potential for karst-related issues was concluded for the Right Embankment.

GEOLOGICAL AND GEOTECHNICAL INVESTIGATION - OVERVIEW

Early in 2023 a geological and geotechnical investigation was undertaken to better understand the karst risk to the Right Embankment. The investigation included geological mapping, a lineament study, scan lines of the right abutment, 25 geotechnical boreholes drilled up to 70 feet into bedrock, packer testing, downhole geophysics, surface geophysics and laboratory testing of rock samples.

The investigation supplemented previous geotechnical investigations which had been completed at the site in 2015 through 2020 and included boreholes, laboratory testing, downhole geophysics and surface geophysics.

Geological Mapping Study

Geologic mapping along the right abutment and the area north the abutment was completed using base maps generated from Burem and McCloud TN USGS 7.5-minute topographic quadrangles and a 2-foot contour interval map developed from Tennessee LiDAR tiles.

Lithology, mineralogy, orientation and characteristics of structural discontinuities were recorded at each map station. Map station locations were recorded using a hand-held, Wide Area Augmentation System (WAAS)-enabled Global Positioning System (GPS) and the Avenza Maps application. Karstic observations were recorded during mapping. The geological map produced for the area is shown in Figure 5.

Stratigraphic and Lithological Findings

The mapping showed that the rocks exposed in the right abutment and the area north of the abutment are consistent with stratigraphy of the lower Ordovician Knox Group and middle Ordovician Chickamauga Group in northeast Tennessee. The following lithologic descriptions are based on observations made during geologic mapping:

Sevier Shale
Rocks exposed in the area north of the abutment above the dam, and in the left abutment were consistent with lithologies of the Sevier Shale. This unit occurs as a tan to light brown, fine grained, fissile, calcareous, silty shale that is locally interlayered with dark gray, massive carbonate (limestone) nodules and beds. The carbonate beds are relatively fresh and typically less than 6-inches thick and comprise approximately 10% of the unit. Locally, carbonate beds have calcite filled fractures and open space crystallization of calcite (crystals). The shale is moderately to completely weathered, locally occurring as saprolitic silty to clayey residual soil.

Bedding within the shale is closely spaced and locally contorted and the unit is poorly jointed. Jointing is primarily isolated to the limestone nodules and beds.

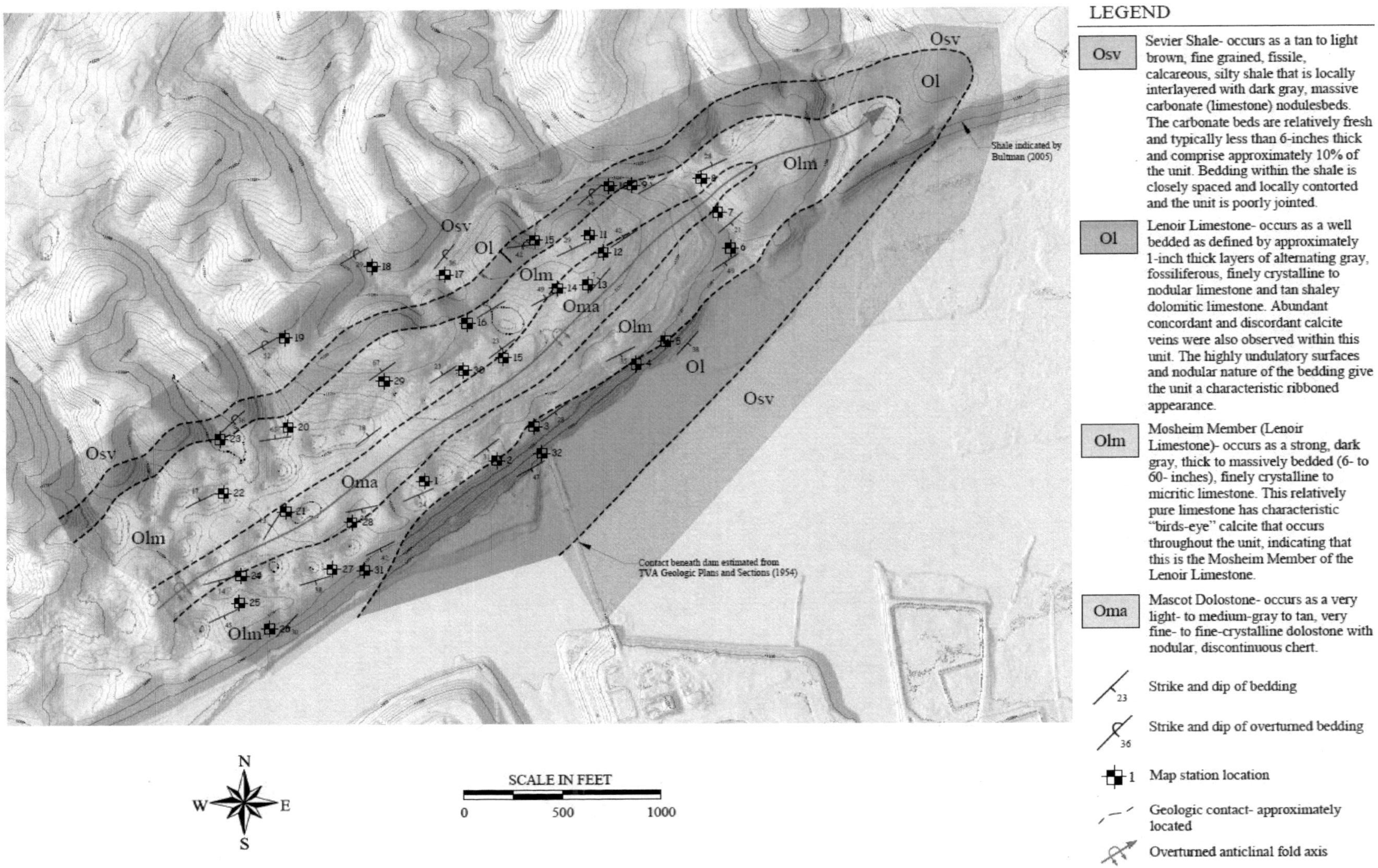

Figure 7: Extract of Geological Map from Field Observations (Petrologic, 2023).

Lenoir Limestone
The Lenoir Limestone directly underlies the Right Embankment and outcrops along the observed rock slope abutment adjoining the embankment. The Lenoir Limestone boundary with the underlying Mosheim Member generally marks the top of the northeast trending ridge immediately north of the right abutment

The unit has well-defined bedding with approximately 1-inch-thick layers of gray, fossiliferous, finely crystalline to nodular limestone interbedded with thin tan shaley dolomitic limestone. Abundant concordant and discordant calcite veins are observed within this unit. The highly undulatory surfaces and nodular nature of the bedding give the unit a characteristic ribboned appearance

The ribboned Lenoir Limestone occurs as fresh to slightly weathered in exposures. Indications of karst development were observed during geological mapping as local development of keyhole karst (i.e., dissolution concentrated along bedding and perpendicular joint) in the limestone beds, but typically, karst features in this unit were not well developed.

Overall, the ribboned limestone had joint spacing generally greater than 3 feet. Where present, joint surfaces occur as single, calcite healed planes that lack persistence (i.e., the length of the discontinuities are not laterally extensive).

Mosheim Member - Lenoir Limestone
The Mosheim Member is a member within the Lenoir Limestone. The Mosheim Member is the basal unit of the of the Lenoir Limestone and was observed during geological mapping within most of the upland topographic surface north of the right abutment (Figure 7).

The Mosheim Member is a strong, dark gray, thick to massively bedded (6- to 60- inches), finely crystalline to micritic limestone. This relatively pure limestone has characteristic "birds-eye" calcite that occurs throughout the unit, which is characteristic of the Mosheim Member of the Lenoir Limestone.

The birds-eye limestone had jointing typically spaced at 1 to 3 feet. A strike joint set was observed to occur subparallel to bedding and is persistent, through-going, and abundant within this unit. Additionally, a dip parallel (dip joint) joint set was observed to occur nearly orthogonal to bedding and is persistent, through-going, and abundant within this unit.

Numerous sinkholes were observed within this unit. Most of the sinkholes observed on the upland topographic area north of the right abutment are controlled by the outcrop areas of the Mosheim Member (**Figure 8 and Figure 9**).

Figure 8: Minor Dissolution in Mosheim Member Along Strike Joint and Dip Joint.

Figure 9: Large sinkhole observed in Mosheim Member north of site

Mascot Dolomite

The Mascot Dolomite is the uppermost part of the Knox Group in the area and is the lithological equivalent of the Newala Formation. The Mascot Dolomite has relatively limited exposures along the highest points of the upland area north of the right abutment (**Figure 7**).

This unit is a light gray to tan, finely crystalline dolostone with abundant chert nodules. Locally, the unit contains thin beds of greenish shale partings. The chert nodules occur at multiple stratigraphic horizons within the Mascot Dolomite.

Lineament Study

A desktop geomorphic evaluation was completed for the Right Embankment. This evaluation consisted of a lineament analysis within a 1.5-mile radius of the dam. Topographic lineaments can be related to local and regional geologic anomalies such as bedrock fracture zones, joints, cleavage, and compositional layering (bedding), faults, and geologic contacts.

Topographic lineaments related to the alignment of karst features exhibited similar orientations to lineaments reflecting regional rock fabric (bedding, joint sets, faults). This suggested structural control of karst development (**Figure 10**). Three main lineament orientations were identified, which were consistent with the discontinuity orientations identified in the structural data collected from the geotechnical investigation.

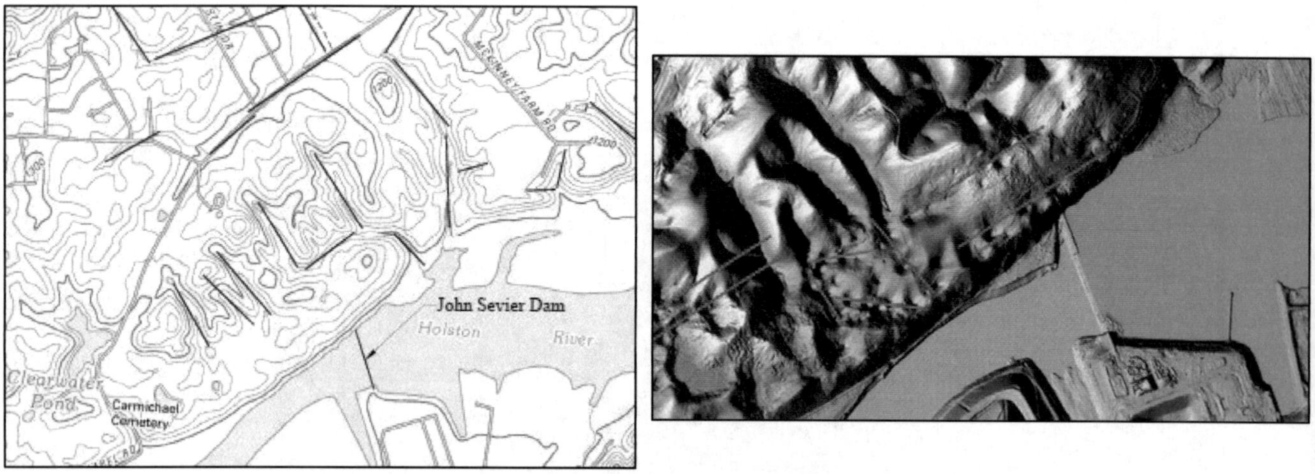

Figure 10: Lineament sets showing consensus with structural control of karst.

Structural Findings of Geological Mapping

The geologic mapping showed the dam is on the southeastern limb of a northeast plunging, overturned anticline with an axis that trends northeast-southwest. The trend of the anticlinal axis is roughly parallel to bedding strike in the area, which is typical within the Valley and Ridge province.

In the immediate vicinity of the dam, bedding strike is consistently to the northeast; however, bedding orientation is variable around the macroscale anticline with the axis north of the dam.

A stereonet generated for bedding over the entire mapped area shows the changing dip characters of a plunging fold, (**Figure 11**). Bedding at the site underlying the JSF is relatively consistent, averaging with a dip of 44° and dip direction of 133°.

Supporting Geotechnical Investigation Data

The geotechnical investigation was completed in 2023 and included coring 25 geotechnical boreholes between 30 and 70 feet into bedrock, including four angled boreholes into the right abutment; packer testing eight borings; downhole geophysics in eight borings, downhole geophysical techniques included 3-arm caliper, spinner flow meter (SFM), gamma, high-resolution acoustic borehole televiewer (ATV), heat-pulse flow meter (HPFM), fluid temperature/conductivity (FTC), high-resolution optical borehole imager/televiewer (OTV); surface geophysics combining a data set generated in 2020 including microgravity, self-potential (SP), seismic refraction tomography (SRT), electrical resistivity tomography (ERT), and microgravity; and soil and rock laboratory testing.

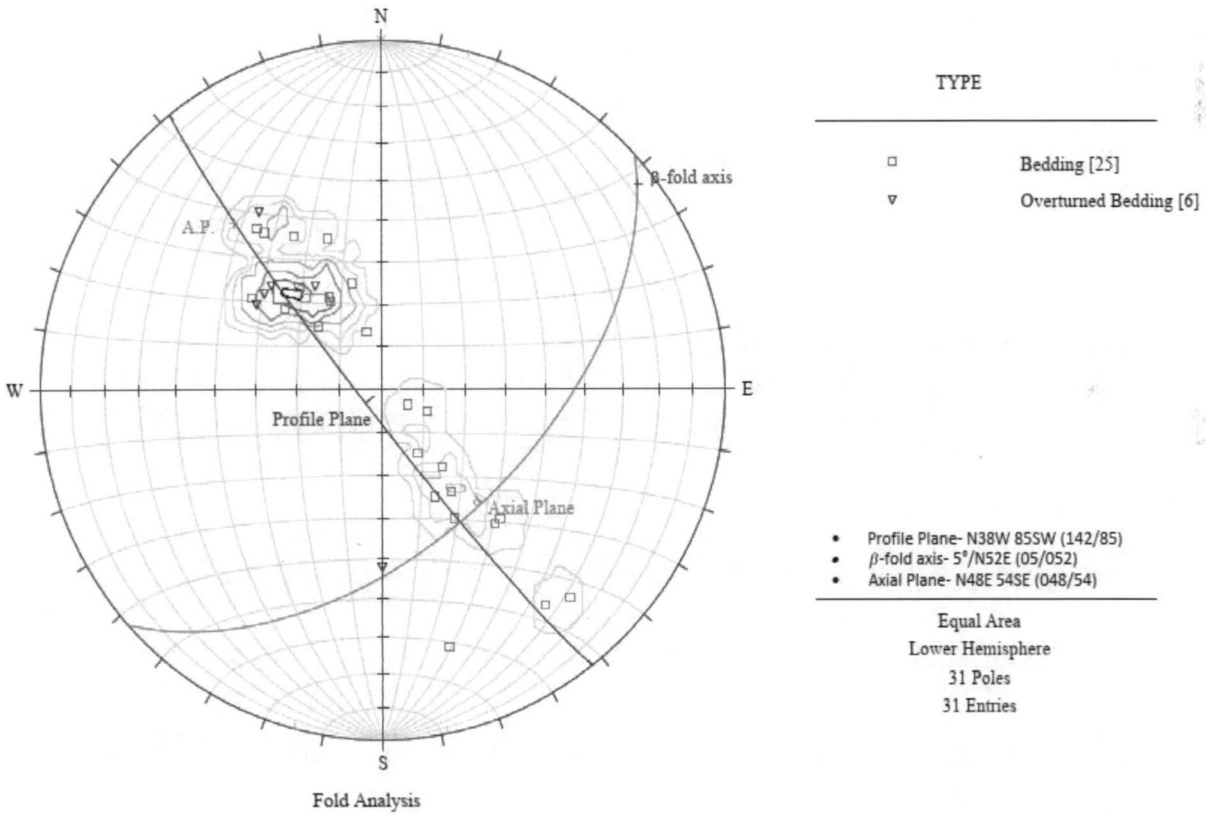

Figure 11: Fabric diagram of bedding surfaces measured during field mapping (Petrologic, 2023) showing plunging anticline. Data are shown as poles to plane.

The geotechnical investigation supported the updated geological interpretation based on geological mapping. Rock core returns showed that the bedrock underlying the site is the Lenoir Limestone. The rock displayed the characteristic, ribbon-like appearance of interbedded limestone and shale.

Drill core confirmed the lithology the Mosheim Member identified during mapping as more micritic limestone and underlying the Lenoir Limestone (**Figure 12**). The Mosheim Member – Lenoir Limestone contact at the site coincided with the contact shown on the geological map.

The Lenor Limestone rock quality designation (RQD), core-run returns, discontinuity sets and rock strength generally showed a high quality, strong rock with a shallow weathering zone. Only two minor voids less than 20 inches were encountered in the Lenoir Limestone. The few borings that encountered the Mosheim Member indicated that this lithology was more susceptible to karst. This correlated well with geological mapping and geomorphology data indicating the Lenoir Limestone is less susceptible to karst compared to the Mosheim member.

Packer testing results showed a majority of holes had very low conductivity (<1 L). Where fracture flow was recorded there was and apparent correlation with the dip

The shale content of the Lenoir Limestone did appear to increase towards the southeast. This is where higher grout takes were recorded in construction drawings near the edge of the Right Embankment.

Core showing Lenoir Limestone – Mosheim Member contact.

Figure 12: Rock Core showing characteristic ribbon texture of interbedded limestone and shale of Lenoir Limestone and massively bedded finely Crystalline to Micrtic Limestone.

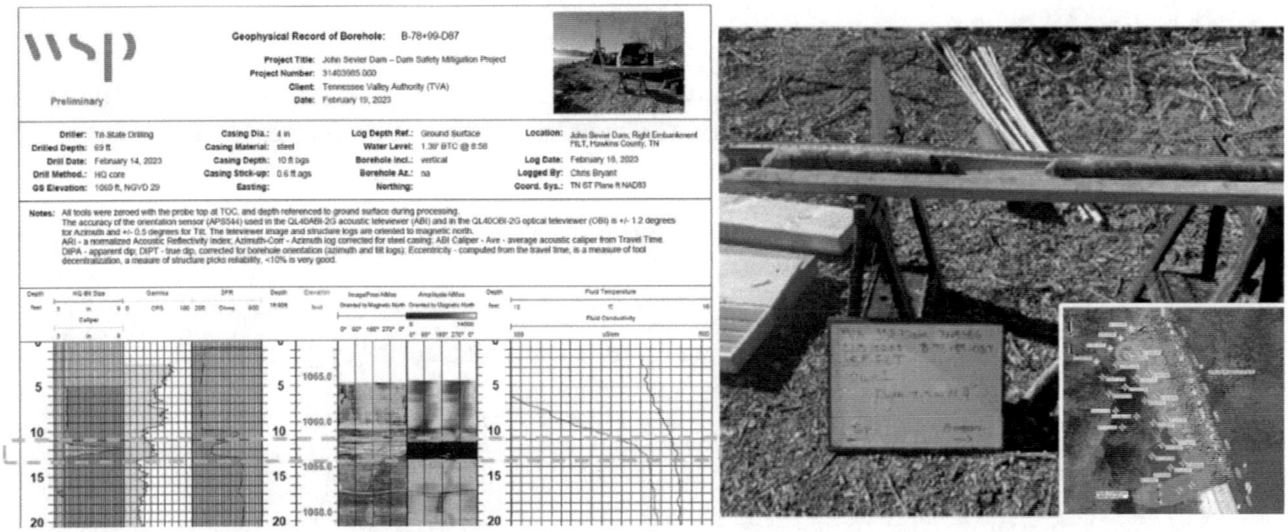

Figure 13: A 1-foot-thick void encountered in a corehole through the Lenoir Limestone.

CONCLUSIONS

Rocks underlying the Right Embankment, exposed in the right abutment, and the area north of the abutment are consistent with stratigraphy of the lower Ordovician Knox Group and middle Ordovician Chickamauga Group in northeast Tennessee.

The Right Embankment is directly underlain by the upper member of the Lenoir Limestone, an interbedded limestone and shale, which is less susceptible to karst dissolution relative to the underlying units exposed in the upper slopes.

The lower member of the Lenoir Limestone is the Mosheim Member, which is exposed along with the Mascot Dolomite on the upper slope and hilltop on the right abutment north of the dam. The Mosheim Member typically occurs as a massive, homogenous limestone. The Mosheim Member and the Mascot Dolomite, are more susceptible to karst dissolution relative to the upper member of the Lenoir Limestone, based on field evidence and geomorphology. The geological map demonstrates the correlation between the distribution of karst features and the mapped geologic units.

The geological mapping was able to provide an updated ground model for the site, that provided context and direct the interpretations of geotechnical investigation. The mapping explained why geomorphic evidence shows karstic terrain near to the dam, but not encountered during drilling.

Overall, the project demonstrated the significant value geological mapping can provide to civil engineering projects.

REFERENCES

Bultman, J.G., 2005. Structure, stratigraphy, and tectonics preserved of the northwestern Bays Mountain synclinorium, Greene and Hawkins Counties, northeast Tennessee: University of Tennessee Master's Thesis.

Petrologic Solutions Inc. March 26, 2023. Letter Report of Geological Mapping and Lineament Analysis.

Rogers, J. 1953. Geologic Map of East Tennessee with Explanatory Text: Tennessee Department of Environment and Conservation, Division of Geology.

TVA. October 27, 1954. Geologic Plans and Sections. Drawing number 41 GE 4 822 K1209.

TVA. 2015. RSOJSFROGCDX000.262015000020 – John Sevier Dam Field Investigation and Laboratory Testing Report.

TVA. 2019. John Sevier Dam – Semi-Quantitative Risk Assessment.

West Virginia Highway 340
An Intensive Rockfall Mitigation in Harpers Ferry National Park

David Crotsley, PG, CPG
Senior Geologist
HDR Engineering, Inc.
1670 Broadway, Suite 3400
Denver, CO 80202
(717) 615-7898
david.crotsley@hdrinc.com

Jody Kuhne, PG, PE
Senior Pursuit Engineer
Access Limited/ GeoStabilization Intl.
Oceano, CA
(828) 779-9482
Jody.kuhne@gsi.us

Prepared for the 73rd Highway Geology Symposium, September, 2024

Disclaimer

Statements and views presented in this paper are strictly those of the author(s), and do not necessarily reflect positions held by their affiliations, the Highway Geology Symposium (HGS), or others acknowledged above. The mention of trade names for commercial products does not imply the approval or endorsement by HGS.

Copyright Notice

ABSTRACT

US-340 is a high traffic volume, two-lane roadway that meanders along the Blue Ridge Mountains and the Shenandoah and Potomac Rivers water gap between Harper's Ferry, WV, and the West Virginia – Virginia state line. Originally constructed in the 50's, the cut slopes along the road and the exposed natural rock slopes of the mountain vary in elevation from 280 feet to 900 feet. The rock slopes exhibit varying degrees of rockfall activity that present potential hazards to the traveling public and require ongoing maintenance by the West Virginia Department of Transportation. The project included a geologic evaluation and rockfall remediation design for three slopes adjacent to US-340 between Chestnut Hill Road (CR-32) and Harpers Ferry Road (VA-671).

Combining the use of photogrammetry; terrestrial and aerial, as well as on slope evaluations, the recommended design included a variety of mitigation measures including simple drapes, hybrid barriers, and roadway level rockfall barriers, among others, to complete the project. As a portion of the project is located within the National Park Service property, and the area of Harper's Ferry relies on the park for commercial and tourism funds. The resulting 22-mile detour during construction and the impacts to the viewshed made the project of particular interest to the community. From the basic to the advanced mitigation techniques, collaboration between the stakeholders, design and construction teams, and the unique characteristics of the Chilhowee Group led to the project being a challenging success.

INTRODUCTION

The project area is located along US Highway 340 in Jefferson County, WV, in the Loudoun Heights region of the Harpers Ferry National Historical Park and west of the West Virginia/Virginia border on the southern bank of the Shenandoah and Potomac Rivers. US 340 is a high-traffic volume corridor serving local, commuter and truck traffic from West Virginia, Virginia, and Maryland. The corridor also experiences seasonal high traffic condition due to its recreational and historical significance in the region. The existing cut slopes are a product of US 340 construction in the mid-1950's and natural erosion along the Shenandoah River. The cut slopes and the exposed rock of natural slopes vary in height from 150 feet to greater than 300 feet above the roadway. The cut slopes in the project study area exhibit varying degrees of rockfall activity that present potential hazards to the travelling public and require ongoing maintenance by the West Virginia Department of Highways (WVDOH). Refer to Figure 1 below for the project location.

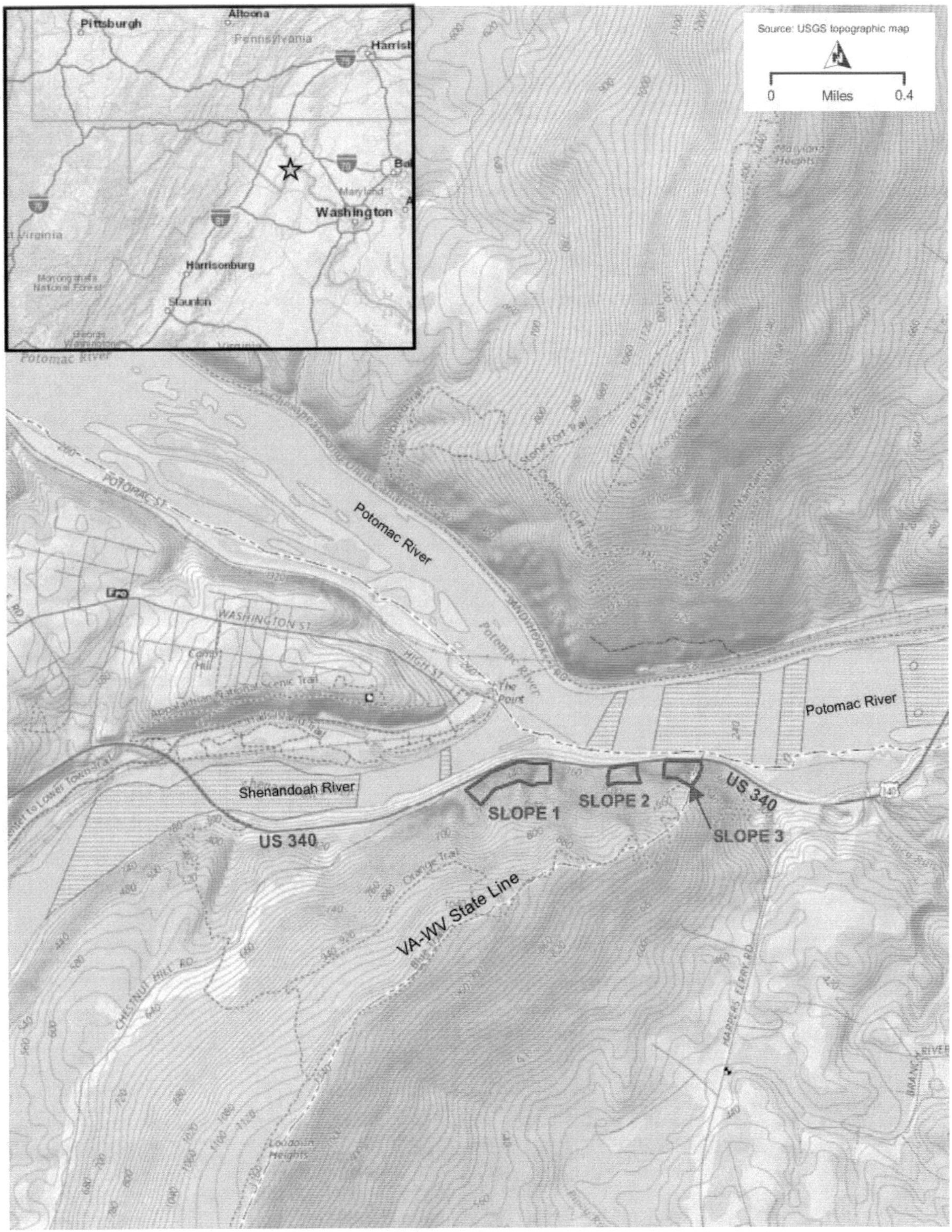

Figure 1 – Project Location Map

SITE GEOLOGY, EXPLORATIONS AND LABORATORY TESTING

Site Geology

The project is located within the Blue Ridge Physiographic Province of the South-Central Appalachian Mountains. Rock formations buttressing the mountains in this area consist of granitoids and paragneisses of Middle Proterozoic, and the bedrock underlying the Blue Ridge-Elk Ridge and Short Hill-South Mountain areas are metasedimentary and metavolcanics of Late Proterozoic to Lower Cambrian.

The project corridor is underlain by the Weverton Formation. Refer to Figure 2 for the Geologic Map of the area. The formation consists of three (3) distinct members along the corridor and include the Buzzard Knob, Maryland Heights, and Owens Creek members. The Buzzard Knob member consists of mature quartzite beds interbedded with sandy metasiltstone, the Maryland Heights member consists of very coarse grained to granular quartzite, metasiltstone and greywacke layers, and the Owens Creek consists of pebble conglomerate, quartzite, and metasiltstone. These formations have been complexly deformed by folding with development of cleavage and overprinted with several phases of deformation. Site evidence suggests at least two (2) periods of ductile formation/folding as indicated by tight to isoclinal folding of the dominant bedding and sub-isoclinal distortion within the dominant foliation pattern. Valley-wall stress-release joints are also present on the slopes and parallel the river valley. A gently dipping strain-slip cleavage pattern is sporadically developed in folds along the dominant bedding.

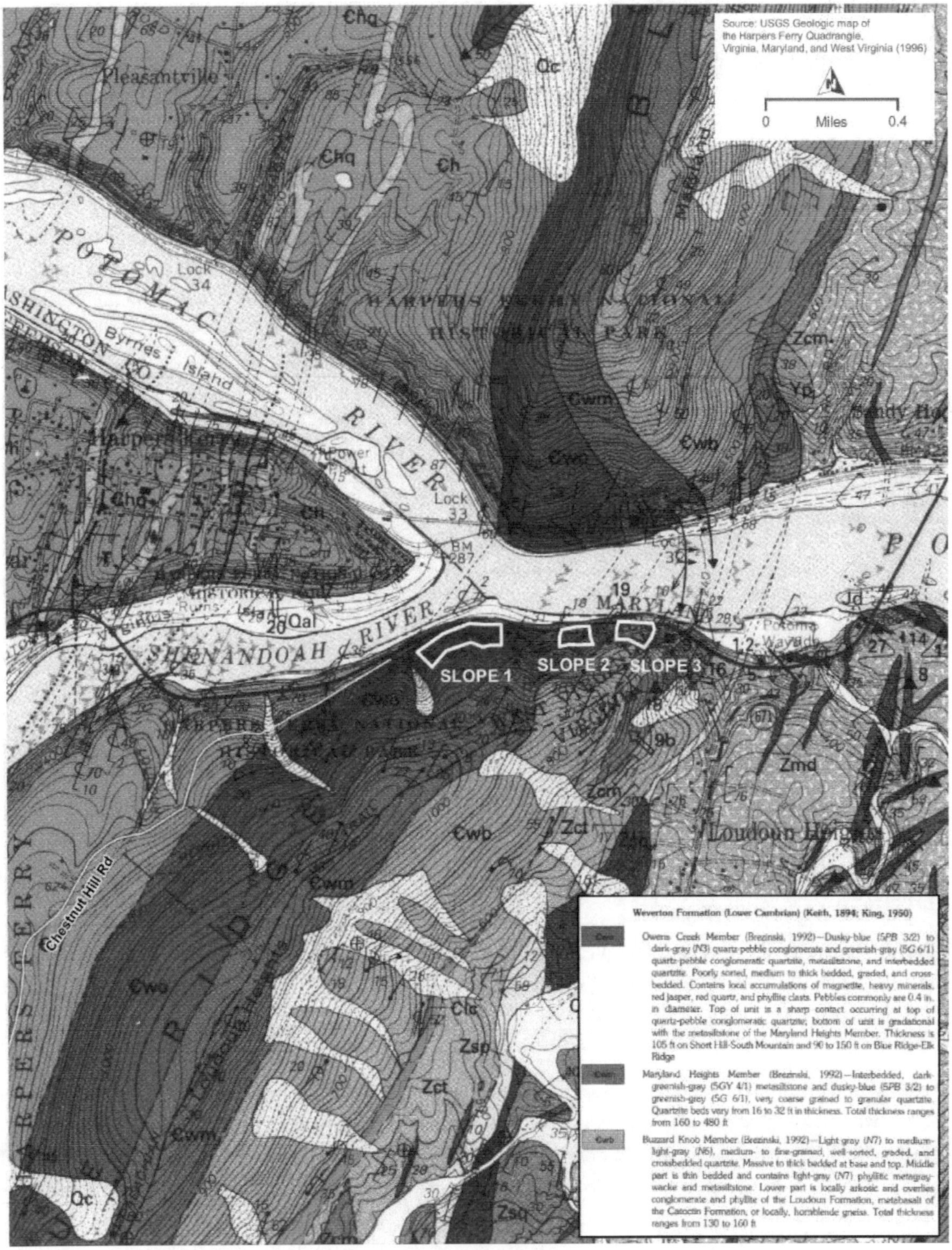

Figure 2 – Geologic Map

Rockfall History

The existing cut slopes within the project study area have historically been prone to rockfalls and slides. Previously, there were no existing rockfall protection measures in place except for small, inconsistent ditch lines that extended along the toe of the slope to the paved travel lane.

Rockfalls and other failures had been documented along US 340 by the WVDOH personnel over the last two decades. The documented rockfall events appear to have historically occurred primarily in the Spring and Winter months and appear to correspond with freeze/thaw cycles and wetter seasons. There had been five (5) rockfall events between Chestnut Hill Road and the state line between January 12 and February 14, 2018, for instance. A majority of the failures that occur along US 340 are minor volume and impact; however, large rockfall events had occurred in the past and posed a constant risk.

On February 6, 2021, a boulder approximately 60 cubic feet in size slid from a previously identified boulder field area down to the roadway causing a vehicular accident. Injuries were non-life threatening; however, this event reinforced the necessity to provide mitigation as soon as possible.

Site Explorations

The complexity of the site required multiple site evaluations utilizing various techniques. An initial aerial imagery survey was conducted in November of 2015, followed by a Light Detection and Ranging (LiDAR) survey that December. Data from the two surveys were processed, merged, and tied with site ground control for use in developing the rock slope conditions and concerns, and to better identify specific areas of interest. Following that effort, a roadway geologic evaluation was performed in January and February of 2016, which included bedrock lithology identification, geologic structure, weathering, water seepage/presence, discontinuity orientation and spacing, areas of slope creep/instability, and the locations of previous/potential rockfalls or other observed failures. Discontinuity mapping was performed utilizing a Brunton compass and/or the mobile phone app. Discontinuities were labeled as joint, cleavage, or bedding, as identified by the consultant's senior geologists, and consisted of strike, dip, and dip direction.

The aerial imagery and LiDAR were collected and processed to generate a point cloud from which discontinuity data of areas inaccessible from the roadway were identified. A follow up aerial effort was performed as part of the Final Design scope of work to refine the data collected during the preliminary phase and eliminate areas with undesirable point cloud density.

In addition to the roadway level evaluation and the aerial mapping, an on-slope evaluation was performed to verify the geologic conditions on the slope, identify high priority areas, evaluate slope stabilization requirements and option feasibility, view obscured slope areas, and confirm slope access conditions for construction. Slope areas were then delineated into primary areas of concern based on the site explorations.

Laboratory Testing

During the upper slope investigation, bulk rock samples were collected from each slope for strength testing. Unconfined compressive strength testing was performed where feasible; however, due to sample sizes Samples MH-1 and MH-2 were tested using a point load methodology to approximate the unconfined compressive strength. Additionally, the direct shear testing was performed along saw cut surfaces of the rock core. A singular core sample was sheared along sawed surface at three (3) normal stresses to determine the residual shear stress.

SUMMARY OF SLOPE CONDITIONS

The structural geology present within each slope area impacted the type of rockfall or failure mechanism. The structure, although highly variable within the project limits due to the abundant folding, presented similar trends in the types of failure mechanisms influencing each slope area. Below is a summary of some of the controlling conditions.

- The Weverton Formation stratigraphy comprising the slopes are part of the larger recumbent folding of the Blue Ridge Mountains and exhibits broad folds extending from below the river elevation to the top of the mountain. Similarly, there is secondary, tighter localized folding within the broader regional folding. In general, folds have a shallow 5-15 degree plunge into the hillside with a NE/SW oriented axis. The stresses along the fold axis tend to be higher, which result in a more jointed rock mass due to the tensile stresses developed in the area. These areas had the highest potential for rockfall activity.
- The bedding dips along the fold crests tended to be near vertical to 80 degrees. The upper and lower limb bedding dip orientations range from 80 degrees to horizontal, and generally dip to the NW and SE. Slopes with a NW face tended to have a higher potential for a global planar block failure with the upper limb daylighting the slope face.
- Tectonic stresses associated with folding activity resulted in cleavage planes developed parallel the axis of the folds and prominent regional NW joint set. Cleavage discontinuities are primarily NE/SW trending with a near horizontal to shall dip to the SW and are typically perpendicular to the fold axis. Rock slopes with a NE face had numerous wedge features that developed between bedding, joints, and cleavage discontinuities. Wedge blocks were controlled by the joint spacing, bedding thickness, and discontinuity orientation.
- Lastly, valley wall stress relief joints that paralleled the existing slopes were a result of the river erosion. These joints were typically steep, 70-90 degrees, with localized planar blocks resulting.

Upon completion of the field mapping and reception of the LiDAR, the consultant developed annotated orthoimagery showing the identified geologic conditions at the site, and the preliminary mitigation recommendations to be applied. Refer to Figures 3 through 6 below for the imagery.

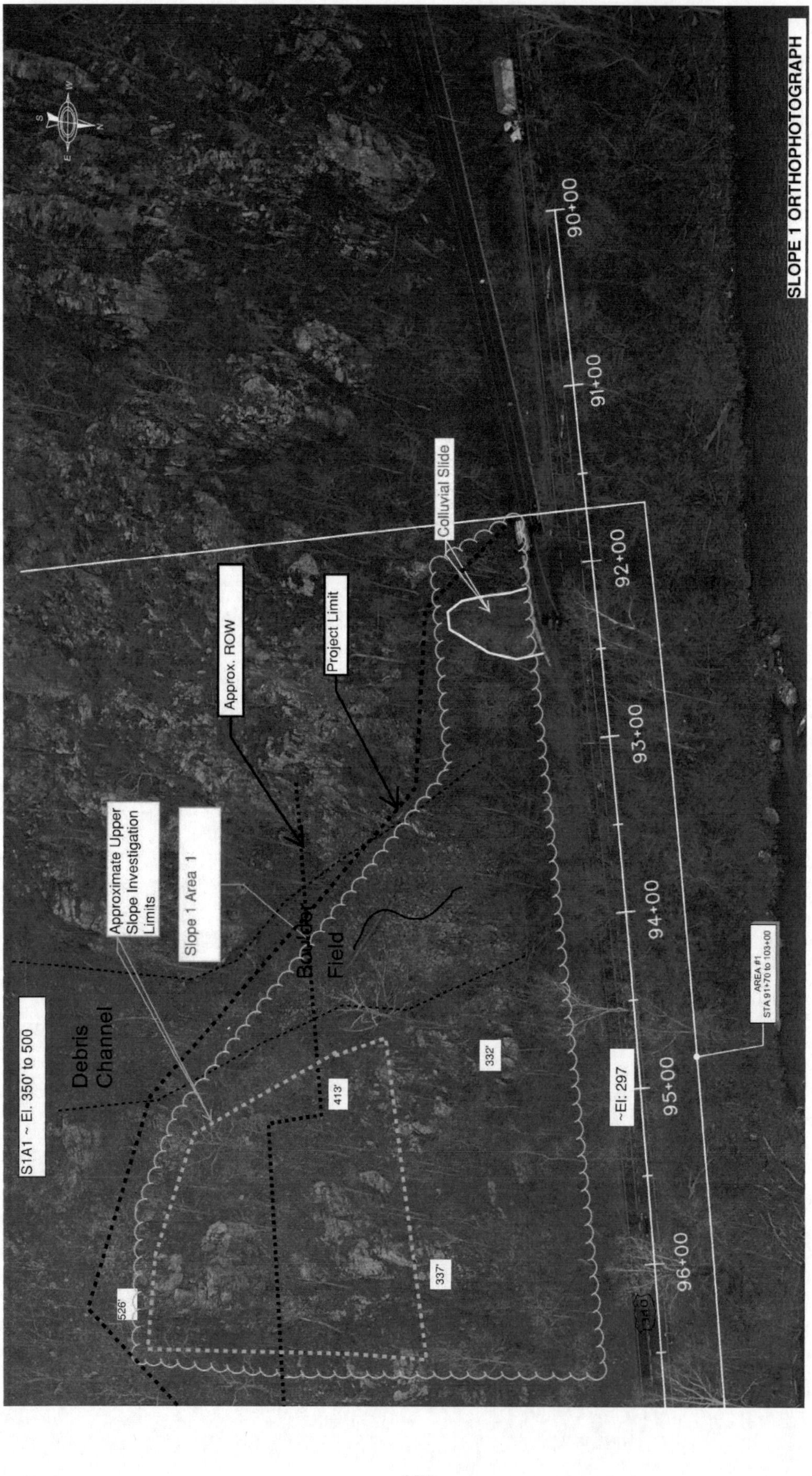

Figure 3 – Geologic Conditions at Slope Area 1

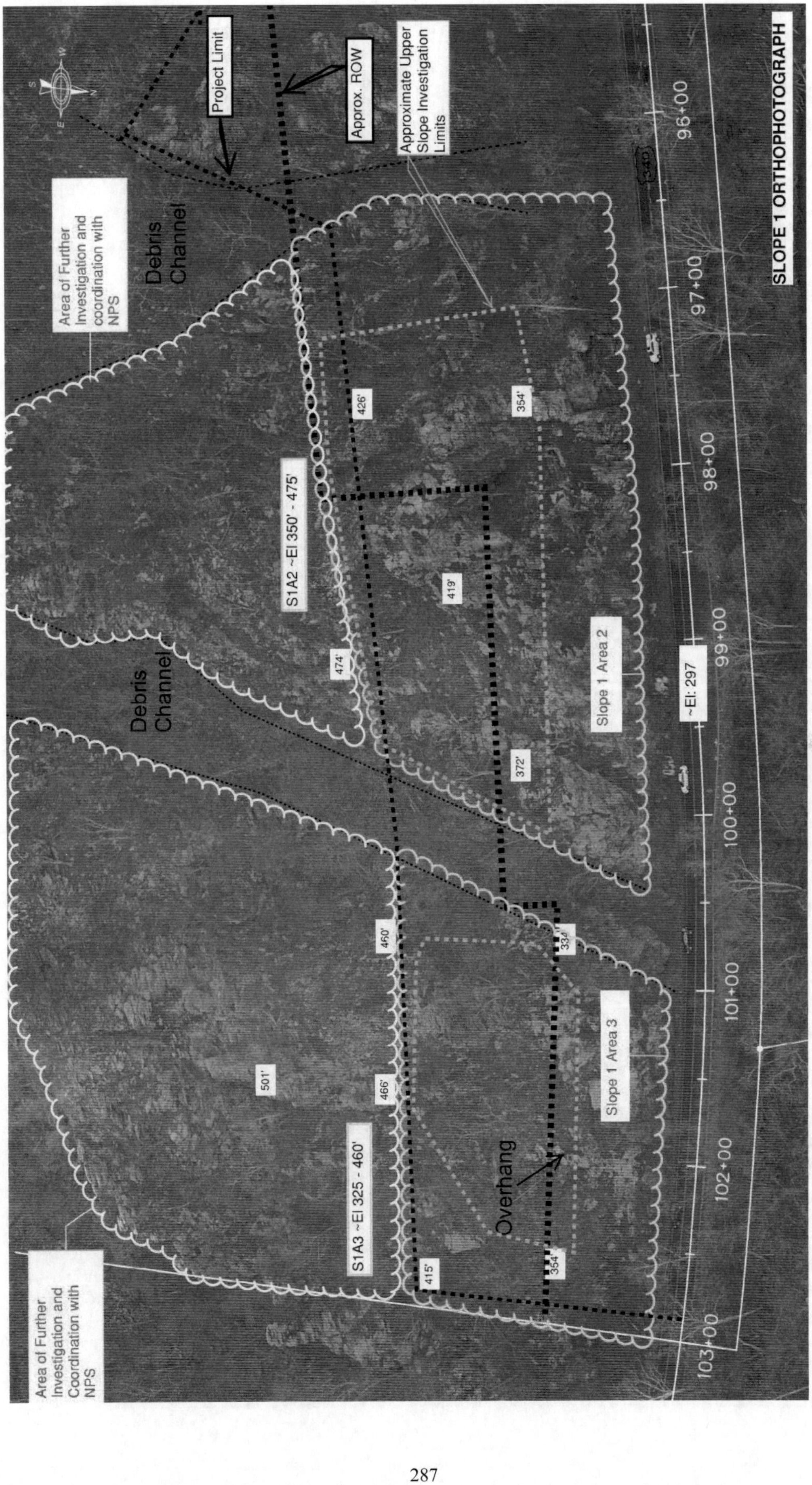

Figure 4 – Geologic Conditions at Slope Area 1

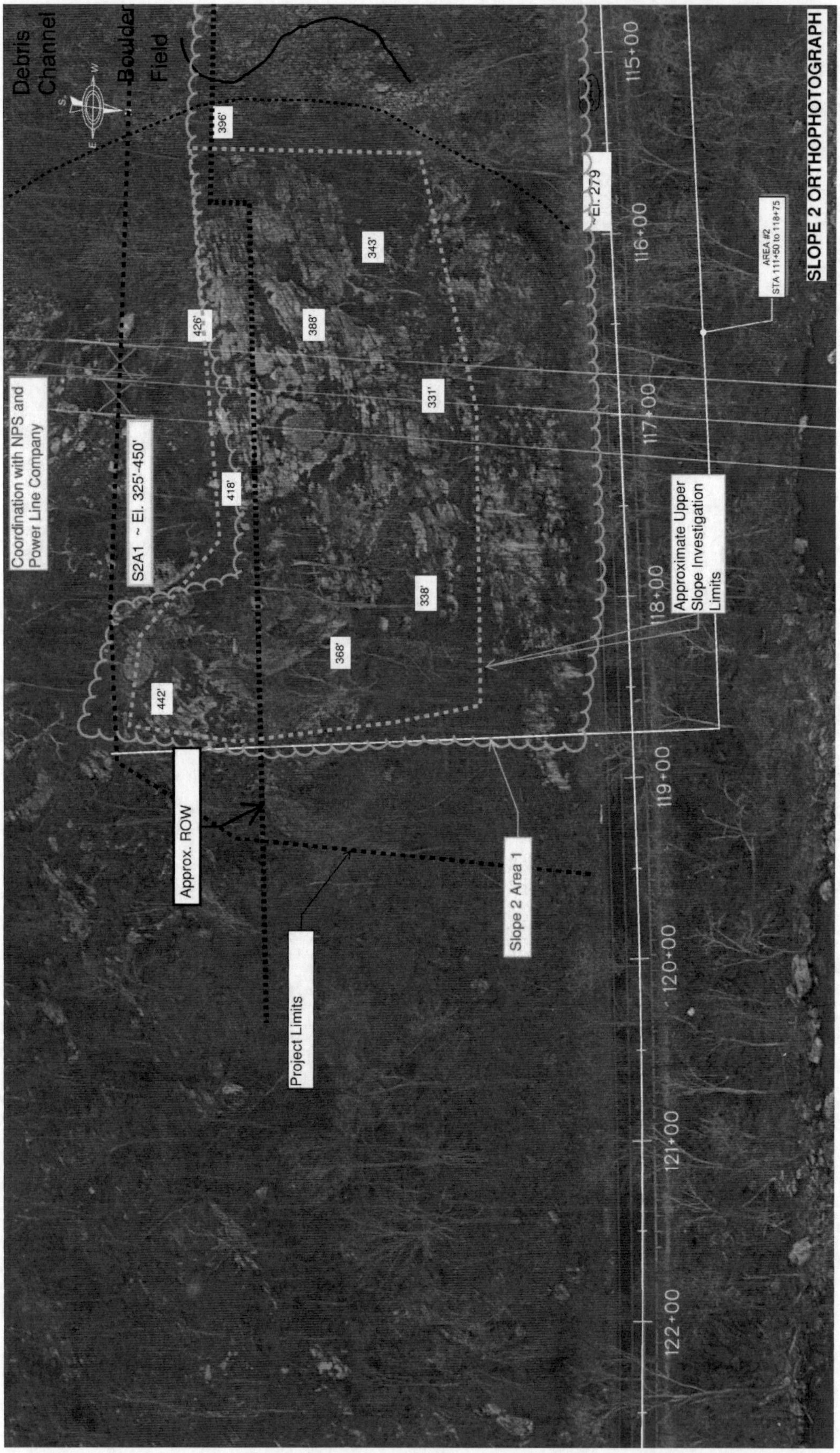

Figure 5 – Geologic Conditions at Slope Area 2

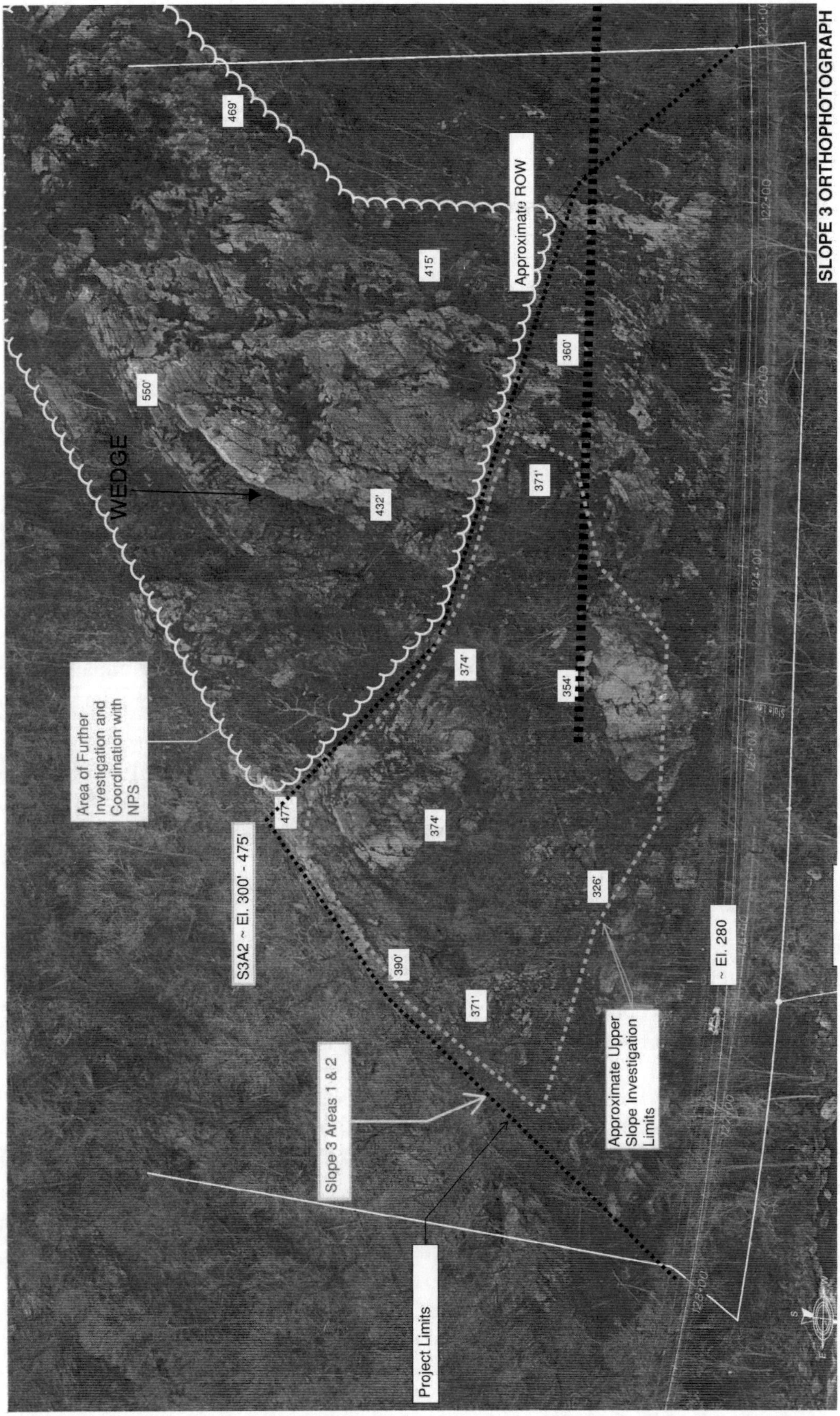

Figure 6 – Geologic Conditions at Slope Area 3

DESIGN CHALLENGES AND CONSIDERATIONS

Preparing a design package for projects like this US 340 project which included multiple stakeholders, potential for a significant detour to maintain traffic, and minimizing impacts to the scenic park was a challenge. Realizing that the construction would require full closure of the roadway for a portion of the project, the consultant's team developed a detour plan to maintain the approximately 30,000 vehicles per day traffic in the area. The detour was approximately 22 miles and would require commercial vehicles to use roads and intersection that weren't intended to see such traffic. As part of the traffic study, the impact to the local tourism industry was considered as well. The Harpers Ferry area is known for river recreation as well as people visiting the park and the town. The final recommendation included a full road closure for the duration of the construction, with an incentive for the Contractor to complete the work in a shorter time than estimated. As a result, the Contractor was able to complete the project ten (10) days early, allowing the roadway to be opened to the public earlier than the advertised date, much to the joy of the local commuters.

In addition to the impact on the local traffic, the consultant also considered the importance of the natural environment in the area. The project is located at the confluence of the Shenandoah and Potomac rivers, surrounded by trees including chestnut oak, poplars, and red maples, and undergrowth including ferns, grasses, and sedges. Approximately 70% of the land in the park is forested. As a consultant, it's our responsibility to respect the need for such areas to maintain the natural beauty. Therefore, a viewshed analysis was performed to evaluate the impact of the proposed mitigation options on the five (5) primary scenic vistas located within the park and the town of Harpers Ferry: Jefferson Rock, St. Peter's RC Church, Maryland Heights Overlook, the C&O Canal trail and the Shenandoah Shoreline. Ultimately, the controlling factor for the design was in the name of safety for the area; however, our viewshed analysis was valuable in conversing with the stakeholders as to what the ultimate impact would be to the area.

Lastly, the design criteria used for the mitigation design was also considered. It's ultimately very difficult and exceedingly expensive to develop a design which provides complete protection from rockfall events, especially in mountainous terrain with complex geology such as the project area, all while minimizing Right of Way Impacts. Therefore, rockfall source locations were discussed with the WVDOH prior to mitigation development, and an acceptable percent passing for rockfall analyses was agreed upon at 10%.

Sub-global and global stability analyses were conducted to assess the stability of the designated rock slope areas. In sub-global stability analysis, it is assumed that rockfall will occur, and the purpose of the analysis is to determine the characteristics of the rockfall and the potential for it to travel beyond a certain designated point of interest, such as a catchment area or barrier. Kinematic analysis was performed to assess the natural discontinuity patterns on a slope relative to the respective slope geometry, and to evaluate the potential and type(s) of failures that may occur.

ROCKFALL AND SLOPE STABILITY ANALYSES

Sub-Global Rockfall Stability

Sub-global rockfall stability was analyzed using a rockfall simulation program to evaluate the potential for rockfall generated on the US 340 slopes to reach the roadway based on the existing slope geometry and noted rockfall dimensions. Based on the field observations and design recommendations per WVDOH guidelines, design rockfall analyses used 1-foot by 1-foot and 3-foot by 4-foot discoidal blocks, weighing 130 and 6220 pounds, respectively.

Slope Area	Station	Source Elevation	Block Size (ft.)	Bounce Height (ft.)	Kinetic Energy (kJ)	Percent Retained
1	98+00	468-440	1x1	0	0	100
			4x3	0	0	100
		560-540	1x1	36	32	97
			4x3	59	1398	88
	100+75	460-420	1x1	0	0	100
			4x3	0	0	100
		597-500	1x1	0	0	100
			4x3	7	1082	98
	102+00	455-420	1x1	0	0	100
			4x3	0	0	100
		589-494	1x1	0	0	100
			4x3	8	980	94
	115+50	416-380	1x1	0	0	100
			4x3	1	728	100
		584-500	1x1	13	19	97
			4x3	11	1388	80
2	123+00	481-420	1x1	17	10	66
			4x3	19	347	1
		562-481	1x1	20	19	82
			4x3	15	764	6
3	125+00	481-420	1x1	5	7	97
			4x3	15	482	91
		562-481	1x1	0	0	100
			4x3	15	1072	60

Table 1 – Summary of Rockfall Analyses

Global Rockfall Stability

Global rock slope stability was analyzed using kinematic and limit equilibrium software, as outlined in the preliminary design report. The purpose of the kinematic analyses was to determine stability of planar or wedge block failure based on different block sizes and orientation, and the potential for toppling failures. Utilizing this information, remediation techniques are evaluated relative to the target factor of safety.

Slope Area	Slope Orientation	Slope Dip	Planar	Wedge	Direct Toppling	Oblique Toppling
1	68	80	25.53	32.63	3.04	9.92
	79		26.95	40.91	1.97	10.14
2	49	80	8.33	29.79	2.86	31.25
	81		31.25	43.35	1.79	0.45
3	49	80	3.95	8.29	7.41	9.2
	93		15.79	23.46	3.62	4.6
	152		6.58	14.08	2.11	4.28

Table 2 – Summary of Kinematic Analyses

ROCKFALL MITIGATION TREATMENTS

Selecting rock slope and rockfall remediation treatments appropriate for the observed and analyzed conditions was critical to the project success. Numerous treatment options are available and were discussed with the WVDOH. As rock slope remediations are historically known to be multi-faceted with numerous conditions to account for, the approach moving forward included a variety of mitigation measures to be applied to the slope. These measures would include scaling, ground mounted rockfall barriers, pinned and draped mesh, attenuator barriers, and bolting. Refer to Figures 7 through 9 for a summary of the remediation recommendations.

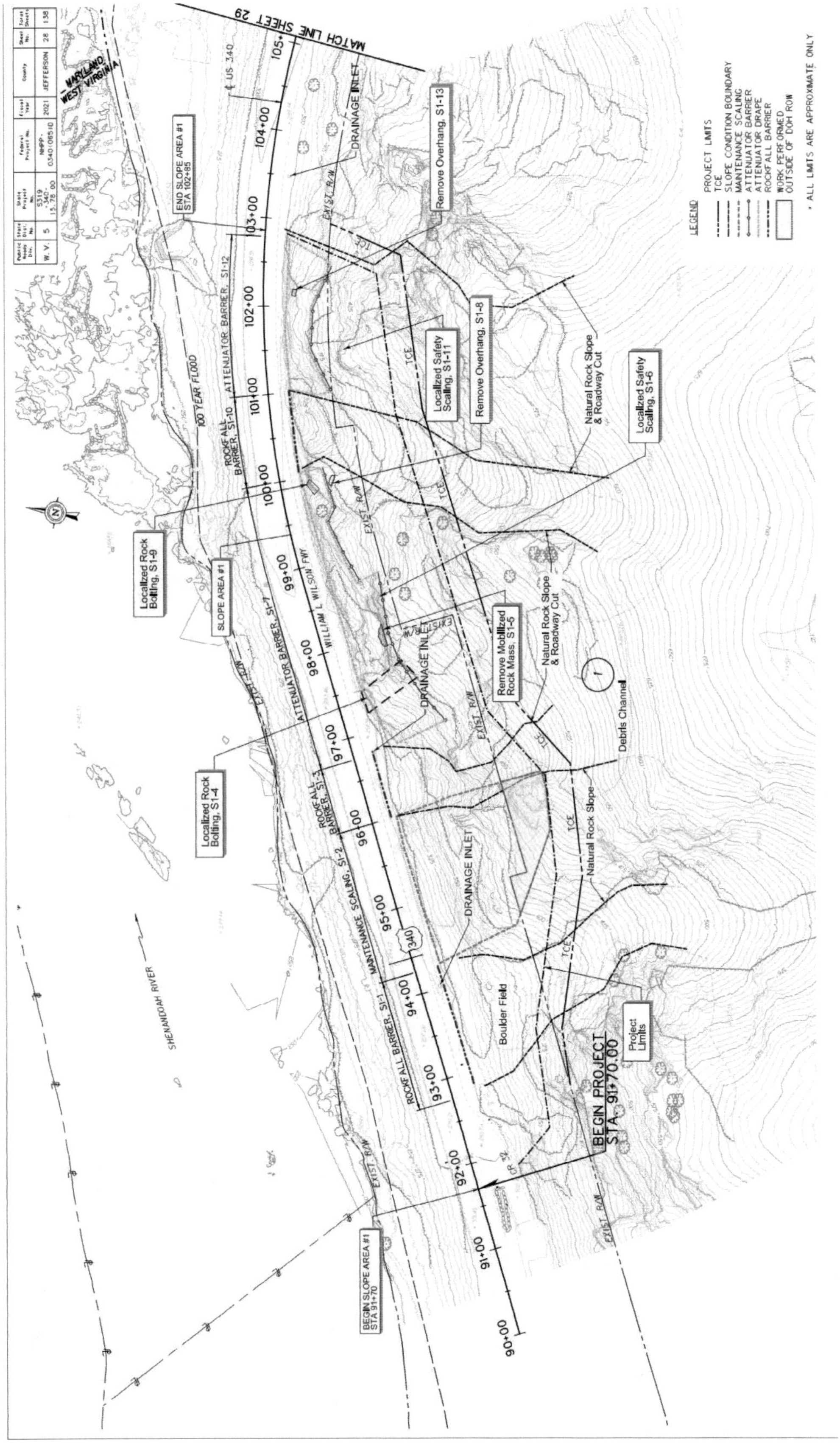

Figure 7 – Slope Remediation Plan Slope Area 1

Figure 8 – Slope Remediation Plan Slope Area 2

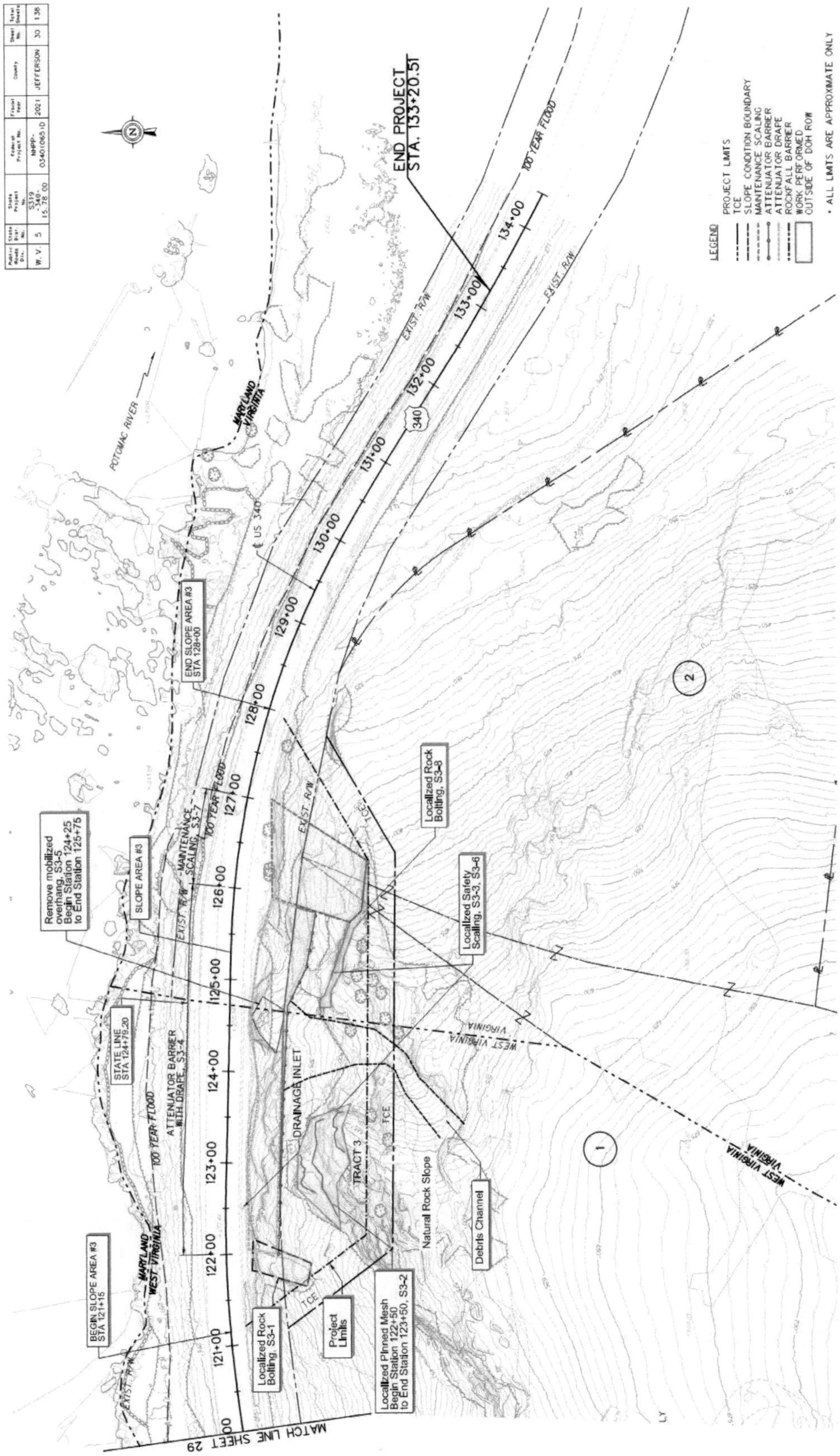

Figure 9 - Slope Remediation Plan Slope Area 3

DESIGN SERVICES DURING CONSTRUCTION

As identified originally in the FHWA Publication HI-99-007, “Rock Slopes”, one of the items uniquely important to rock slope stabilization projects is the need for the project to have built in flexibility. Particularly in areas with significant vegetative cover, design efforts can be particularly difficult determine items such as the extent of scaling, or the length of bolts. The US 340 project was no different.

Design Adjustments

WVDOH contracted another consultant to provide a constructability review, highlight items and areas that may need modification during construction, provide design, inspection and testing services, and gain confidence in the construction timeline. Good communication and timely revisions were made between the prime contractor, the designers, and the subcontractor.

In particular an issue arose during construction in regard to what was referred to as Item S3-5. The area was a local rock overhang, which had the potential to mobilize onto the roadway. During design, the consultant did not permit blasting of the mass, due to the potential for the blast energy to impact surrounding features of the slope. However, the Contractor reviewed the area after scaling and working on the site and determined that conventional means would not be sufficient to obtain the desired outcome. Therefore, an RFI was submitted to the consultant to consider blasting. A project meeting was held to discuss the impact concerns, possible alternatives, and methods for the team to reach the outcome of removing the mass and providing a safer line of sight and eliminate the overhang concern. Between the Contractor, WVDOH staff, and the consultant designers, a specialty contractor was brought in to provide a unique blasting plan which would limit applied pressure to the rest of the site. Upon reviewing and approving the blast plan, the specialty contractor mobilized and performed the blast with excellent results and no negative impacts. Refer to Figures 10 through 12 for imagery around the blast site. Blast size was approximately 60 feet high, and a volume of approximately 4,000 cubic yards. Figure 11 is prior to the additional scaling and bolting performed in the area.

Figure 10 – Pre-blast Condition

Figure 11 – Post Blast Condition

Upon completion of the blasting and successful scaling of loose materials and bolting, the attenuator drape was installed above the crest of the face. Figure 12 below shows the post drape installation condition at the area of concern.

Figure 12 - Area S3-5 after blasting, scaling, rock bolting and attenuator installation.

Another area of the project that was discussed extensively was in regard to Split Rock. Split Rock is a spire towering over the eastern edge of the project corridor. The area is historically stable; however numerous structural features are present that could provide source material for future rockfall events. The intent during design development was to utilize localized pinned mesh material to stabilize the area, as well as rock bolting, if needed. Inherently to be used at the discretion of the on-site staff, the mitigation option was intentionally open ended. Main driver behind this option was the limited understanding of what the area would look like post scaling. As designers, we are required to make assumptions using the best information that we have available. Multiple discussions were had between the Contractor and the design team, including to what extent the project was to be responsible for stabilizing the mountainside. Ultimately, the scaling efforts from the Contractor were successful to the point that additional mitigation was deemed not necessary, and therefore saved time and budget for all involved.

Figure 13 - Overall image of Split Rock, with blast area lower left.

Keys to Success

Rock slope stabilization projects are unique in the world of civil engineering. The convergence of maintaining the visual landscape, providing a suitable safety factor, and providing a constructable design requires clear communication from the beginning of the project during the scoping phase through the end of construction. The US 340 Design Team collaborated for years developing an appropriate design for the project. Once the contract went to bid and was awarded to the Contractor, the WVDOH developed a strategic team to handle incoming questions and concerns from the Contractor, to keep the project on a very difficult timeline. The work of the on site WVDOH personnel and representatives was critical in maintaining the level of design while managing the contractor and mitigation implementation.

REFERENCES

Federal Highway Administration "Rock Slopes Reference Manual", 10-1998.

HDR "Final Phase I Design Study Report", 02-20-2018.

HDR "Final Phase II Geotechnical Engineering Report", 02-2021.

Southworth, S., and Brezinski, D. 2000, Geologic Map of the Harpers Ferry National Park, Virginia, Maryland, and West Virginia, Open File Report 00-297, United States Geological Survey.

Case Histories of Roadside Rockfall Barrier Applications

Mike Koutsourais, PE
Maccaferri, Inc
825 Commonwealth Ave.
Hagerstown, MD 21740
Ph: 301-641-8072
m.koutsourais@maccaferri.com

Lucas Martins
Maccaferri, Inc
825 Commonwealth Ave.
Hagerstown, MD 21740
Ph: 240-707-0633
l.martins@maccaferri.com

Luca Gobbin
Officine Maccaferri
Via Kennedy 10
Zola Predosa (BO, Italy), 40069
l.gobbin@maccaferri.com

Prepared for the 73rd Highway Geology Symposium, September 2024

Acknowledgements

The authors would like to thank the individuals for their contributions in the work described:

Dale Chaychuk – Geofabrics Australasia, Australia
Matteo Lelli – Maccaferri Malaysia, Malaysia
Gary Matthews – Geofabrics Australasia, Australia

Disclaimer

Statements and views presented in this paper are strictly those of the author(s), and do not necessarily reflect positions held by their affiliations, the Highway Geology Symposium (HGS), or others acknowledged above. The mention of trade names for commercial products does not imply the approval or endorsement by HGS.

ABSTRACT

Rockfall hazards pose a serious risk to transportation safety, with the potential to cause fatalities, infrastructure damage, and disruptions in transportation systems. For several decades, rockfall barriers have been a common technique to mitigate geohazards in the United States. Over the years, several of these safety systems have been tested and installed around the world. This study explores the unique challenges in implementing rockfall barriers in roadside applications, emphasizing specific infrastructure considerations that will lead to a comprehensive design of these systems.

Traditionally, capabilities of rockfall barrier systems are measured according to the magnitude of their tested impact energies. The design of these systems should not only consider the capabilities of their maximum impact energies but should evaluate other key aspects such as limited barrier deflection, ease of installation and maintainability. This study will discuss how these critical site-specific considerations affect the design of cost-effective roadside rockfall barrier systems.

Three different case studies from different regions in the United States highlight successful installation of rockfall barriers alongside roadways. Emphasis is given on how the rockfall products unique design features and considerations aided in overcoming site-specific challenges and contributed to the success of these solutions.

Furthermore, this study showcases advancements in smartification of rockfall barriers. Such techniques as remote sensing and warning systems are used to improve the overall effectiveness of rockfall mitigation systems. The integration of these new technologies with traditional rockfall barriers contributes to an integrated approach to geohazard mitigation and is essential for a continued improvement of roadway safety.

INTRODUCTION

Rockfall barriers are widely used to mitigate rockfall risk along civil infrastructures such as roads and railways, mining areas, and construction sites to protect valuable assets and human lives. Rockfall barriers are passive mitigation systems as they do not affect the source of rockfall events, but they arrest the rock masses dissipating their falling velocity and energy, and thus preventing them to reach vulnerable areas. Other examples of passive rockfall risk mitigation systems include debris flow barriers, rockfall embankments, simple draperies, attenuators, etc. Generally, flexible rockfall barriers are used to deal with rockfall events with energy levels up to 9,000-10,000 kJ, (Fig. 1). Beyond this energy level, rockfall embankments are commonly preferred for their ability to absorb very high energy impacts with a limited footprint, and the ability to withstand multiple impacts.

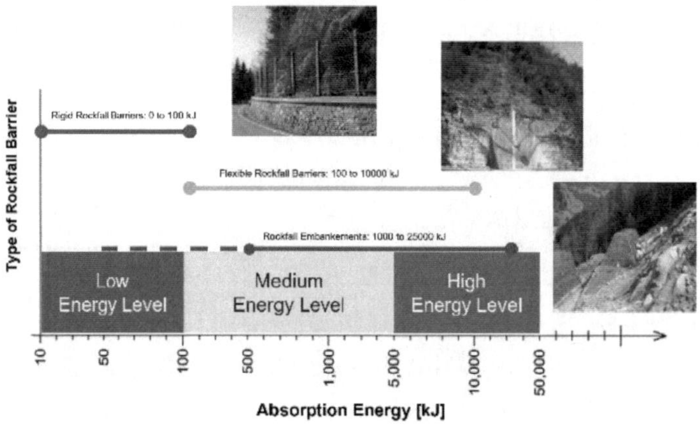

Figure 1: Rockfall Barriers Classification Based On Expected Impact Energy Level

To evaluate the performance of flexible rockfall barriers, the European Organization for Technical Approval (EOTA) issued, in 2008, the ETAG 027 guidelines for the testing and assessment of the performance of rockfall protection kits with maximum energy absorption capacities equal or greater than 100 kJ. In 2018, European Assessment Document (EAD) 340059-00-0106 superseded ETAG 027 and standardized the procedure for carrying out full-scale crash tests on rockfall kits. The UNI 11211-4: 2012, a design guideline for flexible rockfall barriers which considers the performance assessed during the full-scale tests, was issued in January 2012 by the Italian Standard Organization (UNI-Ente Nazionale di Unificazione). In 2018, a revision to this document was released as UNI 11211-4: 2018. EOTA is currently in the process of finalizing and publishing guidelines for testing low energy rockfall barriers less than 100 kJ, EAD 340089-00-0106.

The main benefit of flexible rockfall barriers is their ability to absorb energy and arrest the rock. This is realized only through deformation of the interception panel, allowing time to reduce the forces acting on the barrier (Eq. 1).

$$F = m\,\Delta v / \Delta t \qquad (1)$$

Where F is the force acting against the fence at the time of impact, m is the mass of the block, v is the velocity of the block, and Δt is the time required to arrest the block.

In agreement with the EAD 340059-00-0106 (barriers with energy capacity equal or higher than 100 kJ), and EAD 340089-00-0106 (barriers with energy capacity less than 100 kJ), a rockfall barrier is a "kit" made up of several components, which must be able to stop an impacting block having a certain energy level. The kit is composed by (Fig. 2):

1. Interception structure: made up of principal net and an optional additional layer
2. Support structure: made up of metallic posts (for example, tubular or other steel sections) and base plates
3. Connection components: consisting of metallic ropes, wires and/or bars of different types, junctions, wire rope grips, energy dissipating devices (elements which dissipate energy and/or allow a controlled displacement when activated)

Figure 2: Typical Rockfall Barriers Components

According to the EAD documents mentioned, the barrier foundation system is not considered part of the kit. Due to the uncertainties in modelling the dynamics of a rockfall event impacting a structure made up of several heterogeneous components, it is generally accepted that the performance of rockfall barriers is assessed through full-scale crash tests carried out as per EAD guidelines. Since full-scale crash tests are not able to describe the barrier behaviour for all the real impact conditions, the test can be considered an index test to characterize the performance of rockfall protection solutions.

TESTING

As previously mentioned, rockfall barrier testing is conducted in accordance with the applicable European Assessment Document (EAD). EAD standardized the detailed procedure to carry out full-scale crash tests on rockfall kits and defines the following:

- Shape, minimum dimensions, and density of the impacting block
- Dimension of the tested barrier: it must have at least three (3) spans, and must be impacted in the middle of the center span
- Minimum impact velocity of the block: not lower than 25 m/s (approx. 90 km/h)

- The test field must be able to accelerate the impacting block to the minimum impact velocity; it can be either on a vertical or inclined slope
- EAD 340059-00-0106: Two (2) tests must be performed: the first one involving an impact at the barrier Maximum Energy Level (MEL); the second one consists in two subsequent impacts at the Serviceability Energy Level (SEL = 1/3 of the MEL) of the tested barrier. These two tests must be carried out on two (2) different barriers A and B, having the same geometrical and mechanical characteristics
- EAD 340089-00-016: impact test at one defined energy level

Figure 3: Full-Scale Crash Test On A Flexible Rockfall Barrier

For the kit to pass the MEL test (Fig.3), the barrier must stop the block having MEL without the block touching the ground before the barrier reaches its maximum elongation and without major damage. After the MEL test is completed, the barrier is subjected to the SEL test. Two subsequent impacts are required for the SEL test, each with an energy equal to one third of the MEL. No repairs and components replacements are allowed between the two consecutive tests. Moreover, the second impact can be carried out only if the residual height of the barrier, previously impacted by the first SEL launch, is at least the 70% of the nominal height of the tested fence (before the impact). During the second SEL impact the barrier must withstand the falling block without any requirement on residual height.

During the crash test the following performance parameters are measured:

- Maximum energy capacity of the barrier
- Maximum dynamic elongation of the interception screen (Fig. 4)
- Residual height (h_R): minimum distance between the lower and the upper longitudinal cables, measured orthogonally to the reference slope after the test and without removing the block from the interception structure. h_R is expressed as a percentage of the nominal height of the barrier (h_N), which is the distance between the upper longitudinal cable and the connection line between the base of the posts, before the impact, and measured perpendicular to the reference slope (Fig. 5)
- Forces trasnferred to the foundations
- Lateral gaps between lateral post and interception screen

- Photos and description of the damages occurred during the test

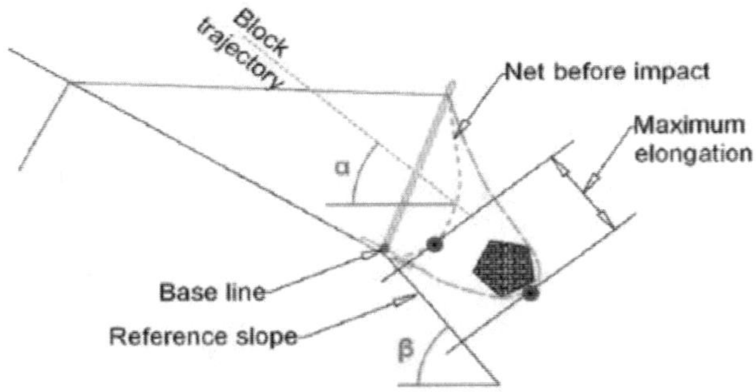

Figure 4: Maximum Elongation of the Interception Screen After Crash Test

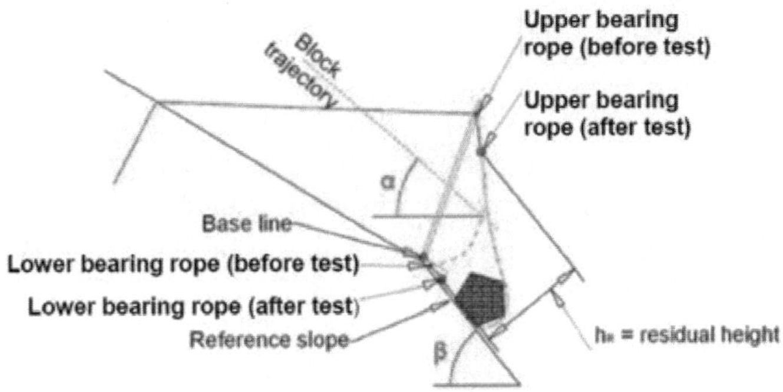

Figure 5: Residual Height of the Barrier After Crash Test

EAD (2018) provides a classification of the barriers in three categories based on the residual height after MEL impact:

- $h_R \geq 50\%\ h_N$, the barrier is classified as Category A
- $30\%\ h_N < h_R < 50\%\ h_N$ the barrier falls in Category B
- $h_R \leq 30\%\ h_N$ the barrier is classified as Category C

DESIGN

The UNI 11211-4:2012 design standard is a design code which describes the methodology to design passive rockfall barriers using a Limit State Design (LSD) approach. In 2018, UNI issued a new version of the mentioned standard: the UNI 11211-4:2018, which supersedes the previous one. Partial factors of safety applied during the design process which account for uncertainties and variability of conditions at job sites are prescribed in the UNI 11211-4:2018.

At the base of passive rockfall protection with barriers, rockfall simulations should be carried out to identify the trajectories of the potentially unstable blocks along the slope. Rockfall trajectory simulations aim at defining the statistical distribution of energy, velocity, height of the bounces, and endpoints of the falling rocks for each cross section under investigation. The input data necessary for the rockfall trajectories analysis are generally geo-mechanical surveys, which are required to characterize the unstable slope areas and the number and the dimension of the potential falling blocks, geological surveys, and finally, topography surveys to identify the geometry of the study area.

Rockfall simulations are generally performed with the aid of commercial software such as RocFall, CRSP, or Rock falls 3D, which may use different calculation approaches. For instance, Lumped Mass Analysis (LMA) has been used extensively. The LMA model applies the normal coefficient of restitution R_n, a parameter that depends on the material property of the ground, and tangential coefficient of friction resistance R_t, an experimental parameter that depends on the slope material and the vegetation. Moreover, rocks are considered dimensionless point masses. Nowadays, new methods have been implemented to offer a more realistic behaviour of the falling blocks. For example, the Rigid Body Impact Mechanics (RBIM) model introduces the effect of the size and shape of the rock and its interaction with the slope. It uses the soil material parameters (R_n, R_t), dynamic friction coefficient (μ: tangent of the friction angle, obtained with experimental data), and the rolling friction (Chai et al., 2013).

According to UNI 11211-4:2018, the design of a rockfall barrier can be done considering an Ultimate Limit State (ULS) or a Serviceability Limit State (SLS) approach. In both cases, the Limit State Design (LSD) approach introduces some partial factors. These factors are load coefficients, which increase the unfavourable actions acting on the barrier, and reduction coefficients, which reduce the resistance of the structure.

The equation at the base of this new design approach is:

$$E_{sd} < E_{barrier} / \gamma_E \quad (2)$$

where:

E_{sd} is the design energy level developed by the block at impact with the barrier.

$E_{barrier}$ is the energy capacity absorption of the barrier, measured during the crash test carried out according to EAD (MEL or SEL).

γ_E is a safety coefficient to be applied, equal to 1.2 in case of design following a ULS approach and equal to 1.0 in case of SLS approach.

E_{sd} can be defined with the classic equation of kinetic energy:

$$E_{sd} = 1/2\ m_d \times V_d^2 \quad (3)$$

Whenever the risk of loss of human lives is deemed to be high, an additional amplification factor between 1.0 and 1.2 should be applied to the design energy level (E_{sd}). Examples include highly trafficked roads and rail lines, and areas near schools and hospitals.

Generally, the spin effect of the falling rock is neglected because it generally contributes only 10-15% of the total kinetic energy; therefore, it can be compensated by introducing partial safety coefficients. In Equation 3, m_d is the design mass of the block; V_d is the design velocity of the block; m_d and V_d are expressed respectively as:

$$m_d = Vol_b \text{ x } g \text{ x } g_m \quad (4)$$

$$V_d = V_t \text{ x } g_F \quad (5)$$

In Eq. 4, Vol_b is the volume of the design block; g is the unit weight of the rock; g_m is an amplification factor expressed as:

$$g_m = g_{VolF1} \text{ x } g_y \quad (6)$$

where:

g_y is a coefficient related to the uncertainties on the unit weight of the rock (generally equal to 1.0);

g_{VolF1} is a safety coefficient depending on the reliability of the rock volume estimation and equal to 1.02 in case of availability of accurate surveys, or equal to 1.1 in case of no site-specific surveys.

In Eq. 5, V_t is the velocity resulting from rockfall trajectories analysis and taken as the 95th percentile of the normal distribution of velocities provided as output in the rockfall trajectories analysis; g_F is an amplification factor expressed as:

$$g_F = \gamma_{TR} \text{ x } \gamma_{DP} \quad (7)$$

where:

γ_{TR} is a safety coefficient depending on the reliability of the simulation which is equal to 1.02 if ground restitution coefficients (R_n, R_t) are derived from back analysis or equal to 1.10 if restitution coefficients are taken from bibliography.

γ_{DP} is a safety coefficient introduced to reflect the quality of the topographic survey used in the rockfall trajectories analysis and equal to 1.02 if a good quality topographic survey is available and equal to 1.10 if a low-medium quality topographic survey is used instead.

UNI 11211-4:2018 also provides guidance on the choice of the height of the barrier to be selected. The total interception height of the barrier at the point of impact should be greater, with a certain factor of safety, than the falling rock trajectory height:

$$h_{tot} \geq h_d + f_{min} \quad (8)$$

where:

h_{tot} is the total interception height of the barrier.

h_d is the design interception height defined as per Eq. 9

f_{min} is a minimum freeboard. The freeboard is to be taken as the highest value between 0.5 m, or the radius of the design impacting block.

$$h_d = h_t \times g_F \quad (9)$$

where:

h_t is the trajectory (i.e. bounce) height resulting from rockfall trajectories analysis and generally taken as the 95th percentile of the normal distribution of heights provided as output in the rockfall trajectories analysis, and g_F is expressed as per Eq. 7.

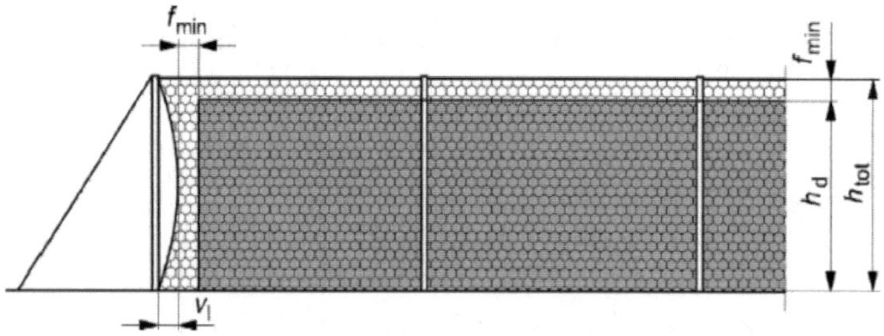

Figure 6: Total interception height (h_{tot}), and freeboard of rockfall barriers as per UNI11211-4:2018

Determining the appropriate barrier length to ensure full protection from the rockfall hazard is also an important design consideration. The barrier length must be extended beyond the area to be protected (area highlighted in dark grey in Fig.6) by a distance not less than the freeboard (f_{min}) added to the lateral gap measured during the full-scale test (Vi). The lateral gap (Vi) is the free gap left between the lateral supporting post and the interception screen which may be recorded at impact for some barriers available in the market. Barriers with no lateral gaps measured during the tests are generally preferred. Lateral gap is a critical performance parameter which allows the engineer to correctly determine the total length of barrier, which should be, in any case, at least three (3) modules (i.e. spans) and approximately 30 m.

In case of more than one row of flexible barriers, UNI 11211-4:2018 prescribes that there should be enough overlap between the two barriers (Fig. 7), also taking into consideration the actual slope morphology, distance between the barriers perpendicular to the alignment, shape of the design boulder and design trajectories.

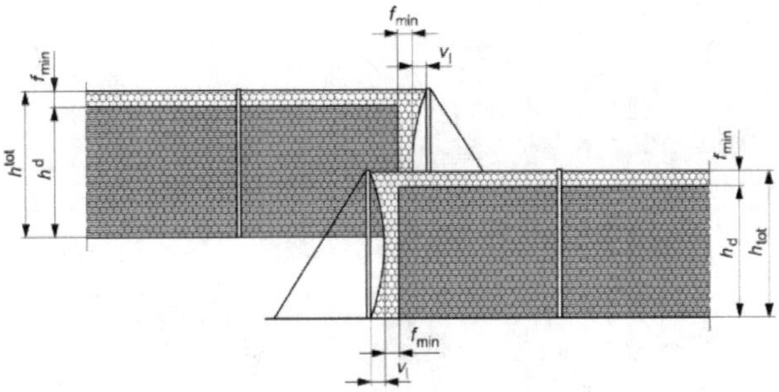

Figure 7: Indication on the overlap length between multiple rows of rockfall barriers as per UNI11211-4:2018

UNI11211-4:2018 also prescribes the minimum distance (d_A) between the barrier alignment and the asset to be protected:

$$\mathbf{d_A = d_{barrier} \times g_D} \qquad \textbf{(10)}$$

where:

$d_{barrier}$ is the maximum deformation measured during the full-scale crash test

$d_{barrier}$ should be taken equal to the maximum deformation during MEL or SEL test depending on whether the rockfall protection kit is designed to withstand a MEL or a SEL impact

g_D is a partial safety factor equal to 1.3

CASE STUDIES

Three case studies from various regions of the United States provide real-world examples of how the technical standards and metrics used in evaluating rockfall barrier systems are applied and how different barriers perform in practice. Each case study highlights unique challenges and solutions in roadside rockfall protection.

Maryland I-68 Sideling Hill

Interstate 68 (I-68) runs from I-79 in Morgantown, West Virginia, east to I-70 in Hancock, Maryland. Also known as the National Freeway in western Maryland, I-68 mainly spans rural areas and crosses numerous mountain ridges along its route. During construction in the 1980's, a section of these mountain ridges was blasted and cut through Sideling Hill, resulting in a pair of long, steep, narrow mountainous slopes approximately 1200 feet long and 360 feet high. Rock mass weathering over the years had posed a serious rockfall risk to the traveling public, prompting the Maryland State Highway Administration to contract Schnabel Engineering for the design of a rockfall remediation solution. Figure 8 shows an aerial view of the Sideling Hill rockfall barriers.

Figure 8: Aerial View of the Sideling Hill Rockfall Barriers

The design called for a rockfall barrier system to provide a maximum energy level capacity of 800kJ. However, due to the tight site constraints as shown in Figure 9, the barrier had to withstand the maximum rockfall energy without deflecting into the roadway, a distance of only 11 feet!

Figure 9: Tight Site Constraint Requiring a High Energy, Low Deformation Barrier

To meet these requirements, a semi-rigid barrier system was proposed by Maccaferri which consisted of unique low deformation High Energy Absorption (HEA) cable nets, manufactured with 1/2-inch diameter cables arranged in a 10 x 10-inch pattern, shown in Figure 10. This heavy net offers a very stiff response to the rockfall impact load, significantly reducing barrier deflection compared to conventional rockfall barriers available on the market. Heavy cantilevered posts, Figure 11, were designed to withstand the increased foundation loads associated with the more rigid impact response of the barrier.

Figure 10: Low Deformation HEA Cable Net Panel with DTWM Backing

Figure 11: Custom-Designed Cantilevered Posts Supporting the High Energy, Low Deformation HEA/DTWM Interception Structure

Construction of the barrier was completed by Geostabilization International (GSI) in 2023, with a total of 1480 linear feet of rockfall fence installed. During the construction of the barrier, the specialized contractor encountered difficulties installing the cable anchors. The natural discontinuities of the site caused some of the holes for the lateral anchors to collapse, making it impossible to insert the cable anchors. To address this issue, the specialty contractor presented to the contractor its tested cable anchor tethers as a solution. These systems are designed to be installed in conjunction with hollow bars, providing a reliable solution for securing the barrier support and bracing cables despite challenging site conditions.

This project serves as an example of how the strength and stiffness of specific products and meticulous engineering can address the unique challenges of roadside rockfall protection.

Idaho I-90

The Idaho Transportation Department required a 35 kJ K-Rail mounted barrier be installed along I-90 between Wallace and Mullan. The system installed was the Maccaferri RB 035, a 35 kJ rockfall barrier that has been tested in accordance with the EAD 340089-00-0106, 2019. This innovative lightweight barrier, Figure 12, is designed for easy and fast installation featuring multiple standard foundation configurations. It can be installed mounted on concrete barriers, directly embedded in the ground, or self-standing.

This solution highlights the adaptability of standard foundations for rockfall barriers and is also a great example of an application where a rockfall barrier protection is necessary and the assurance of having a tested and tried system is essential.

Figure 12: RB 035 (35 kJ) Rockfall Barrier Along I-90

This bonus case study features the same solution implemented on an access road to Camp Pendleton in California.

As depicted in Figure 13, it is clear to see why this is an easy installation barrier. The barrier posts are anchored into the back of the K-rail, enhancing the load transfer into the concrete barrier. Additionally, the lateral anchors are directly secured into the K-rail.

It is worth noting that a standard chain-link fence mounted on a K-rail is often used in similar applications as a lightweight rockfall roadside barrier. These systems, however, are not always tested or rated. The benefits and assurances of having a tested system guarantee the functionality of the barrier and should be considered when specifying systems intended to protect the traveling public.

Figure 13: RB 035 (35 kJ) Rockfall Barrier at Camp Pendleton, CA

Tennessee US 129

This project is situated along U.S. Route 129 near Knoxville, TN. As part of the current roadway project, the TN DOT required the contractor to provide a temporary rockfall protection barrier for workers and the traveling public throughout the duration of the project. TN DOT required that the moveable temporary rockfall barrier have a minimum height of 8-ft and be capable of withstanding repeated impacts of at least 200kJ impact without excessive component damage.

Typically, standard temporary movable rockfall barriers are installed and anchored on top of trench plates, which can be costly and hard to move around. However, the CTR series rockfall barrier offered a more efficient alternative. Despite the need to cover a vast area of 4200 linear feet for the project, the contractor opted for this solution over a movable barrier because of its ease of installation and maintenance.

This temporary rockfall barrier is designed without the need of upslope bracing anchors. The interception structure is a composite cable net that integrates a cable net pattern interwoven into a double twist wire mesh layer. This high strength material can be rolled out and attached to the support cables with shackles which benefit the installation and maintenance of the system.

Figure 14: Temporary Roadside Barrier Along US 129 in Tennessee

SUMMARY

Testing and evaluating the performance of flexible rockfall barriers have historically been conducted in accordance with the European Organization for Technical Approval (EOTA) issued in 2008, the ETAG 027 guidelines. In 2018, EAD 340059-00-0106 superseded this document and standardized the procedure for carrying out full-scale crash tests on rockfall kits for rockfall barriers with maximum energy absorption capacities equal or greater than 100 kJ. EOTA is currently in the process of finalizing and publishing guidelines for testing low energy rockfall barriers less than 100 kJ, EAD 340089-00-0106.

The UNI 11211-4: 2012, issued in January 2012, is a design guideline for flexible rockfall barriers which considers the performance assessed during the full-scale tests. In 2018, a revision to this document was released as UNI 11211-4: 2018.

This paper explains the definitions, tests, and design method for flexible rockfall barriers used as passive protection method. The full-scale test procedure described by EOTA in the EAD documents is a very useful tool for engineers and project owners to evaluate the performance of rockfall barriers in terms of maximum energy absorbtion capacity, maximum deformations, lateral gap, and forces transfered to the foundations. Combined with the design method presented based on UNI11211-4:2018, it allows for designing more effective rockfall protection systems general enough to be applied globally.

In the process of designing or specifying rockfall barriers for roadside applications, it is essential to consider all the metrics of the system, including factors like deformation and its impact on the applied foundation loads, as well as the ease of installation and possible foundation configurations.

REFERENCES

- Conference Proceedings
  1. Brunet G., Giacchetti G., Grimod A., 2013. 'Rockfall barrier behavior under multiple impact events', 64th Highway Geology Symposium, September 9-12, 2013. New Hampshire, USA.
  2. Cerro, M, Giacchetti, G, Lelli, M, Grimod, A & Arul, A 2016, 'ETAG certified rockfall barriers — new design approach according to UNI 11211:4/2012', APSSIM 2016: Proceedings of the First Asia Pacific Slope Stability in Mining Conference, Australian Centre for Geomechanics, Perth, pp. 443-456.
- Federal Government Reports
  1. ETAG 027: Guideline for European technical approval of falling rock protection kits. Brussels, EAD 340059-00-0106, 2018.
  2. UNI 11211-4:2018: Rockfall protective measures – Part 4: Definitive and executive design.
  3. European Organization for Technical Assessment (EOTA) Falling Rock Protection Kits with Energy Level Less than 100 kJ, Brussels, EAD 340089-00-0106, 2019.

Utah HWY 189 Hazard Rock Removal Using Non-Detonating Rock Breaking Cartridges:

A Case Study in Provo Canyon

Jack Fitzgerald, P.E., G.I.T
GeoStabilization International
10225 Westmoor Dr
Westminster, CO 80021
970-370-0433
jack.fitzgerald@gsi.us

Prepared for the 73rd Highway Geology Symposium, September, 2024

Acknowledgements

The author(s) would like to thank the individuals/entities for their contributions in the work described:

Zach McClellan, M.S., P.E. PMP, ENV SP, GeoStabilization International
Fred Sauber – Crew Operations and QC Manager, Access Limited Construction
Graeme Lindsay – Superintendent, GeoStabilization International
Ari Menitove, P.E. – Geological Engineer, Utah Department of Transportation

Disclaimer

Statements and views presented in this paper are strictly those of the author(s), and do not necessarily reflect positions held by their affiliations, the Highway Geology Symposium (HGS), or others acknowledged above. The mention of trade names for commercial products does not imply the approval or endorsement by HGS.

Copyright Notice

ABSTRACT

In early March 2023, blocks of rock fell onto the highway onramp near milepost 7.6 on US-189 in Provo Canyon near Orem, Utah. The event created several relatively large undercut blocks (up to 75 ft in length) along with thinner blocks sitting precariously above the onramp. To limit the danger of these rock masses to the traveling public, Utah Department of Transportation (UDOT) closed the onramp until the unstable masses could be removed.

US HWY 189 cuts through Provo Canyon near Orem Utah; the canyon was formed by the Provo River cutting through Mississippian and Pennsylvanian sedimentary formations. Orientations of the rock units vary throughout the canyon, however there are several areas which present adverse geologic conditions, including bedding dipping into the roadway and near vertical fractures running perpendicular to near horizontal bedding. These adverse geologic conditions created the potential for rockfall hazards and overhung blocks. Traditional scaling techniques alone were judged to be ineffective to remove the blocks, so trim blasting was deemed the best solution.

As an alternative to conventional blasting, a non-detonating rock breaking cartridge (NDRBC) was used to remove the overhang efficiently with limited disruption to the road. An NDRBC is a non-detonating cartridge that is used to break rock via deflagration instead of detonation; this leads to a more controlled process with less fly rock compared to conventional blasting. This paper will present the UDOT block removal project as a case study on the use of NDRBC as a tool for emergency rock downsizing and removal to mitigate highway rockfall hazards.

INTRODUCTION AND PROJECT BACKGROUND

On March 10, 2023, a large block released above the 800N EB to US189 NB Ramp outside of Orem, Utah in Provo Canyon. The failure caused rock to enter the ramp, which UDOT was able to remove with a plow truck, however several precarious features remained above the failure area - a “hanging” undercut block, an adjacent triangular block, and a thin “flake”. Due to the hazard these blocks represented, UDOT closed the ramp to traffic. The annual average daily traffic in this section of US189 is 31,000 vehicles per day (1), and with this ramp closed local traffic was required to either take a detour or an alternate route to enter the highway. Figure 1 shows photos of the rock slope provided by UDOT (2) detailing the slope before and after the failure. The undercut block measured approximately 50’ x 6’ x 6’, and the entire area of concern (including the triangular block and “flake”) was nearly 80’ wide. A near vertical joint set daylighted behind the blocks, indicating it was likely they could be removed without causing further unraveling of the remaining rock above.

Due to the size of the blocks, and access issues to get above the failure area, removal of the blocks by traditional methods of scaling such as hand-scaling and air-bagging were judged unlikely to be efficient means to remove the hazard rocks. As such, non-detonating rock breaking cartridges (NDRBC’s) were used to remove and downsize the blocks in place, leading to an expeditious removal of the hazard and reopening of the highway. GeoStabilization International (GSI) mobilized to the site on March 23, and the ramp was reopened March 31.

Figure 1 – Photos from UDOT (2)

SITE LOCATION AND GEOLOGY

Provo Canyon cuts through the Southern Wasatch Range in North Central Utah, with US HWY 189 following the canyon, connecting Heber City to Orem. The canyon was formed by the Provo River cutting through predominately sedimentary rocks associated with the Pennsylvanian-Aged Oquirrh Group and the Mississippian-Aged Great Blue Limestone and Manning Canyon Shale Formations (2). The project site itself is mapped as the Great Blue Limestone formation, which is described as "Dark-Gray to nearly black, light-to medium-gray weathering, thin- and regularly bedded limestone and shaly limestone with interbedded black and brown shale beds up to 50 feet thick…" (2). Site conditions agree with mapped geology as the cut appears to consist of less weathered black limestone overlain by thinner beds of more weathered limestone, as shown in Figure 2. A thin bed of shale below the limestone layer likely contributed to the initial failure, as it appears to have eroded over time, allowing a loss of toe support to occur. The hazard areas are shown in red in Figure 2.

Figure 2 – Overall Site Photo

ROCK REMOVAL APPROACH

After being notified by UDOT of the issue, GSI personnel began discussing options internally between Engineering and Operations groups. Due to the size of the blocks in question, trim blasting was determined to be the best course of action to remove the hazard partnered with safety scaling before rock removal and a follow up scale afterwards. The near horizontal bedding in the limestone unit allowed for several natural benches to form, and these were covered in previously failed material. There was concern of this smaller-sized material coming loose if ropes were to move across it. Additionally, access to the top of the slope was going to be very difficult, and safety scaling down to the work area was unlikely to be effective with the potential for the loose material mentioned above. It was therefore determined that using aerial lifts would be the safest, most efficient way to tackle the problem.

Due to the location of the blocks relative to HWY 189, the potential for fly rock into the highway was concerning. 20-minute closures of HWY 189 would be permitted during blasting activities, with a goal to keep as little rock from getting into the travel lane as possible. A plow truck was stationed on HWY 189 to remove any fly rock that made it to the highway before reopening to traffic. To maintain live traffic flow on the main highway as much as possible, a temporary rockfall protection system was used during non-blasting activities. This consisted of high tensile strength wire mesh hung between two loaders parked on the outside shoulder of the ramp to prevent smaller rocks from entering the travel lane.

Non-Detonating Rock Breaking Cartridges – Background

Explosives have been used to break and blast rock since the early days of black powder. High explosives are still commonplace today, typically using ammonium nitrate/fuel oil (ANFO) as the blasting agent. For many large-scale blasting projects such as open pit and underground mining, major rock excavation for civil projects, tunneling, and new highway road cut construction, the use of high explosives will likely remain commonplace due its relatively inexpensive cost and familiarity of use. The downsides of conventional blasting include fly rock, air overpressure, ground vibrations, overbreak, and gas emissions; from the standpoint of an emergency rockfall project next to a highway all of these must be considered.

For smaller-scale projects, such as trim blasting and rock downsizing, conventional blasting is oftentimes not the best technique, particularly for discrete blocks or focused areas of a rock slope. Consider a rock slope above a roadway; the potential for fly rock and air overpressure could represent a risk to the public, overbreak could cause further instability in the rock slope, and ground vibrations could damage nearby infrastructure. There are several alternatives to conventional blasting for rock breaking and downsizing including expansive grout, hydraulic splitting, mechanical splitting, and non-detonating rock breaking cartridges (NDRBC's), among others. While each of these alternatives has advantages and disadvantages, NDRBC's represent an option that provides similar results to conventional blasting (i.e., relatively fast, controlled rock breaking) with fewer of the disadvantages.

NDRBC's are self-contained cartridges that operate under the principle of deflagration instead of detonation (4). Essentially, when ignited, a chemical reaction occurs within the cartridge to rapidly produce volumes of gas. Once the reaction begins, the body of the cartridge expands and forms a seal against the walls of the hole drilled in the rock. With the borehole sealed, the gas produced from the reaction enters either existing planes of weakness within the rock, or into any fractures created from the drilling process. The pressure associated with the formation of the gas then exceeds the tensile strength of the rock, causing a tensile failure within the rock and eventually causing the rock to break.

PROJECT SUMMARY

GSI arrived onsite on March 23, 2023, and began safety scaling and an initial inspection of the rock slope the following day. Temporary protection was set up at the base of the slope consisting of high tensile strength mesh and a lightweight geotextile hung between two pieces of equipment. This was used during non-blasting activities such as scaling and drilling to prevent smaller rocks from entering the roadway below, allowing for live traffic to use the main highway. Following the initial safety scale, the planned boreholes were marked out and the crew began drilling. The depth of the boreholes generally ranged from 2' to 6' and were drilled on an approximate 3' by 3' pattern. The first blast was conducted on March 28 and removed approximately 2/3 of the overhung block, with the remainder of the overhung block removed over the next two days in subsequent blasts. Figures 3 through 5 show the progression of each blast, courtesy of a video provided by UDOT. The blasts were followed up with hand scaling to remove any loose material remaining on the slope and to allow for a visual assessment by the site Superintendent and Blaster-in-Charge to evaluate the stability of remaining blocks and determine if any additional blasting was necessary.

Figure 3A – Blast 1 Pre-Blast

Figure 3B – Blast 1 During Blast

Figure 3C – Blast 1 Post Blast

Figure 4A – Blast 2 Pre-Blast

Figure 4B – Blast 2 During Blast

Figure 4C – Blast 2 Post Blast

Figure 5A– Blast 3 Pre-Blast

Figure 5B– Blast 3 During Blast

Figure 5C– Blast 3 – Post Blast

The first blast sequence yielded the least amount of fly rock and most containment of the blasted material within the work area. The second and third blasts both had higher amounts of fly rock than anticipated, which was likely due to the blastholes being closer to the daylighting fractures at the top of the overhung block.

After blasting and scaling were complete, the entirety of the overhang was removed, the triangular block was partially removed, and the "flake" was removed. All remaining blocks in the work area were assessed by the Superintendent before leaving the site. Figure 6 shows the rock slope after work was completed.

Figure 6 – Rock Slope After Completion of Work

CONCLUSION

Non-Detonating Rock Breaking Cartridges represent a useful tool both for rockfall and rock slope projects, particularly for emergency response projects. These devices are quick to install and are very effective for discrete rock breaking and localized trim blasting of rock slopes. In this case, after an initial rockfall event closed an access ramp to US HWY 189 on March 10, GSI began work on March 23 and the ramp was safely reopened to the travel public on March 31. This project demonstrates the benefits of this technology for use in emergency response to hazardous rockfall above a road.

REFERENCES:

1. Utah Department of Transportation, *Traffic Statistics,* https://www.udot.utah.gov/connect/business/traffic-data/traffic-statistics/, Utah Department of Transportation, accessed August 2024

2. Utah Department of Transportation, External Communication, March 2023

3. Solomon, B.J., Constenius, K.K., and Machette, M.N., *Interim Geologic Map of the Orem Quadrangle, Utah County, Utah*. Map, Utah Geological Survey, OFR 567, Scale 1:24,000

4. AutoStem Technologies, *AutoStem Training Manual,* N.D.

Rockfall Hazard Mitigation along I-68 at Sideling Hill, Washington County, Maryland

Brian K. Banks, PG

Schnabel Engineering, LLC
1133 Military Cutoff Rd, Suite 210
Wilmington, NC 28405
240-832-6272
bbanks@schnabel-eng.com

Steve W. Fung, PE

Schnabel Engineering, LLC
300 Red Brook Blvd, Suite 109
410-944-6170
sfung@schnabel-eng.com

Prepared for the 73rd Highway Geology Symposium, September 2024

Disclaimer

Statements and views presented in this paper are strictly those of the author(s), and do not necessarily reflect positions held by their affiliations, the Highway Geology Symposium (HGS), or others acknowledged above. The mention of trade names for commercial products does not imply the approval or endorsement by HGS.

Copyright

ABSTRACT

The Sideling Hill road cut along Interstate I-68 near Hancock, Maryland features two opposite-facing rock cut slopes about 1,600 ft long and 360 ft high. The cut slopes offer an impressive view of interbedded sandstone, siltstone, shale, and coal folded into a tight syncline. Since the initial I-68 roadway and rock slope construction was completed in 1985, slope degradation and differential weathering has led to overhanging rock ledges, loose rock, narrow benches, debris wedges filling the benches, and ultimately a significant increase in the risk of rockfall impacting the I-68 travel lanes. Schnabel Engineering, LLC (Schnabel) performed an initial rockfall hazard investigation in 2012, which included ground-based LiDAR to develop a baseline 3D terrain model as well as slope stability and rockfall hazard analyses to evaluate alternative rockfall risk mitigation concepts. A subsequent LiDAR survey was performed in 2017 for change detection analysis to identify slope areas with significant material loss as well as areas of significant material gain on the benches and catchment areas. Schnabel considered a variety of rockfall risk mitigation options. Maryland State Highway Administration (SHA) ultimately selected a flexible rockfall fence as the preferred risk reduction strategy. Schnabel developed a fence design in collaboration with the proprietary mesh system supplier, Maccaferri. The design consisted of a 9-ft tall fence with steel posts spaced at approximately 25 ft embedded into drilled shafts for fixity, and cable panels for primary interception of rockfalls. Cantilevered posts, wire ropes and lateral anchors were designed to resist the design rockfall force of 800 kJ within the desired deflection limits of 11 ft. Use of upslope anchors were not included to facilitate access behind the barrier for maintenance. The construction project was awarded in January 2023 to Carl Belt, Inc. Construction of the fence was substantially completed in December 2023. The fence option avoided many of the difficult access and safety issues that would have been problematic for alternative rockfall mitigation options. The fence option also protects the iconic views of the I-68 Sideling Hill syncline.

INTRODUCTION

Sideling Hill is a major northeast-trending topographic ridge extending from Pennsylvania through Maryland and into West Virginia. Two opposite-facing rock cut slopes were exposed during construction of I-68 through Sideling Hill in the early to mid-1980's. The site is located a few miles west of Hancock, Maryland, in Washington County (Figure 1).

The Sideling Hill road cut is well known for its impressive exposure of sedimentary rock layers folded into a tight syncline (Figure 2). As one of the most iconic rock exposures in the northeastern U.S, it is a popular destination for geology students and enthusiasts. The visitors center just east of the cut is accessible from I-68 in both westbound and eastbound directions, and it offers free parking, restroom facilities and various views of the cut slopes.

Figure 1 – Site Location (Reference: Google Maps)

Since construction was completed in 1985, slope degradation and differential weathering has led to a significant increase in the risk of rockfall impacting the I-68 travel lanes. Recognizing the increased risk to roadway users, Maryland State Highway Administration (SHA) desired to better understand the conditions contributing to an elevated rockfall potential with an eye toward implementing proactive rockfall risk mitigation strategies.

Figure 2 – North Cut Slope on I-68 at Sideling Hill with Syncline Rock Structure

SHA retained Schnabel Engineering, LLC (Schnabel) to characterize the rockfall hazards, develop risk-reduction recommendations, and design the final rockfall mitigation solution. This study involved engineering geology field mapping, rock slope stability and rockfall hazard analyses, monitoring slope conditions, and development of conceptual designs for rock slope maintenance. SHA used the results of the study to select a preferred maintenance concept to carry forward into final design and construction. SHA's preferred option was a 9-ft tall flexible rockfall fence installed along both the eastbound and westbound shoulders of the roadway. The construction project was awarded in January 2023 and substantially completed in December 2023.

SITE GEOLOGY

Sideling Hill lies in the Valley and Ridge Physiographic Province, a region characterized by folded sedimentary rock strata and alternating valleys and ridges oriented northeast to southwest. Sediments that formed the rock within the Valley and Ridge were deposited during late Cambrian to Mississippian time between about 500 and 320 million years ago (Means, 2010). Folding resulted from compression forces developed in the Earth's crust by the collision of the North American and African tectonic plates during the Alleghenian orogeny between about 320 and 250 million years ago. Folding was accompanied by faulting to accommodate flexural bending of the rock strata. In addition to regional thrust faults, numerous localized faults occur, particularly near fold axes.

The uppermost (i.e., youngest) Valley and Ridge formations are exposed in the I-68 Sideling Hill road cut. These formations include the Purslane and Rockwell Formations of the Devonian/Mississippian-age Pocono Group (Brezinski, 1994). The Purslane Formation is typified by cross-bedded sandstone, conglomerate, and siltstone, with a minority portion of interbedded shale and coal. The underlying and stratigraphically older Rockwell Formation consists of interbedded shale, siltstone and sandstone with a few claystone and coal interbeds, and a glacial diamictite layer near the base of the western exposure. These sedimentary rock strata are folded into a tight, northeast-trending syncline coincident with the regional Sideling Hill ridge. The road cut intersects the syncline roughly perpendicular to the fold axis, resulting in a nearly symmetrical exposure of the eastern and western fold limbs on both sides of the cut.

The sandstones and conglomerates of the Purslane Formation occur in the center of the fold and cap the ridge. The Rockwell Formation is exposed in the lower portion of the cut slopes and forms the outside limbs of the fold. Based on the Maryland Geological Survey's Geology of the Sideling Hill Road Cut website (Brezinski, 1994), the contact between the Purslane and Rockwell Formations can be observed within the eastern and western fold limbs. The contact occurs below grade in the center of the fold.

Numerous faults formed during folding to accommodate flexural bending of the rock strata. Many bedding-parallel thrust faults are exposed in the cut slopes where movement occurred along the bedding planes. This type of faulting is referred to as flexural slip and is similar to the slippage that occurs between cards when a deck of cards is folded. There are also many thrust faults that crosscut bedding. These faults generally strike parallel to the fold limbs and dip more steeply toward the fold axis. Some of these crosscutting thrust faults are offshoots and splays of bedding-parallel faults that appear to step upward irregularly from one bedding plane to another, rising higher in the stratigraphic column in the fold limbs. A few reverse faults completely crosscut bedding for relatively long traces. A major reverse fault was observed on the south face, within the eastern fold limb. This fault exposure was measured to be over 250 ft long.

Seeping groundwater can be observed in the central portions of the rock faces, with greater amounts of seepage occurring in the lower sections. These groundwater seeps originate as rain which infiltrates the permeable rock mass along fractures. Relatively low permeability rock layers (e.g., shale) act as barriers to impede downward groundwater flow. Because these layers are folded into a syncline, they tend to channel groundwater toward the fold axis, which is why the central portions of the cut slopes are associated with greater seepage relative to the outside portions. The amount of seepage varies through the year and in response to precipitation. During the winter, seeping water freezes into ice flows.

CUT SLOPE CONSTRUCTION

The original construction of the Sideling Hill road cut was completed in 1985. The road cuts measures approximately 1,600 ft long and 360 ft high at the center on each side. Based on the as-built drawings (Baker-Wibberley & Associates, 1986), the rock cuts were designed with up to 80-ft-high bench slopes cut as steep as 0.25H:1V, with 20-ft wide benches reverse-sloped at an inclination of 20H:1V. Both the north and south cut slopes have four benches in the tallest (central) portions of the slopes. The rockfall catchment ditches at the base of the slopes were designed to be 38 ft wide with a shallow V-shaped configuration.

Roadway modifications were made circa 1990-1991. Modifications included adding truck climbing lanes in both directions, an exit ramp from the visitor's center on the west-bound side, and an entrance ramp to the rest area on the east-bound side. The visitor center opened in 1991.

The existing geometry of the cut slopes is significantly different from that indicated by the as-built construction drawings. While some differences between current conditions and the as-built drawings results from weathering since construction, we believe that the actual rock cut slope construction varied considerably from what is shown on the construction drawings. For example, the bench widths are narrower than 20 ft in many areas. We found bench widths typically in the 10 ft to 18 ft range in the central portions of the cuts with some locations found to be even narrower. The bench is completely missing in a short section of the third bench on the north slope (Figure 3). Some narrowing of the benches has occurred since construction due to raveling and loss of the bench edges. However, the location of blasting half-casts delineating the originally constructed slope faces suggests that the benches were constructed to be less than 20 ft wide in many areas. Additionally, the catchment area between the toe of the slope and the edge of pavement varies from 27 ft to 45 ft wide at the base of the south slope, and 14 ft to 38 ft at the base of the north slope. There is also some variation in the slope face inclinations from what is shown on the construction drawings.

Figure 3 – Missing Section of Third Bench on the North Slope

ROCKFALL HISTORY AND SLOPE MAINTENANCE

In the nearly 40 years since construction, the Sideling Hill rock cut slopes have significantly degraded. Differential weathering between more competent sandstone and less competent shale interbeds has led to overhanging sandstone rock ledges, narrow benches, loose rock on the bench slopes and bench crests, and debris filling the benches. Several prominent slabs of rock within massive sandstone units have separated from the main rock mass by tension cracks. Figure 4 shows typical slope conditions within the lower-central portion of the north slope.

Fortunately, no significant rockfall events have been reported. Typical rockfall blocks lying on the ground in the catchment ditches and on the benches were found to be blocky, rectangular, and tabular shaped and typically less than about 1 ft wide. However, blocks up to about 3 ft were common in some areas, particularly in the central portion of the slopes. Similar-sized rocks have occasionally been found in the I-68 pavement areas. These rocks are removed as needed by SHA District 6 maintenance crews. No accidents or significant damage has been reported. In 2011, an approximately 2 ft wide rock was found resting on the paved shoulder in front of the eastern end of the south rock cut slope. Reportedly, a divot in the pavement was observed at the edge of the travel lane (i.e., at the solid white line) marking the point of the rock's impact. Evidently, the falling rock bounced backward after impact at the edge of the travel lane and came to rest on the paved shoulder.

Figure 4 – Typical Slope Conditions on the North Slope

We understand from SHA that visual inspections and clearing of rockfall debris from the benches were performed annually for the first approximately ten years after construction. However, after about 1995, regular maintenance tapered off with the last bench-cleaning event occurring in 2002 but only on the north slope. Since then, rockfall debris has been periodically removed from the catchment and pavement areas on an as-needed basis.

ROCK SLOPE HAZARD INVESTIGATION

The rock slope hazard investigation was accomplished in a phased approach, occurring over a period of about 8 years, from 2009 to 2017. After initial site reconnaissance in 2009 and 2011 to check for potential emergency-response conditions, a comprehensive mapping program was performed in 2012 to develop alternative concepts for risk reduction. A follow-up site visit was performed in 2017 to identify changes in the slopes and collect additional data to support final design. This phased approach was possible because there was no need for emergency maintenance and therefore SHA had time to study and monitor the slopes to explore cost-effective alternatives for proactive risk mitigation.

A variety of tasks were performed as part of our field investigations to develop a comprehensive understanding of the rock slope hazard conditions. These tasks included terrestrial LiDAR and UAV-based photogrammetry surveys and engineering geologic mapping of the rock mass and rockfall characteristics across the site. Data from our field investigations were used to identify likely rockfall sources and typical block characteristics, conduct rock slope stability analysis and rockfall hazard modeling, develop viable risk mitigation concepts, and ultimately to design the final constructed solution. Figure 5 shows site conditions in 2017.

Figure 5 – Photo from UAV Photogrammetry Survey in 2017

Engineering Geologic Field Mapping

Schnabel conducted a field investigation to map the engineering geologic features of the rock mass and rock slope faces in April 2012, focusing on identifying evidence of rockfall activity as well as conditions contributing to instability and rockfall hazard potential. We revisited the site in December 2017 to assess changes since 2012.

Typical rockfall block sizes and shapes were measured from rock blocks lying on the ground at the base of the existing rock slopes and on the benches. These rocks were found to be blocky, rectangular and tabular shaped. The size of the rock blocks were typically less than 1 ft wide. However, blocks up to about 3 ft wide were common in some areas such as in the central portion of the slopes. We observed a significant amount of rockfall debris had accumulated on the benches and at the base of the slopes within the catchment ditch, particularly below differentially weathered shale interbeds. The wedge shape of many shale debris accumulations reduces the rockfall containment ability of the benches and increases the overall rockfall hazard of the slopes by increasing the launching potential of rockfall blocks.

Although the entirety of the slopes has the potential to initiate rockfall, more frequent rockfall events are expected from areas with high concentrations of loose and fractured rock. Areas of loose rock were often found at the bench edges. There was evidence the bench edges were raveling, which caused narrowing of the benches and thus a decrease in the rockfall catchment protection offered by the benches. Additionally, raveling of the bench edgesresults in a profile geometry that increases the rockfall hazard. Another negative consequence of bench edge raveling is that narrow benches discourage bench-cleaning activities due to the associated access difficulties.

We observed ample evidence of differential weathering and undercutting of relatively low durability rock layers (e.g., shale and coal) below relatively high durability rock layers (e.g., sandstone and conglomerate) forming overhanging rock ledges. Overhanging ledges were generally found to range up to about 2 ft deep. However, we found overhangs up to 6 ft deep in a few locations. Overhanging rock ledges present an elevated rockfall hazard due to the decreased stability of the unsupported rock mass. The stability of the overhanging rock mass depends mainly on its fracture characteristics and discontinuity orientations. Massive rock above an overhang will be more stable than highly fractured rock. In general, the rock masses immediately above the overhangs were found to be in relatively good condition with respect to large-scale rockfall potential. Overhanging rock ledges were not identified to present an immediate concern with respect to large-scale collapse. However, it is recognized that ongoing differential weathering and undercutting will increase the depth of the overhangs and decrease the stability of the overlying rock masses. Based on field observations, we expect progressive failure of many of the overhangs will involve repeated small-scale rockfall events as opposed to relatively infrequent massive events (Figure 6).

We identified individual or groups of larger (>5 ft) rock blocks that appear to be marginally stable and thus are believed to present a significant rockfall hazard. These larger, marginally stable rocks were partially detached from the slope face, and not clearly supported. On both the north and south slope faces, several marginally stable blocks up to about 15 ft tall were identified on the first, second and third bench faces. Instability is exacerbated by undercutting below some of the blocks. We found evidence that similar-sized rockfall blocks fragmented into smaller blocks as they detach and fall down the slope face.

Figure 6 – Progressive Raveling of Overhanging Sandstone Interbed

We collected representative rock structure orientation measurements from bedding, joints, fractures, and faults across the site for rock slope stability analyses. We also mapped discrete major throughgoing discontinuities, the locations of which correlate well with areas of loose and marginally stable rock blocks. Lithologic boundaries and the locations of major faults and fractures were also mapped digitally by fitting a plane to a discontinuity surface observed in the LiDAR data (see below). Digitally mapped discontinuities were generally consistent with field observations.

Seepage was identified as generally wet areas of the rock face, or in many cases as areas of dripping water. We would expect seepage areas to be associated with an increase in the weathering of the slope face, and thus with an increased rockfall hazard potential. During the winter, water seepage freezes. In areas of greatest seepage, the ice accumulates to cover large areas. Freeze-thaw action in these areas contributes to the on-going weathering and degradation of the cut slopes. These effects are greater on the northern slope face where direct sunlight is expected to cause more frequent cycling between freezing and thawing conditions.

LiDAR Survey & Change Detection Analysis

We performed an initial LiDAR survey in 2012 to create a high-resolution three-dimensional digital terrain model (DTM) of the Sideling Hill rock cut slopes. The DTM was beneficial for documenting existing slope conditions and establishing a baseline for monitoring, and it provided detailed slope geometry for our hazard assessment. We also used the DTM for digital rock structure mapping allowing us the opportunity to collect rock structure orientation measurements from otherwise inaccessible locations. We performed an overlapping LiDAR survey in 2017 to investigate changes to the slopes over time. The LiDAR scans were performed with an I-Site 8800 laser scanner set up in multiple locations to obtain data from a variety of vantage points which allowed us to resolve finer details in the highly irregular slope geometry such as the many protrusions, overhangs, undercuts, and bench surfaces. A total of nine LiDAR scan locations were used, including five on the north slope (to image the south slope), and four on the south slope (to image the north slope). The same LiDAR scan locations were used in the 2012 and 2017 surveys. Based on how precisely the point cloud data from the individual LiDAR scans overlapped, the precision of the survey was within about 0.1 ft in all directions. The resulting 3D terrain model is therefore considered to have an accuracy of 0.1 ft in all directions.

Through change detection analysis between the 2012 and 2017 LiDAR scans, we were able to resolve slope areas with significant ground loss at rockfall sources zones as well as areas with significant material gain representing rockfall debris accumulations on the benches and in the catchment ditches. We produced a series of "heat maps" to assess areas of material loss and gain, and to roughly quantify the change based on the distance between the 2012 and 2017 surfaces. For example, several areas of material loss are indicated in the heat map shown in Figure 7, an approximate 300-ft wide area on the north slope. Differential weathering of several shale beds in this area has resulted in further undercutting of the overlying sandstone beds. Undercutting in the shale caused progressive rockfall in sandstone overhangs (Figure 6).

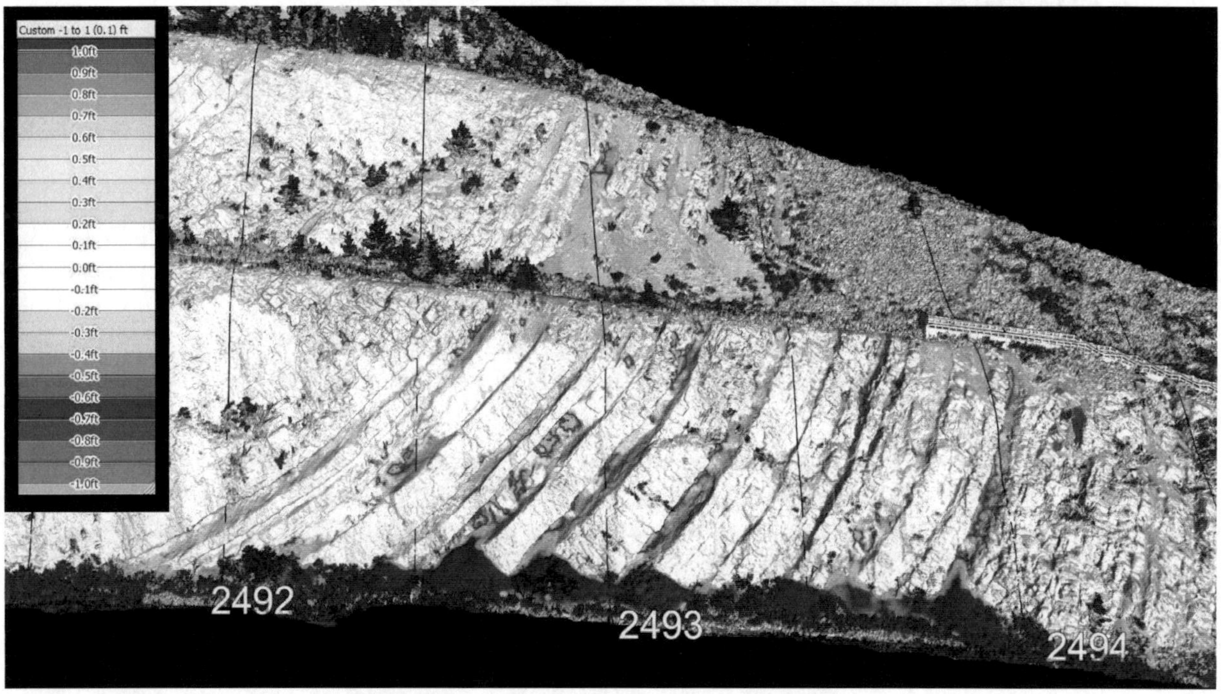

Figure 7 – Change Detection Heat Map

Based on a comparison between the heat maps from multiple vantage points (i.e., the individual LiDAR scan locations) we concluded that much of the material loss from 2012 to 2017 came from differential weathering in the shale layers on the fold limbs. The depth of loss in these shale layers generally ranged up to about 0.5 ft depth and locally up to about 1 ft depth or more. A corresponding increase of material was observed in the debris wedges, with accumulations over 2 ft depth in many areas below these differentially weathered shale layers. Relatively few specific rockfall source zones were identified within the sandstone layers. The sandstone rockfall sources were found to be much more isolated and generally only up to a few feet across and up to about 2 ft in depth, which corresponds to the dimensions of typical sandstone rockfall blocks observed on the benches and catchment areas that are mostly less than 2 ft wide and occasionally up to about 3 ft wide. Most of the sandstone rockfall sources were from the overhanging edges of sandstone layers immediately above undercut shale and coal interbeds (Figure 7). There also appeared to be ongoing raveling and loss of the bench edges generally ranging up to about 1 ft depth, particularly in the areas of regularly interbedded shale and sandstone at the fold limbs.

The more massive sandstone in the central portion of the syncline is relatively unchanged, except for relatively small, isolated rockfall sources and relatively small accumulation of debris on the benches. However, we observed one relatively large rockfall source zone above the first bench on the north slope. The source zone measures approximately 10 ft wide and about 15 ft high. We observed this source zone and the associated rockfall debris in the field. Although the source area is relatively large, the largest rockfall blocks in the debris field are typically less than about 2 ft wide indicating that the rockfall mass broke up as it travelled down the slope.

UAV Photogrammetry Survey

Schnabel also performed an unmanned aerial vehicle (UAV) based photogrammetric survey as a companion to the LiDAR survey to develop a photo-realistic, three-dimensional model of the site. We used the photogrammetry model to better visualize slope conditions and observe inaccessible portions of the slopes to aid with site characterization. An overall perspective view of the photogrammetry model is presented as Figure 8.

Figure 8: Overall Perspective View of the Photogrammetry Model

ROCK SLOPE STABILITY ANALYSES

Rock slope stability analyses were performed to determine areas of potential large-scale rock slope instability based on the geometry and estimated shear strength characteristics of rock mass discontinuities. Through kinematic analysis, we identified several locations with discontinuity orientations indicative of potential large-scale planar and wedge sliding failure modes. However, in all cases our limit equilibrium analysis indicted acceptable factors of safety (i.e., FS > 1.5) which suggests that the identified potential large-scale rock slope failure modes do not pose a significant stability concern. While large-scale failure modes were not predicted, the slope will continue to shed rocks over time because of small-scale failures. Rocks will loosen and dislodge due to weathering and freeze-thaw effects.

ROCKFALL HAZARD MODELING

Schnabel also performed rockfall simulations to identify sections of the slopes where falling rock could impact traffic, and to obtain information about the range of travel distance, bounce height, and energy of simulated rockfall blocks to evaluate rockfall hazard mitigation strategies. We considered a slope geometry reflecting existing conditions (i.e., no bench cleaning and debris wedges left in place) and considered a range of typical rockfall block shapes and sizes based on site observations. We set our analysis point at an assumed rockfall barrier location 10 ft outside of the edge of the travel lane (i.e., solid white line) which considers an 8 ft wide shoulder and an additional 2 ft fence setback.

Our rockfall modeling results indicate that up to about 14 percent of rockfall blocks will reach the assumed fence location which we believe is reasonable based on observations of rocks contained within the benches and catchment areas, the number of rocks observed near the guiderail, and SHA information about the site's rockfall history. For both the north and south slope, results indicate maximum bounce heights and energies of up to about 84 ft and 1,500 kJ, respectively. These results suggest that although the maximum energies can be accommodated, the maximum bounce heights are prohibitively large for standard rockfall barrier installation. The term 'bounce height' refers to the distance above ground a falling rock crosses an analysis point in the rockfall model. It does not necessarily mean that a rock hits the ground in the catchment area and then bounces to the indicated height. Typically, very large bounce heights correspond to rocks that are launched away from the slope and cross the analysis point before hitting the ground. The slope geometry can greatly affect the launching potential of falling rocks. For the Sideling Hill slopes, the geometry of the debris wedges and rounded bench edges contribute greatly to the launching potential at the site. When falling rocks impact these surfaces, vertical-dominated trajectories may change into trajectories with a significant horizontal component, causing the rocks to be launched away from the slope face and travel farther toward the highway. Extreme bounce heights indicated in our analyses may not have been observed in the past. However, the rockfall model reflects the anticipated future potential of rockfall trajectories based on the existing conditions.

RISK MITIGATION CONCEPT OPTIONS

Based on the results of the field investigation, rock slope stability and rockfall hazard analysis, proactive rockfall risk mitigation should consider removal of the debris wedges and provide protection from small-scale rockfall involving blocks up to about 3 ft. Larger rockfall is possible but is considered much less likely and therefore relatively lower risk.

There are many strategies that can be used for rockfall risk mitigation. We considered a wide range of solutions for the I-68 Sideling Hill rock cuts and evaluated each scenario based on its effectiveness at reducing the identified rockfall hazards, as well as other important considerations such as cost and constructability.

Mitigation options can be generally classified as stabilization, protection, and avoidance strategies (Andrew et. al, 2011). Stabilization measures prevent rockfall from occurring by either removing the unstable portion of the slope or providing internal or external support to the rock mass. Some common stabilization methods include scaling, internal stabilization (e.g., rock bolts, rock dowels, shear pins and rock gluing) and external stabilization (e.g., shotcrete and anchored wire mesh). In situations where stabilization is not feasible, rockfall protection measures are used to stop, divert, or control rockfall. The rocks are allowed to fall while preventing them from causing significant damage. Some common rockfall protection methods used in the transportation industry include draped mesh or cable nets, barriers and fences, and providing adequate catchment areas (e.g., mid-slope benches or ditches). Avoidance measures include drastic approaches such as a major slope geometry alteration, changing the roadway alignment or elevation, or tunneling. Consideration of avoidance measures was beyond the scope of our study. Instead, we focused on possible stabilization and protection techniques.
After consideration of a wide variety of potential mitigation scenarios, the following three options were considered as the most viable, cost-effective strategies for reducing the rockfall risk along the I-68 Sideling Hill rock cut slopes.

Option 1: Scaling and Bench Cleaning

Scaling is a common technique used by itself or in conjunction with other slope stabilization or mitigation measures to decrease the rockfall hazard of new and existing rock slopes. Scaling is the process of removing loose or marginally stable portions of the slope that could easily dislodge as rockfall. It is performed with hand tools, mechanical equipment or by small blasting operations called trim blasting. Scaling is a temporary measure that usually needs to be repeated every two to ten years as part of an on-going maintenance program as the slope face continues to degrade. There are many areas of the I-68 Sideling Hill cut slopes that contain loose, marginally stable rocks that could be improved by scaling. Scaling alone will not eliminate the rockfall potential but would reduce it for a while. Vegetation would be removed from the benches and slope faces during scaling to prevent root wedging.

Bench cleaning to remove or reshape the debris wedges during scaling would help reestablish the original reverse-graded bench design for improved rockfall containment. Removing the debris wedges would also reduce the launching potential so that falling rocks would tend to land closer to the slopes instead of being launched so far toward the highway. Scaling and bench cleaning would need to be repeated periodically as the slopes continue to weather and degrade.

Option 2: Rockfall Barrier

Installing rockfall barriers on both sides of the highway is a long-term mitigation scenario. A rockfall barrier is intended to intercept falling rocks before they reach the highway. Barriers are used when the rockfall catchment area is not wide enough to retain falling rocks. A wide variety of rockfall barriers are used for rockfall mitigation, including earthen barriers, concrete barriers, structural walls, ridged and flexible fences, and attenuators. Of all these varieties, we considered a flexible rockfall fence to be the most cost-effective solution for this project. Rockfall fences, are designed to absorb energy through deformation of the fence material and braking elements. Fences come in a wide variety of sizes and energy levels that can be tailored to specific requirements. A variety of fence material is used, including woven wire-rope mesh and interlocking ring nets, depending upon the energy rating of the barrier. The mesh is supported by steel beams typically spaced up to about 40 ft apart. The beams are anchored to a foundation with grouted bolts or embedded into the foundation. The larger, higher-capacity fences incorporate lateral and tieback rope anchorage, and break cables. Periodic maintenance of fences is needed to remove accumulated rock debris.

Option 3: Rockfall Drapery

This option includes installation of rockfall drapes. Rockfall drapes consist of wire mesh or cable netting that is suspended by anchors from the bench edges and hung freely over the bench faces. Only the top of the drapes are attached to the slope, which allows rockfall to occur under the mesh. The draped mesh guides falling rock down along the face to be deposited safely on a bench or within the catchment area. Drapery systems are commonly used on steep rock slopes to control rockfall. They are designed to protect against raveling-type rockfall that involves relatively small-volume slope failures or blocks up to about 5 ft, depending on the strength of the mesh used (Andrew et al., 2011). Drapes that become damaged over time due to rockfall and would need to be repaired. Rockfall drapes may be considered as a long term rockfall protection measure for the I-68 Sideling Hill rock cut slopes in lieu of rockfall fences.

Shotcrete Surface Protection

In combination with any of the three options presented above, we also considered installing surface protection for localized areas that have experienced severe differential weathering of the relatively weak rock layers. The purpose of the surface protection is to slow the degradation and undercutting associated with weak shale interbeds to mitigate the potential for catastrophic failure of the overhanging sandstone rock ledges. Surface protection can be installed as a thin layer of reinforced shotcrete. Before shotcrete is applied, the rock surface must be thoroughly cleaned to remove loose debris and the shotcrete must be fastened to the slope face with anchors or dowels. Wire mesh or fiber reinforcement is required to prevent cracking. Drainage must be installed to relieve the potential buildup of water pressures behind the shotcrete that could damage the shotcrete facing.

Spot Bolting and Anchored Mesh

Spot bolts and/or anchored mesh can be used to stabilize individual rock blocks or groups of rocks that are marginally stable but are not able to be removed during scaling. Spot bolts and/or anchored mesh was not recommended as a stand-alone rockfall hazard mitigation solution, but rather to be considered for use in conjunction with other techniques to further minimize the long-term rockfall hazard potential related to relatively large, marginally stable rock blocks.

Site Access Considerations

Access to the slopes and benches will be extremely difficult for the scaling/bench cleaning, shotcrete surface protection and spot bolting/anchored mesh installation options. Crane assistance could be used for installation on the lower benches. However, access to the upper benches will be problematic. Workers may need to access steep portions of the slopes with ropes and harnesses, and likely with cages, equipment and supplies lowered from the top. Vehicles and equipment may need to access the top of the slope, which may require temporary access road construction to the area behind the slope crests.

FINAL DESIGN

SHA initially advertised a construction project for scaling and bench cleaning. However, the contractor's proposed pricing was much higher than anticipated mostly because of the difficult site access and safety issues associated with the slope height and narrow benches. Because the cost-benefit considerations for scaling and bench cleaning didn't meet SHA's expectations, they opted to pursue an off-slope rockfall risk mitigation strategy instead - namely a flexible rockfall fence - as the preferred alternative.

The barrier configuration selected by SHA for final design was a 9-ft high flexible fence set back 12 ft from the travel lanes along both sides of I-68. The barrier will be about 523 ft long on the north side and 956 ft long on the south side of I-68. The design energy capacity was established as 800 kJ with a maximum lateral deflection of 11 ft. Based on the rockfall bounce heights and energies indicated in our rockfall hazard analysis, the fence is expected to contain at least 90% to 95% of the rock impacts. The fence design incorporates a 1 ft gap below the barrier to facilitate snow plowing. No upslope rockfall barrier anchors were allowed which will facilitate open access behind the barrier for periodic maintenance cleaning of the catchment ditch.

Since the available standard rockfall fence products for proprietary systems requires upslope anchors to resist the design rockfall energy for the specified deflections, a custom rockfall barrier design involving a cantilevered system was required. To resist the design rockfall energy without upslope anchors, posts were designed to be embedded in drilled shafts using the moment capacity of the post. The design required collaboration between the proprietary mesh system supplier (Maccaferri), and the foundation design engineer (Schnabel).

Maccaferri developed the custom rockfall barrier system design to meet the design guidelines outlined above. A primary interception cable panel RB 750 system was designed for posts 25 ft apart. A secondary wire mesh panel layer consisting of double twist wire mesh was used for fly rock protection. Maccaferri uses full scale load test data in conjunction with finite element modelling to develop resultant deflections and forces for project specific custom rock barrier designs such as this.

Schnabel used the resultant forces from Maccaferri to design the rockfall barrier support posts, foundations, wire rope anchors, fairleads, bolts, and welded connections using AASHTO ASD and AISC Steel Construction Manual, 8th Edition (ASD) codes. Based on Maccaferri's design calculations, the maximum horizontal load perpendicular to the rockfall barrier in the middle of the mesh panel is 73 kips, which is distributed between two posts. Therefore, a maximum wire rope lateral anchor force of 40 kips was used to design the lateral anchors.

We considered that the rockfall design energy capacity of 800 kJ is based on a rock mass impact energy on the system and steel support posts. Therefore, we designed the steel posts for a maximum impact stress of $0.9F_y$ – the AASHTO ASD maximum allowable impact driving stress for steel beams. Based on our analysis, we selected a W14X74 ASTM A992 ($F_y = 50$ ksi) wide flange steel beam for the rockfall barrier support post.

The barrier support posts were designed to be installed into pre-drilled holes socketed into the underlying bedrock at least 5 ft to resist the lateral impact forces from the rockfall. We designed the socketed support posts for both axial and lateral loading conditions. The soil parameters and soil stratigraphy used to estimate the axial and lateral resistances were obtained from geotechnical test borings. The horizontal forces and moments will be resisted by the stiffness of the post and the passive resistance of the soil and rock adjacent to the socketed post. The computer program LPILE was used to analyze the lateral capacity of the socketed posts. The computer program calculated the embedded post deflection, internal shear forces, and bending moments within the post foundation.

A 4-inch diameter anchor drilled and grouted into the underlying rock was designed to resist the wire rope lateral design force of 40 kips. The anchors were designed for a minimum bond length of 8 feet into the underlying rock and assumed an ultimate grout-to-ground bond resistance of 116 psi. This minimum bond length is expected to provide an allowable anchor resistance exceeding 40 kips (with a FS > 2.0). The wire rope lateral anchor inclination was designed to be installed at 45° from the horizontal.

CONSTRUCTION

SHA awarded the rockfall fence installation contract to Carl Belt, Inc. with a total bid of \$3.1 million. Approximately \$2.1 million was directly for fence materials and construction whereas the remaining approximately \$1 million was for maintenance of traffic, scree removal, drainage, landscape, and other miscellaneous items. The average cost for the custom flexible rockfall fence was \$1,420 per linear ft. Construction started in April 2023 and final inspection was in December 2023. Figures 9, 10 and 11 show the fence shortly after construction.

Any work within 20 ft of the rock slope required worker protection, although workers in heavy machinery were exempt. The contractor protected workers using high tensile steel wire mesh suspended 50 ft from a steel spreader bar supported by a crane.

CONSLUSION

The flexible rockfall fence option was constructed to provide significant proactive rockfall risk reduction for the iconic rock slopes along I-68 through Sideling Hill. SHA selected the flexible fence option after extensive study and monitoring of the rockfall hazard characteristics and history of rockfall at the site, and after consideration of a variety of alternative options. The new fence was designed to contain the vast majority (but not all) of rockfall expected at the site based on rockfall hazard modeling results and observed site conditions. Selection of the rockfall fence option avoided many difficult site access and safety issues that would have been extremely challenging and expensive for the alternative on-slope maintenance options including scaling, bench cleaning, draped mesh, spot bolts, and anchored mesh. The 9 ft fence height is very small compared to the overall height of the rock cut slopes such that the iconic view of the Sideling Hill syncline is maintained, even when viewed from the closest/right lanes.

Figure 9: View Looking West Through the I-68 Sideling Hill Road Cut Shortly After Fence Installation.

Figure 10: End Post with Double Leg Anchors, Lateral Cables, and Energy Dissipators

Figure 11: Shared Interior Post with Double Anchors, Lateral Cables and Energy Dissipators

REFERENCES

1. Andrew, R.D., Bartingale, R. and Hume, H. (2011). "Context Sensitive Rock Slope Design Solutions." Report FHWA-CFL/TD-11-002, U.S. Dept. of Transportation, Washington, DC.

2. Baker-Wibberley & Associates, Inc., 1986. As-built construction drawings for I-68 through Sideling Hill (Sheet Nos. 5, 26 and 29 of 106) dated December 8, 1986.

3. Brezinski, D.K. (1994). "Geology of the Sideling Hill Road Cut." Maryland Geological Survey Pamphlet Series, <http://www.mgs.md.gov/esic/brochures/sideling.html>

4. Means, J. (2010). “Roadside Geology of Maryland, Delaware, and Washington, D.C.”

Precision Presplitting:

Redefining Accuracy in Rock Blasting

Dr. Anthony Konya
Precision Blasting Services
16800 Little Elkhart Lake Rd
Kiel, WI 53042
(440) 823-2263
Anthony@idc-pbs.com

Alexander Sibley
Precision Blasting Services
16800 Little Elkhart Lake Rd
Kiel, WI 53042
(314) 757-5906
a.sibley@konyamining.com

Prepared for the 73rd Highway Geology Symposium, September, 2024

Disclaimer

Copyright Notice

ABSTRACT

Presplitting is a widely used method of blasting in the mining and construction industries. In recent years a lot of development has gone into the development of empirical equations based on field data to be able to better design the Precision Presplit for various rock types and structural environments. However, a thorough analysis of the mechanics behind presplitting which matches with observed field practices has yet to be developed and treated through chemistry, thermodynamics, and mechanics.

The widely publicized theory of presplitting is that of shockwave collisions between boreholes exceeding the tensile strength of the rock and causing a fracture to occur. This theory did not hold based on basic wave mechanics and the supplies available for performance of the presplit, as shown in this paper. Other authors have suggested alternative theories based on the gas pressurization of the borehole. Recently the concept of hoop stresses as a result of the gas pressurization of the borehole was suggested. No method to analyze the gas pressurization of the borehole and magnitude of the hoop stresses existed. This paper sought to reconcile that and using basic laws from thermodynamics and mechanics of materials has presented a mathematical proof to determine the borehole pressure from a decoupled charge and the magnitude of the hoop stress developed in the rock.

INTRODUCTION

Explosives constitute a crucial element in mining and construction industries, offering an efficient means of rock excavation. The bulk of their application lies in mass rock excavation and production blasting, facilitating the fragmentation of rock. Conventional blasting often causes breakage behind the blastholes' final row, acceptable within the main excavation body, but concerning when occurring near the excavation limits.

Within the construction industry, the final excavation limit, commonly referred to as the 'neat line,' is meticulously engineered to yield a stable slope, minimizing rockfall risks to the public or adjacent infrastructure. These long-term slopes, when subjected to overbreak, undergo accelerated weathering, escalating rockfall risk and necessitating area rework to stabilize the slope.

Moreover, several construction projects demand concrete pouring adjacent to the rock wall for infrastructure development, such as locks and dams. If blasting fails to adequately reach the neat line, mechanical excavation and scaling become necessary, incurring high costs and time consumption. Blasting beyond the neat line brings additional expenses due to the supplementary concrete required for overbreak regions.

In the mining sector, the design of pits follows an overall slope to mitigate large-scale slope failures. The slope design primarily relies on the rock's inherent properties. Poor blasting techniques, leading to fractured rock behind the slope, often necessitate 'laying back' the slope or adopting a shallower slope to guard against failures. This requirement increases waste material mining, amplifying the mine's total cost and diminishing profitability. As mines venture deeper, the necessity for proper slopes with minimal backbreak becomes imperative to reduce slope failure risk, thereby enhancing worker and equipment safety and mine's profitability.

Moreover, bench angles in mines are designed to minimize rockfall risks. Several methods, including bench angle design, catch benches, berms, mechanical scaling of walls, and overbreak control measures, are employed to protect workers, each with their advantages and trade-offs.

The capacity for a mining or construction project to generate smooth walls using explosives is critical for operational economic effectiveness and employee safety. The employment of appropriate presplitting can notably reduce scaling, leading to significant economic savings and safer projects. Although traditional presplit methods work well with hard rock types, their application to weaker rocks often proves ineffective. However, Precision Presplitting has demonstrated efficiency under such conditions, providing near-perfect walls in full-scale construction projects.

The structural properties of the geology being blasted frequently cause backbreak beyond the presplit lines. A potential solution involves bringing borehole spacing closer together. Traditional presplit design employed 'split-factor' to adjust the explosive load based on a linear relationship with spacing. However, this relationship is non-linear, and adherence to the linear model leads to charge overloading.

While the mechanisms behind a presplit formation remain ambiguous and occasionally misunderstood, it's essential to dispel misconceptions. For instance, the shock breakage theory, widely studied and taught, has been shown to be a false concept in numerous studies. A new theory posits that the explosive-generated gases in a borehole produce a hoop stress field, causing the presplit fracture to occur. This theory suggests that small explosive loads could be used depending on the rock type and structural environment, generating a fracture without causing any overbreak to the surrounding structure.

The scope of this paper encompasses an analysis of the shockwave breakage and hoop stress theories, comparing their stress magnitudes against various rock properties, including the Young's Modulus and Tensile Strength. It presents the design of a Precision Presplit round based on the stress magnitudes with empirical research to develop a methodology for designing a Precision Presplit with variations to blasthole spacing and blasthole pressures.

INTER-BOREHOLE FORCES

In a Precision Presplit, or any presplit operation, decoupled charges are utilized, meaning the explosive is not directly in contact with the rock. As such, any shockwave generated from the explosive must transition through an intermediate medium, typically air, before impacting the rock. However, it's critical to understand that the shockwave originating from the explosive does not directly propagate into the air. Instead, the explosive's rapid gas expansion compresses the surrounding air, thereby generating a new shockwave. This resultant shockwave is of significantly lower magnitude than a hypothetical shockwave transiting directly from the explosive to the air.

A detailed analysis of the Hugoniots - curves that express the relationship between pressure, density, and internal energy of a material under shock compression - elucidates this phenomenon. The impedance mismatch between the explosive and the air is too substantial to permit a new wave's generation in the air. Instead, the majority of the wave is reflected back into the explosives. The proof can be solved simply, using base Explosive Engineering methods (*1*), as demonstrated below.

In this treatment, we will analyze TNT being suspended in open air, with TNT having a P_{CJ} of 19 GPa, a density of 102.38 lb/ft^3 (1.64 g/cc), and a Velocity of Detonation of 22801.84ft/s (6.95 km/s). The air will have the following properties: $C_0 = 0.899$, s = 0.939, and a density of 0.0812 lb/ft^3 (0.0013 g/cc). We will analyze the P-u Hugoniot and begin by finding the CJ particle velocity (u_{CJ})

$$u_{cj} = \frac{P_{cj}}{\rho_0 D} = \frac{19}{1.64 * 6.95} = 1.67\ km/s$$

Then, we will use a P-u Hugoniot for air of

$$P = 0.0012u + 0.0012u^2$$

and a P-u Hugoniot for TNT of

$$P = 45.83 - 19.7u + 2.18u^2$$

The solution of the particle velocity (u) results in a complex solution of imaginary numbers. Graphing the equations results in a set of quadratic equations which have no real solutions (Figure 1), this illustrates that the transition of a shockwave from an explosive such as TNT to air is not feasible in base explosive engineering principles.

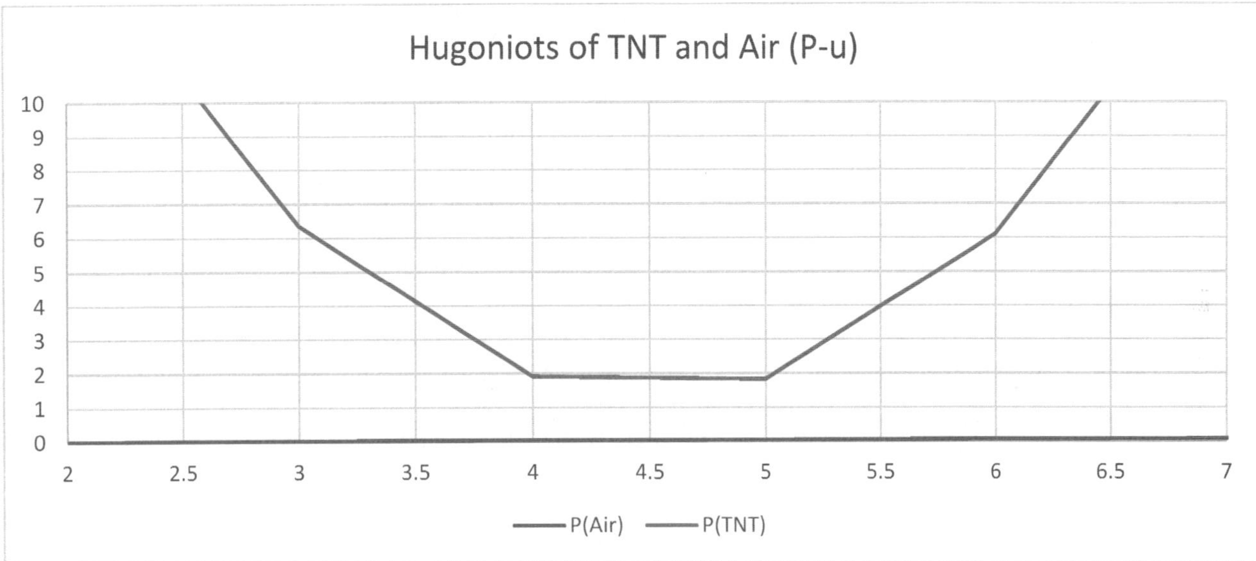

Figure 1 - P-u Hugoniot of TNT and Air

Moreover, the explosive detonation results in the shockwave striking the outer edge of the charge at an angle, which further diminishes any potential or theoretical transmission from the explosive into the air. The confluence of these factors significantly impacts the efficiency of the shockwave transmission and, in turn, the practical applicability of shockwaves being a major mechanism of presplit performance.

What actually transpires when an explosive detonates in air which is surrounded by a rock wall (a decoupled charge in a borehole) is the formation of a new blast wave in the air as a result of the explosive detonating in contact with air. This blast wave, created by the rapid expansion of gases and the compression of air, subsequently travels through the air medium until it strikes the rock. This is a similar phenomena to a bullet exiting a rifle, with hot gases mixing with the ambient air and causing a shockwave to form as the rapidly expanding, pressurized air inside the muzzle pushes on the ambient air (Figure 2).

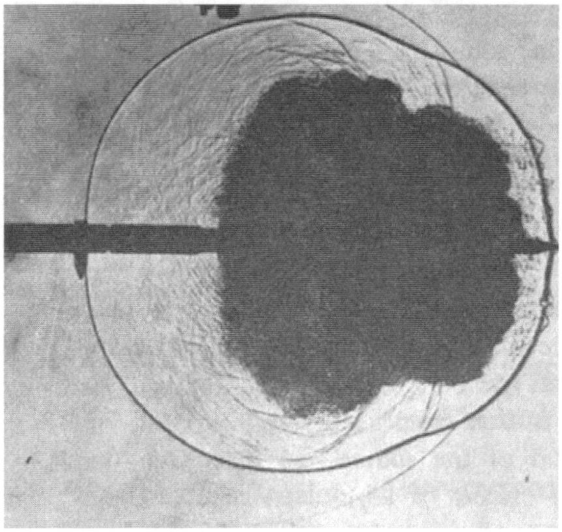

Figure 2 - Gasses from a bullet exiting the muzzle of a rifle, causing a new shockwave to form in ambient air

Clearly, the blast wave generated by the explosive detonation will not traverse directly into the rock mass due to the impedance mismatch between the air and the rock; similar to a blast wave's effects on structures, with damage resulting from the waves pressure and not its transference into the structure. Consequently, the blast wave can essentially be treated as consisting of two initial components: the blast wave overpressure and the blast wave dynamic pressure (*2*). Additionally, with the minimal volume for volumetric expansion, the blast wave will travel at the same rate as the outward expansion of the explosive gases that are generated as a result of the detonation. The blast wave can then be seen as a mechanism which carries the gases to the borehole wall and causes an 'impact' loading which is then followed by a quasi-static loading of the pressure generated by the over-confined, hot gasses. Apropos, the reflected shockwave from the blast wave striking the rock wall will reverberate through the gas mixture theoretically resulting in increases to temperature and pressure, however no experimental evidence exists to show this to be a meaningful impact to the borehole pressure.

Practical observations further reinforce the understanding that the presplit formation is not attributable solely to the shockwave. For instance, it has been demonstrated that presplitting can be effectively accomplished using propellants with similar loads to dynamite, a process that wouldn't be possible if presplit formation was dependent on the high-pressure shockwaves unique to high-explosives like dynamite (*3*).

Furthermore, another empirical revelation contravening the shockwave-driven formation theory is the successful formation of a presplit in boreholes with delayed detonation. If the shockwave collision theory held true, delayed detonations between blastholes should fail to form a presplit as the shockwaves would not collide simultaneously. However, successful presplit formation with such a delay contradicts this, indicating that shockwave collisions are not the solitary or predominant factor in presplit formation.

These practical observations underscore the complexity of presplitting processes and the need to look beyond the shockwave theory to fully comprehend and effectively design presplitting operations.

Further research has yielded insights into the extraordinary conditions inside a presplit blasthole. Upon detonation of a Precision Presplit blasthole, pressures in the range of 1,000 to 100,000 psi (equivalent to 7 to 690 GPa) are generated in less than a millisecond, typically within hundreds of microseconds. These pressures are generated following the initial 'impact' loading of the blast wave and develop from the gas pressure itself, which can be seen from measured borehole pressures (Figure 3).

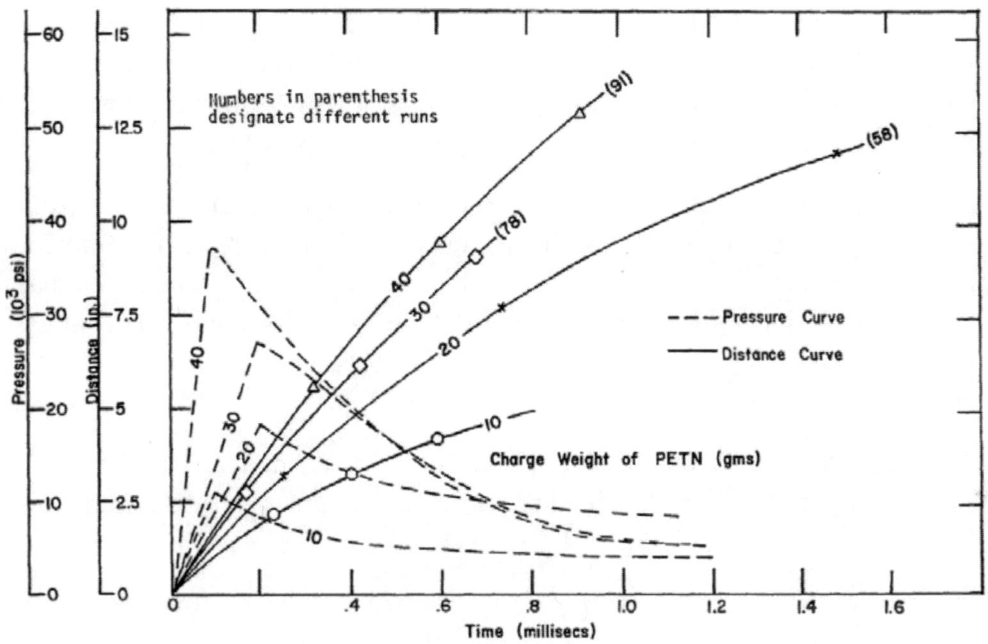

Figure 3 - Borehole Pressures with changes to charge size and stemming (Otuonye, 1981)

This intense pressure is too great to be quickly regulated by heat transfer alone. As such, the primary method for pressure alleviation is deformation of the rock mass, predominantly through fracturing. In presplitting operations, the blastholes are strategically positioned distant from a free-face. This positioning ensures that the pressure regulation can only occur through breakage towards another blasthole or via venting through the top of the hole. Detailed models have been developed for the determination of the borehole pressure which begin in base chemistry and physics and have been validated through real-world experimentation and data (*4*).

Therefore, the challenge for explosive engineers is to design rounds such that the pressure can propagate to the adjacent blasthole faster than the pressure decrease through top venting, considering the effects of stemming. The pressing question that arises is the orientation of fractures and the understanding of the breakage process. Specifically, how do the fractures orient themselves under such conditions, and how does the breakage process function? Answering these questions is vital to optimize presplitting designs and enhance the overall efficiency and safety of rock excavation operations.

HOOP STRESS FIELDS

The succeeding section of this paper introduces the concept of hoop stress fields, a relatively new theory proposed (*5*). The theory suggests that the gases generated by an explosive in a borehole create hoop stress fields, which in turn facilitate the formation of presplit fractures. This implies that the application of very small explosive loads could trigger the formation of fractures, without causing overbreak to the surrounding structure. The feasibility of this approach depends on the specific characteristics of the rock type and the structural environment.

The hypothesis being explored is that the hoop stress field is a function of the gas pressure. The research carried out in this project therefore aims to define the gas pressure in a borehole resulting from both detonating and deflagrating explosives. This investigation aims to determine if it is indeed possible for borehole pressures to generate hoop stress fields.

Moreover, empirical evidence from field trials has consistently shown that presplit blasts without stemming require approximately four times the explosive load, compared to a stemmed borehole, to yield a clean split to the top of the borehole. These observations further reinforce that the gas pressure is a critical function to the performance of the presplit, with significant changes to the gas pressure because of the stemming, while no effects to the shockwave would occur due to the amount or presence of stemming.

There is a clear necessity to develop blasting-specific hoop stress models, which begins with the equation for hoop stress fields in thick walled pressure vessel, this is shown in Equation 1.

$$\sigma_c = \left[\frac{p_i r_i^2 - p_o r_o^2}{r_o^2 - r_i^2}\right] - \left[\frac{r_i^2 r_o^2 (p_o - p_i)}{r^2 (r_o^2 - r_i^2)}\right] \qquad \text{(Equation 1)}$$

Equation 1 is then modified to include blasting specific variables, this derives Equation 2.

$$\sigma_c = \left[\frac{P_g d_b^2 - p_o S^2}{S^2 - d_b^2}\right] - \left[\frac{d_b^2 S^2 (p_o - P_g)}{r^2 (S^2 - d_b^2)}\right] \qquad \text{(Equation 2)}$$

Where σ_c is the Circumferential Hoop Stress

d_b is the diameter of the blasthole

p_o is the ambient borehole pressure

S is the spacing between the boreholes

P_g is the gas pressure

r is the distance from the center of the borehole to a certain point between the boreholes

Multiple assumptions are then made to simplify equation 2. The base case will assume that the pressure outside the borehole is zero. While not technically accurate, the pressure outside the borehole would be very small in comparison to the internal borehole pressure for a majority of situations. The radius 'r' that will be calculated will be halfway between the boreholes, as the gas pressure from each hole is assumed to be enough to break halfway between each set of holes. In this case the radius would equal half the spacing (r = S/2). These assumptions including in equation 2 will result in equation 3 for determination of the hoop stress associated with Presplit Blasting:

$$\sigma_c = \frac{5P_g d_b^2}{S^2 - d_b^2} \qquad \text{(Equation 3)}$$

Equation 3 can then be viewed as the baseline equation for the determination of the hoop stress field which will result in a successful presplit formation between two boreholes.

To implement Equation 3 effectively, it's crucial that the calculated circumferential hoop stress surpasses the tensile strength of the rock mass while remaining below the rock's compressive strength. This balancing act is crucial for ensuring the successful execution of the presplit without incurring additional, unnecessary damage to the rock mass. Specifically, the tensile strength should be surpassed by a margin of 25% or greater when feasible. This margin serves as a buffer to account for variability in rock properties and the inherent uncertainties associated with blasting operations.

While this methodology guides the internal dynamics of the presplit operation, it's important to note that this strategy doesn't directly address the cratering mechanism found at the surface of the boreholes. Cratering, or the formation of a bowl-shaped depression around the borehole due to the explosive action, can lead to undesired rock fragmentation and overbreak. Therefore, managing cratering requires additional considerations and design strategies beyond the purview of this paper.

The mechanics discussed and recommendations on the target strengths help to explain why weaker rock is often difficult to presplit or is over-blasted when the presplit performance is combined with a stronger rock. In Table 1 below, examples of rock strengths are presented, including the compressive and tensile strengths of each rock. A common case for presplitting in sedimentary deposits of the Appalachian Mountains (USA) has shales and siltstones that are interbedded between seams of competent limestones. The limestone has significant 'forgiveness' that when they are attempted to presplit the explosive load can increase be nearly an order of magnitude of 4 without causing additional breakage and exceeding the compressive strength. However, the shales and siltstones normally encounter overbreak, due to the methodology of blasting the limestone extremely aggressively to ensure breakage and minimize mechanical hammering/treatment of the rock face. Simply put, this blasting methodology over-stresses the weaker rocks by using too much explosives for the spacing provided.

ROCK TYPE	COMPRESSIVE STRENGTH OF ROCK (PSI)	TENSILE STRENGTH OF ROCK (PSI)
GRANITE	14,500	900
LIMESTONE	4,350	725
SHALE	750	580
SANDSTONE	2,900	450
SILTSTONE	750	400

Table 1 - Table of Rock Strengths

The author has successfully utilized Precision Presplitting to blast through various seams in a single blast by varying the explosive load and inter-borehole pressures. The pressurization through the hole, and the associated hoop stress fields, can be modified by changing the explosive load throughout the blasthole. This requires the use of the theory of choked gas flow to determine borehole pressure inter-hole, which is beyond the purview of this paper but is regularly completed in the industry.

CASE STUDIES: REAL-WORLD APPLICATIONS OF PRECISION PRESPLITTING

This section introduces a collection of case studies drawn from real-world projects around the United States, primarily those led by the U.S. Army Corps of Engineers. These projects are particularly relevant and valuable due to the comprehensive rock data gathered during their execution. Each case study focuses on successful applications of Precision Presplitting.

To further the understanding of the efficacy of our proposed theories, these case studies retrospectively apply the theoretical framework developed in the earlier sections of this paper. Specifically, they assess the rock properties of each site, recall the Precision Presplitting techniques used, and calculate the magnitude of the presplit hoop stress field that would have been predicted by our theory. They compare these calculated values to the tensile strength of the rock that was actually blasted. The gas pressure has been calculated from the previously published treatise on blasthole pressure calculations (*4*).

Through this process, we aim to validate our theory's predictive power in diverse field settings and conditions, strengthening its practical value for future blasting projects. Additionally, these case studies serve as practical demonstrations of how the theoretical framework can be integrated into real-world project planning and execution, ultimately fostering more efficient and effective rock blasting. Table 2 below shows the theoretical hoop stress fields, calculated from the actual blasting technique successfully deployed on projects to the tensile strength of the rock recorded. This does not take into account the structural geology of the area, which does have other impacts.

ROCK TYPE	HOOP STRESS (PSI)	TENSILE STRENGTH OF ROCK (PSI)
DECOMPOSED GRANITE	837	750 - 900
LIMESTONE	701	650 - 725
SHALE	513	550 - 580
SANDSTONE	484	450
SILTSTONE	353	400

Table 2 - Table of Theoretical Hoop Stress Fields

PRECISION PRESPLITTING IN THE FIELD

This section explores projects where precision presplitting was employed for slope remediation along highways and creating clean neat line vertical walls designed for direct concrete pouring.

Kentucky Locks Downstream Excavation Project

The Kentucky Locks Downstream Excavation Project is an example of precision presplitting in action. This project involved over 100,000 linear feet (30,478.5 m) of presplit faces. Throughout the project numerous geological challenges were faced due to the varied rock formations encountered during the excavation.

The site is comprised of two primary rock formations: the Warsaw Formation and the Fort Payne Formation, both consisting mainly of limestone. The Warsaw Formation located in the upper elevations of the excavation is characterized by weathered limestone with clay and mud seams, large pinnacles, orthogonal jointing, and prominent horizontal bedding planes. These can be seen in Figure 4. In the zone between where the Warsaw meets the Fort Payne Formation, there is an approximately 3 to 5 foot (0.94-1.5 m) thick weathered mud joint. This created a natural bench elevation during the excavation of the project. The Fort Payne Formation consists of a more massive solid limestone with round chert nodules interbedded throughout the formation.

Figure 4 – Warsaw Formation Bedding Planes and Joints

During the development of the blasting program, various load weights of detonating cord and borehole spacings (ranging from 12 inches (30.48 cm) to 24 inches (60.96 cm)) were tested. Initial testing revealed that a load weight of 125 grains per foot of detonating cord with a 1/3-

pound(0.15 kg) booster was sufficient to shear the rock along the desired presplit face without causing damage beyond the neat lines. A stemming length of 24 inches was determined to best confine gas pressure in the borehole while allowing proper venting. These blasting parameters resulted in approximately 80 - 90% half-cast left of the boreholes and with little to no evidence of overbreak.

Throughout the progress of the excavation several geologic factors were encountered where the original blast parameters needed to be adjusted. One of these factors included the increased hardness of the rock. As excavation progressed deeper in elevation the material became harder and harder to break with both mechanical means and explosives. The varying bench depths due to excavation limits and geology required adjustments to the loading weights of the boreholes. Soft seams in boreholes evident in drill logs and horizontal bedding planes resulted in zones of rock where the gas pressures would escape the boreholes and not result in the desired presplit. When these horizontal seams aligned near the base of the stemming zone, uplift in the upper bedding planes led to some overbreak on the surface and increased excavation outside the designed scope as seen in Figure 5.

Figure 5 – Uplift of Upper bedding planes

As the project progressed several adjustments had to be made to overcome these geologic factors. The stemming zone was originally designed at a maximum length of 24 inches (60.96 cm). This length of stemming confined the gas pressure in the boreholes long enough to split the rock without causing damage beyond that of the neat line. When the horizontal joints fell at the bottom of the stemming zone as seen in the red line in Figure 6, the stemming zone was not increased but shifted down to below these joints to where the green line is seen in Figure 6. The stemming was not filled to the top of the borehole as this would have increased the stemming zone length. In turn the stemming was able to stay at the same 24 inches (60.96 cm) in length and this still confine the gas pressure into the proper portion of the borehole and not uplift the rock on the surface.

Figure 6 – Stemming Location in Presplit

Two other parameters were also adjusted as the blasting progressed. These included the presplit load and the borehole spacing. As the borehole depth increased on deeper benches the minimum presplit load was increased to 150 grains per foot (from the 125 grains per foot) with the same 1/3 – pound (0.15 kg) booster as the base charge. It was seen that on these deeper benches the performance was reduced with the lighter loads and increasing these loads, however slightly, allowed for sufficient gas pressures to achieve the desired results.

Through evaluating each variable and geological factor, the development of custom hole loads and design adjustments resulted in neat walls with minimal to no significant breakage beyond the designed excavation limits. The project overcame many geological challenges through meticulous explosive engineering and in-field test blasting.

Interstate 24 Slope Stabilization Project

The Interstate 24 project involved slope stabilization to mitigate rockfall hazards due to weathering and erosion. The high wall faces which had originally been presplit during the roadway's construction had deteriorated over the years. This resulted in an unsafe overhang where a softer rock seam had eroded as seen in Figure 7. The overhang caused dangerous rockfalls to highway motorists. This project removed the rock faces that were unstable and creating these overhangs on the side of the highway.

Figure 7 – I24 Rock Overhand

This project has several challenging factors largely that every blast hole had to be loaded from hanging on ropes or a man basket from the ground. This work was all conducted alongside and above an active highway with very short allowable times for traffic shutdowns. The Unstable and tight working areas provided for the need for precise explosive engineering in the development of the blasting plans.

Some of the geologic challenges included highly laminated rock that was fractured and weathered from the environment. This was seen throughout the drilling where different rock zones were prevalent. With reviews of the drill logs customized blast hole loading and placement was needed to be designed. The most difficult challenge was the placement of the blast holes. The new presplit face needed to meet the existing undercut face without producing a ledge.

To address the overhang and undercut rock face, a non-linear presplit line was required. Various 3D profiles were generated to map the undercut on the face. Unlike a normal presplit that generally follows a straight line or the curve along an excavation these blast holes needed to contour exactly to the weathered undercut. The 3D profiles allowed for the precise placement of boreholes along that contour of the natural weathered undercut as seen in Figure 8. This pattern ensured a straight vertical face with no resultant bench for material to bounce onto the highway.

Figure 8 – Borehole Placement along Undercut

When reviewing the drill logs and field examinations of the rock faces custom borehole loads were developed. In zones of weaker, softer, or fragmented rock a reduced load was used as small as 25grains of detonating cord stretching over this area. Where increased burden over 4 feet (1.22 m) was seen additional holes were placed in front of the presplit. In these increased burden areas the presplit was then loaded to the full design weight and the additional holes were loaded with decoupled charges made to ensure the rock was well fragmented as it fell along the side of the highway.

These adjustments in blast design successfully stabilized the overhang, creating a stable vertical face without hazardous benches. The results were effective in mitigating rockfall hazards and ensuring motorist safety.

These case studies demonstrate the effectiveness and versatility of precision presplitting in various construction settings. The techniques and necessary adjustments for field conditions, geological factors, and project requirements in the blasting programs highlight the importance of precision presplitting to achieve desired outcomes while minimizing unintended damage.

CONCLUSION

This paper has ventured into the uncharted territory of the mechanisms behind Precision Presplitting, providing novel theoretical insights and practical guidance for explosive engineers and rock blasting practitioners. Our investigation started by dissecting the blast wave formation process and its propagation, highlighting the vital understanding that the explosive-induced shockwave doesn't directly transition into the rock mass, but instead, triggers a new blast wave in the air.

We've unpacked the nuances of how this blast wave interacts with the borehole wall, resulting in an air-rock interface and causing the formation of an innovative hoop stress field theory. Our work revealed that this field, generated by explosive gases inside a borehole, holds the key to generating a presplit fracture, even with very small explosive loads.

Further, we developed an equation that connects explosive load to the hoop stress produced, allowing for a more precise prediction of the presplitting outcome. Our model illustrates that achieving a successful presplit requires the circumferential hoop stress to exceed the rock's tensile strength without reaching the compressive strength, thus avoiding unwanted rock mass damage.

Several case studies examples were supplied, applying our theoretical framework retrospectively to validate its predictive power. The promising alignment between our calculations and the actual outcomes in diverse field settings bolsters our confidence in the proposed theory's practical applicability.

In conclusion, our findings promise to transform the practice of Precision Presplitting, by offering a rigorous, science-based approach that replaces the current trial-and-error methods. The theory presented in this paper, with its equations and guidelines, empowers blasting practitioners to plan and execute their operations more accurately, efficiently, and safely. It also provides a solid foundation for further research into the optimization of rock blasting techniques, thus paving the way for continuous advancements in this field.

REFERENCES:

1. Cooper, P. (1996). *Explosive Engineering* . VCH Publishers.
2. Kinney, G., & Graham, K. (1962). *Explosive Shocks in Air.*
3. Konya, C., Barret, D., & Smith, J. E. (1986). Presplitting Granite Using Pyrodex, A Propellant. *Proceedings of the Twelfth Conference on Explosive and Blasting Technique*, 159-166.
4. Konya, A. (2019). *The Mechanics of Precision Presplitting.* Missouri University of Science and Technology.
5. Konya, A., & Konya, C. (2017). Precision Presplitting. *Proceedings of the Society for Mining, Metallurgy, and Exploration.*

Role of Instrumentation and Monitoring in a Geotechnical Asset Management Program

Darren Beckstrand, C.E.G., Senior Associate Geologist

Landslide Technology
10250 SW Greenburg Rd. Suite 111
Portland, Oregon 97223
Darren.Beckstrand@ccilt.com

Prepared for the 73rd Highway Geology Symposium, September 2024

Disclaimer

Copyright

ABSTRACT

Many state DOTs have begun investigating, implementing, or have recently completed Geotechnical Asset Management (GAM) programs. These programs have tackled the efforts in several ways, from pilot programs through specific problematic corridors to statewide assessments of entire highway systems. Several DOTs have inventoried and assessed the condition of hundreds or thousands of assets. A freshly implemented program may reveal dozens or hundreds of assets that appear worthy of immediate attention alongside ongoing chronic issues. Using a targeted program of reconnaissance, monitoring, and instrumentation can help to understand if the assets are actively failing or if the monitored features are in a steady state. Targeted, site- or corridor-specific instrumentation and monitoring programs can inform the planning and budgeting process and focus mitigation efforts where and when required.

INTRODUCTION

Geotechnical Asset Management (GAM) and associated principles have become better understood and have gained acceptance within geotechnical DOT divisions for the past decade. Several state DOTs have implemented some type of GAM system to manage one or more of these critical asset types (Beckstrand et al., 2022; Vessely et al., 2019). These DOTs will benefit by realizing the asset management plan's benefits of *a strategic and systematic process of operating, maintaining, upgrading, and expanding physical assets effectively throughout their life cycle* (AASHTO, 2020).

Most GAM programs implemented to date have focused on the inventory and condition assessment of rock slopes, unstable soil slopes, and retaining walls. These efforts have inventoried and assessed the conditions of thousands of assets across many states. Optimally, the thousands of assets should be monitored and assessed at regular intervals, much like Pavement Management programs measure the condition of thousands of lane miles on a two-year cycle. However, the cost of physically monitoring and assessing thousands of generally stable geotechnical assets does not provide value. Additionally, these GAM programs will often identify dozens or even hundreds of assets that *do* need increased investigation, scrutiny, and monitoring. This paper suggests a phased approach to using GAM programs to identify and prioritize locations within a DOT network.

MONITORING GEOTECHNICAL ASSETS

The idea of monitoring geotechnical assets may conjure up many different scenarios depending on several factors or experiences. Many geotechnical personnel may correctly view monitoring as obtaining data and information from surveys and obtaining readings from piezometers, inclinometers, or other geotechnical instruments. Asset managers may correctly view monitoring as continually measuring and understanding the performance of the highway system and the underlying geotechnical assets. Maintenance personnel may tend to rely on visual monitoring of signs of distressed assets; i.e., 'keeping an eye on that slope'. An ideal GAM program would account for and inform all these monitoring needs. Geotechnical personnel, however, can utilize the GAM concepts and data to select sites along an unstable corridor for enhanced or focused monitoring, including the use of geotechnical instrumentation.

The process of starting up a GAM system is laid out in NCHRP Report 903: Geotechnical Asset Management for Transportation Agencies (Vessely et al., 2019). Several state or provincial DOTs and the FHWA have implemented GAM programs with various complexities, from the essential inventory and pilot program to fiscal forecasting and performance management concepts with near-complete statewide condition assessments of one or more asset type (Beckstrand, Benko, et al., 2017; Beckstrand, Mines, et al., 2017; Farny, 2023; Mines et al., 2023; Oester et al., 2019; Waseem et al., 2022).

Five phases are proposed herein, starting at the 30,000-foot statewide level and getting progressively more detailed in later phases (Figure 1). GAM principles are used to systematically inventory and assess potentially thousands of sites across a highway network. Through the development and application of decision support tools that consider condition, risk, and fiscal concerns thousands of geotechnical assets can be winnowed down to a manageable quantity. This allows the proactive preservation of geotechnical assets and programming via STIP, HSIP, Resiliency, or other funding avenues rather than the traditional worst-first and reactive approach for managing aging geotechnical assets.

Figure 1: Five phases for using GAM programs to select sites for instrumentation and monitoring.

Phase 1: Statewide Monitoring via Inventory, Condition Assessment, and Performance Monitoring

A common approach to starting a statewide monitoring program is to first understand what you have through an inventory and what condition they are in through detailed assessment. These are the basic ingredients of a bare-bones GAM program. Beyond the basics, a risk-informed GAM program accounts for the risks posed by its condition and its deterioration and should be assessed independently from its condition. Risk values based on items such as safety, user and non-user adverse event costs, and risks from uncontrolled deterioration. Risk measured in units of dollars facilitates direct cross-asset comparisons and helps bracket dollar targets for mitigation approaches and benefit-to-cost ratios.

GAM implementation guides and research case histories, cited above, are available that document starting up a GAM program. One of the most important decisions for starting a GAM program is which asset types are included for inventory and assessment. The most common type would be rock cut slopes, with retaining walls, soil slopes and landslides, and culverts also commonly included in GAM programs. A developing consideration is the effects of debris flows, floods, and increased runoff after wildfires and their change of soil properties (Agbeshie et al., 2022).

Monitoring and communicating asset performance through a variety of means are important, but difficult to implement. When successful, they would allow an unbiased and thorough assessment of adverse geotechnical events (rockfalls, landslides, flooding, debris flows, etc.) and deteriorating conditions. At a minimum, all performance information coming into a Geotechnical office from maintenance personnel should be recorded in a systematic way, such as using GIS tools and trackers (Figure 2).

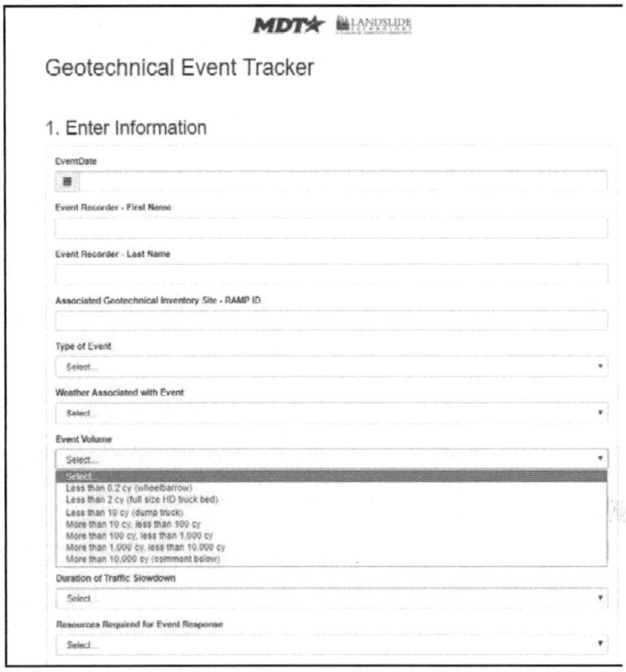

Figure 2: Geotechnical Event Tracker built for the Montana Department of Transportation on an ESRI Platform.

Phase 2: Corridor Identification

Once large-scale assessments have been made, patterns should emerge that highlight where concentrations of poor-condition assets coincide with higher-risk sites. Development and use of Decision Support Tools (DSTs), an example of a performance-based classification scheme shown in Table 1, will help to utilize the volumes of data generated by a GAM program. Likewise, a DOT may bypass entirely the first two phases described here and jump right into a corridor-specific assessment described as Phase 3.

Application of the DSTs by aggregating the risk of adverse events posed by each individual site into one-mile segments and then comparing those to the stated performance objectives will highlight where risk is concentrated. This will provide the advantage of illustrating where not only the Poor condition slopes are located, but also where several large Fair condition slopes contribute to a higher risk corridor. Figure 3 illustrates the process from MDT of using rock slope GAM condition data, assessing risk and comparing it to expectations, and subsequently developing actionable plans and projects that may receive additional monitoring and project development.

Table 1: Example of a Performance-based evaluation system for Montana's Rock Slopes (Beckstrand, Mines, et al., 2017).

RAMP Perf. Class	Road Segment Performance Classification, Likelihood, and Associated Condition Targets*
A	Very high level. Rock slopes pose a very low likelihood (<0.25% annual likelihood per centerline mile) of user delays. *Condition target:* >80% of rock slope area (square-foot basis) in GOOD condition and <2% in POOR.
B	High level. Rock slopes pose a low likelihood of user delays (<0.5% annual likelihood). *Condition target*: >70% of rock slopes in GOOD condition and <5% in POOR.
C	Minimum acceptable level. Rock slopes pose a moderate likelihood of user delays (<1% annual likelihood). *Condition target*: >50% of rock slopes in GOOD condition and <10% in POOR.
D	Unacceptable level. Rock slopes pose a high likelihood of user delays (<3% annual likelihood). *Condition target*: <50% of rock slopes in GOOD condition and <10% in POOR.
F	Failing level. Rock slopes pose an unacceptably high likelihood of user delays (>3% likelihood). *Condition target*: >50% of rock slopes in FAIR condition and >10% in POOR.

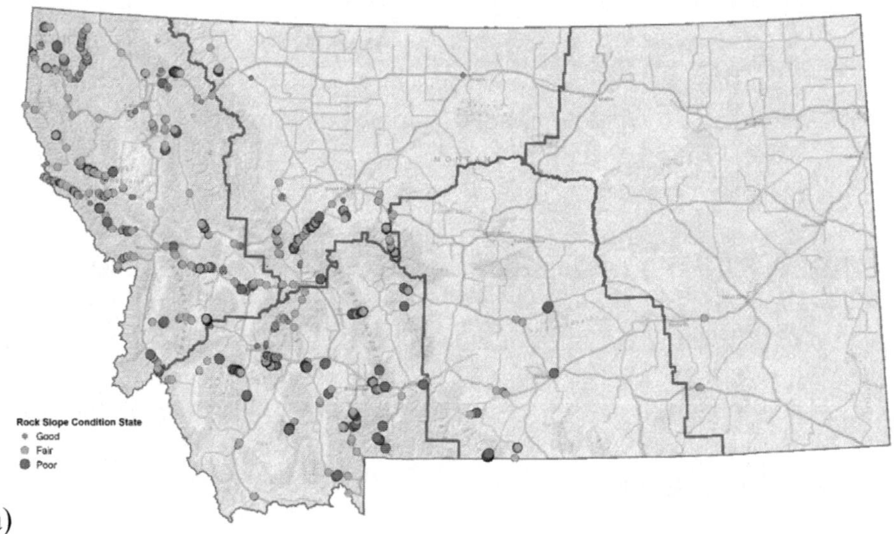

a)

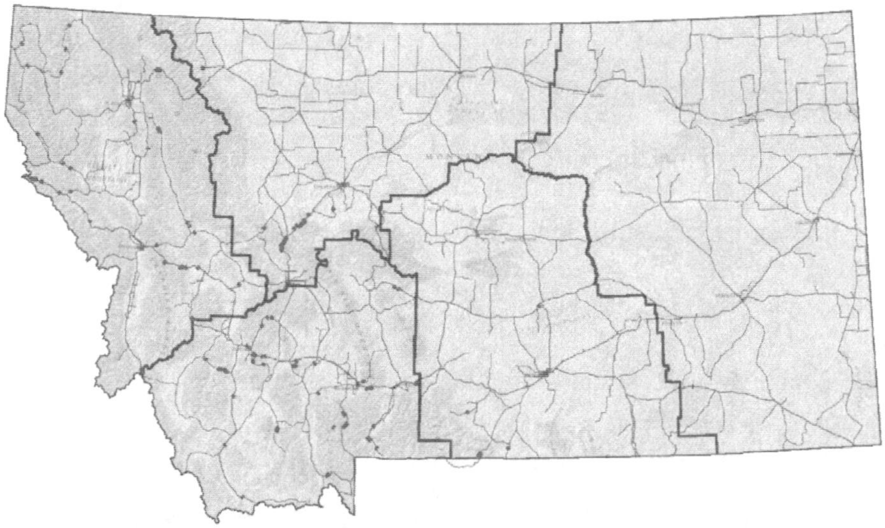

b)

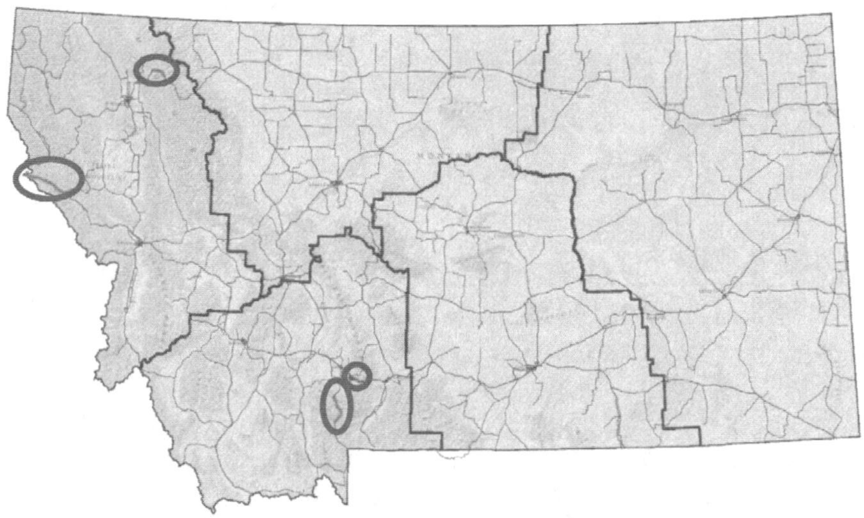

c)

Figure 3: Map series exhibiting a) location of Montana's rock slopes with Good/Fair/Poor conditions; b) one-mile segments (in red) where annual adverse event probability exceeds thresholds; and c) corridors identified for conceptual mitigation designs and potential instrumentation.

Monitoring the statewide conditions and associated risks requires additional, repeat condition surveys. A small number of states, including Tennessee and Montana, have sought to perform repeat surveys on a subset of their GAM assets. This has allowed them to monitor how slopes may have changed over the years, monitoring their deterioration. A key component in a mature GAM system would include monitoring network conditions through time, requiring repeated, large-scale condition assessments.

Techniques for continuous remote monitoring of large areas without unavoidable bias are advancing, but likely only for asset types with certain characteristics. For instance, satellite-based Interferometric Synthetic Aperture Radar (InSAR) will likely return reliable results for large-scale landslides or ground subsidence with sufficient ground reflectors. However, other human-observable evidence of geotechnical issues, such as scars from steep, narrow landslides obscured by dense vegetation; individual rockfall events on steep rock cut slopes; or embankment failures displacing pavements with low coherence, InSAR cannot yet provide full coverage for remote monitoring for GAM programs. However, InSAR techniques and technologies are continually evolving and improving, showing improvements in specific scenarios where it had previously been less reliable (Xu et al., 2023)

Phase 3: Site Identification and Monitoring

A state may study a particular corridor rather than a statewide or larger-scale implementation. The corridor may be selected by a known vulnerability and/or history of adverse geotechnical events. Recent examples include a corridor study in Wisconsin, along the Blue Ridge Parkway, and US 26 in Idaho (Anderson et al., 2022; Banks et al., 2022; Beckstrand, Machan, et al., 2017). Use of GAM principles to document the conditions of, for instance, a corridor with a concentration of geohazards will help prioritize mitigation activities and justify design efforts.

An ongoing corridor project on Utah's US 6 between Thistle and Helper implemented GAM research performed for neighboring states. The project's first phase inventoried and evaluated the condition and risk posed by 112 rockfall and landslide sites across the 47-mile-long corridor. The western portion is situated within weak geology leading to several, predominantly cut-side landslides (Figure 4). The eastern portion is dominated by Price Canyon with resistant, rim-forming sandstone and limestone units undercut with shale and coal-rich beds, leading to road cuts in both colluvium and bedrock leading to rockfall activity (Figure 5).

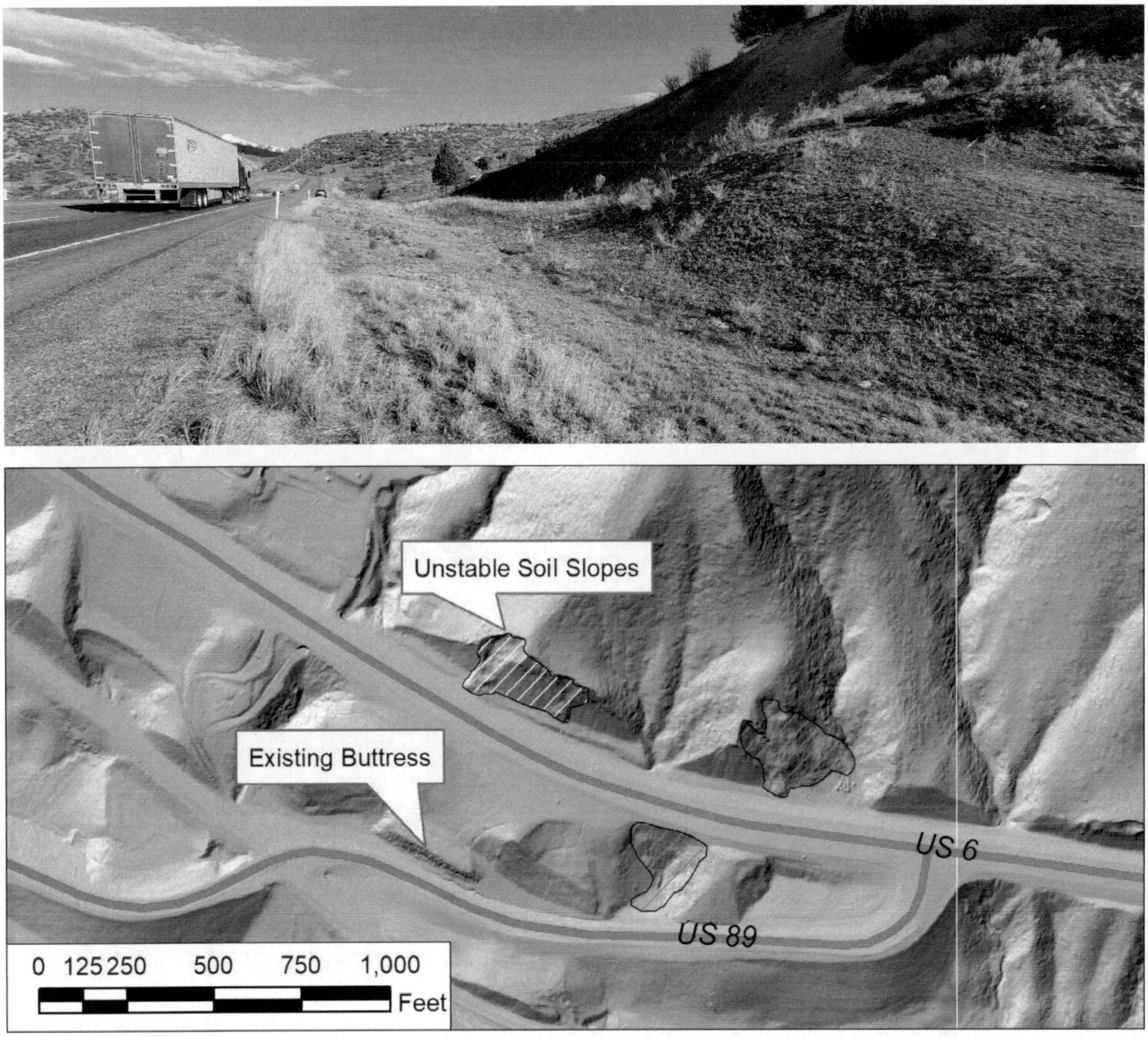

Figure 4: Top: Landslide depositing into and filling roadway ditches, increasing saturation of embankment below. Bottom: Hillshade image of landslide-prone terrain where drainage impairment would negatively impact stability.

Figure 5: Examples of large fallen rock blocks lie on the slopes below these cliff bands. Large rocks are evident in the colluvium excavated in some cuts.

This project identified 95 rock cuts and rock slopes that pose various levels of rockfall risk to the UDOT and road users. Sixty-two of these slopes were in Fair (56) or Poor (6) Condition. Nine slopes scored above 450 points, considered an unfavorable RHRS score. Collectively, the probability analysis predicted more than one adverse rockfall event per year along the corridor. The collective risk exposure posed by rockfall was $1.4 million annually while the cost estimate to improve the slopes to a Good condition was nearly $70 million. The project identified 20 of the 95 rock slopes to receive conceptual mitigation designs and cost estimates in a second phase. UAV-based imagery was collected by UDOT surveyors and subsequently reviewed by geologists to identify where large blocks high above the highway may be prone to failure.

Seventeen landslide sites were identified. Many of these landslide sites were in the western portion of the corridor and affected the cut slopes above the highway, failing into and inhibiting the drainage ditch function and slowly nearing the pavement. Other landslide sites were below the highway, affecting pavement performance. Fourteen of the sites were in Fair (13) or Poor (1) condition. These sites were assessed for high, medium, or low risk using professional judgment for the potential of affecting the travel lanes in the short (1-3 years), medium (3-10 years), or long (10+ years) term. Landslides were also assessed for their ability to negatively affect other nearby facilities. Overall, eight of the seventeen landslides were recommended for site-specific conceptual mitigation costs.

From this project, UDOT can now opt to further focus monitoring and instrumentation efforts at the locations selected for conceptual mitigation. Monitoring can include repeat UAV data collection and change detection at select geohazard sites and installation of inclinometers and crackmeters at select landslide and rockfall sites, respectively. Furthermore, several existing inclinometers were discovered at previously mitigated landslides. A recommendation to read these inclinometers was provided to understand if the prior mitigation efforts were functioning as intended.

Phase 4: Instrumentation Installation and Monitoring

Monitoring the sites to determine if displacement is occurring or otherwise problematic can occur after the completion of a corridor or larger-scale assessment. Instrumentation and monitoring that enables the monitoring of an accelerating displacement rate will also allow a DOT to make time-aware decisions regarding site sequencing or even postponement.

Selection of a remote monitoring approach will depend largely on the scale of the monitoring area and the level of detail required. For GAM systems seeking to monitor identified medium- to large-scale landslides, earth observation satellites, and large-scale lidar data collection and subsequent change detection would likely be applicable. However, these technologies do not provide information related to depth of movement and thus do not have sufficient information to inform engineering analysis and mitigation design. Locations with multiple lidar datasets may have an initial change detection analysis facilitated with tools available from websites such as OpenTopography.Org, which offers change detection analysis via cloud computers.

A phased approach to selecting monitoring methods, consisting of three escalating monitoring levels, is described below. This offers a flexible, but rational approach to selecting which monitoring method to start with and then escalating as evidence of movement is acquired.

Level 1: Visual

Use visual observations and periodic photos obtained from an identical position over time to help establish if deformations are growing. Establishment of reference points and use of tape measures across the deformations of interest will document and quantify the visual observations. Recording widths and dates will allow estimation of movement rates and provide information regarding whether the movement is accelerating or staying at a steady-state creep.

Level 2: Specialized Manual Instruments

After movement is known to be occurring or the consequences of movement are moderate to severe, an elevated monitoring approach is warranted. The use of specialized personnel and equipment to gain additional or more precise information but requiring a site visit by the specialized personnel to take measurements defines this Level. These types of specialized instruments or techniques can include:

- Installation of survey hubs and ground control points for repeat surveys,
- Drilling and installation of standpipe piezometers and/or inclinometers, and
- Precise surface measurements with tape extensometers, laser distance sensors, etc.

Level 3: Automated and Remote Telemetry (3+)

Implementation of a Level 2 monitoring approach may reveal displacement that is not tolerable. Several of the Level 2 techniques can be retrofit with sensors and dataloggers capable of unattended measurements. These include:

- Vibrating wire piezometer sensors installed within standpipe piezometers or directly within a borehole,
- Continuous or discrete in-place inclinometers installed within existing inclinometer casing or within new holes,
- Short or long-range crackmeters, and
- Tilt sensors on structural elements.

These sensors can be connected to dataloggers that are programmed to obtain readings at set intervals. Typically, an instrumentation specialist will need to visit the site to download readings with processing occurring later in the office.

Alternatively, more specialized automated data acquisition systems (ADAS) can be programmed with local radio-based telemetry and connected to cellular networks for a Level 3+ monitoring approach. The data telemetry removes the need for data-collection site visits. The sensors are typically the same as those in a basic Level 3, but the ADAS can be programmed to send SMS or email alerts, perform more complex calculations with the instrumentation data, and be connected to data presentation websites.

Case History

Montana DOT has used site-specific instrumentation information on rock slope sites to prioritize rockfall risk mitigation activities. Maintenance personnel called two rock slopes in the Missoula District to the Geotechnical Office's attention following a period of Level 1 visual monitoring.

One site is about 200-feet tall at MP 14 on Highway 135, east of St. Regis while the other 50-foot-tall slope is at MP 70 on Highway 93, near Elmo. Both sites had large tension cracks near the crest of the slope and were believed to have been visually enlarging. MDT opted to monitor the two sites with a combination of manual and electronic instrumentation in a combination Level 2 and Level 3 monitoring approach. Level 2 manual monitoring consisted of tape extensometer anchor points and point cloud change detection with Level 3 electronic vibrating wire crackmeters attached to dataloggers. Data was recorded at hourly intervals.

After one winter of monitoring and observation of higher-than-expected displacements at the Highway 135 site, the Level 3 sensors were supplemented with a datalogger-connected camera and connected to an ADAS with cellular data capabilities. The added capabilities were needed due to the confirmation of movement and the position of the displacing block's position 200 feet above the highway, justifying the Level 3+ approach. The Highway 93 crackmeters revealed one primary episode of displacement followed by slowing creep displacement.

Data plots for over five years are shown in Figure 6. Tape extensometer points and photogrammetric change detection exhibited similar trends. Comparing photos of the Highway 135 site from the Level 3+ monitoring year over year highlighted the ongoing displacement with visual confirmation that the data was accurate. Due to the monitoring results, MDT began the rockfall mitigation design process following an acceleration in displacement in the early 2021 winter months. Mitigation consisting of scaling, dowel installation, and ditch improvements was constructed in 2022. Soon after construction was complete, the crackmeters were reinstalled and have recorded a significantly reduced displacement rate in the time since.

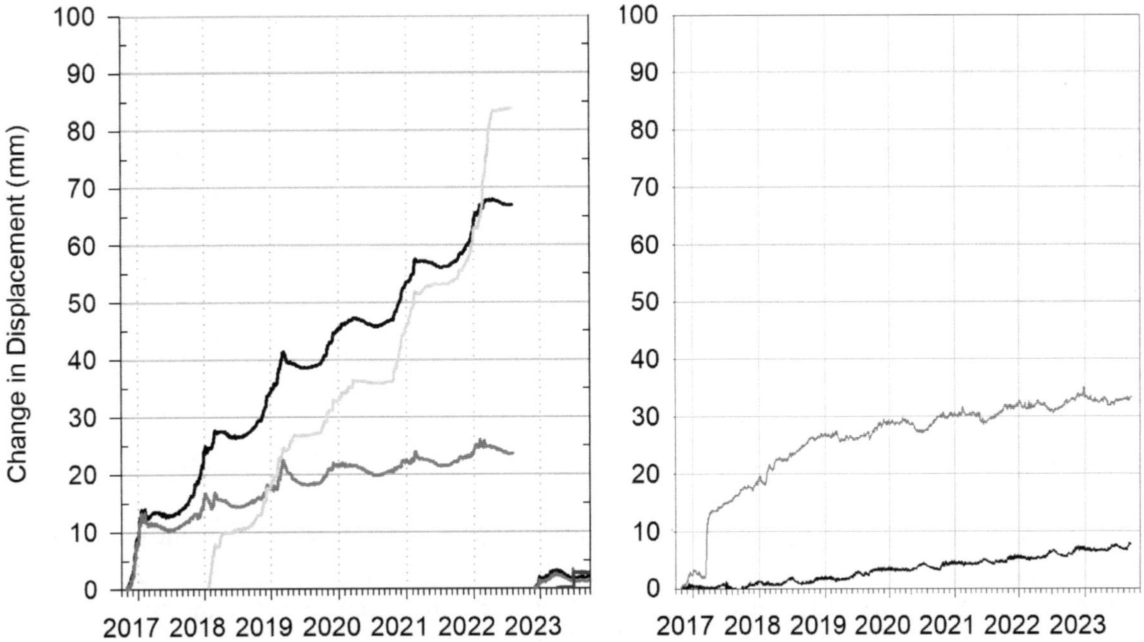

Figure 6: Crackmeter time-displacement plots for Highway 135 (left) and Highway 93 (right). Rockfall mitigation was performed at Highway 135 in mid-2023 and crackmeters were reinstalled thereafter and show much slower displacement.

Mitigation at Highway 93 is still being considered. A more detailed corridor study revealed that the monitored site had generally low RHRS scores and favorable condition assessments, but also possessed the lowest sight distance on the corridor, suggesting that should a rockfall event occur and reach the road a vehicle may not have sufficient reaction time.

Phase 5: Incorporate into Design

The data and information gathered in the previous Phases will provide a great deal of information for the design of mitigation measures. The data will highlight where critical unknowns still exist for the selected mitigation approach. For instance, the data may be sufficient for a geosynthetic-reinforced subgrade improvement for a weak embankment, but additional soil data and subsurface displacement measures may be required for a tieback anchor design.

CONCLUSIONS

In conclusion, the implementation of a Geotechnical Asset Management (GAM) program is crucial for the effective management and maintenance of geotechnical assets within state DOTs. This paper has outlined a phased approach to geotechnical monitoring within a GAM framework, emphasizing the importance of systematic inventory, condition assessment, and targeted monitoring to prioritize and address geotechnical risks effectively.

The initial phase involves a comprehensive statewide inventory and condition assessment, that forms the foundation of the GAM program. This phase ensures a thorough understanding of the existing assets and their conditions, providing a baseline for subsequent risk evaluations. Incorporating risk assessments based on condition and deterioration helps prioritize assets that require immediate attention, thereby optimizing resource allocation.

The second phase focuses on corridor identification, using Decision Support Tools (DSTs) to analyze large datasets and identify high-risk areas. This targeted approach allows for more efficient monitoring and mitigation planning, addressing the most critical sites within the transportation network. By highlighting corridors with concentrated risks, DOTs can develop more effective mitigation strategies and allocate resources where they are most needed.

Site-specific monitoring and instrumentation, as detailed in the third and fourth phases, are essential for understanding the behavior of geotechnical assets over time and potentially, in real-time. Implementing various levels of monitoring, from visual inspections to advanced automated systems, provides critical data to inform decision-making and design efforts. The case histories presented in this paper demonstrate the practical application of these monitoring techniques, showcasing how they can effectively identify and address geotechnical risks.

Finally, incorporating the collected data into the design phase ensures that mitigation measures are based on accurate and comprehensive information. This data-driven approach enhances the effectiveness of the mitigation efforts, reducing the likelihood of geotechnical failures and improving the overall resilience of transportation infrastructure.

In summary, a well-implemented GAM program offers a strategic and systematic approach to managing geotechnical assets, ensuring their safety and functionality throughout their lifecycle. By prioritizing high-risk areas and employing targeted monitoring and mitigation strategies, state DOTs can optimize their resources and improve the resilience and reliability of their transportation networks. The phased approach outlined in this paper provides a robust framework for achieving these goals, demonstrating the critical role of instrumentation and monitoring in a successful GAM program.

REFERENCES

AASHTO. (2020). *Transportation Asset Management Guide: A Focus on Implementation* (2nd ed., p. 324p) [Web]. https://store.transportation.org/Item/CollectionDetail?ID=217

Agbeshie, A. A., Abugre, S., Atta-Darkwa, T., & Awuah, R. (2022). A review of the effects of forest fire on soil properties. *Journal of Forestry Research*, *33*(5), 1419–1441. https://doi.org/10.1007/s11676-022-01475-4

Anderson, S., Carter, R., Dowling, C., Vessely, M., BGC Engineering USA Inc., Wisconsin Highway Research Program (WHRP), Wisconsin Department of Transportation, & Federal Highway Administration. (2022). *Geotechnical Asset Management for Slopes, Wisconsin Project ID: G21-06* (01917542; p. 56p) [Digital/other]. https://wisconsindot.gov/documents2/research/0092-21-06-final-report.pdf

Banks, B., Bauer, J., Abdelhadi, M., Ramey, T., & Highway Geology Symposium. (2022). *Unstable Slope Inventory along the Blue Ridge Parkway, NC* (01852529). pp 204-229. https://www.highwaygeologysymposium.org/wp-content/uploads/71st_HGS-OPT.pdf

Beckstrand, D., Benko, B. A., Mines, A. E., Pierson, L. A., Thompson, P. D., Kimmerling, R. E., Landslide Technology, Thompson (Paul), PanGEO, Alaska Department of Transportation and Public Facilities, & Federal Highway Administration. (2017). *Statewide Geotechnical Asset Management Program Development: Final Report for Rock Slopes, Unstable Soil Slopes and Embankments, Retaining Walls, Material Sites* (01660107; p. 378p). http://www.dot.state.ak.us/stwddes/research/assets/pdf/stp4000-126a.pdf

Beckstrand, D., Black, B., George, B., Kimmes, N., Mines, A., & Ruud, D. (2022). *Design Practices for Rock Slopes and Rockfall Management* (01851045; Issue 588). Transportation Research Board. https://trid.trb.org/view/1989287

Beckstrand, D., Machan, G., Dirringer, S., Shaw, B., & Enright, S. (2017). Avoiding the Slide—Planning for Transportation Modernization in Challenging Terrain. *Proceedings - 3rd North American Symposium on Landslides*, 777–786.

Beckstrand, D., Mines, A., Black, B., George, B., Thompson, P. D., & Stanley, D. (2017). *Rockfall Hazard Process Assessment* (FHWA/MT-17-008/8239-001; p. 373). Montana Department of Transportation. https://www.mdt.mt.gov/other/webdata/external/research/docs/research_proj/rockfall/FINAL_REPORT_OCT17.pdf

Farny, N. J. (2023). A Case Study on the Implementation of the Unstable Slope Management Program. *Transportation Research Circular*, *E-C283*, pp 120-124.

Mines, A. E., Beckstrand, D. L., Bunn, M. D., Landslide Technology, Missouri Department of Transportation, & Federal Highway Administration. (2023). *Development of a Geotechnical Asset Management Collection and Rating Program for Missouri Department of Transportation* (01878780; Final Report, p. 97p) [Digital/other]. https://spexternal.modot.mo.gov/sites/cm/CORDT/cmr23-004.pdf

Oester, N., Group, R., Ortiz, T., & Highway Geology Symposium. (2019). *Implementation and Application of Geotechnical Asset Management in Colorado* (01735578). 12p. https://www.highwaygeologysymposium.org/wp-content/uploads/70th_HGS-OPT.pdf

Vessely, M., Robert, W., Richrath, S., Schaefer, V. R., Smadi, O., Loehr, E., & Boeckmann, A. (2019). *Geotechnical Asset Management for Transportation Agencies, Volume 2: Implementation Manual* (01691948; Issue 903). Transportation Research Board. http://www.trb.org/Main/Blurbs/178602.aspx

Waseem, A., Tappenden, K., St. Michel, G., SKIRROW, R., & Transportation Association of Canada. (2022). *Alberta Transportation Geotechnical Asset Management (GAM) Framework Development, GAM Planner Application and Pilot Study* (01865918). 1 PDF file, 1.8 MB, 24p. https://www.tac-atc.ca/sites/default/files/conf_papers/am4_waseema.pdf

Xu, Y., Lu, Z., Bürgmann, R., Hensley, S., Fielding, E., & Kim, J. (2023). P-band SAR for ground deformation surveying: Advantages and challenges. *Remote Sensing of Environment*, *287*, 113474. https://doi.org/10.1016/j.rse.2023.113474

Modern Borehole Logging & Database Management – A Geodata Lifecycle

Declan Vanderhor, B.E Geotechnical and Mining Engineering
TabLogs, Denver, USA and Maroochydore, Australia
Declan@tablogs.com

Prepared for the 73rd Highway Geology Symposium, September 2024

Disclaimer

Copyright Notice

Abstract

The geotechnical engineering sector has seen significant advancements in borehole logging and data management processes, transitioning from manual methods to sophisticated digital systems. This paper explores the historical evolution and current innovations in these processes, highlighting the impact of modern technologies, particularly Software as a Service (SaaS) solutions. The shift towards digital logging has enhanced the accuracy and accessibility of data, enabling real-time analysis and improved decision-making. Advanced data management systems offer features such as data integration, security, and user-friendly interfaces, that are crucial for managing complex datasets. SaaS platforms provide scalable, cost-effective solutions, offering unparalleled accessibility and regular updates. The adoption of these technologies has led to increased efficiency and cost-effectiveness, emphasizing the importance of data integrity in achieving successful project outcomes. This presentation will summarize the key findings and lessons learned from the industry's evolution, underscoring the critical role of precise data collection and sophisticated management systems in modern geotechnical studies.

Introduction

The geotechnical sector has undergone a profound evolution, particularly in soil investigations and borehole logging, transitioning from manual, error-prone methods to sophisticated digital systems. Historically, site investigations were often neglected or underappreciated by the construction industry, as evidenced by Ashton and Gidado (2001), who found that fewer than 1% of construction firms in the UK conducted any form of risk assessment regarding uncertain ground conditions. This widespread neglect of thorough site investigations frequently led to costly project delays, structural failures, and unforeseen risks, clearly demonstrating the need for better approaches to managing geotechnical risks.

Over time, these issues underscored the importance of accurate and reliable site investigations, prompting a shift in the construction industry. Today, the appreciation for site investigations has evolved significantly, with geotechnical data now seen as a key requirement for construction projects. Modern technologies—particularly Software as a Service (SaaS) solutions—have transformed borehole logging and data management, providing real-time data, improving accuracy, and enabling faster, more informed decision-making at every stage of a project, from initial proposals to final design.

This paper explores the evolution of these processes, highlighting how technological advancements have addressed the past deficiencies and driven a more data-informed approach to geotechnical engineering.

Background and Objectives

Soil investigations provide vital data that underpin geotechnical studies, influencing everything from initial proposals to final reporting. Traditionally, this sector has faced numerous challenges, including data loss, inconsistent record-keeping, and difficulties with data retrieval. Paper-based methods often resulted in fragmented or incomplete data due to mismanagement, transcription errors, or damage caused by environmental factors during fieldwork. Furthermore, the manual entry of field data delayed its availability for analysis and decision-making, often compromising the project's timeline and the ability to make real-time engineering adjustments.

To mitigate the challenges posed by manual logging, modern digital methods have been developed to replace traditional paper-based systems. These advancements allow for more accurate, consistent, and timely data capture and retrieval, ensuring that geotechnical data is reliable and can be accessed when needed without the risk of transcription errors or data loss. This shift has greatly enhanced the reliability of geotechnical investigations by streamlining the processes of data management and reporting (*4*).

The complexity and large volume of data in geotechnical projects further underscored the need for advanced data management systems capable of handling diverse datasets effectively. These systems integrate field data into centralized, accessible databases, enabling real-time data transmission and analysis, which reduces the risk of project delays and enhances decision-making accuracy.

This paper aims to explore the historical development of borehole logging and data management systems, focusing on how modern technologies, particularly Software as a Service (SaaS) solutions, are transforming the geotechnical industry by increasing efficiency, accuracy, and responsiveness.

Evolution of Borehole Logging

Borehole logging has transitioned from rudimentary, manual methods to sophisticated, automated systems. Initially, geotechnical engineers relied on manual logs and analog data storage, which were often susceptible to loss or misinterpretation by data officers in the office who would have trouble reading hastily scrawled field notes (or notes covered in dirt from the site). The advent of digital logging prevents this issue, providing reliable and easily accessible data to engineers at all levels within an organization.

For instance, the use of nuclear logging at the Kakrapar nuclear power plant in India (*5*) illustrated the application of advanced geophysical methods for accurate subsurface characterization, which is crucial for understanding soil properties such as density, moisture content, and shear strength. These properties are fundamental to soil mechanics, influencing critical engineering decisions about foundation design, slope stability, and load-bearing capacity. In Lagos, Nigeria, geophysical logs were instrumental in delineating aquifers and understanding the complex geology of the area (*6*). Beyond hydrogeological assessments, accurate logging data also provided insights into soil stratification, compaction, and the identification of weak zones that could impact soil behavior under stress.

These cases demonstrate the crucial role of accurate data collection in ensuring not only the safety and efficiency of infrastructure projects but also the proper application of soil mechanics principles, which are vital for designing stable and resilient structures in diverse geotechnical environments.

Moreover, digital borehole logging enables real-time data acquisition and analysis. This capability allows engineers to make informed decisions promptly, thereby reducing project delays and costs. The integration of tools like gamma-ray logging and resistivity measurements has further enhanced the precision of subsurface investigations, providing a more comprehensive understanding of geological conditions (*6*).

Key Properties Logged in Digital Borehole Logging

Digital borehole logs capture a wide range of critical subsurface properties, which include:

1. **Lithology:** Digital logs provide detailed information about the different soil and rock layers encountered during drilling. This includes descriptions of grain size, mineral composition, and texture, which are essential for understanding site conditions.

2. **Moisture Content and Porosity:** Measuring water content and porosity helps determine the permeability of the subsurface materials, which is vital for groundwater studies and foundation design.

3. **Strength and Compressibility:** Cone penetration tests (CPT) and standard penetration tests (SPT) assess the strength of the subsurface

materials, helping engineers determine load-bearing capacities and potential settlement issues.

Engineering Decisions Informed by Digital Borehole Logs
Digital borehole logs provide comprehensive data that enables engineers to make critical decisions on various aspects of construction and design:

- **Foundation Design:** The strength, compressibility, and lithology data help in determining the most appropriate foundation type—whether shallow, deep, or piled foundations—for a given site.
- **Slope Stability and Landslide Risk:** By understanding the mechanical properties of soils and rock, engineers can assess slope stability and design appropriate mitigation measures to prevent landslides or other forms of slope failure.
- **Groundwater Management:** Moisture content, porosity, and permeability data help in designing dewatering systems and managing groundwater flow, particularly in areas with high water tables or complex hydrogeology.
- **Environmental and Contaminant Assessment:** Lab results confirming log findings assist in identifying potential contaminant plumes or hazardous materials that may need to be managed during excavation and construction.

- **Real-Time Feedback and Decision-Making**
  One of the most significant advantages of digital borehole logging is the real-time feedback it provides. As data is collected, it is transmitted directly to senior engineers in the office, who can immediately analyze it using specialized software. This instant feedback loop allows for rapid adjustments in field operations—whether that means altering drilling techniques, refining design parameters, or identifying potential risks early in the process.

  For example, if unexpected weak soil layers are identified during drilling, foundation designs can be quickly reassessed to accommodate these conditions, thereby avoiding costly delays and ensuring that the construction proceeds smoothly. This immediate transfer of data helps in making faster, more informed decisions, significantly reducing project delays and cutting costs.

Modern Data Management Systems

The development of advanced data management systems has revolutionized the geotechnical industry. These systems offer a range of features designed to streamline data collection, storage, and analysis. Key features include:

- Data Integration: Modern systems support the integration of various data types, including geotechnical, geophysical, and environmental data. This integration provides a broader understanding of site conditions, essential for complex projects like urban infrastructure development (*7*).

- Real-Time Analysis: The ability to analyze data in real-time is a significant advantage, enabling immediate responses to emerging issues. This feature is particularly valuable in time-sensitive projects, such as emergency response and construction (*4*).
- Data Security and Compliance: With increasing concerns over data security and regulatory compliance, modern systems incorporate robust security protocols to protect sensitive information. These measures ensure that data integrity is maintained, and industry standards are met (*4, 7*).
- User-Friendly Interfaces: The complexity of geotechnical data often requires specialized knowledge to interpret. User-friendly interfaces, including graphical data visualization tools, make it easier for non-specialists to understand and use the data, thereby facilitating better communication among project stakeholders (*4*).
- Adaptability and Automation: Modern systems are highly adaptable, capable of adjusting to new data types and project requirements. Automation of routine tasks, such as data processing and reporting, increases efficiency and reduces the potential for human error.

A noteworthy example is the use of GIS-based borehole data management systems for 3D subsurface modeling, which provides a detailed representation of geological structures. This approach has been successfully implemented in projects across various regions, including Chennai, India, and Austin, Texas, demonstrating its versatility and effectiveness (*7*).

The Impact of SaaS Solutions

SaaS solutions have emerged as a cornerstone of modern geotechnical data management, offering a range of benefits that traditional software cannot match. These cloud-based platforms provide scalable, cost-effective, and accessible solutions for managing complex datasets (*8*).

- Cost-Effectiveness: SaaS platforms generally require lower initial investment compared to on-premises software, making them accessible to a broader range of organizations, from small firms to large corporations.
- Scalability: One of the most significant advantages of SaaS solutions is their scalability. These platforms can easily accommodate growing data volumes and expanding user bases, ensuring that they can support projects of varying sizes and complexities.
- Accessibility: Cloud-based systems offer unparalleled accessibility, allowing users to access data from anywhere with an internet connection. This feature is particularly beneficial for collaborative projects that involve multiple stakeholders in different locations.
- Regular Updates: SaaS providers frequently update their platforms, ensuring that users have access to the latest features and security measures. This continuous improvement cycle helps organizations stay ahead of emerging challenges and opportunities.

Results and Lessons Learned

The evolution of soil investigations and borehole logging has significantly impacted the geotechnical industry, leading to greater efficiency, accuracy, and cost-effectiveness. However, the journey has also highlighted critical lessons, particularly the importance of data integrity (*3*). Inaccurate or poorly managed data can compromise project outcomes, emphasizing the need for meticulous data collection and management practices.

The use of SaaS solutions has demonstrated substantial time savings and improved data accuracy. Clients have reported saving up to 20 hours per week in data handling tasks (*4*), which allowed them to focus on more critical aspects of their projects. This efficiency gain underscores the value of adopting advanced data management technologies in the geotechnical field.

Conclusion

The geotechnical industry is in the midst of a transformative shift, driven by advancements in technology and data management practices. The adoption of modern borehole logging methods and SaaS solutions has enhanced the accuracy, efficiency, and reliability of geotechnical data. These improvements are essential for making informed decisions and ensuring the success of projects. As the industry continues to evolve, maintaining data integrity and leveraging advanced technologies will be crucial for addressing the challenges and opportunities ahead.

References

1. Ashton, P and Gidado, K (2001) *Risk associated with inadequate site investigation procedures under design and build procurement systems.* In: Akintoye, A (Ed.), 17th Annual ARCOM Conference, 5-7 September 2001, University of Salford. Association of Researchers in Construction Management, Vol. 1, 961-9

2. Alhalaby, N.M.H. and Whyte, I L (1994*) The impact of ground risks in construction on project finance.* Risk and reliability in ground engineering: Proceedings of the conference organized by the Institution of Civil Engineers, held in London on 11th 12th November 1993. London, Thomas Telford, 54-67

3. Hytiris, N, Stott, R. McInnes, K (2014) *The importance of site investigation in the construction industry: a lesson to be taught to every graduate civil and structural engineer.* Worls Transactions on Engineering and Technology Education, Vol. 12, No. 3. http://www.wiete.com.au/journals/WTE%26TE/Pages/Vol.12,%20No.3%20(2014)/15-Hytiris-N.pdf

4. TabLogs. (2024). *Case Study: Efficiency Gains in Geotechnical Data Management.* TabLogs White Paper Series.

5. Bhowmick, S., Pund, A., Suresh Kumar, B. (2024). Importance of Borehole Logging Studies in Design of Foundation for Civil Engineering Structures. In: Jose, B.T., Sahoo, D.K., Shin, E.C., Choudhury, D., Joseph, A., Pai, R.R. (eds) Proceedings of the Indian Geotechnical Conference 2022 Volume 1. IGC 2022. Lecture Notes in

Civil Engineering, vol 476. Springer, Singapore. https://doi.org/10.1007/978-981-97-1737-8_37

6. Jimoh, R.A., Bankole, O.M., Ahmed, K. *et al.* Use of geophysical logs in hydrogeological studies and borehole designs: case study of Apapa coastal area, Lagos, Nigeria. *Appl Water Sci* 8, 191 (2018). https://doi.org/10.1007/s13201-018-0804-9
7. B., D.P., Dodagoudar, G.R. *An integrated geotechnical database and GIS for 3D subsurface modelling: Application to Chennai City, India. Appl Geomat* 10, 47–64 (2018). https://doi.org/10.1007/s12518-018-0202-x
8. Ground Engineering. *Lessons learned. The importance of ground investigation* (2022) https://www.geplus.co.uk/opinion/lessons-learned-the-importance-of-ground-investigation-01-12-2022/

SGAM - Smart Geotechnical Asset management

Alessandro Brunetti
NHAZCA S.r.l.
Via Vittorio Bachelet, 12
00185 Rome
+39 06 95065820
alessandro.brunetti@nhazca.com

Andrea Chessa
NHAZCA S.r.l.
Via Vittorio Bachelet, 12
00185 Rome
+39 06 95065820
andrea.chessa@nhazca.com

Marco Deana
NHAZCA S.r.l.
Via Vittorio Bachelet, 12
00185 Rome
+39 06 95065820
marco.deana@nhazca.com

Maria Elena Di Renzo
Sapienza Università di Roma
Dipartimento di Scienze della Terra
Piazzale Aldo Moro, 5 – 00185 Rome
mariaelena.direnzo@uniroma1.it

Paolo Mazzanti
Sapienza Università di Roma
Dipartimento di Scienze della Terra
Piazzale Aldo Moro, 5 – 00185 Roma
paolo.mazzanti@uniroma1.it

Prepared for the 73rd Highway Geology Symposium, September 2024

Disclaimer

Copyright Notice

ABSTRACT

SGAM (Smart Geotechnical Asset management) is a semi-automated decision support system that integrates cutting-edge technology, data fusion algorithms, and satellite Earth Observation data. Its comprehensive approach involves extensive data collection, which emphasizes the importance of data quality for hazard analyses.

In this paper, the focus is on hazard assessment, which requires a comprehensive study of the available data and how they can be integrated to produce a meaningful result. Remote sensing analyses, including InSAR, assess the impact of geological processes on infrastructure, thereby enhancing hazard analysis.

The hazard analyses conducted in the framework of the SGAM project are customized to suit various types of processes that pose threats to infrastructure. As the project progresses, the next challenge will be to effectively incorporate considerations of exposure and vulnerability into the assessment framework. Addressing this challenge will require a comprehensive and integrated approach, utilizing advanced technologies and collaborative efforts with infrastructure managers to ensure a holistic understanding of infrastructure resilience across diverse contexts and conditions.

INTRODUCTION

Natural hazards, such as earthquakes and landslides, pose significant threats to infrastructure from a global perspective, resulting in substantial socioeconomic losses. Statistics reveal that annually, approximately 0.5% of worldwide assets are exposed to these natural calamities (1). Common consequences include damage to roads, bridges, and other vital infrastructure, as well as disruptions to traffic flow and loss of life and property. Various strategies aim to mitigate the impact on linear assets, among which is the enhancement of knowledge and comprehension of asset data (2), necessitating a detailed understanding of the interaction between geohazards and infrastructure.

Assessment of infrastructure resilience at a regional scale is becoming increasingly crucial in planning studies aimed at vulnerability reduction and resilience enhancement. While existing studies often concentrate on single hazards or micro-scale assessments, the combined impact of multiple hazards at a regional scale is considerably more significant. One specific challenge lies in evaluating infrastructure resilience in multi-hazard areas and its correlation with social dimensions (3) (4). Geotechnical monitoring plays a pivotal role in transportation asset management by providing valuable insights into the condition and functionality of geotechnical structures such as foundations, embankments, and slopes. The integration of geotechnical monitoring with data analysis, decision support systems, and risk assessment tools, named as Smart Geotechnical Asset Management (SGAM), ensures continuous monitoring and proactive risk management (5).

This paper offers a comprehensive introduction to the methodology developed for the SGAM platform. The primary goal of SGAM is to conduct risk assessments for linear infrastructure, with a specific emphasis on guiding asset maintenance strategies. The platform is designed to leverage state-of-the-art technology and data-driven approaches to enhance the overall resilience and performance of critical infrastructure.

This paper presents an introduction to the core principles and methodologies that underpin the SGAM project. It focuses specifically on the hazard assessment aspect and offers an exploration of the framework developed within SGAM. The overview aims to clarify the processes involved in assessing the risks posed by various geohazards to critical infrastructure, while the hazard assessment methodology is described in detail to provide a better understanding of SGAM's approach to enhancing infrastructure resilience.

SGAM METHODOLOGY AND APPLICATION

SGAM is a semi-automated decision support system designed for asset management and predictive maintenance. Its mission is to improve the financial resilience of assets through the strategic deployment of data fusion algorithms and satellite Earth Observation (EO) technologies. The project takes a comprehensive multi-hazard perspective, considering the impact of natural hazards on linear infrastructure.

The project focuses on landslides, floods, geotechnical challenges such as subsidence and settlement, and seismic hazard. The initial phase involves extensive data collection to lay the foundation for subsequent hazard analyses. The efficacy of these analyses depends on the quality and validity of the collected data, which requires a rigorous assessment process to ensure the reliability of hazard evaluations.

The accuracy of risk assessment in SGAM depends on a detailed analysis of factors such as the exposure value and the vulnerability of the asset being evaluated. Structural data are also considered, implying collaboration with infrastructure managers to obtain specific information. This collaborative approach ensures that the risk assessment is not only precise but also well-informed, drawing on the expertise of those responsible for managing the assets.

SGAM has a distinctive feature of integrating remote sensing analyses to evaluate the interference of geological processes with infrastructure. The project leverages advanced techniques such as Satellite InSAR (Interferometric Synthetic Aperture Radar) to capture ground movements, enabling a comprehensive evaluation of their impact on assets. The integration of ground movement information with hazard analyses through data assimilation results in a comprehensive multi-hazard analysis and risk evaluation (Figure 1).

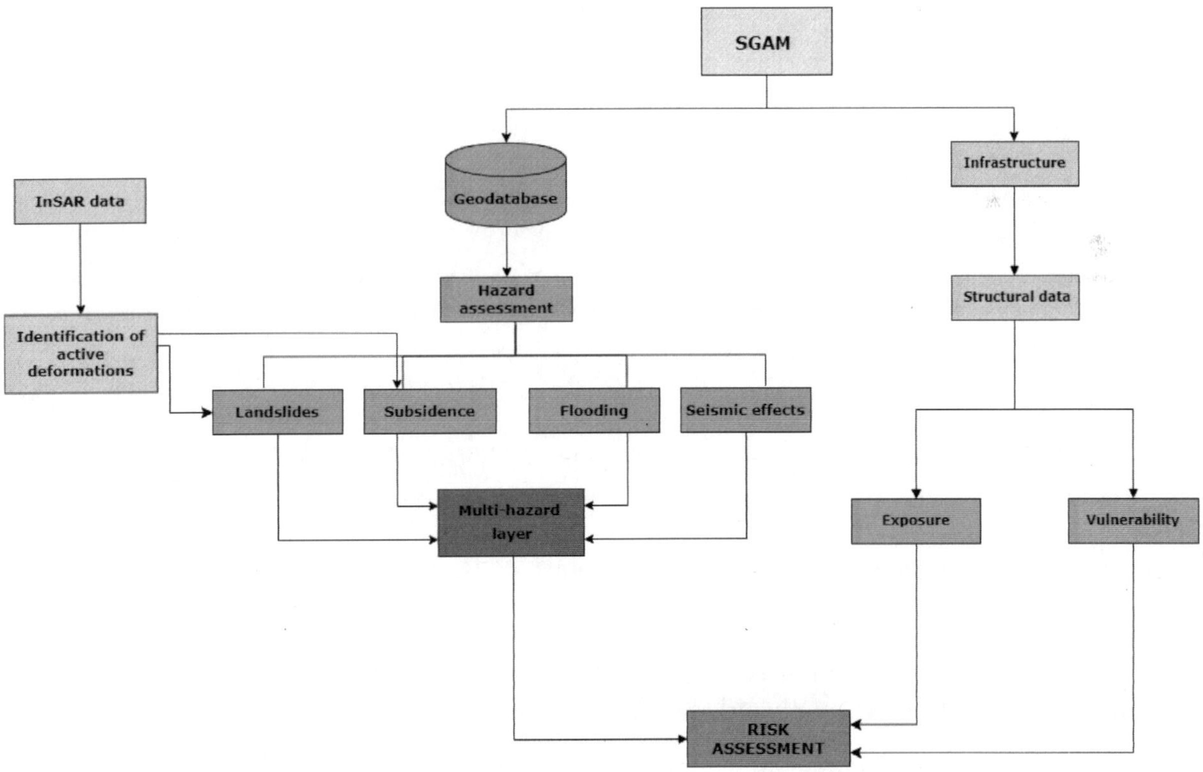

Figure 1 - Framework of the SGAM project, outlining the main steps.

Geodatabase

A proprietary geodatabase has been developed and implemented on the platform to collect the primary data necessary for subsequent analyses. This repository offers detailed information on various geohazards, including earthquakes, landslides, floods, and more. The geodatabase contains valuable information on various datasets, including their type, spatial coverage, data accessibility, download availability, data format, license type, and update frequency.

After a thorough search, 114 links to geohazard datasets were identified (global extent). The quantitative analysis shows that earthquakes are the most common geohazard in the database. They account for 24% of the datasets. Landslides and floods, considered as a unique category, account for 15% of the hazards. A further 15% fall into the category of multiple hazards, describing more than one geohazard.

Italy has the largest spatial coverage in the geodatabase at 38%, followed closely by global datasets at 38% and those focused on Europe at 20%. This analysis reveals potential gaps in hazard data for regions such as Africa, the Americas, and Eastern countries (Figure 2).

The accessibility of data emerges as a crucial aspect of geodatabase analysis. A comprehensive review shows that 93% of the databases offer open access to their datasets, with or without the requirement of user registration. Of the freely accessible datasets, 86% permit commercial use, indicating their potential application for commercial purposes.

The geodatabase mainly uses vector and raster formats, which are widely employed in Geographic Information System (GIS) environments for compatibility and ease of management. However, a challenge arises in the frequency of updates, as many datasets lack information about their update frequency. A scalable and automated approach is necessary to access and update geohazard resources.

Standardizing information related to different geohazards is critical for dealing with data sources that have varying attributes. This standardization ensures that all datasets within the geodatabase are structured consistently, easing integration, and ensuring compatibility between them. Moreover, standardized information facilitates interoperability with different data sources and enables seamless management in various environments.

To integrate and standardize the information, a few steps are necessary due to the heterogeneous nature of these datasets. Decoding data and identifying common attributes that form the basis for evaluating hazard levels are crucially aided by spatial geoprocessing.

The data homogenization and integration process of the SGAM project is controlled by a semi-automated approach, with expert users overseeing the entire workflow. The tool allows for the upload of different vector datasets, and the identification of relevant attributes is a critical step where user expertise is paramount. The uploaded data then undergo a series of geoprocessing steps to produce consistent and comprehensive layers.

This user-friendly system streamlines the overall process and enhances efficiency in data management and analysis. The integrated geodatabase layers prove instrumental as an information substrate for infrastructure management, offering insights into geological and structural conditions. When multiple independent sources corroborate the same information, it strengthens confidence in its accuracy and validity. This multi-faceted approach to reliability evaluation fosters greater trustworthiness in the database content and enhances its utility for decision-making processes.

The geodata stored in a local database, combined with data from other modules, provide the necessary information to assess the impact of geological processes on the user's asset. This approach is designed to account for differences in data granularity across various regions or aspects of the study area, ensuring a thorough evaluation.

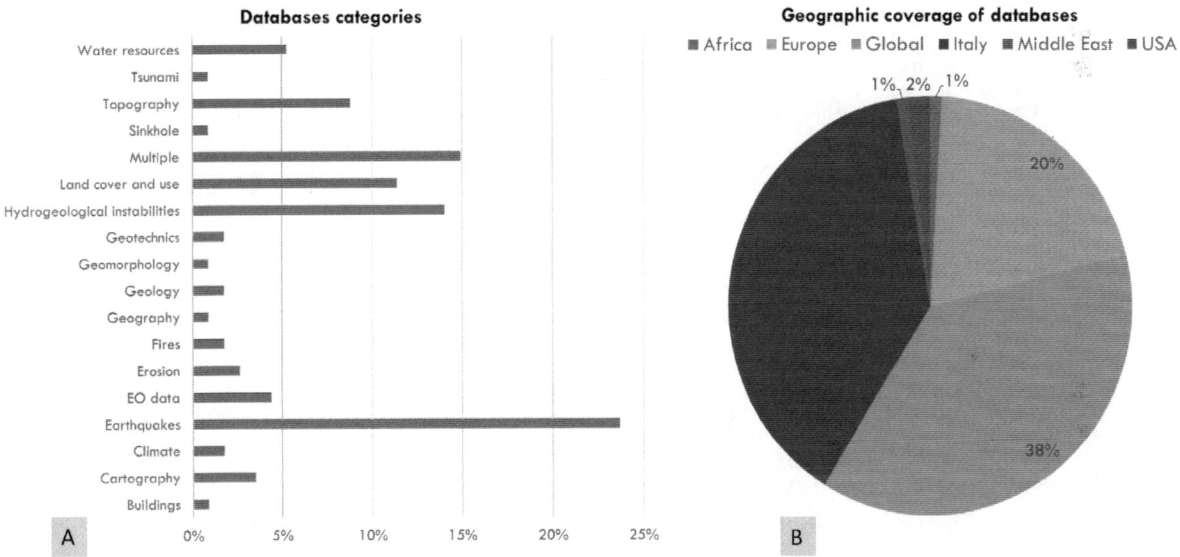

Figure 2 - Pie charts related to the identified datasets. Picture A shows the distribution in terms of categories, graph B shows the distribution of spatial coverage.

Hazard Assessment

By using advanced technologies such as remote sensing, Geographic Information Systems (GIS), and numerical modelling, hazard assessments can provide detailed hazard maps and predictions of potential hazards. This improves the development of effective mitigation strategies and land-use planning, which are essential for safeguarding human life, infrastructure, and the environment.

The SGAM project identifies geohazards that are pivotal for ensuring infrastructure safety. Among these, the threat of landslides is particularly significant. To evaluate the probability of landslides occurring in designated areas, a susceptibility analysis was undertaken, leveraging available inventories. However, the analysis focuses solely on susceptibility due to constraints in ancillary data availability, omitting temporal occurrence considerations. The assessment distinguishes between two kinematic types of movement: slow and fast landslides. This differentiation arises from the varying degrees of interference these movements pose to linear infrastructure, highlighting the potential for divergent impacts on safety and functionality. A data-driven approach, using machine learning (ML) techniques, was employed to analyse the presence or absence of landslides. The objective was to infer the multivariate combination of causative and predisposing factors across stable and unstable sectors within the investigated area. Efforts were previously made to enhance the reliability of the landslide database, given the dependency of machine learning algorithms on the quality and quantity of input data.

To assess hydraulic hazards of infrastructures, including flooding, erosion, and local erosion related to potential scours around piles, a different approach is adopted. Field observations during bridge inspections are utilized to evaluate each category. The assessments are integrated to yield a comprehensive score for each structure across the three risks: flooding, erosion, and local erosion. Using individual scores, the overall risk level for each structure can be determined.

The methodology used to develop soil susceptibility mapping for liquefaction across the entire territory is based on identifying and reclassifying various geological and seismic attributes. These attributes allow for an assessment of the likelihood of a given area experiencing liquefaction phenomena. This approach facilitates the development of a comprehensive classification of the area, offering valuable insights into its susceptibility to liquefaction hazards. A cut-off threshold is then identified to partition the data into two distinct zones based on their maximum expected acceleration values.

In the hazard assessment, Satellite InSAR data, particularly Persistent Scatterers (PS), are used to validate and reinforce the preceding susceptibility analysis. The European Ground Motion Service (EGMS) has made it easier to include ground deformation data. In the context of landslides, these data act as a proxy for activity status and are instrumental in identifying areas that may pose a threat to infrastructure by revealing movements along slopes. These steps were specifically applied to slow movements, as the interferometric technique has limitations that prevent its application to fast deformations. Based on the results of the susceptibility analysis, attention was focused on areas with higher levels of hazard, ranging from medium to very high. A velocity threshold was established using PS data to identify areas that are actively experiencing movement and are correlated with landslides.

PS data are the primary component for subsidence hazard assessment. Like landslide hazard assessment, attention is directed towards areas with slopes below a specified threshold, usually 5 degrees. This criterion helps to identify and analyze movements associated with the subsidence process.

The hazard analysis results are spatialized over a large area, requiring a consistent amount of input data for accurate assessment. It is crucial to evaluate how these processes intersect with infrastructure. The initial step involves segmenting the linear infrastructure and assigning a hazard value to each segment (Figure 3), considering each hazard analysis separately.

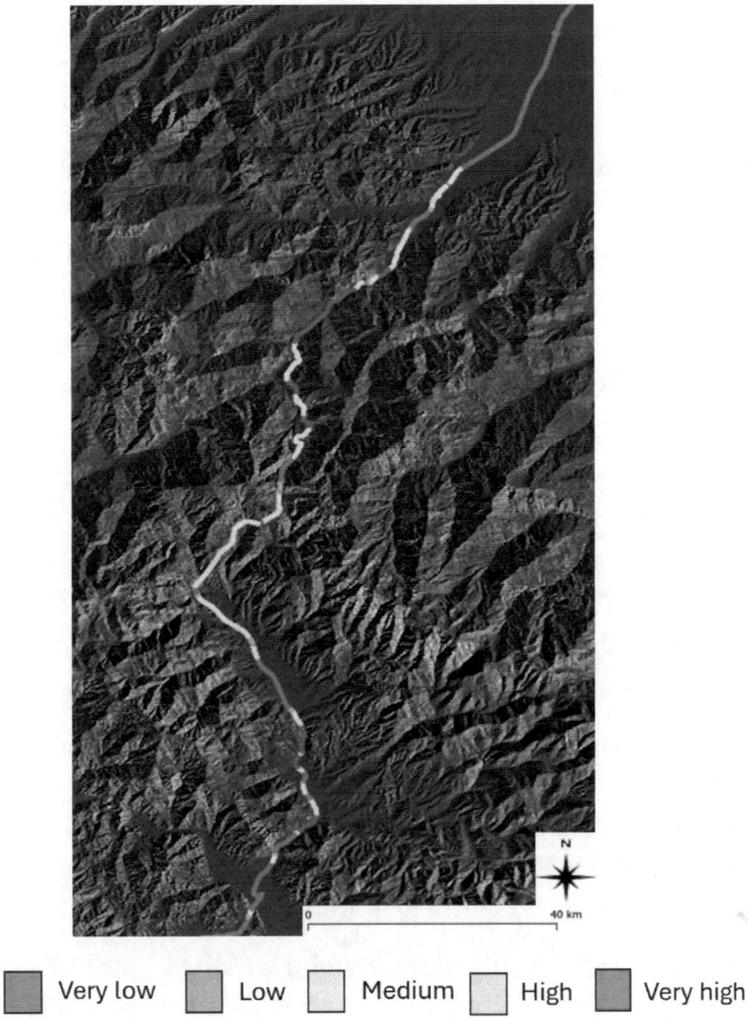

Figure 3 – Map of hazard value for linear infrastructure.

For hazards such as landslides and subsidence, where InSAR data supports monitoring, the integrated hazard value results from susceptibility analysis and PS information. This integration allows for a more comprehensive understanding of the risks posed to linear infrastructure by these geological phenomena, enabling informed decision-making and targeted mitigation strategies.

CONCLUSION

This paper describes the outlines of the SGAM platform that aims to conduct risk assessments for linear infrastructure.

The platform focuses on guiding asset maintenance strategies and is a semi-automated decision support system designed for asset management and predictive maintenance. SGAM employs data fusion algorithms and satellite Earth Observation technologies to take a multi-hazard perspective.

SGAM integrates a geodatabase and remote sensing data derived from advanced analyses, such as Satellite InSAR, to evaluate geological processes' interference with infrastructure. This integration results in a comprehensive multi-hazard analysis and evaluation.

Standardizing information across different geohazards is essential for consistency, integration, and compatibility within the geodatabase. SGAM's approach to data homogenization and integration is semi-automated, with a user-friendly system that streamlines the process and increases efficiency.

Moving forward, the project aims to adopt 'smart' methods that can be easily tailored to different types of data for hazard analysis. The emphasis on customization is crucial to accommodate the distinct features of various hazards and regions, ensuring the adaptability and applicability of the developed methodologies. This approach can streamline the hazard analysis process and improve the efficiency and effectiveness of decision-making in infrastructure monitoring.

The incorporation of Satellite InSAR (Interferometric Synthetic Aperture Radar) data is a dynamic element in SGAM. The use of InSAR technology provides valuable data for assessing the activity status of natural phenomena; it is particularly useful for phenomena such as landslides and subsidence, providing insights into ongoing ground movements.

Assessing infrastructure vulnerability and exposure is a challenging task due to its inherently localized nature. Vulnerability includes various factors such as structural integrity, material composition, and environmental conditions, which can significantly vary from one asset to another. Developing standardized vulnerability assessments that adequately capture the diverse array of risks faced by infrastructure is difficult due to this localized complexity.

In contrast, assessing exposure is usually more straightforward as it primarily involves identifying the physical presence and proximity of infrastructure to potential hazard sources. Although exposure may vary depending on the specific hazard being considered, the number of distinct scenarios requiring evaluation is relatively limited.

Despite the challenges, the SGAM project acknowledges the crucial role of infrastructure managers in providing insights for vulnerability and exposure assessments. Infrastructure managers can contribute valuable information that enhances the accuracy and comprehensiveness of vulnerability assessments by leveraging their detailed knowledge of asset characteristics and operational conditions. Establishing effective channels of communication and collaboration with asset managers is essential for obtaining the necessary data and ensuring a comprehensive understanding of infrastructure resilience across diverse contexts and conditions.

In conclusion, SGAM not only provides a robust methodology for current infrastructure risk assessment but also lays the groundwork for future advancements. SGAM is at the forefront of innovative approaches to geotechnical asset management due to its commitment to smart and customizable methods. The project's success is not only attributed to its methodologies but also its potential to significantly contribute to the broader field of infrastructure resilience in the face of natural hazards.

REFERENCES:

1. Koks, E. E., J. Rozenberg, C. Zorn, M. Tariverdi, M. Vousdoukas, S. A. Fraser, J. W. Hall, and S. Hallegatte. "A Global Multi-Hazard Risk Analysis of Road and Railway Infrastructure Assets." *Nature Communications 10*, no. 1 (June 25, 2019): 1-10.

2. Argyroudis, Sotirios A., Stergios Mitoulis, Mike G. Winter, and Amir M. Kaynia. "Fragility of Transport Assets Exposed to Multiple Hazards: State-of-the-Art Review toward Infrastructural Resilience." *Reliability Engineering & System Safety* 191 (November 2019): 1-16.

3. Arvin, Mahmoud, Parisa Beiki, Seyed Jafar Hejazi, Ayyoob Sharifi, and Nasrin Atashafrooz. "Assessment of Infrastructure Resilience in Multi-Hazard Regions: A Case Study of Khuzestan Province." *International Journal of Disaster Risk Reduction* 88 (April 2023): 1-13.

4. Joshi, Dheeraj, Wataru Takeuchi, Nirmal Kumar, and Ram Avtar. "Multi-Hazard Risk Assessment of Rail Infrastructure in India under Local Vulnerabilities towards Adaptive Pathways for Disaster Resilient Infrastructure Planning." *Progress in Disaster Science* 21 (January 2024): 1-12.

5. Mazzanti, P. "Toward Transportation Asset Management: What Is the Role of Geotechnical Monitoring?" *Journal of Civil Structural Health Monitoring* 7, no. 5 (October 17, 2017): 645-656.

Spatial Analytics, Remote Sensing and Advances in Technology as They Apply to Highway Geological Investigations and Critical Infrastructure Asset Management.

Robert Hendricks, PhD
Elevate
3403 Willowbrook Drive, Richardson TX 75082
214-458-4660
robert@elevateuas.com

Prepared for the 73 rd Highway Geology Symposium, September, 2024

Disclaimer

Copyright

ABSTRACT

Highways have long been considered critical economic and strategic infrastructure to state and federal institutions. Generally, attention is given to paved assets in populated or connecting populated areas, unpaved assets are critical in more rural and remote areas of the United States and the developing world. As such their development and management is an important part of financial resource planning. The assessment of the area adjacent and below these assets, paved or unpaved, is part of this planning program.

Advances in imaging and remote sensing technology as allowed for the collection of data pertinent to the development and management of these assets, quickly, accurately and more completely. Small Unmanned Aerial Systems (sUAS or "drones") are increasingly being used for the acquisition of this data. The creation of three-dimensional models that include integrated surficial and subsurface data allows for site analysis and design in a digital space that can be integrated into existing geospatial databases and easily shared with others. By developing a spatial model in a digital format analysis of the data using machine learning and artificial intelligence is possible. As this technology matures it use in the geosciences and civil engineering will increase but is dependent on accurately modeling the physical asset.

This paper will provide case studies of the use of sUAS and other imaging and remotes sensing data acquisition technologies as they apply to critical infrastructure management projects. It will also explore the results of the use of machine learning softwares in the analysis and cataloging paved and unpaved asset distress.

INTRODUCTION:

Highways have long been considered critical economic and strategic infrastructure to state and federal institutions. From the transportation of goods to providing a means for people to safely and efficiently move from place to place, they impact our lives on a daily basis. Since the inception of the United States Highway system in the 1920s to the evolution of the Interstate Highway System in the 1950s, they have become a mechanism of economic growth and a source of national pride both in the United States as well in other developed nations in the world.

Generally, attention is given to paved highway assets in developed nations and populated regions or connected populated areas. While paved highways may demand attention, unpaved roads and highways are just as critical. Worldwide over fifty percent (57%) of road length is unpaved and in the United States unpaved roads account for approximately thirty five percent of road length. Unpaved highway assets are critical in more rural and remote areas of the United States and the developing world. With road density lower in these areas these unpaved assets can be the only physical link these communities have to the rest of the world. As such, their development and management are an important part of financial resource planning for nations (developed or developing) and communities.

The management of these unpaved assets is beset by a number of challenge and unique conditions. Firstly, unpaved assets are, generally, a rough surface that makes traditional analysis difficult. The remote location and low density of infrastructure make mobilizing and reviewing these roads significantly more costly. Many of the roads are within rural to even extreme terrain meaning that they are exposed to environmental conditions such as washout, rockfall and other natural processes. Like traditional highways, the analysis and management of the road requires data from not just the road but the regions adjacent to the road. Collecting this data, given the characteristics noted above, can be difficult if not impossible.

Advances in imaging and remote sensing technology has allowed for the collection of data pertinent to the development and management of these assets, quickly, accurately and more completely. Advances in battery and communication technology have allowed for Small Unoccupied Autonomous Systems (sUAS or "drones") to become more accessible and reliable. This has allowed for a cornucopia of new applications for this technology. In a similar context, technological advancements have allowed for remote data collection technology to be reduced in size and weight. When this is coupled with the sUAS a powerful data collection tool is created. This article will focus on the application of remote sensing and sUAS based remoted sensing methodology to the analysis of both paved but specifically unpaved assets.

Spatial remote sensing techniques can be used to create accurate, three-dimensional, digital models of the current conditions of an object, space or area. Both passive (orthophotography) and active (light detection and ranging or "LiDAR") methods are available depending on site conditions and applications. Fielding these devices on a sUAS allows for increases in data density over traditional aerial methods as the above ground level (AGL) of the craft is lower. In addition, the sUAS are able to travel at lower speeds as well as hover and maneuver in smaller spaces. This makes them ideally suited for collection of data in remote and rugged terrain, typically associated with unpaved highways and assets. Additionally, the ability of LiDAR to penetrate vegetation allows for surface model creation in areas with tree cover or high grass. Both collection methods are limited to a "top-down" method of data collection. These payloads can be mounted on gimbals. With this orientation and by flying the sUAS in a vertical pattern imaging of vertical objects such as trusses, dams, and cliffs can be accomplished.

The sUASs abilities mentioned above also make it a possible (even preferable) platform for the fielding of geophysical systems for data collection. The decreased AGL allows for more subtle variations to be noted as well as providing reduced background interference. Closely spaced transects are able to be traversed autonomously using preprogrammed flight paths with limited delay between transect collection due to maneuverability. This also reduces variations in transects caused by human error during collection. Data is improved by decoupling the sensor or transport device from the ground where variations in the transect surface (rock, potholes, etc.) can cause noise as they are traversed. Obstacles can be avoided by both traveling around them or over them.

For spatial and geophysical data collection sUAS remote sensing allows for the gathering of data in areas that are difficult impractical or unsafe for traditional methods. In areas with steep slopes this method can eliminate the need for people to traverse the area, reducing fall risk and the risk posed by falling debris to the work crew or the public. This is true in areas of low lying terrain as well where the risk of uneven ground and wet or boggy conditions causing injury is reduced. Remote sensing has even been employed in aquatic environments. While autonomous watercraft exist even aerial sUAS can be used to suspend data collection devices (like echo sounders or sampling devices) in lakes, rivers and oceans, eliminating the need for people to venture into the water. In general, this technology allows for the user to remain at a safe standoff distance from the hazards, be it traffic or site conditions, that might cause injury or harm to themselves or others.

The development and accessibility of machine learning and artificial intelligence is poised to improve and hasten the science of geology and civil engineering. It will allow for efficient analysis of large amounts of data that can lead to a better insight into a site's history and potential future. While this technology is promising, using a computer algorithm to conduct an analysis on a physical object does have challenges. The main one is converting the physical object into a digital format that the algorithm can understand, i.e. how does a computer "see" the physical object. The high resolution of the remote data collections above allows for a digital model, or "clone", to be created. This model can then be used to train the algorithm to the users needs. By maintaining the spatial aspect of this data, the analysis can efficiently be applied back to the physical world. There are currently a number of softwares available for this type of analysis. Most are focused on structures and paved assets though there are those under development for the analysis for unpaved structures.

CASE STUDIES:

The following are case studies of where this technology has been applied. While some examples are in the use of this technology to highway geology, others are examples that are applicable though not directly conducted on highways or their adjacent areas.

Highway Realignment

Realignment of a highway resource can be a complex endeavor. As in general construction, the physical attributes of the site change on a daily, if not hourly, basis, not just due to intended events but also to unintended and natural events (accidents, floods, rockfall, etc.). Maintaining an awareness of the site conditions is critical for the project planning and for communication of the site activities. During a highway realignment, a sUAS mounted camera was used to collect imagery of the site on a periodic basis. These images were used to create orthomosaics as well as generate point clouds and digital terrain models (DTMs). Volume changes were computed by subtracting previous image sets from current image sets (Figure 1).

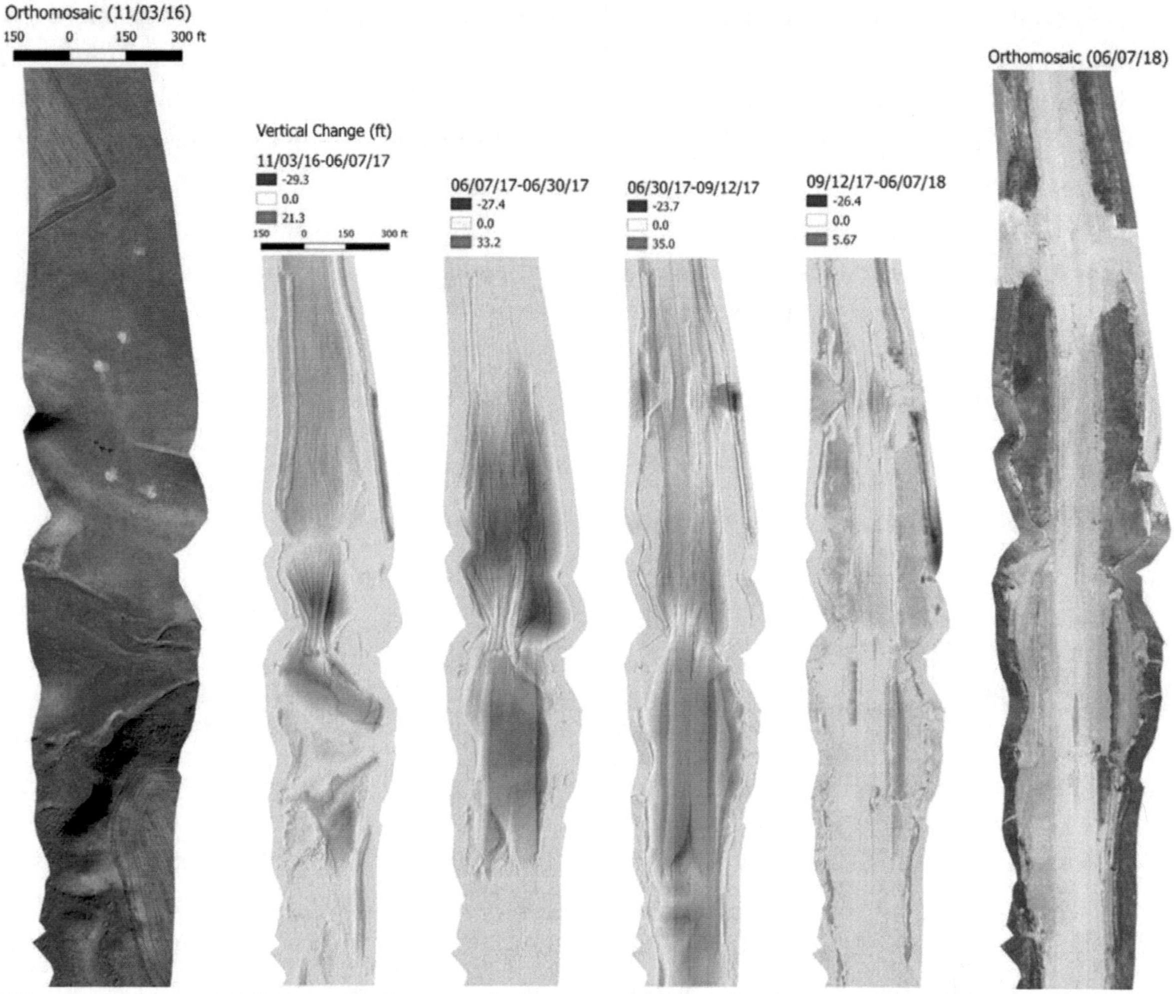

Figure 1: Shown left is an orthomosaic image of the site prior to start of the realignment construction. The images in the center are the temporal analysis of the site as the construction progressed. An image of the construction progress is shown to the right.

The remotely collected data was able to be shared with the project team in a digital format raising awareness of the project conditions. The images also served as a record of site conditions allowing for potential issues to be identified or for progress goals to be updated and communicated. The ability to quickly collect a visual of site conditions, without impeding progress, and share that with stakeholders allowed for improved project communication and planning of the project and the impact to the public.

Critical Infrastructure

The Chiniak Highway is located on Kodiak Island Alaska in the Pacific Ocean. This road is the only land access for residents of Chiniak with all other transport occurring by sea or air. Storms, fog and rough seas can limit the ability to travel by these methods, making the road the only viable option. It is therefore critical that it remains in good condition and safely operational. Numerous geohazards, primarily from coastal erosion, were identified and targeted by the Alaska Department of Transportation Public Facilities. To assess and monitor these hazards a sUAS was used to collect high resolution imagery and LiDAR over a number of seasons. Ground classified pointclouds, DEMs, and orthomosaics, were used to conduct a temporal analysis ("change over time") of areas identified as at risk by the State. Measures the vertical difference between the ground surfaces was identified with observable variations in the scale of inches. This analysis was highly effective for monitoring surficial changes across project sites and targeting areas for remediation (Figure 2).

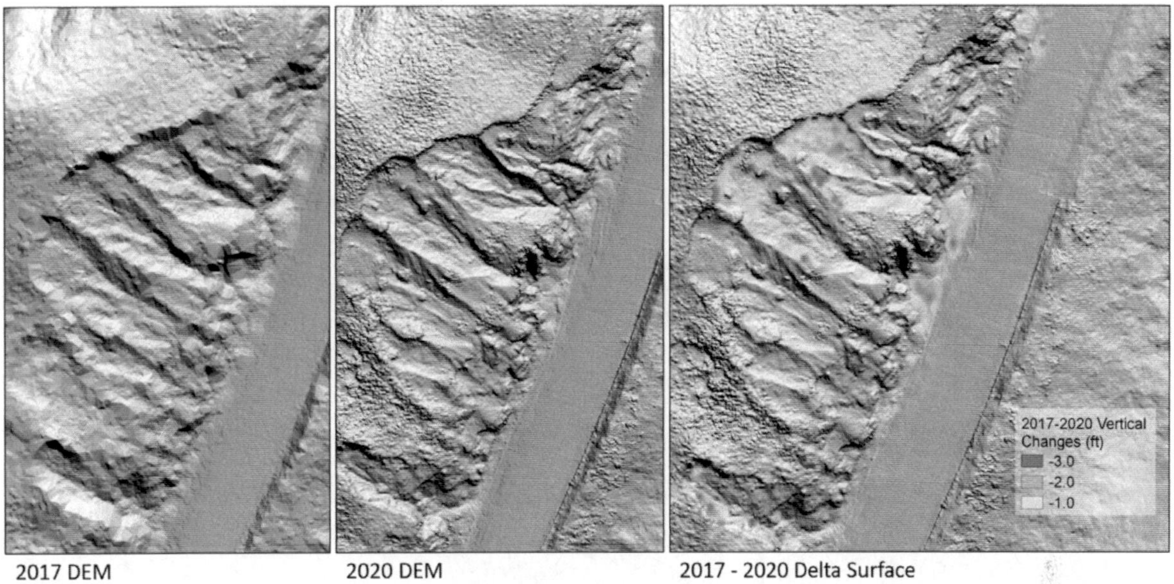

Figure 2: Using ground classified point clouds from two different periods measures the vertical difference between the two ground surfaces.

The sUAS with a two-person crew was able to collect data across the project sites that included terrain that would be difficult to traverse on foot. They were able to remain in a safe location, away from traffic, physical hazards as well as wild animals while the collection process was on-going. Further, the collection did not impact the road operation. The high resolution and accuracy of the data allowed for a relevant temporal analysis to be conducted. It also eliminated human bias in the locating of spatial data points to be collected. The difficulties of the remote location and difficult terrain were overcome by the use of sUAS and remote data collection methods that allowed for a large amount of high-resolution data to be collected quickly, accurately and safely.

Multimethod Digital Model

The spatial characteristics of the data collected with remote systems allows for multiple datasets to be combined into a single model, linked by the spatial data. This allows for the creation of models that show multiple attributes of a site and how they relate to one another.

The focus of this project was a building that was the site of a tea garden and social club in the 1920s-1930s that provided cover for an illicit, subsurface casino and host to other illegal activities at the time. The building was renovated in the 1950s into an administration building and cafeteria for a seminary college. As part of the building's illicit past, an "escape tunnel" is located in the subsurface on the western side of the building, leading from what was the casino (now a cafeteria kitchen) to a wooded area adjacent to the building. It was speculated that additional, un-documented escape tunnels existed as well. To determine the possible existence and location of additional escape tunnels as well as documenting the conditions of the historic site multiple remote sensing methods were used. The site was imaged with both sUAS photography as well as sUAS LiDAR. Terrestrial LiDAR was used to image the exposed interior of the historic structure as well as the interior of the tunnel. Both the sUAS and the terrestrial LiDAR were integrated into a single model that shows both surface and subsurface features. sUAS based geophysical data was collected from the site to identify variations that could indicate other potential underground tunnels. This data was integrated into the overall model.

Figure 3: Shown above is sUAS collected LiDAR of the site.

Figure 4: Complete surface (disseminated for visual purposes) and subsurface LiDAR data shown in a single model.

Figure 5: sUAS geophysical data merged with the spatial data from the LiDAR collections. The relationship between the tunnel and the geophysical response was clear.

The combination of models allowed the client to view, in a digital 3D space, both the surface and subsurface features in a single location. The relationship between the different datasets, such as the position of the tunnel in the subsurface and the variations in geophysical field surrounding it, provided insight into the interpretation of the geophysical results and the potential for other features in the subsurface. The use of remote sensing techniques allowed for collection of the data with minimal impact on the facility's operations. This type of insight is only possible when the data can be linked by spatial attributes.

CONCLUSION:

Remote sensing technology allows for the collection of data quickly, accurately and more completely. sUAS have become easier to operate as well and decreased in cost, make their use more ubiquitous in these applications. The creation of models that integrate data from multiple sources allows for a more complete understanding of a site. The digital nature of this data collection method allows analysis using machine learning and artificial intelligence. Finally, by collecting data from a standoff distance site safety is increased and the impact to the area minimized.

Advanced Multi-Sensing Techniques for Geotechnical Asset Management

Rakesh Salunke, Ph.D.
Postdoctoral Research Scientist
Department of Civil and Environmental Engineering
Jackson State University
1400 J.R. Lynch Street, JSU Box 17068, Jackson, MS, 39217-0168
(601) 979-0553
rakesh.salunke@jsums.edu
https://orcid.org/0000-0003-2466-1121

Sadik Khan, Ph.D., P.E
Associate Professor
Department of Civil and Environmental Engineering
Jackson State University
1400 J.R. Lynch Street, JSU Box 17068, Jackson, MS, 39217-0168
(601) 979-6373
j00797693@jsums.edu
https://orcid.org/0000-0002-0150-6105

Ian La Cour, P.E.
Geotechnical Engineer
Materials Division, Geotechnical Branch,
Mississippi Department of Transportation (MDOT)
412 E. Woodrow Wilson Ave., Jackson, MS 39216, USA
ilacour@mdot.ms.gov

Prepared for the 73rd Highway Geology Symposium, September, 2024

Acknowledgements

This study was supported by State Study 316, funded by the Mississippi Department of Transportation (MDOT) and the authors thank MDOT's continued support in their research efforts. The contents of this paper do not necessarily reflect the views of the funding agency. The authors would like to acknowledge their fellow members of the CREATE research team at the Jackson State University for their help with collecting some portions of the data used in this study.

ABSTRACT

Geotechnical asset management(GAM) is not mandated nationwide in the United States, and as a result, there is a lack of guidelines for practicing advanced techniques for implementing GAM. There is a growing need to incorporate advanced methods for temporal monitoring of surface deformation and subsurface anomalies of geotechnical assets. To this end, this study proposes a practical solution utilizing advanced non-destructive techniques, such as terrestrial Light Detection and Ranging (LiDAR) scanning, Uncrewed Aerial Vehicle (UAV) imagery, and Electrical Resistivity Imaging (ERI), for rapid and accurate characterization of geotechnical assets. Implemented within State Study 316, sponsored by the Mississippi Department of Transportation (MDOT), this research targeted highway embankment slopes prone to landslide failures. The geohazard susceptibility of geotechnical assets in the central MS region is exacerbated by extreme rain events and expansive clay soil prevalent in the region.

Over a two-year period, two representative highway embankment slopes in central Mississippi along Interstate-20 were temporally investigated using LiDAR, UAV, and ERI techniques. The study's findings demonstrated the efficacy of LiDAR and UAV technologies in capturing high-quality images and dense point cloud data, facilitating georeferenced topography and surface profile generation. Comparing the digital elevation models from periodic UAV photogrammetry and LiDAR surveys detected centimeter-level changes and surficial movements in the embankment slopes. ERI testing provided invaluable insights into subsurface anomalies and moisture-logged zones, alerting to potential future deformation risks and slide-slip surfaces. Combining these methodologies facilitated real-time asset performance data evaluation at the Geotech asset's surface and sub-surface levels, supporting proactive monitoring and preventive measures for asset managers. Results underscore the efficiency of LiDAR and UAV technologies in rapidly screening asset distresses, including slope movements, settlements, and deformations.

In conclusion, the study furnishes a practical methodology for proactive asset characterization, integrating advanced non-destructive techniques to bolster highway infrastructure safety and resiliency. These findings provide valuable insights for transportation agencies to refine asset management practices and mitigate geotechnical risks. Although this study focused on highway embankments and slopes as representative geotechnical assets, the developed methodology can be adapted for other geotechnical assets with appropriate adjustments.

INTRODUCTION

Traditional in-situ evaluation methods are outdated and expensive, leading to unexpected asset failures. Implementing warning systems and frequent geo-structure health monitoring is crucial to mitigate risks on highway slopes and embankments. Manual inspections and in-situ instrumentation base dinspection methods lack spatial data, and run the risk of landslides and asset failures going undetected. Frequent geo-structure monitoring is essential for risk mitigation. Remote sensing is increasingly popular for infrastructure monitoring, but boots-on-the-ground inspections are still the norm to identify vulnerable geo-structures after extreme events. However, these manual inspections are costly and time-consuming, which can delay prompt action. Additionally, conventional asset management typically involves creating inventories or databases by conducting physical condition surveys on individual assets (*1*).

Advanced near-surface remote sensing techniques like light detection and ranging (LiDAR) and uncrewed aerial vehicles (UAV) based photogrammetry provide high-quality geospatial and temporal topographic data for assessing geotechnical assets. Repeated LiDAR surveys enable timely tracking changes, ensuring high-quality inventory (*2,3*). Periodic investigations using drones inform the status of slopes with high fidelity (*4*). Highway and road slopes can be reliably inspected using UAVs and inform about risks from landslides (5) and rock falls (*6*). Utilizing point cloud data from LiDAR and high-resolution drone images allows for investigating and evaluating critical areas like geotechnical assets (*7,8*). Temporally spaced LiDAR point cloud surveys at the exact location over time help identify subtle changes in geotechnical assets and enable risk mapping (*2*). LiDAR offers accurate topographical information and Digital Elevation Models (DEM) (*9*) that help mitigate risks by warning about high landslide risk areas (*10*).

While LiDAR and UAV-mounted sensor technologies are excellent for high-quality surficial information, electrical resistivity imaging (ERI) is an excellent alternative for non-destructive subsurface investigation. ERI can effectively determine soil moisture variation, deformation, voids, conductivity, soil composition, failure depth, and slope surface stability (*11–14*). While traditional borehole sensors have their merits in point monitoring, they have limitations, such as limited spatial coverage, reduced sensitivity, and susceptibility to environmental factors (*14,15*).

This study implemented advanced multi-sensing technologies for comprehensively characterizing geotechnical assets required for a robust geotechnical asset management (GAM) program. 3D laser scanning with LiDAR equipment was employed to obtain dense point cloud data of critical highway embankment side slopes. Subsequently, topographical surfaces were generated, and surface profiles were created for comparative analysis of failed and undamaged areas of the slopes. The subsurface condition of the slopes was studied using the ERI technique. The surface profiles were compared with resistivity imaging profiles collected during the same period. The results revealed that LiDAR and UAV-based surficial monitoring can detect potential issues on the surface, especially when there are underground anomalies such as excess water-logged zones. Georeferencing the DEMs enables the creation of baseline models, allowing subsequent investigations and slope scanning to compare and identify any subsequent deformations. Finally, using the findings of this study as a guide, the rapid characterization of geotechnical assets process flow is outlined. Parallels can be drawn from the proposed approach to rapidly characterize any type of geotechnical asset. By implementing these practices, transportation agencies can proactively prevent significant performance degradation and associated impacts, ensuring the safety and long-term sustainability of the geotechnical assets.

METHODS

Surficial and subsurface characterization of geotechnical assets was carried out using advanced multi-sensory tools, UAV, LiDAR, and ERI. The multifaceted data was collected from two sites over two years between 2021 and 2022. The temporal variation of the surficial topography and subsurface tomography were evaluated.

Site Description

Two representative embankment side slopes along I-20 were selected for this study, as presented in Figure 1. Slope 1 is located along eastbound I-20. It is a low-pitch slope (5.5 H:1V) with an elevation difference of about 23ft between the toe and crest of the slope. It has experienced shallow slide failures at the crest and the middle of the slope.

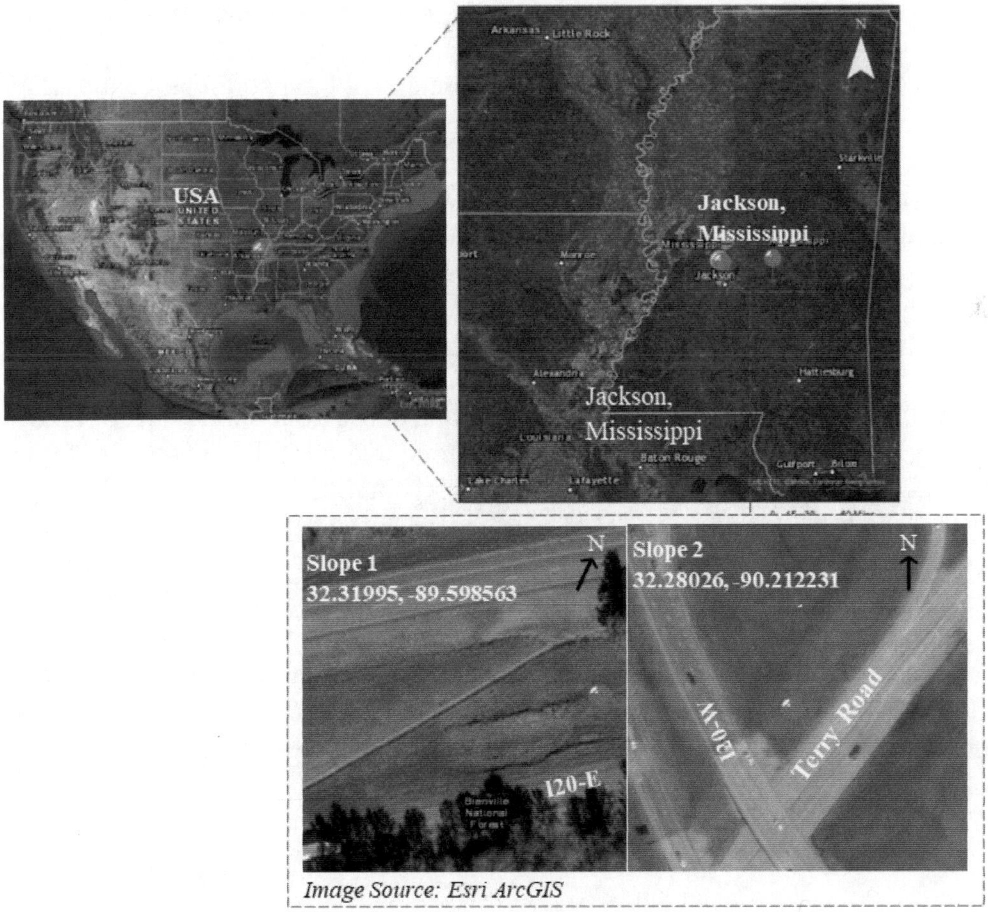

Figure 1 - Study Slope Locations Around Jackson MS

Slope 2 is 15 ft. high with a grade ranging between 3.5H: 1V to 4H: 1V. It is situated along the bridge exit embankment along westbound I-20. The slope had previously encountered a shallow landslide at the south end close to the bridge abutment, which was remediated by reinforcing the soil with steel piles.

Advanced multi-sensing techniques

UAV Technology

Uncrewed aerial vehicles (UAVs), also known as drones, offer a fast and accurate way to collect high-resolution image data, making them a valuable tool for inspecting and assessing geotechnical assets. This study used a drone mounted with high-resolution RGB and thermal cameras to capture images. The UAV flight missions were carried out at different seasons in tandem with ERI and LidAR surveys. The data collection and processing procedure for rapid landslide characterization using UAV technology is outlined in Figure 2.

Ground control points were established with the help of the Mississippi Department of Transportation (MDOT). The collected images and GCP text files were processed, aerial triangulation process was employed with multiple tie points to stitch the images together and generate digital elevation model (DEM) and digital terrain models (DTM). Ground control points (GCPs) with identified coordinates from the field were utilized to georeference the DEM. The state plane coordinate system for Mississippi was used to georeference all images, allowing the superimposition of DEMs, creation of surface profiles, and change detection. Subsequently, the images were orthorectified and stitched to create an orthomosaic representation of the entire slope site, with 3D models developed for specific seasonal surveys.

LiDAR technology

Terrestrial 3D laser scanning was performed and dense point cloud data of highway slopes experiencing surficial deformations were captured overtime. The procedure for using LiDAR technology for rapid landslide characterization is outlined in Figure 2. The laser scanner was placed at several stations along the slope surface, and several point clouds were captured and later registered together to form a large point cloud of the entire slope surface. Due to the open area terrain, spherical targets attached to poles were placed strategically so that at least two targets were common to each scanning station. This arrangement allowed for easier registration of point clouds by picking two common points between each scanning station. Once registered, the point clouds were processed to extract surface profile informatioin.The UAV and LiDAR data collection and processing workflow is outlined in Figure 2.

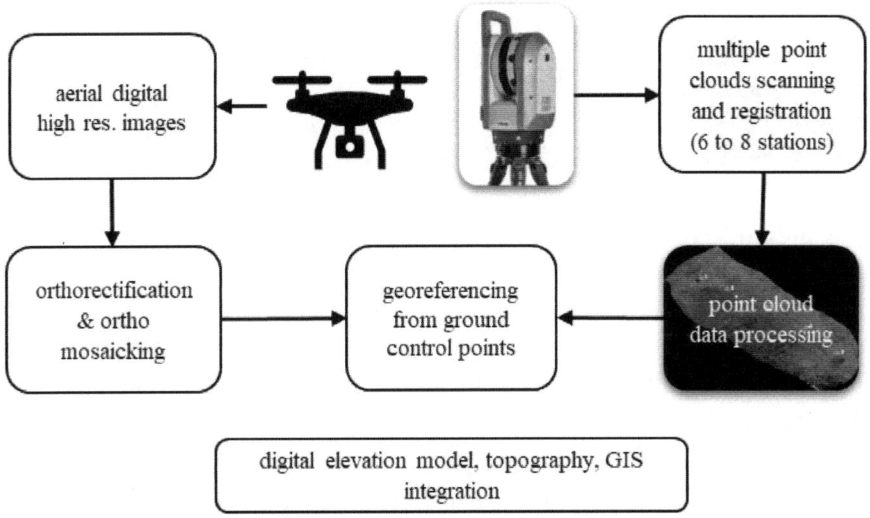

Figure 2 - LiDAR and UAV Workflow For Surficial Characterization Of Geotechnical Asset

DEM alignment and profile creation

The point cloud was georeferenced using ground control points with known MS state plane projected coordinates, allowing for superimposition of the point clouds collected at different seasons. The raw point clouds were processed by segmenting the cloud and eliminating unwanted points at the edges of the scan. Then, a ground extraction algorithm was used to eliminate vegetation and other unwanted above-surface points, and only bare ground points were extracted. The point clouds were dense, making the surface creation process computationally expensive. Therefore, the point clouds were sampled to increase the minimum distance between points to 0.1ft (0.03m) ~ 0.2 ft. (0.06m). Surfaces were generated from the sampled point clouds using the Triangulated Irregular Network (TIN) method. The topographic surface views were generated at 1 ft. major and 5 ft. minor contour intervals.

The georeferenced point clouds and surfaces were geolocated using the MS state plane coordinate system. The topographic surfaces of were superimposed, which was feasible due to the careful georeferencing of the point clouds. Multidirectional alignments were created along the superimposed surfaces in the direction of the slope and the transverse direction. Surface profiles along the downward direction of the slope and in the transverse direction were generated. The slope surface profiles at different seasons were compared to identify any variations. The LiDAR data collection and processing procedure is outlined in Figure 3. Comparative analysis of surface profiles and resistivity imaging data provided insights into soil movement with changing weather patterns and subsurface moisture.

ERI Technology

ERI is a non-destructive technology for investigating soils, providing rapid assessment of larger subsurface areas. It measures electrical resistivity to offer valuable insights into geological structures, moisture distribution, and other parameters without disturbing the subsurface soil. Numerous studies have highlighted the effectiveness and advantages of ERI in different geotechnical applications (*16–19*)

Gallipoli et al. (2000) (*20*) used ERI to detect regions impacted by landslides in southern Italy. ERI can reliably provide subsurface information, efficiently cover large areas, offer cost-effectiveness, identify site heterogeneity and areas with high moisture content, and enable fast data processing.

ERI investigations along two lines of the slopes were accompanied with the LiDAR and UAV surficial surveys. ERI surveys captured the subsurface resistivity distribution data. Then ERI inversion was carried out which is a process of inverting the resistivities through an iterative process based on least-squares inversion. Field-measured apparent resistivities were compared to calculated resistivities, and were adjusted iteratively until the root mean squared (RMS) error falls to a low acceptable value close to 5 to 10%. The typical ERI workflow followed in this study is explained in Figure 3.

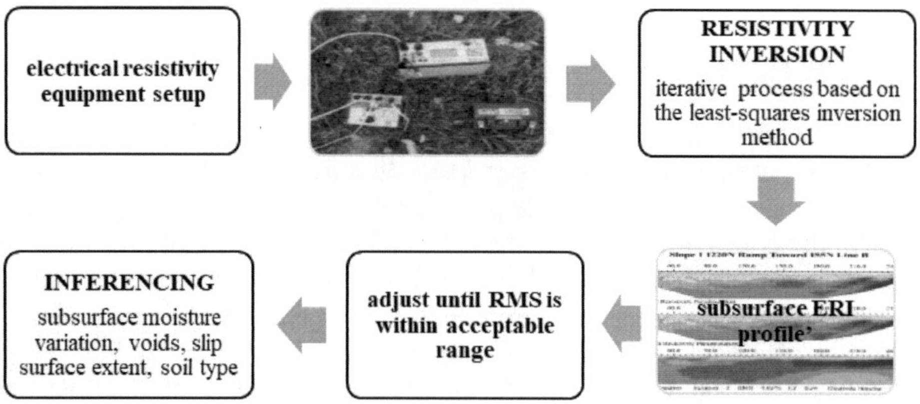

Figure 3 - ERI Workflow for Subsurface Characterization of Geotechnical Asset

RESULTS

Slope 1 Monitoring Results

UAV photogrammetry results

Drone survey missions were conducted during the fall of 2021 and fall/winter of 2022 over slope 1. The captured aerial imagery was meticulously processed as described in the methodology section to ensure accurate results. The DEM and orthomosaic representations for Fall 2021 and Fall/Winter 2022 can be observed in Figure 4 (a) and (b). The fall 2021 orthomosaic image and the DEM in fall 2021 (Figure 4a) show cascading slide failures at the crest and middle of the slopes. The UAV imagery-based DEM developed in the fall of 2022 indicated that the landslide failure had widened from 158' up to 222', spanning horizontally across most of the slope width. Furthermore, from Figure 4 (b) , In Fall 2022, a new layer of heaving soil can be observed between the two cascading layers of slides that didn't exist in the prior year. The 3D model representations and surface profile views are presented in Figure 5. The surface profile was obtained by following the same surface creation procedure explained in the methodology section of this paper. In addition to the contour views and surface profiles, 3D model representations of the slope were developed.

LiDAR results

Terrestrial LiDAR scanning data at the slope 1 site were collected during the fall of 2021 and fall of 2022. For fall 2021 data, ground control points were not available, leading to a non-georeferenced point cloud. Nevertheless, the DEM generated for Fall 2021 was superimposed by overlapping the fixed points at the external boundary of the landslide and active movement area. The topographic surface views for Fall 2021 and Fall/Winter 2022 were generated at 1 ft. major and 5 ft. minor contour intervals, visually depicting the terrain's elevation changes over time. Figure 6 (a) displays the topographic surface views from point clouds obtained from 3D laser scans in fall 2021 and 2022. Additionally, surface profile views for Fall 2021 and Fall/Winter 2022 were created. Figure 6 (b) and (c) present the surface profile views, offering valuable insights into the slope's cross-sectional variations during the two seasons.

ERI results

ERI tests were conducted along two lines, A and B, on the failed slope 1, as displayed in Figure 7. The ERI tests covered a length of approximately 270 ft. with electrodes spaced at 5 ft. intervals. The ERI testing results for Lines A & B are presented in Figure 8(a) and (b). For Line A, the results show the presence of high resistivity layers up to approximately 15 ft. depth, spanning the horizontal distance between 30 ft. and 230 ft. These high resistivity layers are indicative of deformed soil. The soil deformation appears to have been due to creep-induced movement.

Additionally, the ERI results for Line A reveal low resistivity areas directly below the loose soil, suggesting the presence of perched water conditions resulting from rainfall infiltration. These perched water zones may have contributed to slope instability and further slope movement by weakening the soil and triggering localized instabilities. Similarly, Line B exhibits a similar ERI pattern but with lower resistivity values than Line A. This lower resistivity may indicate different soil properties or variations in the extent of soil deformation and perched water conditions along this slope section.

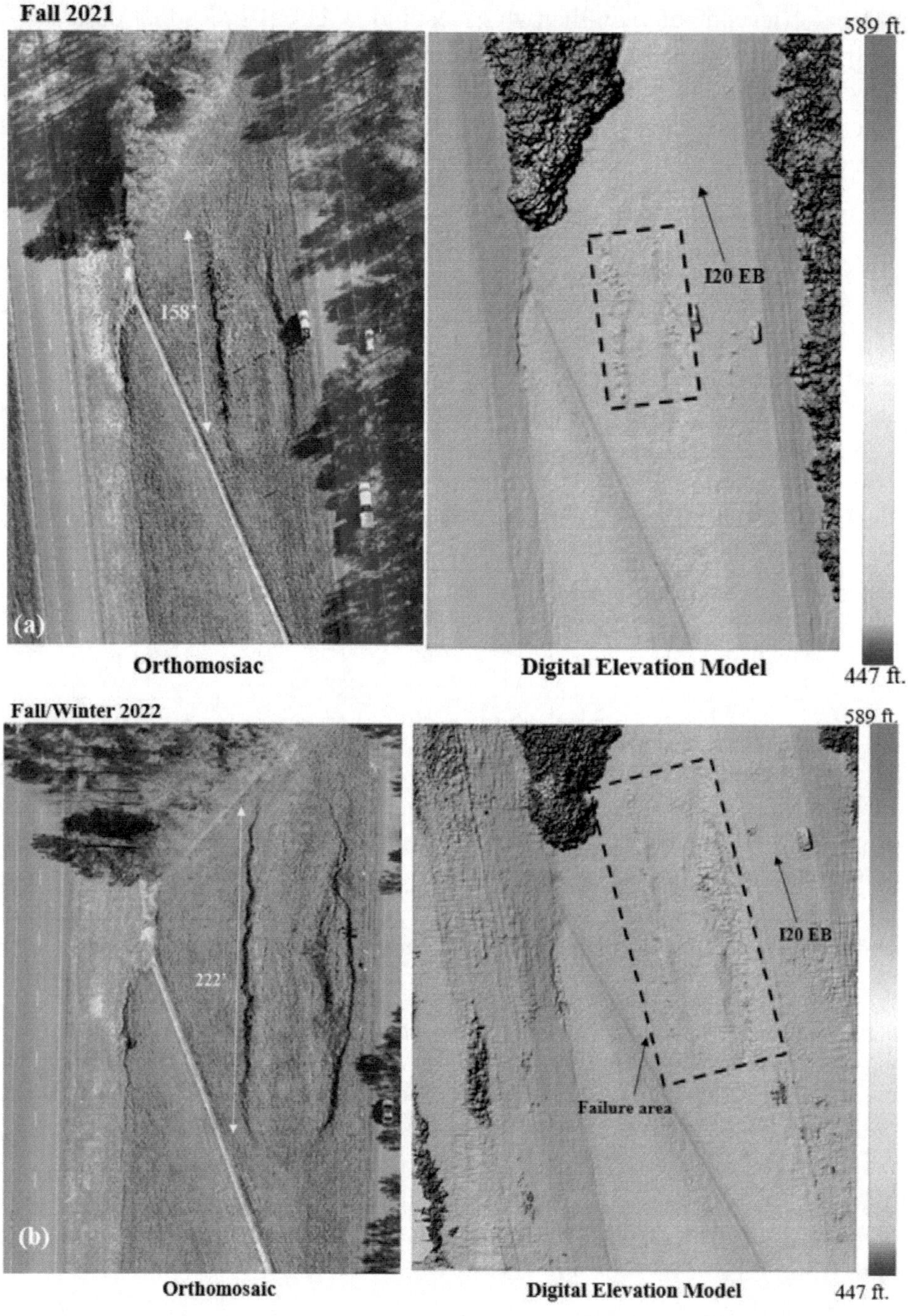

Figure 4 - Slope 1 Orthomosaic & Digital Elevation Models (a) Fall 2021 (b) Fall/Winter 2022

Figure 5 - Slope 1 3D Model & Superimposed Surface Profile (Fall 2022)

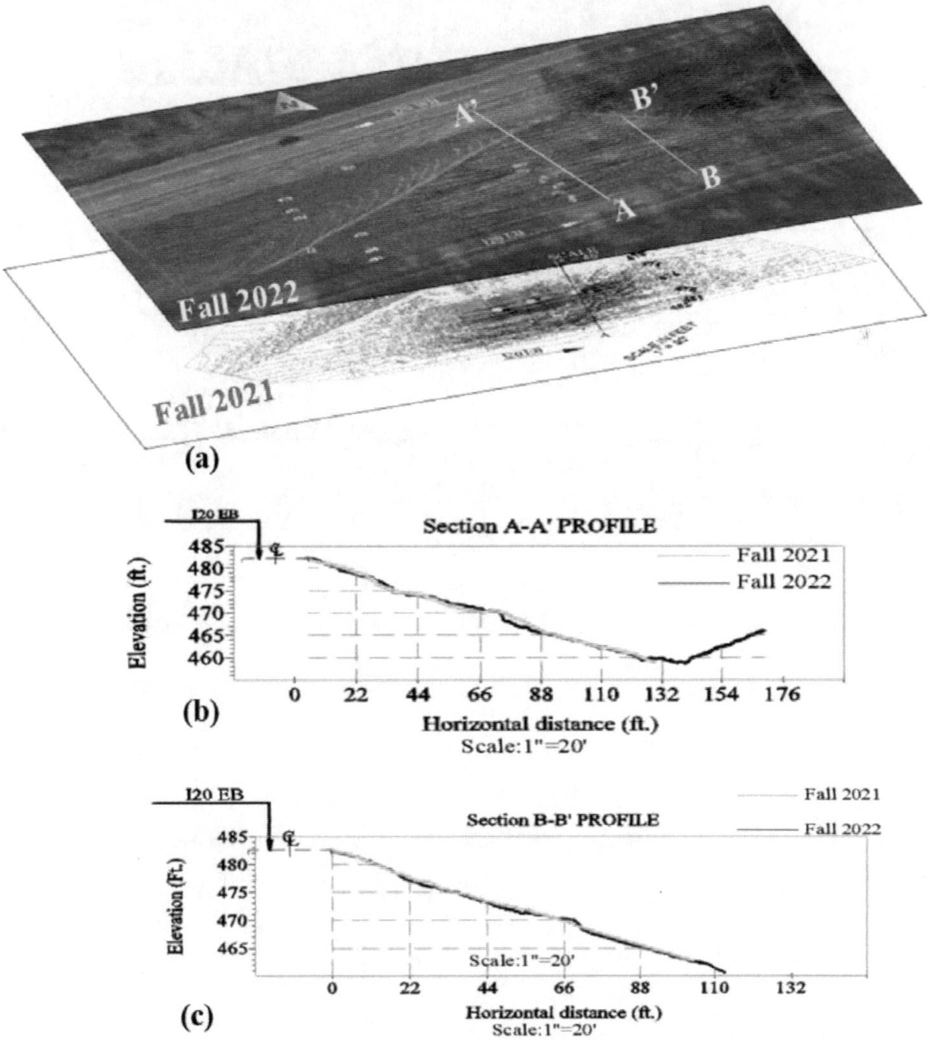

Figure 6 - Slope 1 LiDAR Point Cloud Results (a) Surface Topographies Of Different Seasons (b) Stacked Surface Profile A-A' (c) Stacked Surface Profile B-B'

Figure 7 - ERI Test Lines at Failed Slope 1

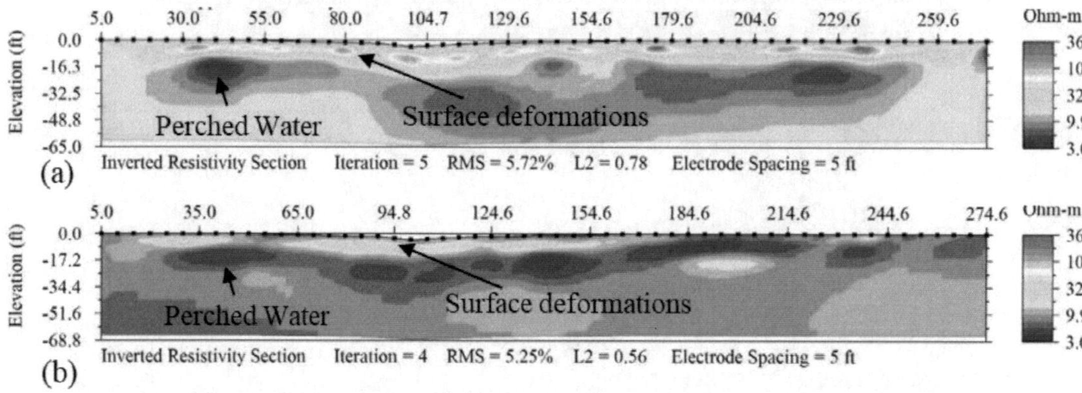

Figure 8 - ERI Test Results of Failed Slope 1: (a) Line A (b) Line B

Slope 2 Monitoring Results

UAV photogrammetry results

A comprehensive drone survey was conducted at slope 2 during various seasons, spanning from summer 2021 to fall 2022. The survey yielded valuable data, leading to the creation of digital elevation models (DEMs) and orthomosaic digital image representations for each season. The photogrammetry outputs for Summer and Fall 2021 are presented in Figure 9 and Figure 10, respectively. The overlapping imagery data captured during fall 2021 and fall 2022 helped develop good representative 3D models and point clouds, as depicted in Figure 11 and Figure 12. The procedure for creating the surface profiles using drone imagery-based point clouds was similar to that for LiDAR point clouds.

LiDAR results

Following the surface creation procedure outlined in the methodology section, topographic surface views were generated at 1 ft. major and 5 ft. minor contour intervals. Figure 13 (a-d) portray the topography of slope 2 developed from Terrestrial LiDAR scanning data collected during different seasons, spanning from summer 2021 through fall/winter 2022. The topographic surfaces of slope 2 created seasons were superimposed, allowing for the exploration of the surface profile variations over time. The stacked surface profiles spanning from summer 2021 through fall/winter 2022 are presented in Figure 14.

Recalling that slope 2 was reinforced with pipe at the crest close to the bridge abutment, and the adjacent areas were built without any reinforcements. Surface profile results in Figure 14a showed that the reinforced slope area closer to the bridge abutment is stable and not experiencing any significant movement over time. This is indicative of the fact that the H-pile stabilization of the slope is indeed working as designed. However, at the unreinforced area of slope 2, hints of surficial deformations and elevation changes were found, as displayed in Figure 14.

ERI results

ERI testing results for slope 2 provided valuable insights into the subsurface conditions and soil movement over time. The testing was conducted along a 165-foot (50-meter) length of the slope's crest and middle (lines A and B), as depicted in Figure 15. Figure 16 (a - e) present the ERI results obtained from testing Line A during various seasons from spring 2021 through fall/winter 2022. Similarly, Figure 17 (a - e) illustrates the ERI results for Line B for the same period. The ERI results revealed the presence of a high resistivity zone at shallow depths up to a depth of 6 ft at the left side of the test lines closer to the bridge abutment. This zone of high resistivity is observed both at the crest (Line A) and the middle (Line B) of the slope between the 5 ft. (1.5m) horizontal distance mark up to 54 ft. (16.5m) mark, which can be observed in Figures 18 & 19.

Interestingly, beyond the 6 ft. depth, resistivity drops significantly to extremely low levels, measuring less than 5 Ohm-m. This low resistivity zone extends across the crest and middle of the slope, just below the unsaturated soil zone, specifically at the 54 ft. (16.5m) horizontal distance mark. The high resistivity is usually a sign of loose soil particles with larger air voids that resist electric currents. Therefore, the slope area needs further monitoring for signs of land movement. Furthermore, the low resistivity zones identified in the ERI results under the potentially deformed soil suggest the presence of perched water zones. As water enters the loose soil, it gets trapped in the highly impermeable clay mass, creating perched water zones. The perched water zones create a bathtub effect, enabling a slipping surface for the soil above and causing landslides and associated hazards. The last two test results for Spring and Fall 2022 (Figure 16 d&e, and Figure 17 d & e) indicate that the perched water zones expanded closer to fall/winter 2022 in the 8 to 25' depths, which is a cause for concern.

The high resistivity areas have stayed uniform over time and have not changed, which is encouraging and does not trigger immediate maintenance efforts. Even though LidAR and drone results for slope 2 were inconclusive of any damage, the ERI subsurface investigation provided valuable insights to put this slope on further monitoring status.

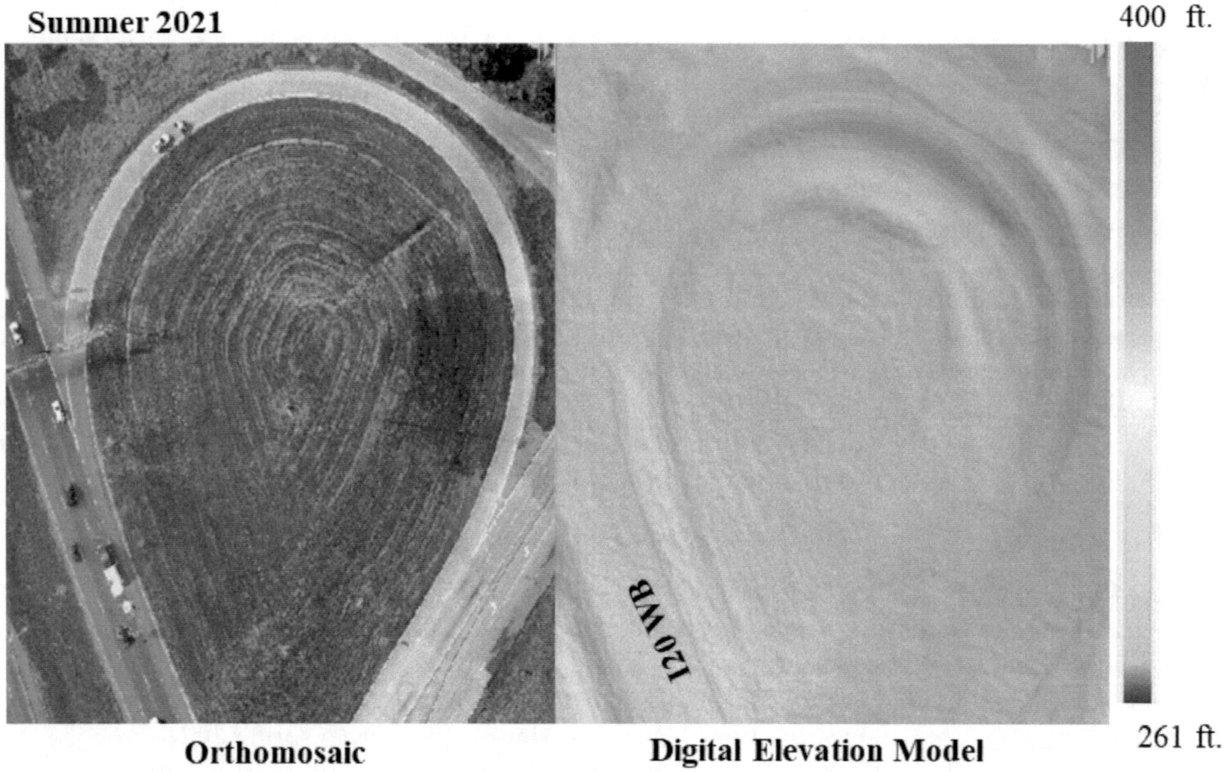

Figure 9 - Slope 2 Digital Elevation Model from UAV Imagery in Summer 2021

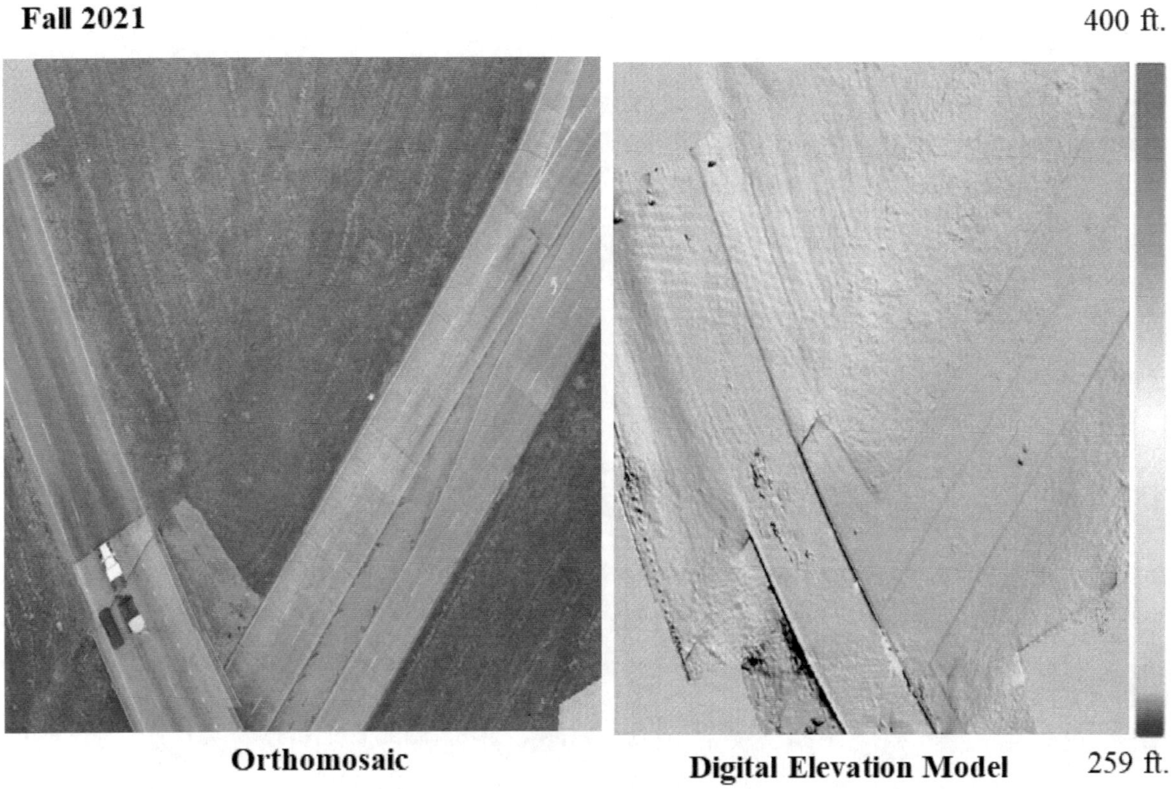

Figure 10 - Slope 2 Digital Elevation Model from UAV Imagery in Fall 2021

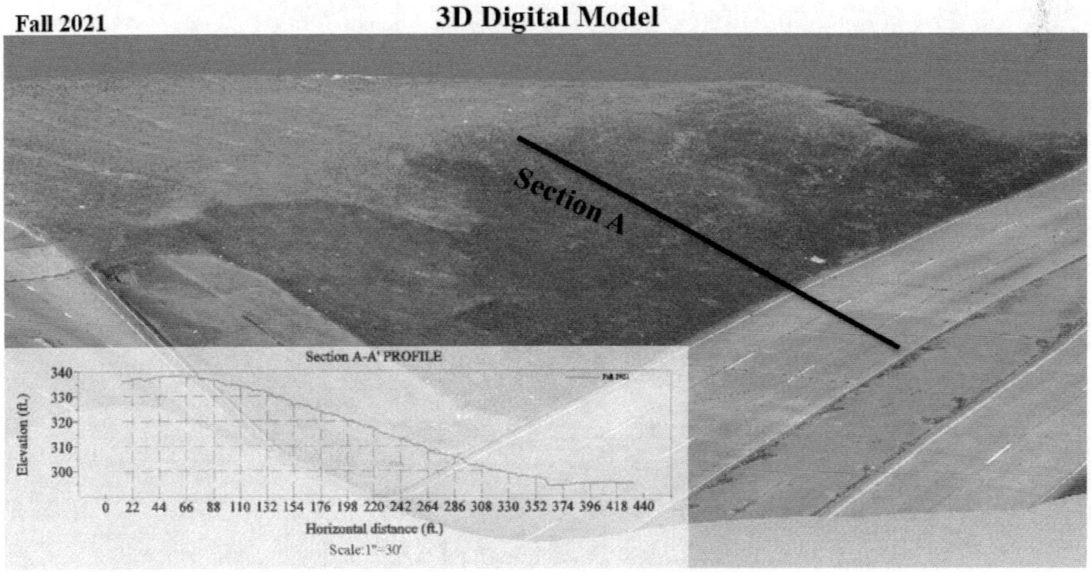

Figure 11 - Slope 2 3D Model from UAV Imagery in Fall 2021

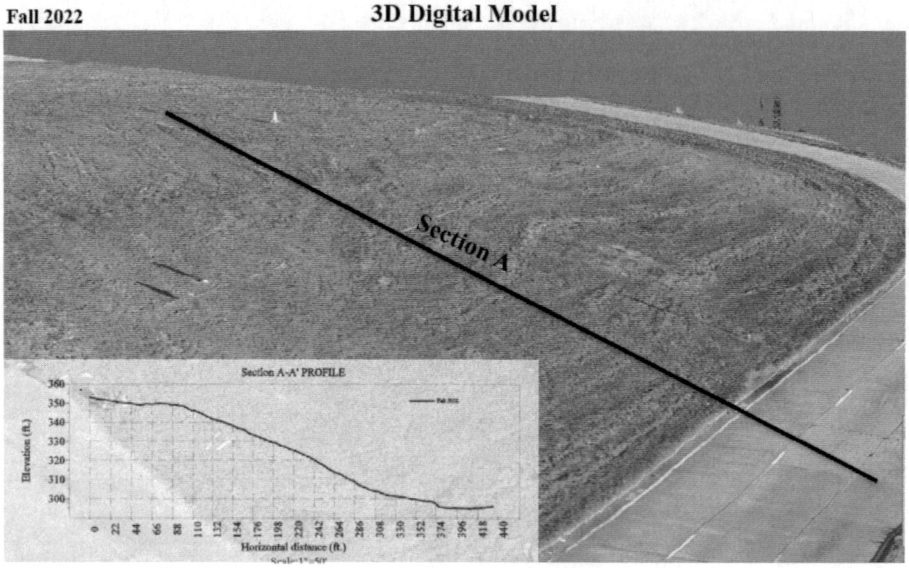

Figure 12 - Slope 2 3D Model from UAV Imagery in Fall 2022

Figure 13 - Slope 2 LiDAR Point Cloud Surface Topographies at Different Seasons: (a) Summer 2021, (b) Fall 2021 (c) Spring 2022 (d) Fall 2022

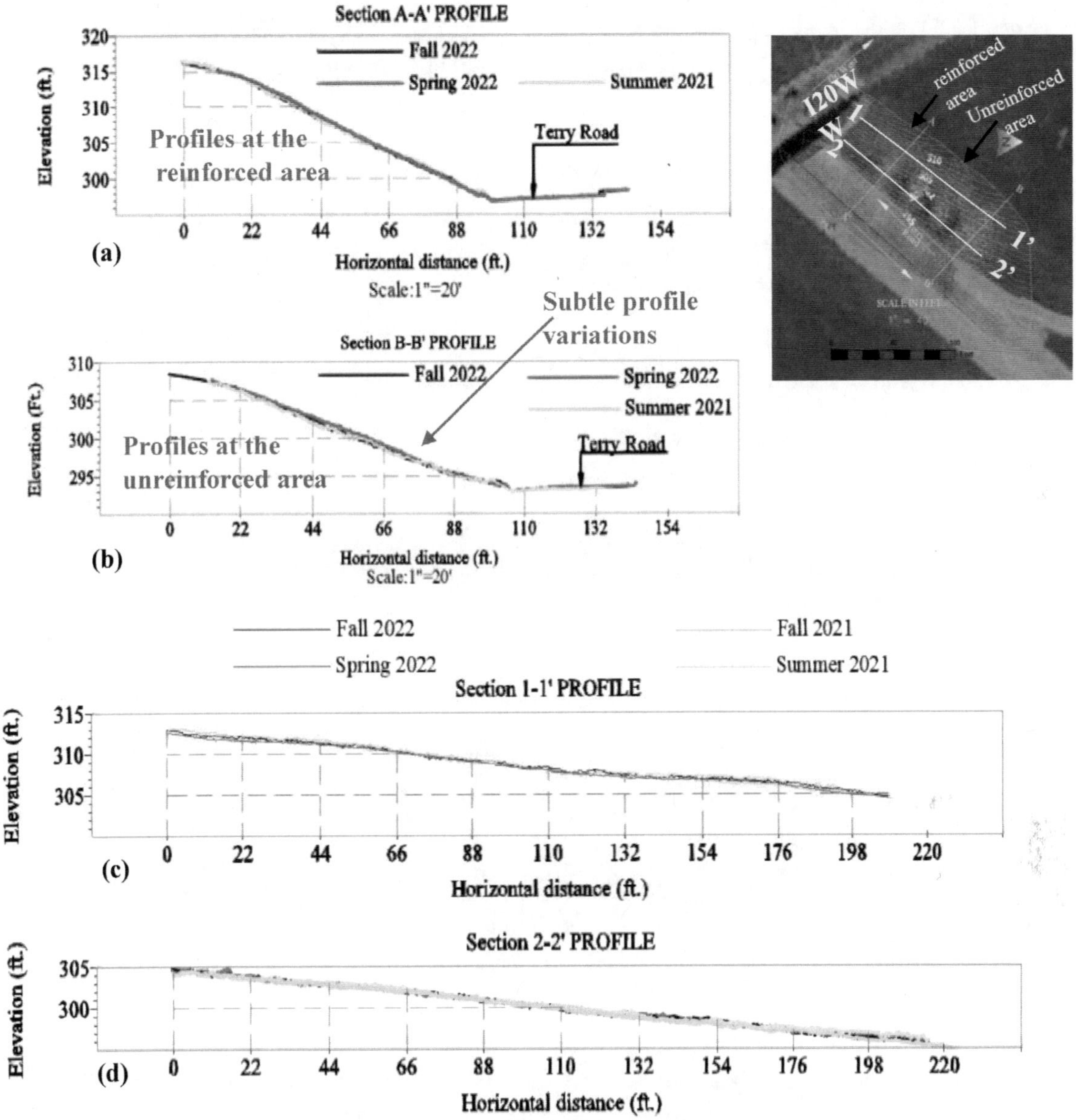

Figure 14 - Seasonal Variations in Slope 2 LiDAR Point Cloud Surface Profiles: (a) Slope Profile Section A-A', (b) Slope Profile Section B-B' (c) Transverse Profile Section 1-1' (d) Transverse Profile Section 2-2'

Figure 15 Slope 2 ERI Test Lines

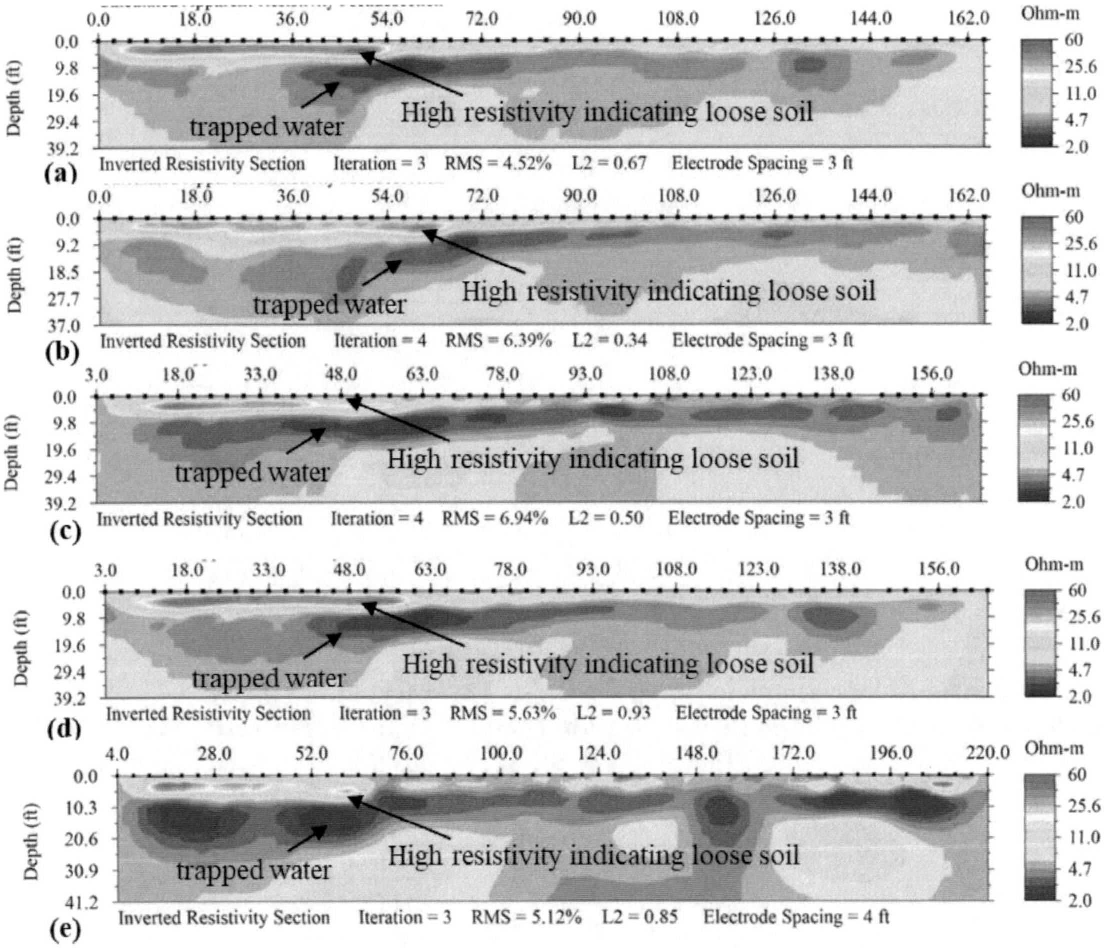

Figure 16 ERI Line A of Slope 2: (a) Spring 2021, (b) Summer 2021, (c) Fall 2021, (d) Winter 2021/Spring 2022, (e) Fall/Winter 2022

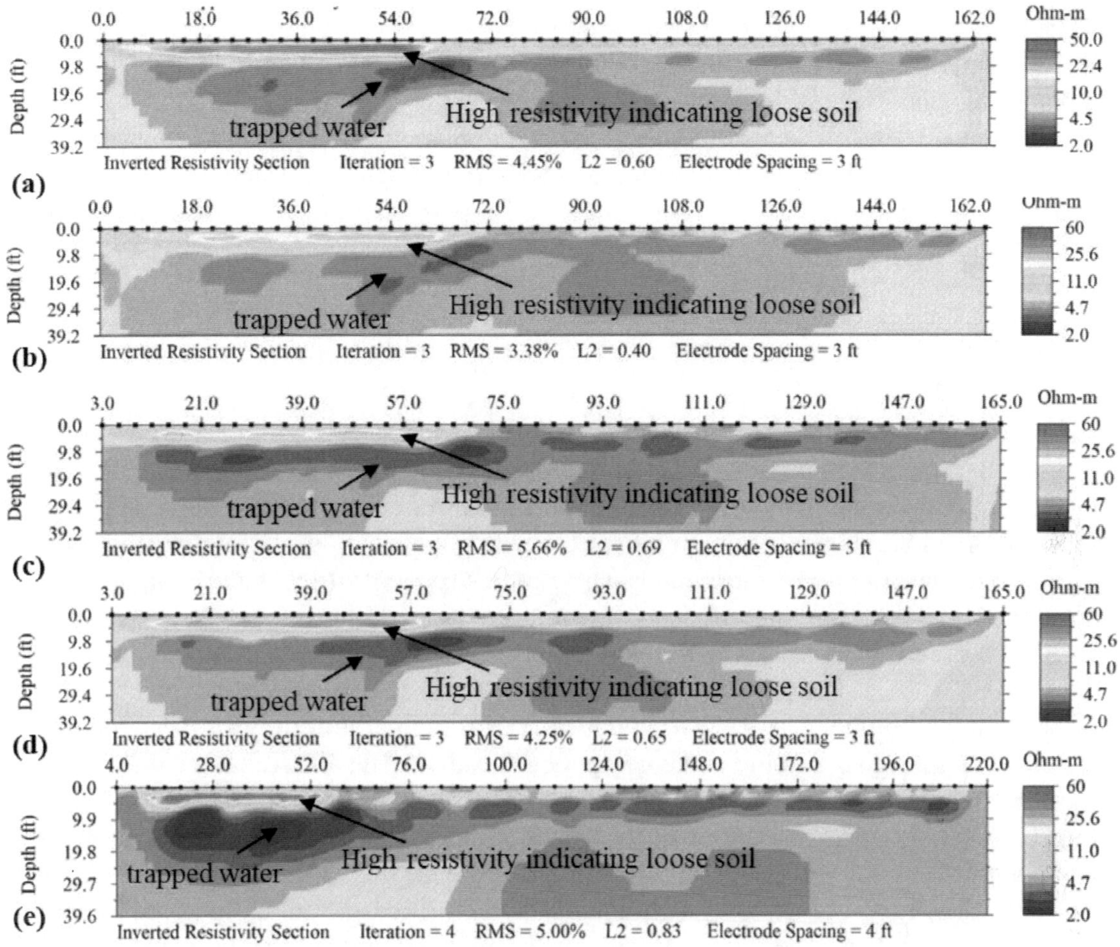

Figure 17 19 ERI Results Line B of Slope 2: (a) Spring 2021, (b) Summer 2021, (c) Fall 2021, (d) Winter 2021/Spring 2022, (e) Fall/Winter 2022

DISCUSSION

The resulting evidence from periodic investigations of the slopes in this study affirms that LiDAR and UAV serve as fitting technologies for initial rapid screening for landslides, slope movements, settlements, and soil deformations in general. Slope 2 results proved that although UAV and LiDAR results show a stable slope, the subsurface tomography from ERI showed signs of future issues due to perched water creating slide slip surfaces underground.

The proposed approach can be further strengthened by an interdisciplinary evaluation of the results of rapid characterization, combining expertise from geotechnical engineers, transportation engineers, government agencies (such as DOTs), geologists, and other relevant experts. This collaboration aims to interpret the collected data comprehensively and assess the stability of the slopes or any given geotechnical asset. Potential hazards and geotechnical risks could be identified based on the information gathered during the evaluation process.

The authors advocate integrating the proposed characterization of geotechnical assets using advanced multi-sensory tools into a comprehensive geotechnical asset management (GAM) plan. This plan will guide the framework for addressing slope-related challenges and potential emergencies. It will establish clear protocols to respond effectively to changes in slope behavior and unforeseen events. Regular monitoring of implemented measures would help gauge their effectiveness and update the GAM as needed to adapt to changing conditions and ensure the long-term stability of the assets.

CONCLUSION

This study leveraged multi-sensor technology tools to establish an innovative rapid characterization of Geotechnical Asset conditions, specifically focusing on highway slopes subject to landslides. Using 3D laser scanning with LiDAR equipment allowed the acquisition of high-density point cloud data from an embankment with a history of surficial deformations and failures. This data generated topographical surfaces and surface profiles, facilitating a comparative analysis between the failed and undamaged areas of the slopes. Additionally, the study correlated the surface profiles with resistivity imaging profiles collected simultaneously.

The findings from the various investigations performed at the study slope sites (Slopes 1 and 2) offer crucial insights into the slope behavior and subsurface conditions over time.UAV photogrammetry, LiDAR, and ERI testing provided valuable data for assessing the evolving conditions of slopes 1 and 2. The drone surveys captured the slope's changing topography, showcasing significant voids at the slope's toe and near the bridge abutment. The LiDAR results revealed stable conditions at the slope section close to the bridge abutment, indicating that the installed piles have worked to stabilize the previously failed slope. At the unreinforced area of slope 2 however, hints of surficial deformations and elevation changes were visible.

The ERI findings of slope 2 indicated the depths of shallow slide slip surfaces. They identified perched water zones beneath the unsaturated soil zones, raising concerns about potential slope instability in the future. The drone surveys and LiDAR scanning provided detailed data for slope 2, showing stable conditions and the effectiveness of the pile stabilization system. The ERI testing revealed high resistivity zones indicative of loose soil particles resisting electric currents and low resistivity zones suggesting the presence of perched water. Slope 1 exhibited cascading slide failures and an expanding landslide failure from the drone and LiDAR results. The ERI testing indicated high resistivity layers at shallow depths, suggesting deformed soil, and low resistivity zones, indicating perched water conditions, both contributing to slope instability.

The results revealed that initial rapid characterization using LiDAR and UAV sensors provides enough information to detect potential geotechnical issues. The topographic and surface profile variations were compared with subsurface resistivity imaging profiles collected during the same period. This comparison revealed patterns between surficial and sub-surficial dynamics of highway slopes prone to landslides.

Overall, the combination of drone surveys, LiDAR scanning, and ERI testing has been instrumental in understanding the changing conditions and potential hazards associated with the monitored and failed slopes. These findings are vital for geotechnical assessments, engineering evaluations, and designing appropriate slope stabilization and remediation measures. By implementing these practices, transportation agencies can adopt a proactive approach to prevent significant performance degradation and mitigate associated impacts, ensuring the safety and long-term sustainability of geotechnical assets.

REFERENCES

1. Kimmerling RE, Thompson PD. Assessment of retaining wall inventories for geotechnical asset management. *Transp Res Rec*. 2015;2510:1-6. doi:10.3141/2510-01
2. Wolf RE, Bouali EHY, Oommen T, et al. *Deliverable 2-A: Candidate Remote Sensing Techniques for the Different Transportation Environments, Requirements, Platforms, and Optimal Data Fusion Methods for Assessing the State of Geotechnical Assets*.; 2014.
3. Nobahar M, Salunke R, Alzeghoul OE, Khan MS, Amini F. Mapping of Slope Failures on Highway Embankments using Electrical Resistivity Imaging (ERI), Unmanned Aerial Vehicle (UAV), and Finite Element Method (FEM) Numerical Modeling for Forensic Analysis. *Transportation Geotechnics*. 2023;40:100949. doi:10.1016/j.trgeo.2023.100949
4. Kader U, Altunel AO. *Monitoring the Stability of Highway Cut Slopes Utilizing Drone Photogrammetry*.; 2021. http://igd.mersin.edu.tr/2020/
5. Tan L, Deng S, Huang H, He R. Research on intelligent inspection of high slopes on Yuqian Road based on UAV. In: Mikusova M, ed. *International Conference on Smart Transportation and City Engineering (STCE 2022)*. SPIE; 2022:94. doi:10.1117/12.2658188
6. Yaacob MLM, Rasib AW, Majid Z, et al. Investigation of Rock Slope Stability Using Drone-Based Thermal Sensor. *Journal of Advanced Geospatial Science & Technology*. 2022;2(1):67-91. Accessed July 18, 2024. https://jagst.utm.my
7. Luo H, Paal SG. Machine Learning–Based Backbone Curve Model of Reinforced Concrete Columns Subjected to Cyclic Loading Reversals. *Journal of Computing in Civil Engineering*. 2018;32(5). doi:10.1061/(asce)cp.1943-5487.0000787
8. Marx A, Chou YH, Mercy K, Windisch R. A lightweight, robust exploitation system for temporal stacks of UAS data: Use case for forward-deployed military or emergency responders. *Drones*. 2019;3(1):1-11. doi:10.3390/drones3010029
9. Jaboyedoff M, Oppikofer T, Abellán A, et al. Use of LIDAR in landslide investigations: a review. *Natural Hazards*. 2012;61(1):5-28. doi:10.1007/s11069-010-9634-2
10. Hatta Antah F, Khoiry MA, Abdul Maulud KN, Abdullah A. Perceived Usefulness of Airborne LiDAR Technology in Road Design and Management: A Review. *Sustainability*. 2021;13(21):11773. doi:10.3390/su132111773
11. Jongmans D, Garambois S. Geophysical investigation of landslides : a review. *Bulletin de la Société Géologique de France*. 2007;178(2):101-112. doi:10.2113/gssgfbull.178.2.101
12. Hen-Jones RM, Hughes PN, Stirling RA, et al. Seasonal effects on geophysical–geotechnical relationships and their implications for electrical resistivity tomography monitoring of slopes. *Acta Geotech*. 2017;12(5):1159-1173. doi:10.1007/s11440-017-0523-7
13. Khan S, Amini PEF, Masoud Nobahar PE. *MDOT State Study 286-Performance Evaluation of Highway Slopes on Yazoo Clay FINAL REPORT*.; 2020. Accessed March 20, 2024. (Report No. FHWA/MDOT-RD-20-286). Mississippi Department of Transportation. https://rosap.ntl.bts.gov/view/dot/56232/dot_56232_DS1.pdf
14. Nobahar M, Salunke R, Khan S, Amini F. Development of Soil Moisture Content and Soil Matric Suction Model Based on Field Instrumentation and Electrical Resistivity Imaging (ERI) for Highway Slopes Constructed on High Expansive Clay Soil. *Geotechnics*. 2022;2(3):671-705. doi:10.3390/geotechnics2030033

15. Gałuszka A, Migaszewski ZM, Namieśnik J. Moving your laboratories to the field – Advantages and limitations of the use of field portable instruments in environmental sample analysis. *Environ Res*. 2015;140:593-603. doi:10.1016/j.envres.2015.05.017
16. Samouëlian A, Cousin I, Tabbagh A, Bruand A, Richard G. Electrical resistivity survey in soil science: a review. *Soil Tillage Res*. 2005;83(2):173-193. doi:10.1016/j.still.2004.10.004
17. Hossain S, Kibria G, Khan S. *Site Investigation Using Resistivity Imaging*. Taylor & Francis Group; 2018.
18. Koehn WJ, Tucker-Kulesza SE, Lebow V, Rahimi S, Bernhardt ML, Wood CM. *Enhanced Analysis of Landslide Failure Mechanisms in the Ozark Plateau Region with Electrical Resistivity Tomography.*; 2019.
19. Manzur SR, Hossain MS, Kemler V, Khan MS. Monitoring extent of moisture variations due to leachate recirculation in an ELR/bioreactor landfill using resistivity imaging. *Waste Management*. 2016;55:38-48. doi:10.1016/j.wasman.2016.02.035
20. Gallipoli M, Lapenna V, Lorenzo P, et al. Comparison of geological and geophysical prospecting techniques in the study of a landslide in southern Italy. *J Environ Eng Geophys*. 2000;4:117-128.